Chemical Compounds and their Properties

Chemical Compounds and their Properties

Edited by **Clayton Murphy**

Willford Press

New York

Published by Willford Press,
118-35 Queens Blvd., Suite 400,
Forest Hills, NY 11375, USA
www.willfordpress.com

Chemical Compounds and their Properties
Edited by Clayton Murphy

International Standard Book Number: 978-1-68285-244-6 (Hardback)

Printed in China.

Contents

Preface

This book aims to highlight the current researches and provides a platform to further the scope of innovations in this area. This book is a product of the combined efforts of many researchers and scientists from different parts of the world. The objective of this book is to provide the readers with the latest information in the field.

This book elucidates the physical and chemical properties of chemical compounds. It will aid an advanced understanding of chemical compounds through chemical bonds, resonance structures and chemical equations. It demonstrates theoretical and conceptual practices in laboratory and commercial production. It shall serve as a valuable reference for experts, chemical engineers, students and anyone else interested in this field.

I would like to express my sincere thanks to the authors for their dedicated efforts in the completion of this book. I acknowledge the efforts of the publisher for providing constant support. Lastly, I would like to thank my family for their support in all academic endeavors.

Editor

Preparation of Cobalt (III) complexes with trans-1,2-diaminocyclohexane N,N,N',N'-tetra acetatic acid (CDTA) and ethylenediamine tetraacetic acid (EDTA)

Abida Perveen, Teimoor Nezamoleslam and Iftikhar Imam Naqvi

Department of Chemistry, University of Karachi, Kaarchi-75270 Pakistan.

The preparation and isolation of complexes trans 1,2 diaminocyclohexane N, N, N', N'- tetraacetato cobalt (III) [Co(CDTA)] - and ethylenediamine-tetraacetato cobaltate (III) [Co(EDTA)]- by direct oxidation of cobalt (II) with peroxodisulfate ion in the presence of Ag+ ion as a catalyst is being reported. The stoichiometry of the complexes has been confirmed through spectroscopic techniques i.e. UV-Visible, Infrared and Atomic Absorption Spectroscopy. The complexes were found to have similar analytical values, further to confirm the procedure the complex of Co (III) EDTA was first prepared, isolated and analyzed to confirm the structural formula, Co (CDTA)- was also studied as comparatively. The approach was set to characterize compounds as expected to have the molecular formula, K[Co(CDTA)} – xH_2O in crystal form and [Co(CDTA}- in ionic form.

Key words: Isolated complexes, redox reactions, infrared (IR) spectroscopy, peroxodisulfate protonation.

INTRODUCTION

The nature and the number of bonding sites in multidentate ligands is one of the most interesting and important aspect of present day coordination chemistry. Woodruff et al. (1973) have worked on the structures of metal-EDTA complexes in the solid state and have tried to produce some answers but they could not render to the structures of the metal complexes in aqueous solution. The problem has been carefully considered by Chisholm and Clark (1973) and concluded that the substantial number of EDTA complexes of bivalent metals contained quinquedentate ligand that is, could be represented by $MEDTA[H_2O]^{2-}$ with one uncoordinated carboxylate group. Jolley et al. (1992) studies support for this idea comes from the observation of a common pk value for these complexes which is related to protonation of the free carboxylate group rather than involving any breakage of a metal-carboxylate bond. Offsetting this argument is the study of Nobuyuki and Hirosin (1965), which confirmed that definite spectral changes accompany the protonation of the copper(II) and cobalt(II) complexes and this is more easily rationalized in terms of at least some six coordinated EDTA compound.

Micheal and Robert (1992) studied the interaction of multidentate cobalt (II)-chelate complexes with ferricyanide ion. According to them, a cyanide-bridge that is, intermediate bridge is formed between $CoCyDTA^{2-}$ and $Fe(CN)_6^{3-}$ in a manner exactly analogous to that for the $CoEDTA^{2-}$ - $Fe(CN)_6^{3-}$ reaction. CyDTA was chosen because the stereochemistry of the cyclohexyl ring plays an important part in reactions of complex of this ligand as reported by Sivaswaroop et al. (1991). Number of inner-sphere reactions involving cyano bridging have been reported by Loussouarn et al. (1992). In such kind of reactions, when at least one of the product ions is substitutional labile, the decomposition (or dissociation) of the successor binuclear complex is rapid and conclusion concerning the structure and reactivity of that species must be drawn from indirect evidence as reported by Suzuko et al. (1991).

The self exchange of Co (II) and Co (III) EDTA complexes was first studied by Adamson and Vorres (1956) at pH 2.0 using a 60°C tracer. Schneider and Sheldrick (2002) described an experiment for the light induced electron transferred reaction between [Co (EDTA)] $^{2-}$ and $S_2O_8^{2-}$ by employing $[Ru(bpy)_3]^{2+}$ as a photocatalyst, but this reaction hardly proceed in the absence of $[Ru(bpy)_3]^{2+}$. Number of other studies have been undertaken related to complex reaction of EDTA and CDTA complexing agents (Wilkings and Yelin, 1967; Nishida and Kimura, 1987; Albert and Wilkinson, 1970; Ohashi and Yamamoto, 1977; Nakamoto, 1978; Marr and Rocckett, 1972; Fornaro and Coichev, 1999; Tur'yan and Kohen, 1995; Horacio and Coichev, 2006). Present study reveals the structure of cobalt (III) complexes and their ratios verification through different spectral techniques.

MATERIALS AND METHODS

Chemicals

Cobalt sulfate, Potassium Peroxodisulfate, Silver Nitrate, Ethylenediamine tetraacetic acid -EDTA, Trans-1,2-Diaminocyclo hexane-CDTA N,N,N',N'-tetraaceticacid, Sodium Chloride, Ethyl Alcohol, Cobalt(II) Chloride. All the chemicals were of the analytical grade.

Apparatus and instrumentation

Spectrophotometric techniques were used for the element detection in complex. All types of glass wares, flask, beaker, pipette, cylinder etc used were made up of Pyrex and good quality from SCHOTT DURAN. Iron Magnet heater was of IKAMAG_RTC. Double distilled deionized water and oven having CSW-300 system was used. Balance (Analytical Balance) was of mettler HIOT. Shimadzo UV-Visible Spectrophotometer (type 160 A) was used for determination of cobalt (III). Atomic absorption Spectrometer (modle Shimadzo) was used for the determination of elements present in the organic part of complex, that is, carbon, hydrogen, nitrogen and oxygen.

Preparation of reagents

Solutions preparation was done in deionized water. The reagent cobalt (II) sulfate, peroxydisulfate, silver nitrate and ligands (that is, CDTA or EDTA) were mixed in solid form and then dissolved in deionized water to get a clear colored solution by help of magnet stirrer and a little heat. Standard solution of cobalt for atomic absorption spectroscopy measurement, of cobalt (II) having 1000 ppm concentration was prepared by dissolving an appropriate amount of cobalt (II) chloride in measuring flask with deionized water and making up to the mark. Three standard solutions pertaining to concentrations of 1, 2, and 3 ppm were prepared from stock solution cobalt (II) chloride. To establish a concentration of cobalt solution to 400 ppm, calculated amount of compound was dissolved and made up to the mark with deionized water to prepare the solution of concentration beyond 3.5 ppm. The stock solution was diluted to 200 times.

Methodology

0.01 moles of cobalt (II) sulfate, 0.012 moles of ligands (that is, CDTA and EDTA), 0.003 moles silver nitrate and 0.05 moles of potassium peroxodisulfate were dissolved in accurately measured 80 ml of deionized distilled water at room temperature. The dissolution was facilitated using some heat with stirring to get a clear violet-pink solution. Reaction mixture was kept in water bath between temperatures from 85 to 90°C, while stirring for about two hours such that all peroxodisulfate got decomposed. The color of solution was changed from pink to reddish brown. 0.003 moles of sodium chloride was added to hot solution, to remove silver ion as silver chloride. Then the solution was cooled and filtered and that of total volume of filtrate was about 70 cm^3. Ethyl alcohol was added to the filtrate in the same volume and it was kept in controlled temperature for overnight to precipitate out and these precipitate were dried to in oven to crystallization. On the very next day reaction mixture was filtered to isolate the pink precipitate which was washed with excess amount of alcohol, re-crystalization was followed by drying in oven at temperature ≥ 90°C.

Analysis

Determination of cobalt

Cobalt (III) complex was examined on Shimadzu 160A UV visible spectrophotometer. The value of λ_{max} was calculated with in two different ranges of 200 to 800 nm and at of 300 to 650 nm to observe the wavelength in which the absorbance is maximum. Different maximum absorbances were observed at 380 and 540 nm. Quantitative determination of cobalt element was analyzed by 2380 Perkin – Elmer Atomic Absorption Spectrophotometer. Hallow cathod lamp of Co^{59} and Air-Acetylene flame (air as oxidant and acetylene as a fuel) were applied and the wavelength was adjusted on 240.7 nm to obtain the maximum absorbance reading for cobalt. Stock solution (containing 1.658 g of complex) was prepared as of unknown solution, three standard solutions of cobalt(II) chloride contain 1, 2 and 3 ppm cobalt were prepared to get linear baseline graph of absorbance vs. concentration as the standard procedure as shown. The observations and the graph are shown in Table 1 and Figure 1 respectively.

Elemental analysis of organic part of the molecule

To characterize the structure of the compound, it was analyzed by infrared spectrophotometer. The previous reports have suggest that the method of Nujol mull sampling method for compound. Instead alkali halide disk, spectra was obtained with KBr pellets.

RESULTS AND DISCUSSION

The preparation and isolation were carried out as described in experimental conditions. The crystal pink chloride compounds were isolated and identified by spectroscopic techniques. Spectrophotometric analysis show two different λ_{max} for cobalt III in the UV visible range at 380 and 540 nm as have shown in the Tables 2 and 3 and Figures 2 to 7. The silver catalyzed, direct

Table 1. Atomic Absorption Spectroscopic Data for Co-CDTA.

Solutions	Concentration in ppm	Absorbance
Standard No. 1	1	0.0060
Standard No. 2	2	0.0120
Standard No. 3	3	0.0170
Standard No. 4	×	0.0015

The amount of cobalt = X × 50 / 100 × 100 / 2.5 × 500 / 1000 = mg, the unknown concentration 'X' from the graph is 1.925 mg per liter.

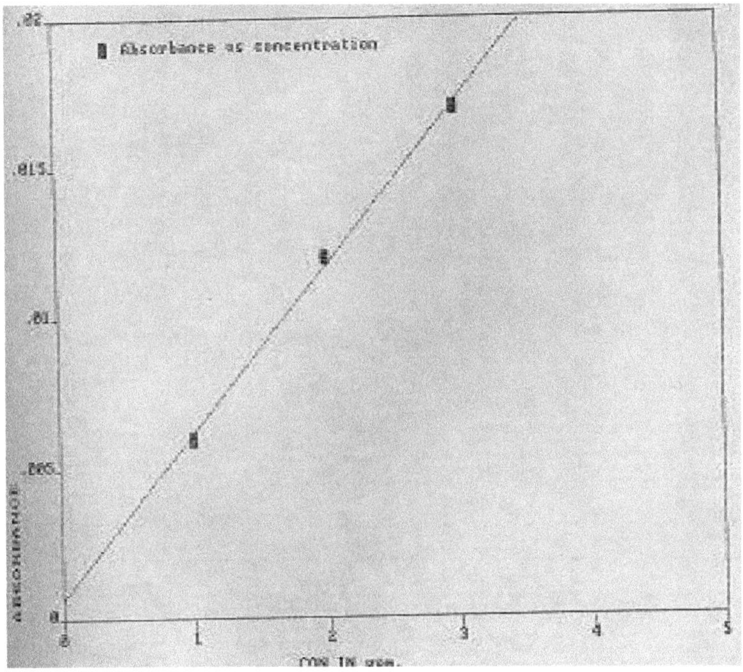

Figure 1. CoCDTA graph of concentration vs. absorbance by Atomic Absorption Spectroscopy, the regression of polynomial line (6.667E-04) + (5.500E-03) × x.

oxidation of cobalt II by peroxodisulfate to prepare [Co (III) CDTA]$^-$ complex has been studied. The prepared and isolated complexes are expected to contain cobalt metal and other components as carbon, hydrogen, nitrogen and oxygen. The stoichiometry of the complex has been confirmed through spectroscopic techniques that is, UV-Visible, infrared and atomic absorption spectroscopy. Complex was found to have similar analytical values as reported. To understand more and confirm the procedure the complex of Co (III)-EDTA was first prepared and isolated and analyzed to confirm the structure formula, as shown in Figures 6 and 7.

The [Co (CDTA)]$^-$ was then studied the same manner as [Co (EDTA)]$^-$ complex. Elemental analysis for the value of cobalt was estimated to have the same % value as reported before. Accordingly the compound is expected to have the molecular formula, K [Co (CDTA)]-xH_2O in crystal from and [Co (CDTA)]$^-$ in ionic form (in solution form). The molecular formula of K [Co (CDTA) }-xH_2O is supposed to be correct, having one hexadentate ligand bonded to one central Co atom [CoOOCH$_2$) NCH (CH$_2$)$_4$CHN (CH$_2$COO)$_2$}]$^-$. The number of hydrated molecules of water is supposed to be more probably two than one molecule. The presence of three molecules of water in the hydrated molecule is not supported by our results as discussed below.

If x= 1 (one molecule of water), the molecular weight of K [Co (CDTA)], H_2O will be 458.35 g/ mole.

Thus the expected percentage of cobalt content would be 12.86%. If number of water molecules present in the hydrated form of complex is two, x = 2, the molecular weight of K [Co (CDTA), H_2O would be 476.38, in which

Table 2. UV –Visible Data of Absorption of CoCDTA in Filtrate and ppt.

CoCDTA in Filtrate				*CoCDTA in ppt*			
λ	**ABS**	**λ**	**ABS**	**λ**	**ABS**	**Λ**	**ABS**
650	0.038	640	0.048	650	0.022	640	0.033
630	0.065	620	0.089	630	0.051	620	0.079
610	0.122	600	0.165	610	0.118	600	0.167
590	0.214	580	0.268	590	0.222	580	0.280
570	0.321	560	0.369	570	0.335	560	0.378
550	0.411	540	0.440	550	0.407	540	0.419
530	0.453	520	0.440	530	0.412	520	0.383
510	0.403	500	0.349	510	0.338	500	0.281
490	0.290	480	0.236	490	0.223	480	0.169
470	0.192	460	0.157	470	0.126	460	0.097
450	0.130	440	0.116	450	0.081	440	0.081
430	0.120	420	0.144	430	0.097	420	0.133
410	0.190	400	0.257	410	0.191	400	0.268
390	0.321	380	0.347	390	0.344	380	0.375
370	0.328	360	0.289	370	0.352	360	0.307
350	0.261	340	0.264	350	0.271	340	0.278
330	0.331	320	0.506	330	0.350	320	0.553
310	0.931	300	1.766	310	1.013	300	1.946

Table 3. UV –Visible data of absorption of CoEDTA in filtrate and ppt.

CoEDTA in filtrate				**CoEDTA in ppt**			
λ	**ABS**	**λ**	**ABS**	**λ**	**ABS**	**λ**	**ABS**
600	0.172	590	0.221	600	0.188	590	0.248
580	0.275	570	0.327	580	0.314	570	0.383
560	0.376	550	0.419	560	0.449	550	0.510
54	0.449	530	0.464	54	0.559	530	0.590
520	0.453	510	0.419	520	0.583	510	0.542
500	0.367	490	0.308	500	0.472	490	0.391
480	0.254	470	0.210	480	0.319	470	0.256
460	0.174	450	0.148	460	0.205	450	0.158
440	0.133	430	0.141	440	0.127	430	0.118
420	0.171	410	0.222	420	0.138	410	0.183
400	0.293	390	0.368	400	0.254	390	0.320
380	0.406	370	0.391	380	0.340	370	0.304
360	0.351	350	0.310	360	0.242	350	0.185
340	0.297	330	0.339	340	0.167	330	0.218
320	0.486	310	0.853	320	0.398	310	0.856
300	1.636			300	1.829		

the expected percentage of cobalt contents are to be 12.37%. In the case of x = 3 the molecular weight of K [Co (CDTA)], $2H_2O$ be 476.38 and the expected percentage of cobalt content would be 11.92% as in Table 4.

In the present study the data pertains to the 12.28% of cobalt present in complex. This leads to believe that a molecular formula with two of hydrated water molecules in the coordination sphere of the complex is correct, and also presence of one K^+ ion in the crystal form of the

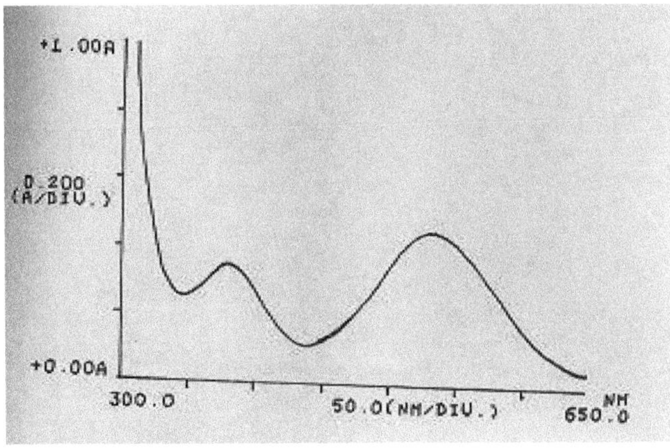

Figure 2. Graph show absorbance vs. wavelength (λ) nm for CoCDTA in filtrate.

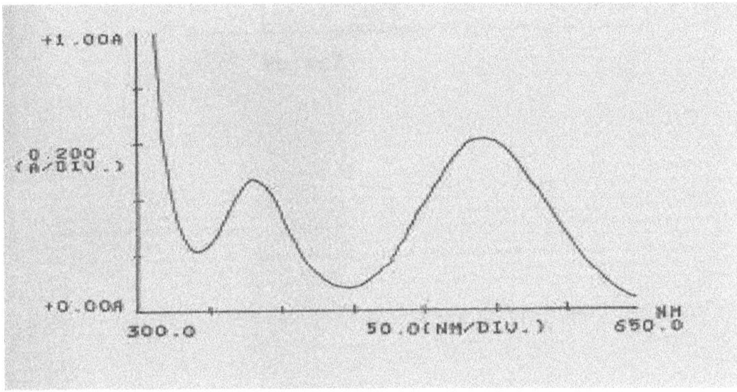

Figure 3. Graph shows plot of absorbance vs. wavelength (λ) nm for CoCDTA in ppt.

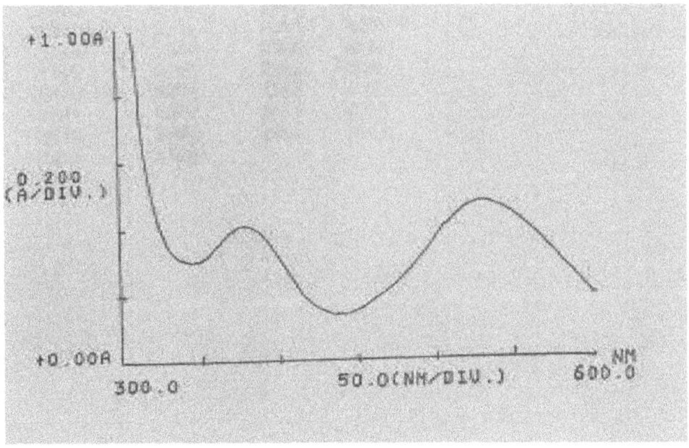

Figure 4. Graph shows plot of absorbance vs. wavelength (λ) nm for CoEDTA in filtrate.

compound is substantial, therefore the proposed molecular formula for the complex of cobalt (III) with CDTA is K [Co (CDTA], $2H_2O$ where-as the complex of cobalt (III) with EDTA is K [Co {EDTA}]$2H_2O$.

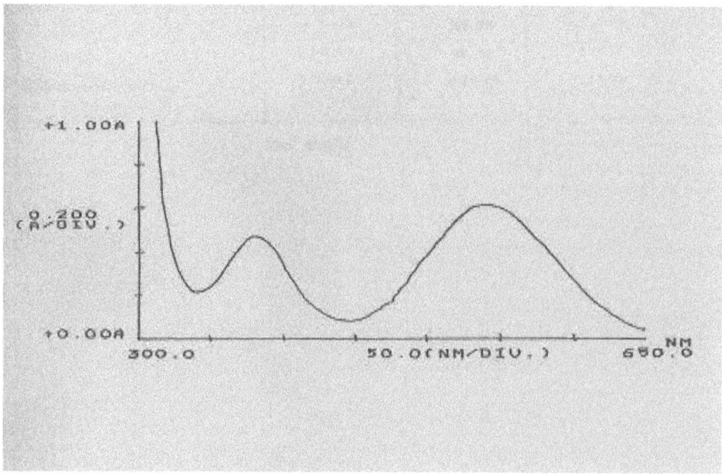

Figure 5. Graph shows plot of absorbance vs. wavelength (λ) nm for CoEDTA in ppt.

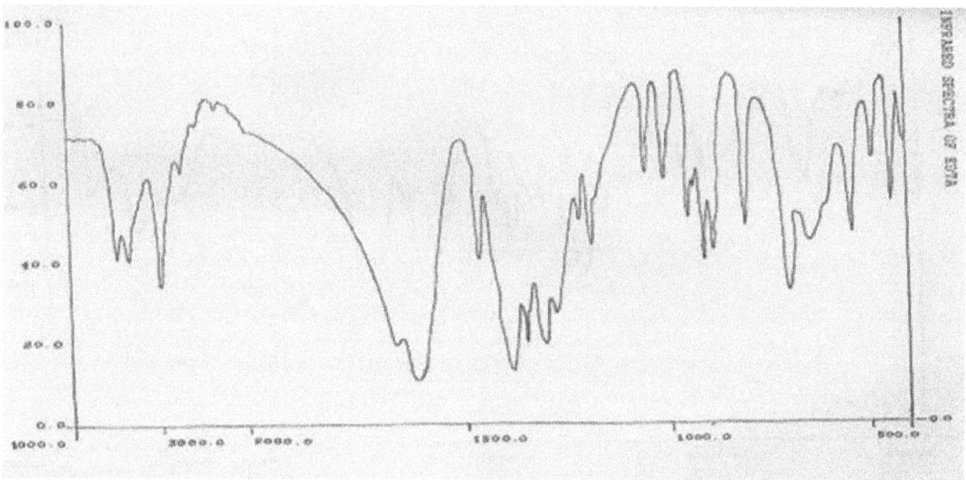

Figure 6. Infrared spectra of EDTA.

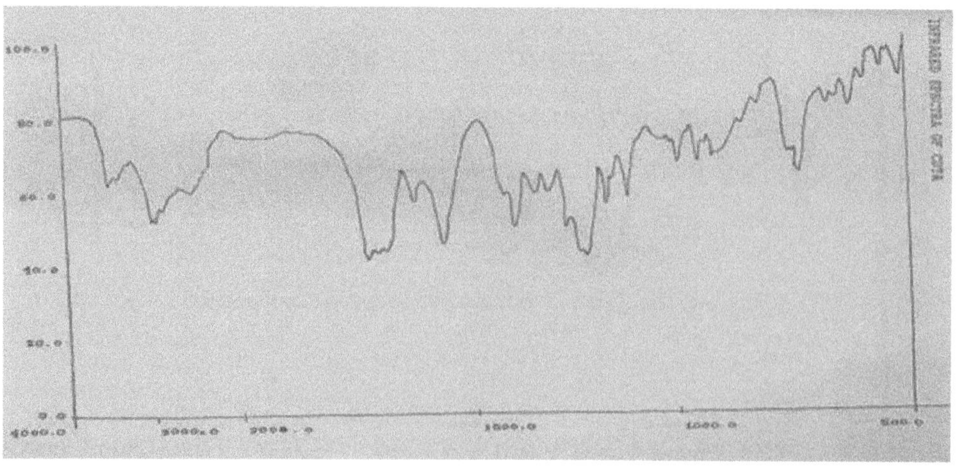

Figure 7. Infrared spectra of CDTA.

Table 4. Percentage of Cobalt in Complex Theoretically and Experimentally Supposed to have Molecular Formula K [Co (CDTA)] xH_2O, So $x \leq 2$.

Method	X Molecules	Wt. of complex (g)	Wt of Cobalt (g)	Percentage of Cobalt
Theoretically	1	458.35	58.93	12.86
Theoretically	2	476.38	58.93	12.73
Theoretically	3	494.40	58.93	12.92
Experimentally	x	1.568	0.1925	12.28

ACKNOWLEDGEMENT

The authors gratefully thank the Dean of Faculty of science, University of Karachi, for providing the financial assistance during the course of this work.

REFERENCES

Albert FC, Wilkinson C (1970). The element of the first transition series. Advanced Inorganic Chemistry, FR.S, 3rd Ed. 877-878.

Chisholm MH, Clark HC (1973). Cationic acetylenic platinum (II) compounds and their derivatives. V. Comparative study with .pi.-olefinic complexes. Inorg. Chem. 12(5):998-1008.

Fornaro A, Coichev N (1999). Iron (III) complex- catalyzed reduction of Tris (Oxalato) cobaltate (III) by L-Ascorbic acid. J. Coord. Chem. 46:519-525.

Horacio M, Coichev N (2006). Kinetic studies of the oxidation of L-ascorbic acid by tris (oxalate) cobaltate in the presence of CDTA metal ion complexes. J. Braz. Chem. Soc. 17:364-368.

Jolley WH, Stranks DR, Swaddle TW (1992). Kinetics of the aqueous cobalt (II)/cobalt (III)/EDTA system at variable pressure. Inorg Chem. 31:507.

Loussouarn A, Ouadi A, Morandeau L, Remaud P, Giles P, Gestin JF, Webb J (1992). Synthesis of new semi rigid agents of Samorium-153. Tetrahedron Letts. 44:3539.

Marr G, Rocckett BW (1972). Practical Inorganic Chemistry, 1st Ed Van Nostrand Reinhold London. 25-26:147-149.

Micheal DJ, Robert JB (1992). Kinetics and mechanism of metal ion substitution on a cobalt (III) complex containing a pendant phenanthroline. Inorg. Chem. 31:808 -815.

Nakamoto K (1978) Infrared and Raman Spectroscopy of Inorganic and Coordinated Compounds, 2 Ed John Willy of Sons N.Y; New York, P. 313.

Nishida S, Kimura M (1987). Light induced electron transfer reactions, kinetics of the oxidation of EDTA Cobaltate(II) complexes by peroxodisulphate ion, induced by irridation with visible light of aqueous solution containing trans2,2' bipyridine-ruthenium(II) ion. Bull. Chem. Soc. Jpn. 60:2367.

Nobuyuki T, Hirosin O (1965). Stability constant of some substitution inert cobalt (III) complaxes. Bull. Chem. Soc. Jpn. 38:1054.

Ohashi K, Yamamoto K (1977). The Silver (I) ion catalyzed oxidation of EDTA cobaltate (III) by the peroxodisulfate ion. Bull. Chem . Soc. Jpn. 50:3049.

Schneider TR, Sheldrick GM (2002). Substructure solution with *SHELX*. Acta Crystallogr. Sect. D. 58:1772.

Sivaswaroop P, Kumar KA, Rao PVK (1991). Oxidation of Trans 1,2 Cyclohexanediaminetetraacetate cobalt (III) by peroxodisulphate catalyzed by Ag(I) in Acetate buffer. Bull. Chem. Soc. Jpn. 64: 641-666.

Suzuko YN, Aiko N, Noriko M, Masaru K (1991). Kinetic Study on the quenching of the photo excited $[Ru(bipy)_3]^{2+}$ with $[Co(EDTA)]^-$ in the presence of EDTA, CDTA and $C_2O_4^{2-}$. Bull. Chem. Soc. Jpn. 64:1866-1875.

Tur'yan Y, Kohen R (1995). Formal redox potentials of the dehydro-L-ascorbic acid/L-ascorbic acid system. J. Electro. Anal. Chem. 380:273-282.

Wilkings RG, Yelin R (1967). Kinetics evidence pertaining to the structure of the cobalt (II) Ethylenediaminetetraacetate (EDTA) complex in aqueous solution. J. Am. Chem. Soc. Jpn. 89:5497.

Woodruff WH, Margerum DW, Mlano MJ, Pardue HL, Santini RE (1973). Kinetics and mechanism of the oxidation of cobalt (II) aminopolycarboxyalte complexes by hypobromous acid and hypobromite ion. Inorg. Chem. 12:1498 -1501.

Synthesis of cholesteryl-(3,4-dihydroxy-5-methoxy tetrahydro-furan-2-yl) methanone

Soheila Sedaghat

Department of Chemistry, Islamic Azad University, Tehran North branch, 19136-74711, Iran.
E-mail: s_sedaghat@iau-tnb.ac.ir.

Various biological membranes are synthesized from sugars-based compounds. The hydrophobic groups of them made up the amphiphilic properties in the structure of membrane that consisted fatty acids, and long chain alkyl groups. It had established that cholesterol was an integral component and it affects the fluidity of membranes. We prepared the cholesteryl-(3, 4-dihydroxy-5-methoxy-tetrahydro-furan-2-yl)-methanone from reaction of N-methoxy-N-methyl amide as an effective acylating agent, Weinreb amide, with cholesteryl chloride. Spectroscopic methods elucidated the structure of these compounds.

Key words: Cholesterol, Weinreb, membrane, mannose.

INTRODUCTION

The biomembranes had been built from a water-induced self organized architecture of amphiphilic molecules. They are made from polar hydrophilic head groups and non-polar hydrophobic part like cholesterol. In prokaryotes, the triterpenoid compounds had been the membrane stabilizers (Dannenmuller et al., 2000). The efficiency of the stabilizers had been assigned to fit their dimensions and to their amphiphilic character locking them in the membrane (Pozzi et al., 1996). In the previous studies, it suggested that cholesterol were inserted perpendicularly to the surface of the membrane and that its inclusion improves the order of the chain (Sedaghat et al., 2004).

The present study shows that a new convenient strategy of synthetic route leads to the preparation of the triterpenoid compounds in good overall yields (Bourgeois and Gueyrard, 2005). In order to prepare the compound, we used the reaction of N-methoxy-N-methyl amide (Duvold, 2000) as a polyhydroxylated moiety with organolithium species in tetrahydrofuran solvent (Francais, 1991) to form ketones type bond between the hydrophilic and hydrophobic parts of molecule.

EXPERIMENTAL

The present study was a combinatorial synthesis method for preparation of biocompatible compound. All the reagents were purchased from Merck Chemical Company. The proton nuclear magnetic resonance (HNMR) and carbon NMR (CNMR) spectra (in $CDCl_3$, internal standard Me_4Si) were recorded on a Bruker 300 and 500 MHz spectrometers. The analytical thin layer chromatography (TLC) was performed on silica gel plate (Merck). Infrared spectra were recorded on a Perkin-Elmer 297 instrument. Reactions were monitored by TLC.

1-(6-Methoxy-2, 2-dimethyl-tetrahydro-furo [3, 4- d] [1, 3] dioxol-4-yl)-ethane-1, 2-diol

A mixture of D-mannose (5 g, 27.77 mmol), 2,2-dimethoxypropane (17 ml), acetone (16.5 ml), methanol (16.5 ml) and concentrated HCl (0.5 ml) was refluxed for 2 h. The cooled mixture was diluted with water (50 ml) and concentrated under reduced pressure at below 30°C. Methanol (50 ml) and concentrated HCl (1.25 ml) were then added. The reaction mixture was stirred at room temperature for 3 h, neutralized with a saturated solution of sodium hydrogen carbonate, and evaporated to dryness. The residue was taken in hot chloroform and solution was filtrated through the celite pad. The solvent evaporated to give methyl-2, 3-O-isopropylidene-mannofuranoside (Figure 1) as yellow oil (5.6 g, 85%).

6-Methoxy-2, 2-dimethyl-tetrahydro- furo [3, 4- d] [1, 3] dioxole-4-carbaldehyde

Sodium periodate (5.46 g, 25.52 mmol) dissolved in minimum of water was added to a solution of methyl-2, 3-O-isopropylidene-

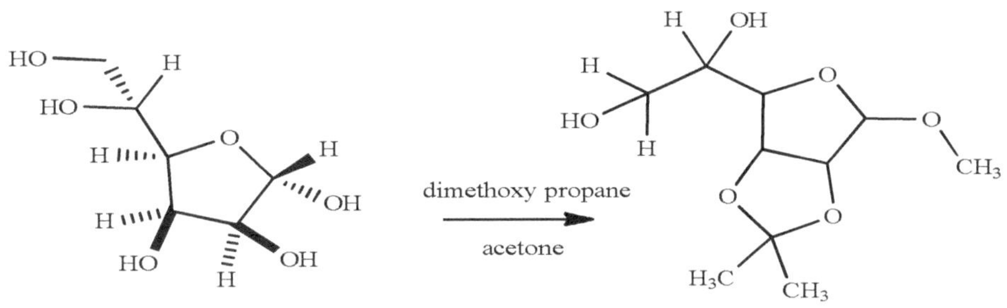

Figure 1. H-NMR (300 MHz, $CDCl_3$): δ = 4.87 (s, 1H), 4.72 (d, 1H), 4.49 (d, 1H), 3.12 (brs, 1H), 4,30 (dd, 1H), 3,57 (dd, 1H), 3.61 (dd, 1H), 3.32 (s, OCH_3, 3H), 1.37 (s, CH_3, 3H), 1.22 (s, CH_3, 3H); C-NMR (500 MHz, $CDCl_3$): 113, 107, 64.8, 80.4, 79.5, 70.7, 85.1, 54.9, 26.3, 25.1; FTIR: 3448, 1638, and 1084 cm^{-1}.

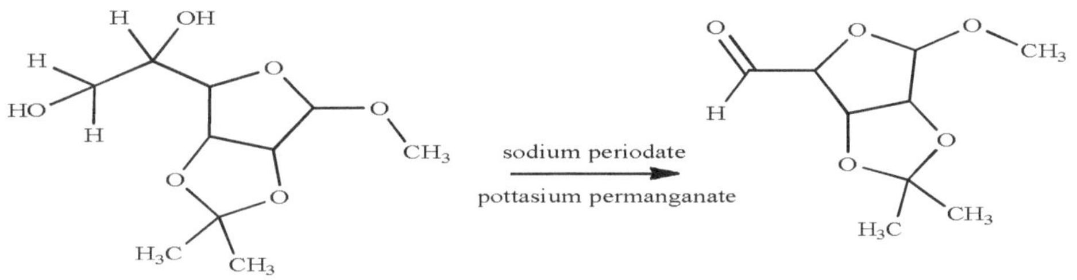

Figure 2. H-NMR (500 MHz, $CDCl_3$): δ = 9.5 (s, H fermyl), 5.1 (s, 1H), 5.08 (d, 1H), 4.63 (d, 1H), 4.38 (d, 1H), 3.36 (s, OCH_3, 3H), 1.44 (s, CH_3, 3H), 1.30 (s, CH_3, 3H); C-NMR (500 MHz, $CDCl_3$): 200.7, 109.1, 89.4, 84.4, 80.7, 55.6, 25.8, 24.8; FTIR: 3448, 1638, and 1088 cm^{-1}.

mannofuranoside (5.43 g) in methanol (150 ml). The mixture was stirred for 2 h at room temperature and the solvent was evaporated under reduced pressure (Duvold, 1995). The residue was taken in acetone and the solution was filtered through a celite pad. Thefiltrate was evaporated to dryness to give the intermediate aldehydes (Figure 2) (4.1 g, 75%).

6-Methoxy-2, 2-dimethyl-tetrahydro- furo [3, 4- d] [1, 3] dioxole-4-carboxylic acid or Methy-2, 3-O-isopropylidene-a-D-lyxofuranosiduronic acid

The intermediate aldehydes dissolved in water (120 ml) and a solution of NaOH (0.23 g) in water (5 ml) and sodium permanganate (7.332 g, 46.4 mmol) in water (200 ml) were added at 0°C. The mixture was stirred at room temperature for 16 h. The excess of potassium permanganate was removed with a solution of H_2O_2 (33%) and the mixture filtered through a celite pad (Bourgeois and Gueyrard, 2005). The filtrate concentrated to 100 ml and washed twice with ethyl acetate (100 ml). The aqueous layer was acidified to pH 3 at 0°C with HCl and extracted with ether (2 × 100 ml). The organic layer was washed with a saturated sodium chloride solution, dried (Na_2SO_4), filtered and evaporated under reduced pressure to yield acid as a colorless oil (3.2 g, 68%) (Figure 3).

6-Methoxy-2,2-dimethyl-tetrahydro- furo [3, 4-d] [1, 3] dioxole-4-carboxylic acid methoxy-methyl-amide

The acid compound (3 g) dissolved in dichloromethane (30 ml) and 2.31 g of carbonyldiimidazole was added portion wise over 10 min at 22°C. The solution was stirred at room temperature for 30 min. N,O-dimethylhydroxylamine hydrochloride (4 g) was dissolved in water (17 ml) and 10 N sodium hydroxide (7 ml) was added to it. The solution was extracted with dichloromethane (3 × 17 ml) and the extract dried by $MgSO_4$. After stirring for 3 days, the solution was washed with 0.5 M citric acid, sodium bicarbonate solution, and evaporated in vacuum to give the compound as colorless oil (1.6 mg, 54) (Figure 4).

[17-(1,5-Dimethyl-hexyl)-10,13-dimethyl-2,3,4,7,8,9,10,11,12,13,14,15,16,17-tetradecahydro-1H-cyclopenta[a]phenanthren-3-yl]-(6-methoxy-2, 2-dimethyl-tetrahydro-furo [3, 4-d] [1, 3] dioxol-4-yl)-methanone

Lithium and di tert - butyl bromide were added to a flask and after 45 min of stirring, the flask cooled by cold acetone and ice. Cholesteryl chloride solution added to it and to continue stirring for 3 h until the lithium cholesteryl formed. The compound 4 (1.6 mg) were mixed into the flask and to continue stirring it for 3 h under the 0 degree. The organic layer washed, dried, filtered and evaporated under reduced pressure to yield the yellow solid (1.1 mg, 54%) (Figure 5).

RESULTS AND DISCUSSION

The purpose of the study was to synthesis the sugar-based membrane (Barton et al., 1994; Blunk and Bierganns, 2006; Tirnaksiz and Kalsin, 2005) and according to the results, carbon-carbon bond was formed by Weinreb method (Just and Martel, 1973; Nahm and Weinreb, 1981). The synthesis took place as a modified functional group in ribosyl portion of compound.

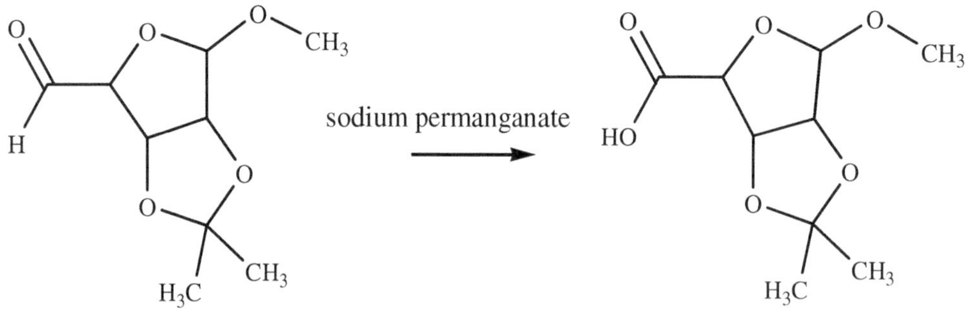

Figure 3. H-NMR (500 MHz, $CDCl_3$): δ = 5.1 (s, 1H), 5.07 (d, 1H), 4.9 (d, 1H), 4.67 (d, 1H), 3.35 (s, OCH_3, 3H), 1.50 (s, CH_3, 3H), 1.36 (s, CH_3, 3H); C-NMR (300 MHz, $CDCl_3$): 177, 113, 107, 83.9, 80.3, 79.1, 54.9, 25.7, 24.7.

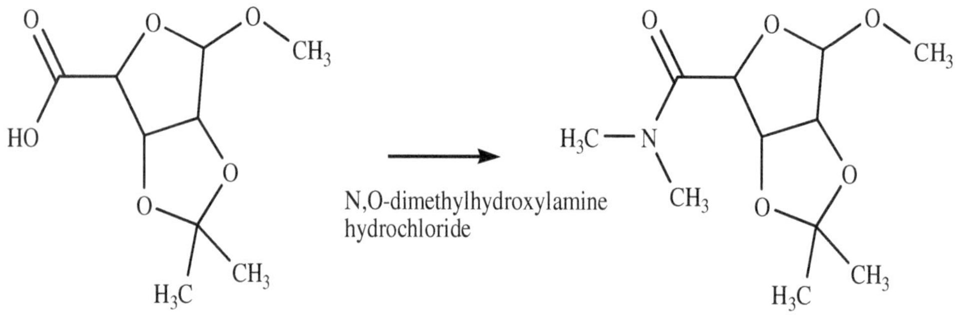

Figure 4. H-NMR (500 MHz, CDCl3): δ = 5.1 (s, 1H), 4.9 (d, 1H), 4.8 (brs, 1H), 4.6 (dd, 1H), 3.7 (s, 3H, OCH3), 3.38 (s, 3H, OCH3), 3.12 (s, 3H, OCH3), 1.43 (s, OCH3 , 3H), 1.35 (s, CH3, 3H); FTIR: 3400, 1638, 1062, 913, and 742 cm^{-1}.

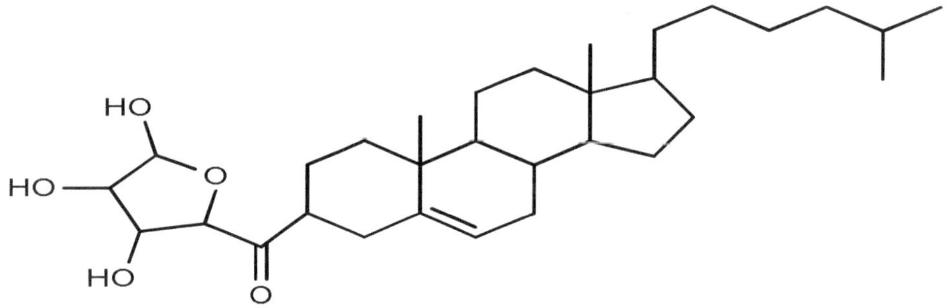

Figure 5. H-NMR (500 MHz, CDCl3): δ = 5.1 (s, 1H), 4.9 (d, 1H), 4.8 (brs, 1H), 4.6 (dd, 1H), 3.7 (s, 3H, OCH3), 3.12 (s, 3H, OCH3); FTIR: 3400, 1638, 1385, 1062, 913, and 742 cm^{-1}.

It started by protecting hydroxyl groups on sugar ring and these part of chain synthesis increased the yield of reaction (Petru, 2005). We applied the amide form of side carbohydrate to connect the cholesteryl group (Dual, 1995). This synthetic protocol represents a useful tool for the access to natural sugar derivatives of biological interest as well as to labeled ribose compounds for biosynthetic studies.

Conclusion

The synthesis is based on modifying functional group in ribosyl side group starting from protecting by hydroxyl groups on sugar ring and subsequent reaction of functional moieties. In a key step, an aldehyde is oxidized to the corresponding acidic group by means of an oxidant. This synthetic protocol represents a useful tool for the access to natural sugar derivatives of biological interest as well as to label ribosyl compounds for biosynthetic studies.

REFERENCES

Barton DHR, Gero SD, Quiclet-sire B, Samadi M (1994). Stereocontrolled radical reaction in carbohydrate and nucleoside chemistry. Tetrahedron Asymmetry 5:2123-2136.

Blunk D, Bierganns P (2006). New speciality surfactants with natural structural motifs. New J. Chem. 30: 1705-1709.

Bourgeois MJ, Gueyrard D (2005). Exploration of Synthetic Pathways to Unprecedented TerpenylThioglycosides. Letts. Org. Chem. 2:665-669.

Dannenmuller O, Arakawa K, Eguchi T, Kakinuma K, Blanc S, Albrecht AM, Schmutz M, NakataniY, Ourisson G (2000). Membrane Properties of Archeal MacrocyclicDiether Phospholipids. Chem. Eur. J. 6:645-654.

Duvold T (1995). Benzyl 2,3-o-isopropylidene- b- D- ribo- 1,4-pentodialdofuranoside as a D-ribose Chiron for the synthesis of terpenyltetraols and aminotriols. Tetrahedron Letts. 36:3153-3157.

Duvold T, Rohmer M (2000). Synthesis of 3b-(5-D-ribityl) cholestane, a putative biological precursor for fossil 3-alkylsteranes. Tetrahedron Letts. 41:3875-3878.

Francais GW (1991). Approaches to the synthesis of terpenylcarbohydrate. Acta Chem. Scandinavica 45:18-23.

Just G, Martel A (1973). C-nucleosides and related compounds ,synthesis of D,L-3,4-Isopropylidene -2,5- anhydroallose: A novel periodate cleavage. Tetrahedron Letts. 17:1517-1525.

Nahm S, Weinreb S (1981). N-methoxy N-methylamide as effective acylating agents. Tetrahedron Letts. 22:3815-3818.

Petru ID (2005). Chemical synthesis of cholesteryl b-D-galactofuranoside and pyranoside. Carbohydr. Res. 340:2052-2058.

Pozzi G, Birault V, Werner B, Dannenmuller O, Nakatani Y, Ourisson G (1996). Single-Chain Polyprenyl Phosphates Form “Primitive” Membranes. Angewandte Chemie 35:177-180.

Sedaghat S, Désaubry L, Streiff S, Ribeiro N, Michels B, Nakatani Y, Ourisson G (2004). A Novel Type of Membranes Based on Cholesteryl Phosphate. Chem. Biodivers. 1:124-128.

Tirnaksiz F, Kalsin O (2005). A topical w/o/w multiple emulsions prepared with Tetronic 908 as a hydrophilic surfactant: Formulation, characterization. J. Pharm. Pharmaceut. Sci. 8:299-315.

Synthesis of trans-3,4-dibenzyl-2-oxo-tetrahydrofuran

Jack I. R. and Orubite K. O.*

Department of Chemistry, Rivers State University of Science and Technology, P. M. B. 5080, Port Harcourt, Nigeria.

Trans-3,4-dibenzyl-2-oxo-tetrahydrofuran (6) a substituted tetrahydrofuranoid lignan was synthesized in 70% yield via the Stobbe condensation between (1) benzaldehyde and (2) diethyl succinate. The structure was confirmed by the spectral data such as infra-red, proton magnetic resonance and mass spectrum. It has two similar but non equivalent aromatic groups which appeared at δ: 6.48 - 6.76 and 7.06 - 7.25 as multiplets.

Key words: Lignans, tetrahydrofuranoid, Stobbe.

INTRODUCTION

Lignans belonging to the substituted tetrahydrofuran class represent a unique group not only owing to their distribution pattern (Rao, 1978) but also due to their utility as synthons in the synthesis of 2,4-diaryl substituted furofurans (Jack and Fekarurhobo, 2003). Although, most lignans occur in plants (Jack et al., 2003), a few have been isolated from the urine of humans and animals such as baboons, vevet monkeys and rats (Cooley et al., 1981; Set-Chell et al., 1980).

The physiological importance of these lignans as synergists, anti-hypertensive agents and germination inhibitors cannot be overemphasized (Chakraverty et al., 1979; Sih et al., 1976). Isolation and synthesis of some lignans belonging to this class have been previously reported (Jack and Fekarurhobo, 2003; Jack et al. 2003) using a diaryl dioxo diester.

In this paper, an alternative approach leading to the formation of a lignan belonging to the substituted tetrahydrofuranoid class is described and discussed. The advantage of this method over the other is that the reactions are smoother with fever mixture of products caused by isomerisation and the overall yield of the butyrolactone is better.

MATERIALS AND METHODS

Infra-red spectra were recorded on Pye-Unicam SP1050 Spectrometer. Proton magnetic resonance was run on a Varian HA 100 instrument using tetramethylsilane (TMS) as an internal standard in $CDCl_3$ with chemical shift value given in δ (ppm). Mass spectra were obtained on an Airborne Electronic Terrain Map System (AETMS) 9 double focusing spectrometer at 250°C and 70 eV.

Thin layer chromatography (TLC) was performed on silica gel GF 254 plates using ethyl acetate: cyclohexane (1:1) as solvent system. All reagents and solvents were purified before use by refluxing over calcium hydride and redistillation.

Compound (3)

1.92 g (0.08 moles) NaH suspension was put into 1 litre three neck flask and washed with dry pentanes (70 ml x 3). The flask and its content were dried on the pump for a few minutes to remove the pentane. 100 ml dry diethyl ether was added into the flask which was flushed with nitrogen for twenty minutes. To this ethereal suspension of NaH was added dropwise, 6.96 g (0.04 moles) diethyl succinate in 40 ml dry diethyl ether and stirring (with the flask surrounded by cold water) for about two hours until the sodium salt was formed as a sludge. An ethereal solution of 8.48 g [(0.08 moles) benzaldehyde in 80 ml diethyl] ether was added into the flask dropwise while the mixture was stirred continuously for about 4 h. The mixture was acidified with 3 M HCl to pH 1 and then extracted with diethyl ether (80 ml x 3). The ether extract was washed with water (100 ml x 2) and dried over anhydrous Na_2SO_4. Evaporation of the solvent gave an oil which could not crystallize when treated with ethanol, acetone and acetone ethyl acetate mixture separately.

Yield: 9.2 g (78 %) IR (film) (CH) 3012-2900, (COOH) 2945-2700, (arom.) 1608 cm^{-1}; ^{1}H n.m.r ($CDCl_3$) (COOH) 12.00 s, (Ar<u>H</u>) 7.00-6.94 m, (C<u>H</u>) 6.62-6.50 m; mass spectrum M^+ 295.

2,3-bis-benzyl succinic acid (4)

Into a 250 ml two neck round bottom flask containing 100 mg of

*Corresponding author. E-mail: kaineorubite@yahoo.com.

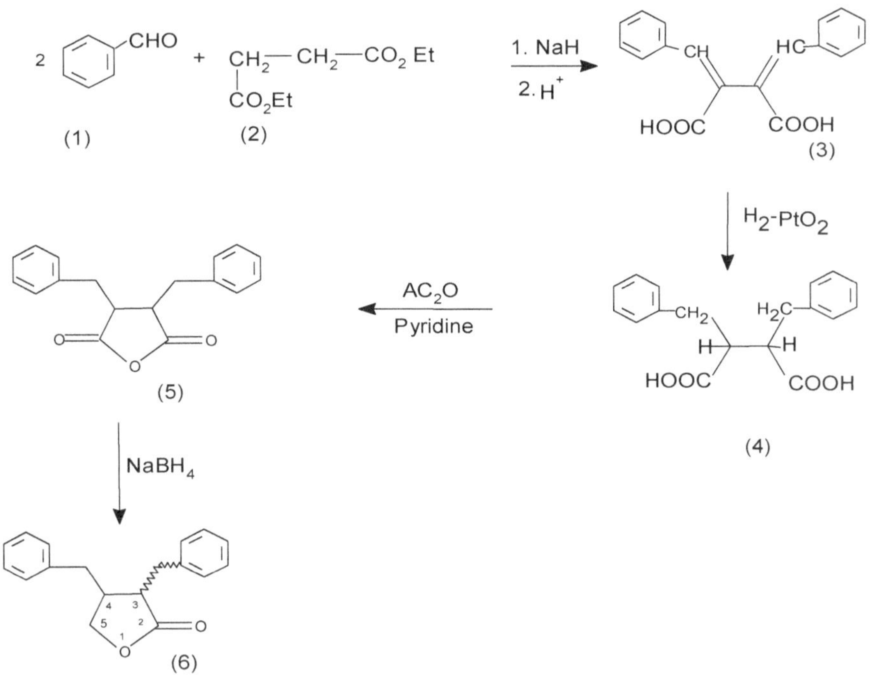

Scheme 1. Refluxation of dibenzyl succinic acid (4) with acetic anhydride and pyridine to yield the anhydride (5) which was partially reduced with sodium borohydride in isopropanol to give the lignan

platinum oxide catalyst was put 5.2 g (0.02 moles) of compound (3) in 100 ml absolute ethanol and hydrogen gas was passed into the flask at room temperature over 7 h, stirring by using a stirrer. TLC of the mixture after 7 h showed that all the starting material had reacted.

The mixture was filtered to remove the platinum and the filtrate evaporated to give an oil which was carried forward to the next reaction.

Yield: 4.6 g (88 %) IR (film) (CH) 3100-2890, (COOH) 2700-2500, (arom.) 1605 cm^{-1}; ^{1}H n.m.r. ($CDCl_3$) (COOH) 12.04 s, (ArH) 7.10-6.90 m, (CH) 5.80-5.54 m, (CH_2) 4.70-4.50 d; mass spectrum M^+ 299.

The anhydride (5)

3 g of compound (4) in 20 ml dry pyridine was refluxed with 80 ml acetic anhydride in a 200 ml round bottom flask for 5 h. The flask was allowed to cool to room temperature and the content poured into 200 ml of ice-cold water in a beaker. After allowing to stand for 30 min the mixture was extracted with ethyl ethanoate (50 ml x 3), washed with IM HCl (25 ml x 3) and finally with water (30 ml x 2). After drying over anhydrous Na_2SO_4, the solvent was evaporated to give an oil which failed to crystallize on standing in the fridge over night and also after treatment with n-hexane.

Yield: 2.22 g (78.6 %) IR (film) (CH) 3050-2910, (anhydride) 1866, (arom.) 1610 cm^{-1}; ^{1}H n.m.r. ($CDCl_3$, ArH) 7.40-6.80 m, (CH) 5.10-4.80, (CH_2) 3.10-2.90; mass spectrum M^+ 281.

Trans-3,4-dibenzyl-2-oxo tetrahydrofuran (6)

1.5 g (5.4 mmoles) of (5) was suspended in 40 ml isopropanol in a 200 ml flask and 100 mg (2.7 mmoles) $NaBH_4$ was slowly added, and was stirred for 36 h. 50 ml of aqueous ether was added and the pH of the medium was brought to 6 with dilute HCl with vigorous stirring.

The organic layer was removed, washed with water (25 ml x 3), dried over anhydrous Na_2SO_4, filtered and evaporated to give a gum which defied many attempts to crystallize it.

Yield: 980 mg (70 %) IR (film) (CH) 3090-2900, (γ-lactone) 1770, (arom.) 1608 cm^{-1}; ^{1}H n.m.r. ($CDCl_3$) (ArH, 5H) 7.06-7.25 m, (ArH, 5H) 6.48-6.76 m, (ArCH_2, 4H) 2.52 m, H-3 (1H) 2.92 m, H-4 (1H) 2.96 m, H-5 (2H) 3.87, 4.13 dd; mass spectrum M^+ 267.

RESULTS AND DISCUSSION

The reaction involved the classic Stobbe condensation using an ethereal solution of two moles of benzaldehyde (1) with one mole of diethyl succinate (2) in the presence of sodium hydride. Acid hydrolysis gave the unsaturated substituted succinic acid 3 which could not crystallize in attempt to purify it. Catalytic hydrogenation of (3) over platinum oxide in absolute ethanol at room temperature gave 2,3-bis-benzyl succinic acid (4) in over 80% yield. When the dibenzyl succinic acid (4) was refluxed with acetic anhydride and pyridine, it afforded the anhydride (5) which was partially reduced with sodium borohydride in isopropanol to give the lignan (6) as a gum having R_f 0.58 using silica gel GF254 in a system ethyl ethanoate: cyclohexane (1:1) v/v (Scheme 1). There was also another minor sport overlapping the major spot having R_f 0.58. This may presumably account for the difficulty encountered in crystallizing the lignan (6) which may be

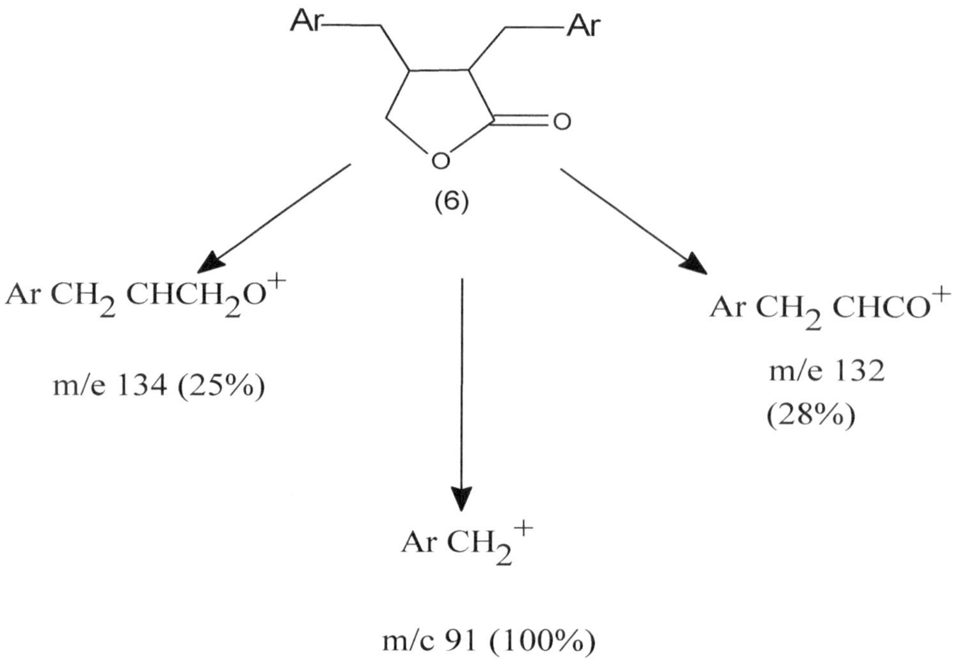

Scheme 2. Establishment of the molecular formula $C_{18}H_{18}O_2$ for the lignan (6) from the mass spectrum giving the parent peak at m/e 266 with other attendant prominent peaks at m/e 134 and 132 corresponding to the fragment ions $ArCH_2$ CH CH_2O^+ and Ar CH_2 $CHCO^+$ respectively. The base peak is the benzyl cation Ar CH_2^+ at m/e 91.

considered as a mixture.

In the infra-red spectrum of the lignan (6), the presence of an aromatic and a five member γ-lactonic groups at v_{max} 1608 and 1770 cm^{-1} respectively were evident. This can be distinguished from the anhydride (5) which has an R_f 0.46 in the same solvent system above and gave a band at v_{max} 1866 cm^{-1} (Jack and Orubite, 2009). The anhydride (5) also gave a molecular ion peak M^+ at m/e 280.

The molecular formula $C_{18}H_{18}O_2$ was established for the lignan (6) from the mass spectrum which gave the parent peak at m/e 266 with other attendant prominent peaks at m/e 134 and 132 corresponding to the fragment ions $ArCH_2$ CH CH_2O^+ and Ar CH_2 $CHCO^+$ respectively (Scheme 2). The base peak is the benzyl cation Ar CH_2^+ at m/e 91.

The proton magnetic resonance spectrum of the lignan (6) revealed signals characteristic of two similar but non equivalent aromatic groups which centred at δ6.48-6.76 and 7.06-7.25 as multiplets with each group integrating for five protons. In addition, the AB- parts of two ABX-systems which centred at δ4.13 (J_{AB} = 9, J_{AX} = J_{BX} = 7) and δ 2.96 (J_{AB}= 14,J_{AX}=5,J_{AB} = 6) could be detected arising from the partial structures of O CH_2-CH-CH- and Ar CH_2CH- with the methylene protons next to the oxygen appearing as double doublet at δ3.87 and 4.13. Consequent on these facts, the structure of this lignan is given as (6) in Scheme 1.

REFERENCES

Chakraverty DP, Roy S, Sinha-Roy SP, Majumber S (1979). Podotoxin-a germination inhibiting lignan from Zanthoxylum accananthopodium. Chemistry, Industry London 667.

Cooley G, Perrant PD, Kirk DN, Wyn S (1981). Chemical synthesis of the first lignan to be found in humans and animals. Tet. Lett., 22: 349-350.

Jack IR, Fekarurhobo GK (2003). Synthesis of Aptosimone- a 2,4-diaryl mono epoxy lignanolide. Int. J. Sci. Tech., 1(2): 1-3.

Jack IR, Fekarurhobo GK, BobManuel RB (2003). Isolation and Synthesis of 3,4-bis (3',4'-methylene dioxy phenyl) butyrolactone from Crassostrea gasar. Int. J. Sci. Tech., 1(2): 23-25.

Jack IR, Okorosaye-Orubite K (2009). Some constituents of the bark of *Gmelina arborea. Scientia Africana* 8(1): 26-32.

Rao CBS (1978). Chemistry of Lignans Andra Univ. Press Ch. 4 95.

Set-Chell KDR, Lawson AM, Mitchell FL, Adlercreutz H, Kirk DN, Axelson M (1980). Lignans in man and in animal species. Nature 247: 740-742.

Sih CJ, Ravikumar PR, Buckner FHC, Whitlock H (1976). Isolation and synthesis of Pinoresinol diglucoside. A major antihypertensive principle of Tu Chung J. Am. Chem. Soc., 98:5412.

4

Hydrogenation of 2H-azirine by aluminium lithium hydride over palladium catalyst

Kuevi U. A., Atohoun Y. G. S. and Mensah J. B.*

Laboratoire de Chimie Théorique et de Spectroscopie Moléculaire (LACTHESMO) Université d'Abomey-Calavi, 03 BP: 3409 Cotonou, Bénin Republic.

The hydrogenation of 2H-azirine by the aluminium lithium hydride ($AlLiH_4$) in the presence of palladium used like catalyst, was studied by Hartree Fock method. For the modelling of reaction, the catalyst was represented by one atom of palladium; we took one molecule of 2H-azirine for two molecules of aluminium lithium hydride. The reaction gave a saturated derivative of 2H-azirine (C_2H_3N) called aziridine (C_2H_5N). On the basis of the analysis of the results obtained, the probable mechanism of reaction was proposed.

Key words: 2H-azirine, aziridine, aluminium lithium hydride, palladium, modelling of reaction, Hartree Fock.

INTRODUCTION

The 2H-azirine, is an unsaturated heterocyclic molecule comprising a nitrogen atom and two carbon atoms, with a double C=N bond. The chemical compounds which contain a cycle azirine in their structure are extremely reactive and they seem intermediary in certain chemical reactions like the rearrangement of Neber (McCarty 1970). Its saturated derivative (aziridine) is an intermediate and monomer in the preparation of cationic polymers which are used to improve wet strength of paper, in fuel-oil and lubricant refining, as flocculating agents and in protective coatings, in textile finishing and for adhesives, polymer stabilizers and surfactants (Lewis, 1993; Scherr et al., 1995).

Hydrogenation is an important reaction which intervenes in the hydrotreating process of the heterocyclic molecules contained in the residues of oil, under a strong hydrogen pressure over catalysts. During this process, the hydrogenolysis of carbon-heteroatom bonds such heterocyclic molecules often occurs (Bachelier et al., 1984; Breysse et al., 1984; Bimal et al., 1999; Yermakov et al.,1985; Kasztelan et al., 1984; Murena et al., 2004).

In our present study, we modelled hydrogenation of 2H-azirine over a palladium catalyst by using as hydrogenating reagent the aluminium lithium hydride. This reaction has been studied because experimental data exist on the same system in literature. Compared with our study, the difference is that the experimental catalyst is palladium on carbon as support, (Knouants, 1988). The objective of our study was to confront experimental results with our theoretical results in order to propose the mechanism of reaction. The calculation methods used are based on the Hartree-Fock theory and the basis set is LanL2DZ.

*Corresponding author. E-mail: menfolben@yahoo.fr.

MATERIALS AND METHODS

Theory

The calculating method used in our work is the Hartre-Fock method (Abarenkov et al., 1989; Rivail, 1994). According to this method, the wave function is the functional of spin orbitals. The electronic energy of the ground state is obtained by applying the variation theorem. The energy has the form: $E_{HF} = V + \langle hP \rangle + \frac{1}{2}\langle PJ(P) \rangle - \frac{1}{2}\langle PK(P) \rangle$ where V is the nuclear disgust energy, P is the density matrix, $\langle hP \rangle$ is the one–electron (kinetic plus potential) energy, $\frac{1}{2}\langle PJ(P) \rangle$ is the classical coulomb repulsion of the electrons, and $-\frac{1}{2}\langle PK(P) \rangle$ is the exchange energy resulting from quantum (fermion) nature of electrons. The basis set used is LanL2DZ.

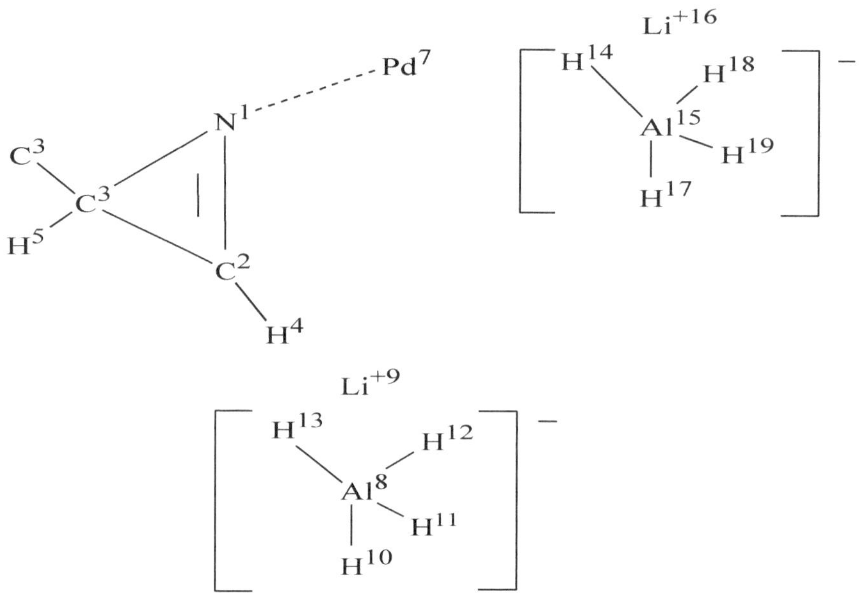

Figure 1. Chemical system studied.

Calculation program

The calculations were performed with Gaussian-98W program (Klark, 1990; Frisch et al., 1998), which carries out the quantum calculations based on *ab initio* and semi empirical methods. The program data are the geometry of systems, their multiplicity and charge. The curves and the drawings of the system studied were performed with Excel 2003 and ChemDraw Ultra 6.0, respectively. The calculations were carried out in the " Laboratoire de Chimie Théorique et de Spectroscopie Moléculaire (LACTHESMO)" of "Université d'Abomey-Calavi" in Benin Republic.

Modelling of the reaction

Two stages characterize the reaction. The first is the adsorption of the 2H-azirine (C_2H_3N) on palladium and second is the reaction between two molecules of aluminium lithium hydride and the adsorbed 2H-azirine molecule. The approach of both molecules $LiAlH_4$ was made according to two different directions in comparison with the adsorbed reactive.

The whole chemical system studied is shown in Figure 1. Two coordinates of reaction were taken into account. They are the N^1H^{14} and C^2H^{13} distances. The rapprochement of both molecules of hydride across previous both coordinates led to deep transformations of system.

In these work, we made the assumption that the active phase of catalyst used by Knouants (1988) would be palladium, and for this reason we modelled this active site of their catalyst by a palladium atom.

RESULTS

Both coordinates of reaction have as initial values 4.000 Å and 3.877 Å for N^1H^{14} and C^2H^{13}, respectively. During the process, these coordinates of reaction were gradually decreased and the parameters whose evolutions were followed in particular are interatomic distances, angles between chemical bonds and Mulliken atomic charges.

According to both coordinates of reaction, we were interested in the evolutions of the interatomic distances C^2N^1, C^3N^1, $Al^{15}H^{14}$, Al^8H^{13} and N^1Pd^7 which are represented on the Figure 2. We also took in consideration angles $N^1C^2C^3$, $N^1C^3C^2$ $H^6C^3H^5$, and $H^{14}N^1C^2$ whose evolutions are represented by the Figure 3.

The analysis of the curves of the Figures 2 and 3 made it possible to determine the ruptures, formations and transformations of chemical bonds occurred during the reaction.

With regard to the atomic charges of the system, the analysis of Mulliken populations also contributes to the determination of the ruptures, formations and transformations of chemical bonds occurred during the reaction. The variations of the Mulliken atomic charges of the atoms C^2, C^3, N^1, H^{13} and H^{14} are shown in Figure 4, according to the coordinates of reaction. The energy of the system also varied according to both coordinates of reaction, during the process. The variation of energy of the system is given (Figure 5).

DISCUSSION

The curves of Figure 2 show that at the end of the process, certain chemical bonds disappeared while others were formed or transformed. Indeed, the interatomic distances between the aluminium atom Al^8

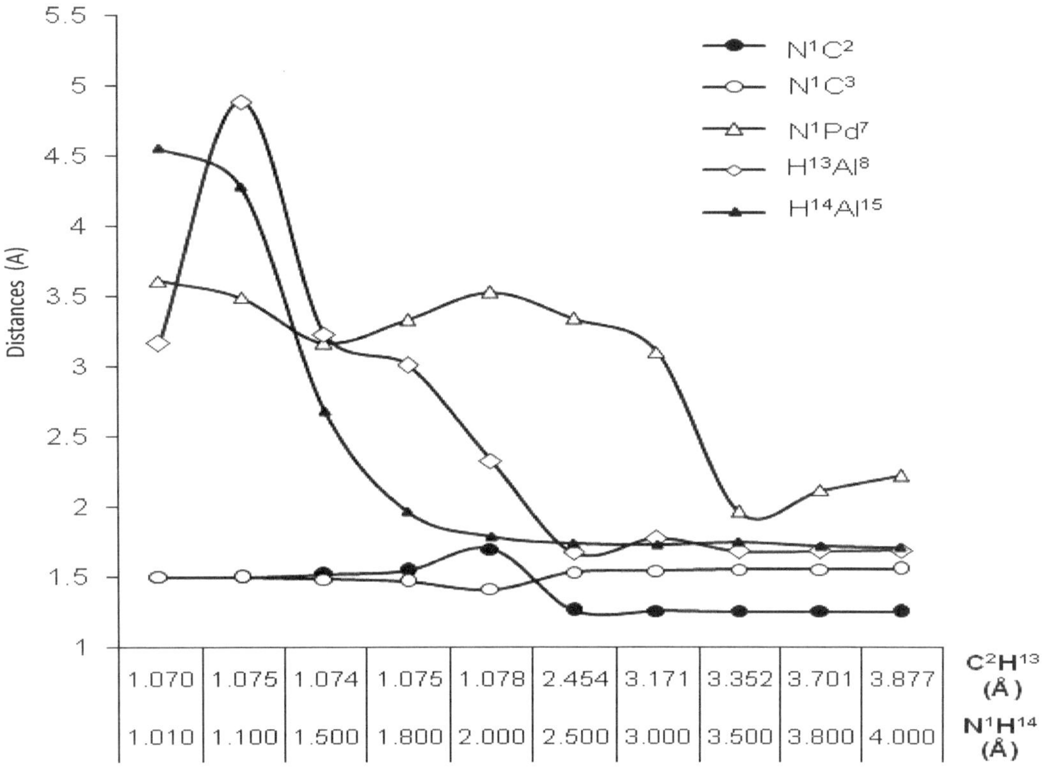

Figure 2. Variations of interatomic distances according to reaction coordinates N^1H^{14} and C^2H^{13}.

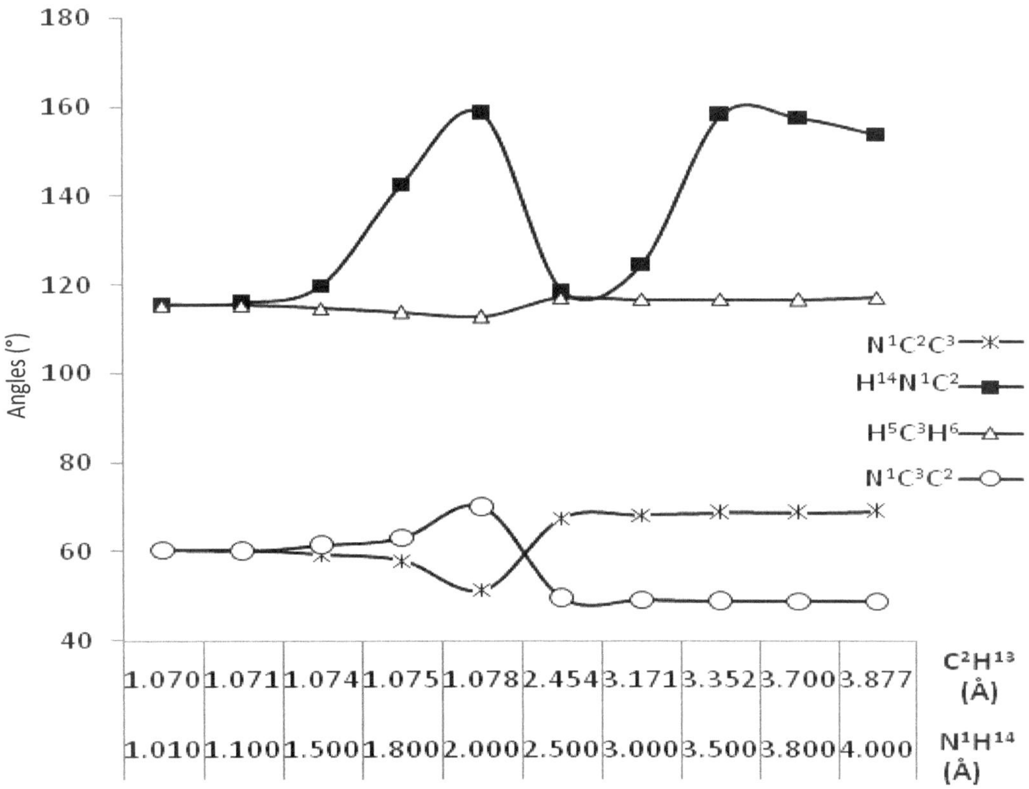

Figure 3. Variations of angles according to reaction coordinates N^1H^{14} and C^2H^{13}.

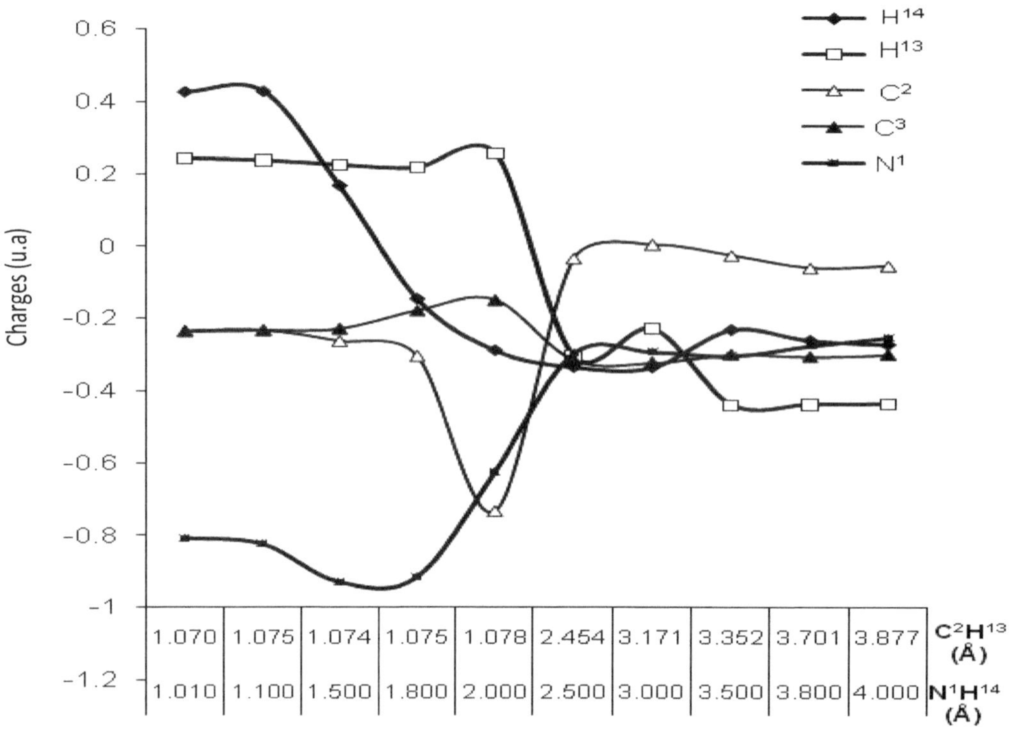

Figure 4. Variations of Mulliken atomic charges according to reaction coordinates N^1H^{14} and C^2H^{13}.

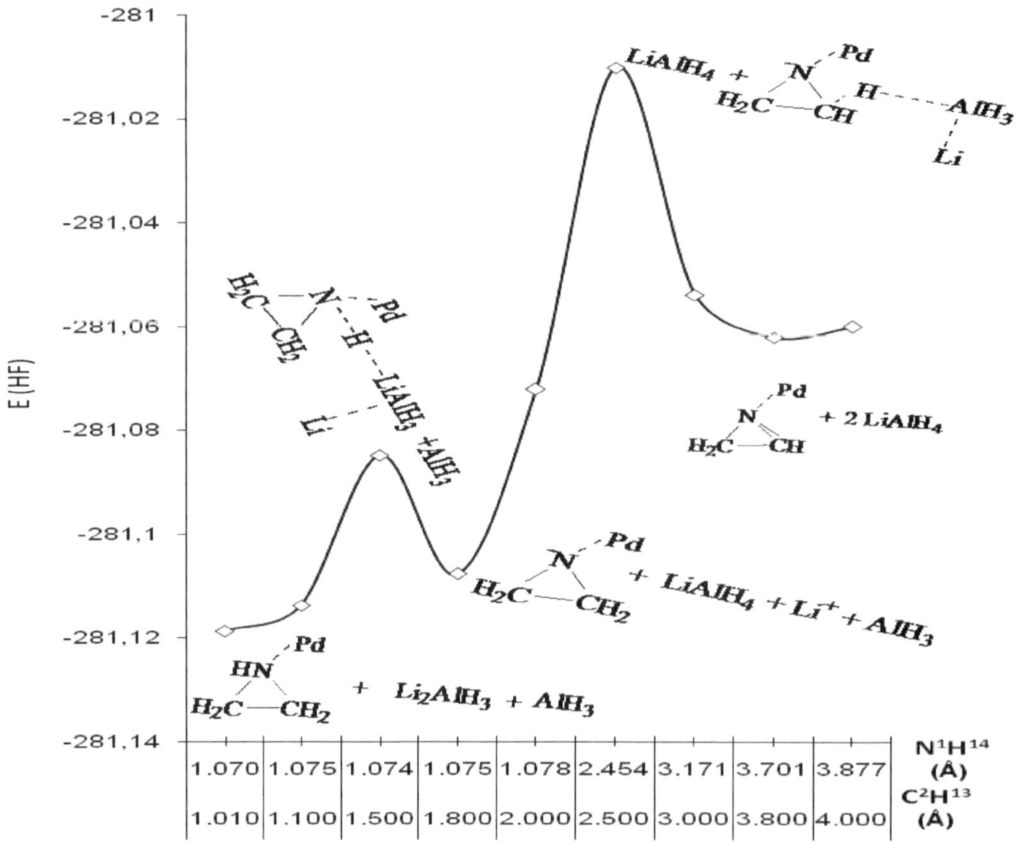

Figure 5. The variation of the energy of the system according to N^1H^{14} and C^2H^{13} coordinate.

Table 1. Calculated vibration frequencies of the reagent (2H-azirine) and the reaction product and standard IR absorption regions.

Bonds	Vibration frequencies (cm^{-1})		
	Before reaction (reagent 2H-azirine)	After reaction (Product aziridine)	Standard IR absorption region
C=N	1645.0	-	1500.0 - 1900.0
C-N	1092.5	1333 and 1333.7	800.0 -1300.0
C-H	3330.0 - 3505.0	3320.0 - 3334.0	2700.0 - 3800.0
N-H	-	3700.0	2700.0 - 3800.0

and the hydrogen atom H^{13}, between the aluminium atom Al^{15} and the hydrogen atom H^{14}, and between the palladium atom Pd^7 and the nitrogen atom N^1, exceed became very close at the end. This means that during the reaction, the double chemical bond N=C was transformed into a simple chemical bond between the two atoms N^1 and C^2. At the end of reaction, the final optimized values of N^1H^{14} and C^2H^{13} are 1.010 Å and 1.070 Å, respectively In addition, angles $N^1C^2C^3$ and $N^1C^3C^2$ of respective initial values 68.956° and 48.715° passed to approximately 60.300°. This result confirms the transformation which took place on the level of chemical bond N^1C^2 during the process. The very weak variation of the value of angle $H^5C^3H^6$ makes it possible to say that the carbon atom preserved its state of hybridization sp^3, during the process.

At the end of the reaction, the absolute values of the dihedral angles $H^{14}N^1C^2C^3$ and $H^{13}C^2N^1C^3$ are 105.886 and 111.135, respectively. These values are equal neither to 0° nor with 180°, which means that the atoms H^{13} and H^{14} are not in $C^3C^2N^1$ plan.

At the beginning, atoms N^1 and C^2 were bound by a double bond and carried negative electronic charge. But at the beginning as at the end of the reaction, the charge carried by atom N^1 is definitely more important than that carried by C^2 because of its electronegativity. The atom C^3, whose charge was equal to - 0.3 ua at the beginning of reaction, carries at the end of the reaction, the same charge of value -0.24 ua as atom C^2. This last observation confirms well that two atoms C^2 and C^3 are found finally in the same state of hybridization (sp^3). At the end of the reaction, hydrogen atoms H^{13} and H^{14} carry positive charge + 0.43 ua and + 0.24 ua, respectively. While fixing itself on N^1, H^{14} compensates for approximately 50% the N^1 electronic charge so that the remaining negative charge on N^1 corresponds only to its free electronic doublet. As for H^{13}, its positive charge compensates for exactly the negative charge of C^2, this last carbon atom do not have a free electronic doublet.

Figure 5 representative variation of the energy of the system according to the coordinates of reaction N^1H^{14} and C^2H^{13}, presents two maximum to 2.500 and 1.500 Å for N^1H^{14} and 2.454 and 1.074 Å for C^2H^{13}. Between the two preceding values of C^2H^{13}, the system was relaxed concerning this coordinate of reaction. C^2H^{13} then takes the optimized value of 1.075 Å T; value very close to that 3.000 Å. There are no more chemical bonds between these atoms. The interatomic distances like C^2N^1 and C^3N^1 which were different at the beginning of the process published in the literature for a chemical bond C–H (Potapov et al., 1981). At the second maximum of the curve of energy, the system was also released compared to the second coordinate of reaction N^1H^{14} whose optimized value was stabilized to 1.010 Å. This observation means that chemical bond N–H was formed, insofar as this value is of the same order of magnitude as for the length published in the literature for a N–H bond (Potapov et al., 1981).

Ultimately, the two maximum presented by the curve (Figure 5) correspond in two transition states associated with the formations of C^2H^{13} and N^1H^{14} bonds, respectively.

In order to clarify the structure of the product of the reaction, the vibration frequencies of the reagent (2H-azirine) and the product were calculated and consigned in the Table 1, which contains also certain data on the standard IR absorption regions (Silverstein et al., 1991).

The analysis of these results shows that between 1500.0 and 1900.0 cm^{-1}, a vibration frequency (1645.0 cm^{-1}) characteristic of C=N bond contained in the molecule of reagent. This frequency disappeared with the profit from frequencies 1333.0 and 1333.7 cm^{-1}, characteristic of the C–N bond contained in the product (Table 1). All this confirms the opening of the π bond between C^2 and N^1 atoms and thus the transformation of C=N into C–N. The frequency 3700.0 cm^{-1} which appeared only at the end of reaction, shows that it formed an N–H bond during the reaction. All these are only one confirmation that the results of the study carried out higher on the variation of the geometrical and different parameters.

On the basis of various stages of the reaction process indicated on Figure 6 of the geometrical parameters variations and the atomic charge of Mulliken and also on the basis of vibration frequencies of various chemical bonds on the level of the principal reagent and the product of the reaction, a probable mechanism of the reaction was proposed (Figure 6).

Conclusion

By methods of quantum chemistry and the hydrogenation

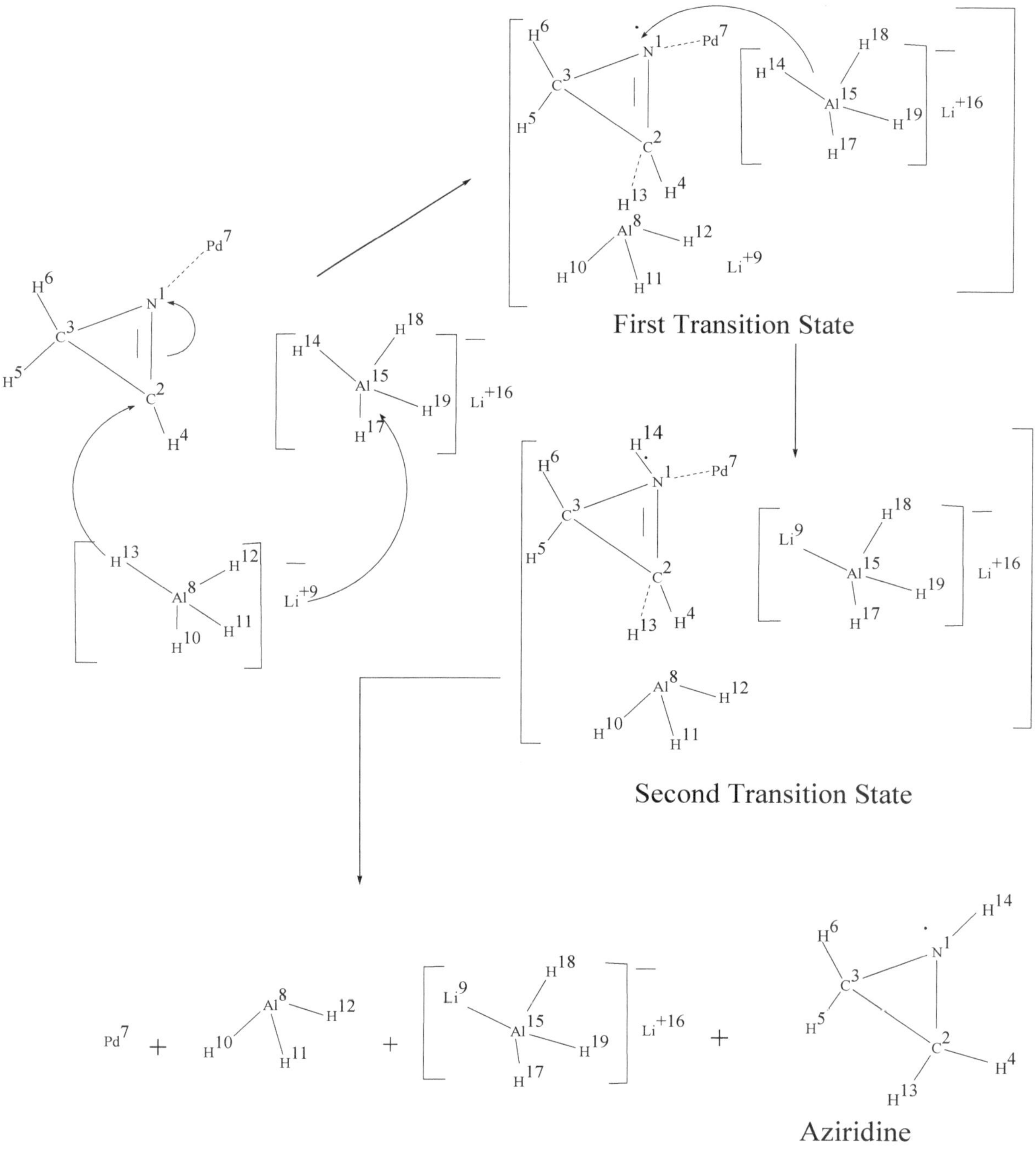

Scheme of reaction: C_2H_3N + $2LiAlH_4$ $\rightarrow$ C_2H_5N + $Li[LiAlH_3]$ + AlH_3

Figure 6. Probable mechanism of Hydrogenation of 2H-azirine by aluminium lithium hydride over palladium catalyst.

of 2H-azirine by aluminium lithium hydride over palladium catalyst was studied. The catalytic site was modeled by a palladium atom, insofar as the same reaction was carried out in experiments by other authors who used palladium supported by carbon as catalyst. According to the data of the literature and our results, it appeared that two molecules of aluminium lithium hydride reacted by providing each one to the molecule of 2H-azirine a hydrogen atom to give a molecule of aziridine, according to a mechanism confirmed by several studied parameters.

The results of our work made, once more, the proof of

the utility of the quantum chemistry in the field of the molecular reactivity, in the sense that it makes it possible to highlight steps of chemical conversions that the experiment alone cannot elucidate.

REFERENCES

Abarenkov IV, Bratsev BF, Touloub AV (1989). Natchala Kvantavoï Himii, Edition Vichaya Chkola ,Moscou, pp. 72-79.

Bimal KB, Khaled JB, Dilip RW, Maghar SM, Ajay KB (1999). Microwave-assisted rapid and simplified hydrogenation, J. Org. Chem., 64: 5746-5753.

Breysse M, Frety R, Vrinat M (1984). Unsupported cobalt molybdenum sulphide catalysts. Part II: Characterization and evolution of physicochemical properties during catalytic reaction, Appl. Catal., 13: 165-178.

Bachelier J, Duchet JC, Cornet D (1984). On the promoting effect in sulfided NiMo/Al_2O_3 catalysts as studied by chemisorptions, J. Catal., 87(2): 283-291.

Frisch MJ, Trucks GW, Schlegel HB, Scuseria GE, Robb MA, Cheeseman JR, Zakrzweski VG, Montgomery JA, Stratmann RE, Burant JC, Dapprich S, Millam JM, Daniels AD, Kudin KN, Strain MC, Farkas O, Tomasi J, Barone V, Cossi M, Cammi R, Mennucci B, Pomelli C, Adamo C, Clifford S, Ochterski J, Petersson GA, Ayala PY, Cui Q, Morokuma K, Malick DK, Rabuck AD, Raghavachari K, Foresman JB, Cioslowski J, Ortiz JV, Stefanov BB, Liu G, Liashenko A, Piskorz P, Komaromi I, Gomperts R, Martin RL, Fox DJ, Keith T, Al-Laham MA, Peng CY, Nanayakkara A, Gonzalez C, Challacombe M, Gill PMW, Johnson BG, Chen W, Wong MW, Andres JL, Head-Gordon M, Replogle ES, Pople JA (1998). Gaussian 98 (Revision A.1), Gaussian Inc., Pittsburgh PA,. Kasztelan S, Toulhoat H, Grimblot J, Bonnelle JP (1984). A geometrical model of the active phase of hydrotreating catalysts. Appl. Catal., 13: 127-159.

Klark T (1990) Kompioutornaya Himia, Edition MIR, Moscou, pp. 120-122.

Knouants IL (1988). Sovietskaya entsyklopédya, Tome1, Moscou., p. 76.

Lewis RJ (1993). Hawley's Condensed Chemical Dictionary, 12th Ed., New York, Van Nostrand Reinhold, pp. 490-491.

McCarty CG (1970). The Chemistry of the Carbon-Nitrogen Double Bond, S. Patai, Ed. Interscience, New York, p. 447.

Murena F, Gioia F (2004). Catalytic hydrotreating of 2,4-DDT and 4,4'-DDT, J. Hazard Mater., 112(1-2): 151-154.

Potapov V, Tatarintchik S (1981). Chimie organique, Edition Mir, p. 37.

Rivail JJ (1994) Éléments de chimie quantique à l'usage des chimistes. InterEditions / Editions du CNRS, Paris, pp. 169-193.

Scherr G, Steuerle U, Fikentscher R (1995). Imines, cyclic. In: Kroschwitz, JI, Howe-Grant M, eds, Kirk-Othmer Encyclopedia Chem. Technol., 4th Ed., New York, John Wiley, 14: 2-40.

Silverstein RM, Bassler GC, Morrill TC (1991). Spectrometric identification of organic compounds, Fifth Edition, John Wiley & Sons, Inc.New York, p. 93.

Yermakov YI, Startsev AN, Burmistrov VA, Shumilo ON, Bulgakov NN (1985). Sulphide catalysts on silica as a support. VI. On the structures of the active phase and active centers of hydrodesulfurization catalysts, Appl. Catal., 18: 33-46.

5

Separation and determination of trace amounts of Pb(II)ions in water samples using organic solution-processable functionalized-nano graphene

Ali Moghimi

Department of Chemistry, Varamin (Pishva) Branch Islamic Azad University, Varamin, Iran.
E-mail: alimoghimi@iauvaramin.ac.ir, kamran9537@yahoo.com.

A novel and selective method for the fast determination of trace amounts of Pb(II) ions in water samples has been developed. The first organic solution-processable functionalized-nano graphene (SPFGraphene) hybrid material with porphyrin–graphene nanohybrid, 5-4 (aminophenyl)-10, 15, 20-triphenyl porphyrin and its photophysical properties including optical triphenyl porphyrin (TPP) and grapheme oxide molecules covalently bonded together through an amide bond (TPP-NHCO-SPFGraphene) were used as absorbent for extraction of Pb(II) ions by solid phase extraction method. The complexes were eluted with HNO_3 (2M)10% $V.V^{-1}$ methanol in acetone and determined the analyte by flame atomic absorption spectrometry. The procedure is based on the selective formation of Pb(II) at optimum pH by elution with organic eluents and determination by flame atomic absorption spectrometry. The method is based on complex formation on the surface of the ENVI-18 DISK™ disks modified porphyrin–graphene nanohybrid, 5-4 (aminophenyl)-10, 15, 20-TPP and grapheme oxide molecules covalently bonded together through an amide bond (TPP-NHCO-SPFGraphene) followed by stripping of the retained species by minimum amounts of appropriate organic solvents. The elution is efficient and quantitative. The effect of potential interfering ions, pH, TPP-NHCO-SPFGraphene, amount, stripping solvent, and sample flow rate were also investigated. Under the optimal experimental conditions, the break-through volume was found to about 1000 ml providing a preconcentration factor of 600. The maximum capacity of the disks was found to be 398 ± 3 μg for Pb^{2+}. The limit of detection of the proposed method is 5 ng per 1000 ml. The method was applied to the extraction and recovery of Pb in different water samples.

Key words: Pb(II), Solid phase extraction (SPE), Octadecyl slica disks, flame atomic absorbtion spectrometry(FAAS), porphyrin–graphene nanohybrid, 5-4 (aminophenyl)-10, 15, 20-triphenyl porphyrin (TPP), grapheme oxide molecules covalently bonded together through an amide bond (TPP-NHCO-SPFGraphene).

INTRODUCTION

Trace amounts of metals are present in natural biosphere. Presence of some of these metals in very low concentrations and certain oxidation states are necessary. Higher concentrations and other oxidation states might be toxic and dangerous. The direct determination of trace metals especially toxic metal ions such as Cobalt, tin, arsenic, lead, antimony, and selenium from various samples requires mostly an initial and efficient pre-concentration step (Leyden et al., 1976). This preconcentration is required to meet the detection limits as well as to determine the lower concentration levels of the analyte of interest (Jones et al., 1983). This can be performed simply in many ways including liquid and solid phase extraction techniques (Nambiar et al., 1998; Caroli et al., 1991). The application of solid phase extraction technique for preconcentration of trace metals from different samples results in several advantages such as the minimal waste generation, reduction of sample matrix effects as well as sorption of the target species on the solid surface in a more stable chemical form

(Alexandrova and Arpadjan, 1993).

The normal and selective solid phase extractors are those derived from the immobilization of the organic compounds on the surface of solid supports which are mainly polyurethane foams (Arpadjan et al., 1997), filter paper (Leyden and Luttrell, 1975), cellulose (Gennaro et al., 1983) and ion exchange resins (Saber et al., 2005; Shin et al., 2005; Mazlum et al., 2003; Mahmoud and Soliman al., 1997). Silica gel, alumina, magnesia, and zirconia are the major inorganic solid matrices used to immobilize the target organic modifiers on their surfaces (Unger and Porous, 1979) of which silica gel is the most widely used solid support due to the well documented thermal, chemical, and mechanical stability properties compared to other organic and inorganic solid supports (Boudreau and Cooper, 1989). The surface of silica gel is characterized by the presence of silanol groups, which are known as weak ion exchangers, causing low interaction, binding and extraction of the target analytes (Kvitek et al., 1982). For this reason, modification of the silica gel surface with certain functional groups has successfully been employed to produce the solid phase with certain selectivity characters (Bruening et al., 1991). Two approaches are known for loading the surface of solid phases with certain organic compounds and these are defined as the chemical immobilization which is based on the chemical bond formation between the silica gel surface groups and those of the organic modifier, and the other approach is known as the physical adsorption in which direct adsorption of the organic modifier with the active silanol groups takes place (Unger and Porous, 1979). Selective solid phase extractors and preconcentrators are mainly based on impregnation of the solid surface with certain donor atoms such as oxygen, nitrogen, and sulfur containing compounds (Mahmoud, 1997a, 1997b; Tong et al., 1990; Dadler et al., 1987). The most successful selective solid phases for soft metal ions are sulfur containing compounds, which are widely used in different analytical fields. Amongst these sulfur containing compounds are dithiocarbamate derivatives for selective extraction of Pb(II) (Mahmoud, 1998, 1999) and preconcentration of various cations (Leyden et al., 1976; Narin et al., 2000; Choi and Choi, 2003; Tabarzadi et al., 2010; Tahaei et al., 2008; Ohta et al., 2001; Cuculic et al., 1997; Tehrani et al., 2005; Thurman and Mills, 1998; Pawliszyn, 1997; Izatt et al., 1996; Hagen et al., 1990; Krueger and Fild, 1995; Yamini and Ashraf-Khorassani, 1994; Shamsipur et al., 1999; Shamsipur et al., 2001; Brunner et al., 2003; Zelder et al., 2004; Boll et al., 2005; Nayebi et al., 2006; Moghimi et al., 2007; Moghimi et al., 2008; Moghimi, 2007a, 2007b) and 2- mercaptobenzothiazol-modified silica gel for online preconcentration and separation of silver for atomic absorption spectrometric determinations (Qiaosheng et al., 1998; Taylor et al., 1995). Ammonium hexa-hydroazepin-1-dithiocarboxylate (HMDC)-loaded on silica gel as solid phase preconcentration column for atomic absorption spectrometry (AAS) and inductively coupled plasma atomic emission spectrometry (ICP-AES) was reported (Alexandrova and Arpadjan, 1993).

Mercapto-modified silica gel phase was used in preconcentration of some trace metals from seawater (Moghimi et al., 2010). Sorption of Pb (II) by some sulfur containing complexing agents loaded on various solid supports (Tajodini and Moghimi, 2010) was also reported. 2-Amino-1-cyclopentene-1-dithiocaboxylic acid (ACDA) for the extraction of silver(I), Pb(II), and palladium(II) (Moghimi et al., 2006), 2-[2-triethoxysilyl-ethylthio] aniline for the selective extraction and separation of palladium from other interfering metal ions (Narin et al., 2000) as well as thiosemicarbazide for sorption of different metal ions (Campderros et al., 1998) and thioanilide loaded on silica gel for preconcentration of palladium(II) from water (Narin et al., 2000) are also sulfur containing silica gel phases.

In the present report, we wish to describe a proper concentrative method for the assessment of trace levels of Pb in different water samples. To the best of our knowledge, octadecyl silica membrane disks modified by (TPP-NHCO-SPFGraphene) have not been used for Pb isolation and preconcentration, previously this work was on the development of a rapid, efficient, and highly sensitive method for selective extraction and concentration of ultra trace amounts of Pb^{2+} ions from aqueous media using octadecyl silica membrane disks modified by porphyrin–graphene nanohybrid, 5-4 (aminophenyl)-10, 15, 20-triphenyl porphyrin (TPP) and grapheme oxide molecules covalently bonded together through an amide bond (TPP-NHCO-SPFGraphene) and Flame Atomic Absorption Spectrometric (FAAS) determination.

EXPERIMENTAL

Reagents

All acids were of the highest purity available from Merck and were used immediately when received. Methanol and Chloroform were of HPLC grade from Merck. Analytical grade nitrate salts of lithium, sodium, potassium, magnesium, calcium, strontium, barium, zinc, cadmium, lead, cobalt (II), and Pb (II) were of the highest purity. Ultra pure organic solvents were obtained from Merck (Darmstadt, Germany), and high purity double distilled deionized water was used throughout the experiments.

The stock standard solution of Pb^{2+} was prepared by dissolving 0.1000 g of the Pb powder in 10 ml concentrated nitric acid and diluted to 1000 ml with water in a calibrated flask. Working solutions were prepared by appropriate dilution of the stock solution from Merck (Darmstadt, Germany).The synthesis of the TPP-NHCO-SPFGraphene, was illustrated in Figure 1a and b.

Synthesis of TPP-NHCO-SPFGraphene

The first organic solution-processable functionalized-nano graphene (SPFGraphene) hybrid material with porphyrins. The synthesis of the porphyrin–Graphene nanohybrid, 5-4 (aminophenyl)-10, 15, 20-triphenyl porphyrin and grapheme oxide molecules covalently bonded together through an amide bond (TPP-NHCO-SPF Graphene, Scheme 1 and 2) was carried out

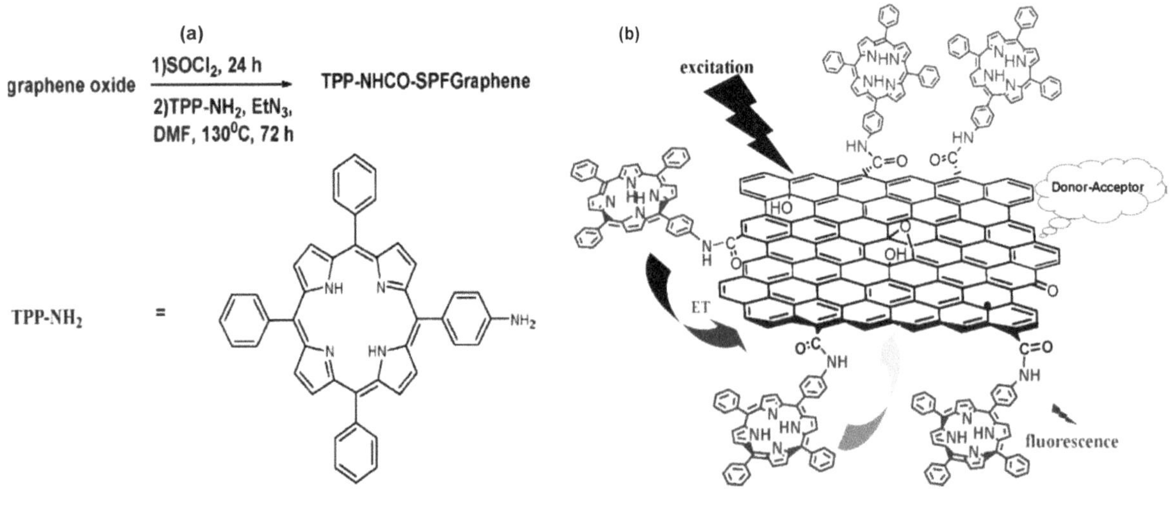

Figure 1. (a) Synthesis scheme of TPP-NHCO-SPFGraphene, (b) Schematic representation of part of the structure of the covalent TPP-NHCO-SPFGraphene.

using an amine-functionalized prophyrin (TPP-NH_2) and nano graphene oxide in N,N-dimethylformamide (DMF), following standard chemistry. Large-scale and water-soluble nano graphene oxide was prepared by the modified Hummers method (Becerril et al., 2008; Liu et al., 2008; Hummers and Offeman, 1958).

Results of atomic force microscopy characterization have confirmed that this grapheme material can be easily dispersed at the state of complete exfoliation, which consists of almost entire single-layered nano graphene sheets in H_2O (Becerril et al., 2008; Liu et al., 2008; Hummers and Offeman, 1958).

TPP-NH_2 and Nano graphene oxide molecules are covalently bonded together by an amide bond. Much care had been taken to make sure all the unreacted TPP-NH_2 had been removed using extensive solvent washing, sonication, and membrane filtration. Details are given in the experimental part. The attachment of organic molecules to nano graphene oxide has made TPP-NHCO-SPFGraphene soluble in DMF and other polar solvents (Becerril et al., 2008; Liu et al., 2008; Hummers and Offeman, 1958).

Apparatus

Determination of Pb^{2+} contents in working samples were carried out by a Varian spectra A-200 model atomic absorption spectrometer equipped with a high intensity hallow cathode lamp (HI-HCl) according to the recommendations of the manufacturers. TPP-NHCO-SPFGraphene, (40 mg) were packed into an SPE ENVI-18 DISK™ disks. Solid phase extractions were carried out by glassy membrane disks, ENVI-18DISK™ 47 mm diameter × 0.6 mm thickness containing octadecyl silica bonded phase (30 µm particles, 70 A° pore size) obtained from Supelco in conjunction with a standard Millipore 47 mm filtration apparatus equipped with a vacuum pump. The pH measurements were carried out by an ATC pH meter (EDT instruments, GP 353).

Sample extraction

Extraction were performed with glassy membrane disks, ENVI-18 DISK™ 47 mm diameter × 0.6 mm thickness containing octadecyl silica bonded phase (30 µm particles, 70 A° pore size) from Supelco. The disks were used in conjunctions with a standard Millipore 47 mm filtration apparatus connected to water aspirator (Moghimi, 2007).

Sample treatment

The water samples were filtered through 45 µm nylon filters. Sampling vessels were polyethylene bottles soaked in 1 ml^{-1} HNO_3 overnight and rinsed twice with deionized water. The analysis must be done within 2 days of sample collection to limit the risk of interconversion of Pb(II). Then, 5 ml of methanol was added to a 90 ml portion of each before analysis. The surface of the ENVI-18 DISK™ disks was modified with TPP-NHCO-SPFGraphene, and therefore could retain Pb^{2+} ions properly. Instead of 10 mg of TPP-NHCO-SPFGraphene, an appropriate volume of an organic solvent (5 ml) was miscible with water. The most suitable solvent under the experimental conditions was HNO_3 (2M)10% $V.V^{-1}$ methanol in acetone.

Disk cleaning and conditioning

A disk was placed in the apparatus and was washed with 10 ml of methanol to remove all contaminants arising from the manufacturing process and the environment. Then, the disk was dried by passing air through it for several minutes. To insure optimal extraction of the analytes of interest, the disk was again washed with 10 ml of methanol, immediately followed by 10 ml of water, without letting the surface of the disk to get dried. This step pre-wets the disk surface prior to extraction. Improper performance of this step causes slow flow–rate and poor analyte recoveries. It is important to avoid any air contact with the surface of the disk before the addition of the sample.

Sample addition

After complete homogenization, accurate volumes of the sample solutions (100 ml portions) were transferred to the top reservoir of the disk apparatus. At the same time, the solution was drawn through the disk by applying a mild vacuum. Application of vacuum was continued until the disk was completely dried (about 5 min).

Table 1. The effect of presence of TPP-NHCO-SPFGraphene on extraction percent of Pb(II)[a].

TPP-NHCO-SPFGraphene	pH	Extraction percent of Pb(II)[a]
Absence	2-6	0.03(7.5)[b]
Presence	2-6	98.9(2.1) to 65(2.8)

[a] Initial samples contained 10 µg of Pb(II) in 100 ml of water; [b] Values in parentheses are RSDs based on five individual replicate analyses.

Table 2. Effect of different eluting solvents on percentage recovery of Pb(II) adsorbed on the disk [a].

Stripping solution	Recovery (%)		
	2 ml	5 ml	10 ml
HNO_3 (2M)10% V.V^{-1} methanol in acetone	82.9 (2.6)[b]	92.8 (2.6)	98.7 (2.0)
Acidified methanol[c]	54.5 (2.3)	83.2 (2.2)	83.6 (2.9)
Ammoniacal methanol[d]	54.4 (2.5)	87.5 (2.6)	86.0 (2.8)
Nitric acid (1M)10% V.V^{-1} methanol	82.7 (1.6)	99.6 (1.5)	99.8 (2.3)
Acetonitril	36.6 (4.8)	46.5 (5.5)	69.0 (2.9)
Formic acid (1M)10% V.V^{-1} methanol	55.6 (1.2)	68.3 (2.0)	78.5 (2.8)
Hydrochloric acid (1M)10% V.V-1 methanol	52.7 (1.9)	92.0 (2.6)	90.2 (2.0)
Hydrochloric acid (1M)10% V.V^{-1} methanol	51.6 (2.5)	65.3 (2.5)	97.3 (1.6)
Methanol	52.8 (1.9)	85.2 (2.3)	82.4 (2.0)
Ethanol	64.0 (2.5)	85.6 (2.1)	86.9 (1.6)

[a] Initial samples contained 10 µg of each Pb in 100 ml water; [b] Values in parentheses are RSDs based on five individual replicate analysis; [c] Acidified solvents obtained by the addition of 0.1 M HCl; [d]Ammoniacal solvents obtained by addition of 0.1 M NH_3.

Analyte elution

In order to elute the analyte selectively, exactly 5 ml of acidified solvents 0.1 M HCl in methanol was passed through the disk and collected into a 5.0 ml volumetric flask under the extraction funnel. It was found that ultra pure alcoholic organic solvents were the best eluting agents. The concentration of Pb(II) in the eluates were then determined by FAAS using an external calibration graph.

RESULTS AND DISCUSSION

Evaluation of the role of the TPP-NHCO-SPFGraphene,

Some preliminary experiments were performed for investigation of absence or presence of TPP-NHCO-SPFGraphene, on the quantitative extraction of Pb(II). It was concluded that the membrane disk itself does not show any tendency for the retention of Pb(II), but introduction of 100 ml portions of aqueous Pb(II) samples containing 10 µg of Pb(II) and 10 mg of TPP-NHCO-SPFGraphene, which leads to its satisfactory retention (Table 1). The latter case is most probably attributed to the existence of a considerable interaction between Pb(II) and the TPP-NHCO-SPFGraphene. It should be mentioned that, the formation of stable complexes between Pb(II) and TPP-NHCO-SPFGraphene, at pH = 2 is probably due to an ion pair formation mechanism. However, at pH higher than 2 the retention and percentage recovery of Pb(II) are negligible.

Choice of eluent

In order to select the most appropriate eluent for the quantitative stripping of the retained Pb(II) on the disks, 5 ml of various non organic (each containing 10% V.V^{-1} methanol) and different organic solvents were tested. The results were tabulated in Table 2. As can be seen, the best eluting solvents were found to be 5 ml of HNO_3 (2M)10% V.V^{-1} methanol in acetone, resulting in quantitative elution of Pb(II) from the disk. It should be emphasized that the presence of HNO_3 (2M)10% V.V^{-1} methanol in acetone in any kind of employed solvents helps to better the contact of eluent with hydrophobic surface of the disk.

The effect of the pH

The pH of the sample solutions were adjusted to different values between 2 to 9 by the addition of hydrochloric acid or a suitable buffer such as sodium acceate-acetic acid or sodium dihydrogen phosphate- disodium hydrogen phosphate, and then solutions passed through the disks. Eventually, the metal ions were stripped by pure methanol or ethanol solutions followed by flame atomic absorption determination of the eluted Pb(II). Then, percentage recovery at various pH values was determined (Figure 2). According to the results shown in Figure 2, up to pH 4.0 to 4.5, complete recoveries are obtained. However, at higher pH values, percentage recovery decreases. This is due to fact that in an acidic solution, the protonation of TPP-NHCO-SPFGraphene occurs and there is a weak tendency for retention between Pb(II) and TPP-NHCO-SPFGraphene, whereas at higher values (pH > 5), Pb(II) reacts with hydroxide ions to produce $Pb(OH)_2$. Therefore, sodium acceate-acetic acid buffer with pH = 4.5 was used for the preconcentration step. Other solvents used for dissolving TPP-NHCO-SPFGraphene, were methanol and ethanol.

The influences of these solvents on the recoveries as a function of pH are compared and shown in Figure 2. Meanwhile, other organic solvents were not tested because of their restricted solubility and formation of two phases with aqueous solutions, and incompatibility with flame having higher pH values (> 7) were not tested because of the possibility of the hydrolysis of octadecyl silica in the disks (Moghimi, 2007). Pb (II) ions can be retained quantitatively by the modified membrane disk through the pH range from 4.0 to 4.5 However, at lower pH value (< 4.0), nitrogen atoms of the TPP-NHCO-SPFGraphene could be protonated and the stability of complex is reduced.

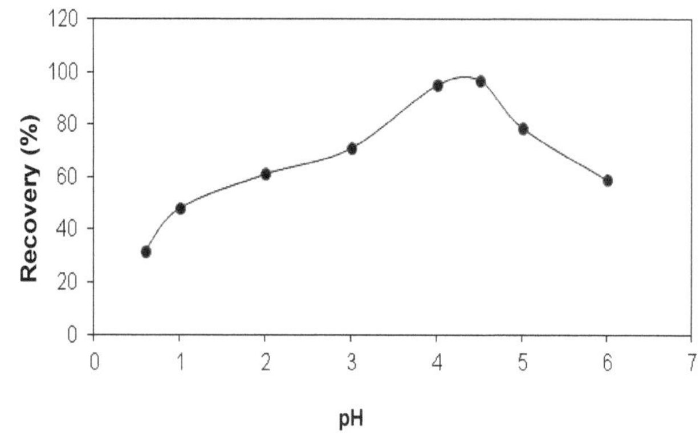

Figure 2. Influence of sample pH on the percentage recovery of Pb(II).

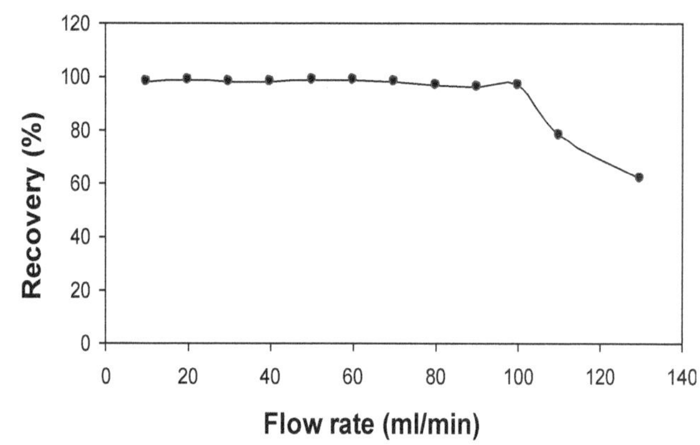

Figure 3. The effect of the flow-rate on extraction percent of Pb(II).

Effect amount of counter anion

In order to investigate the effect of counter ion on the recovery Pb^{2+} ions by the modified disks, different counter anions were tested (Table 3), it is immediately obvious that the nature of the counter anion strongly influences the retention of Pb ions by the disk. The results revealed that the TPP-NHCO-SPFGraphene, behaves in pH range of 4.0 to 4.5 (Moghimi, 2007) so that the Pb ions are retained as ion pair complexes by the membrane disks. As seen, acetate ion is the most efficient counter anion for the SPE of Pb(II) ions. The influence of the concentration of sodium acetate ion on Pb recovery was investigated, and the results are shown in Table 3.

As seen, the percent recovery of Pb^{2+} increased with the acetate concentration until a reagent concentration of about 0.1 M is reached, beyond which the recovery remained quantitative.

Moreover, acetate ion acts as a suitable buffering agent, while it effectively contributes to the ions pair formation; thus, in the Solid phase extraction (SPE) experiments, there was no need for the addition of any buffer solution.

Table 3. Percent recovery of Pb from the modified membrane disk in the presence of 0.01 M of different counter anions [a].

Counter anion[a]	Recovery (%)
Cl^-	15.9
Br^-	20.8
ClO_4^-	35.7
SCN^-	47.6
Picrate	74.7
Acetate	98.8

[a]Initial samples contained 10 µg of Pb(II) in 100 ml of water.

The influence of flow-rate

One of the most important parameters affecting solid phase extraction is the speed of the process. Hence, the effect of flow-rates on extraction efficiencies was investigated. It was found that in the range of 10 to 100 $ml.min^{-1}$, the retention of Pb(II) was not considerably affected by the sample solutions flow-rates and leads to reproducible and satisfactory results (Figure 3). Thus, the flow-rate was maintained at 89 $ml.min^{-1}$ throughout the experiment.

Table 4. Influence of the TPP-NHCO-SPFGraphene amount on the recovery of Pb(II) ions [a].

TPP-NHCO-SPFGraphene amount (mg)	Recovery of Pb(II) (%)
2	32.20(2.3)[b]
5	45.40(2.0)
8	83.24(2.8)
10	95.7(2.8)
15	98.5(2.0)
20	98.6(2.8)

[a]Initial samples contained 10 µg of each Pb in 100 ml water; b Values in parentheses are RSDs based on five individual replicate analysis.

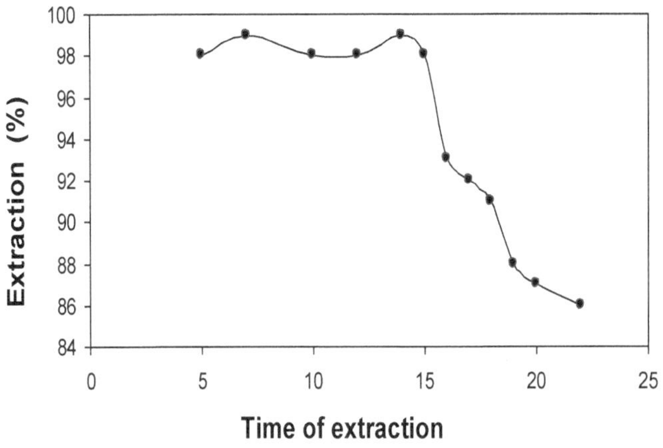

Figure 4. Influence of eluent (5 ml of HNO_3 (2M)10% $V.V^{-1}$ methanol in acetone) type on disk efficiency.

Quantity of the TPP-NHCO-SPFGraphene

The optimum amount of TPP-NHCO-SPFGraphene for the quantitative extraction of Pb(II) was also investigated by adding various amounts of it to the solution (between 2 to 20 mg). The results are listed in Table 4. The experimental results revealed that the extraction of Pb(II) was quantitative using a sample solution containing more than 10 mg TPP-NHCO-SPFGraphene. Hence, subsequent extractions were performed with 15 mg of TPP-NHCO-SPFGraphene.

Disk efficiency

Undoubtedly, one of the major parameters affecting the SPE determinations is the efficiency of the used membrane disks. However, to the best of our knowledge this case has not been discussed elsewhere in similar reports. Under the optimum experimental conditions, it was found out that, each ENV-18 DISK™ disk could perform at least 14 replicate analyses if organic eluting solvents are used. On the other hand, acidic, eluents practically decrease the number of time a disk could be used to perform 10 replicates. These observations are presented in Figure 4.

Analytical performance

When solutions of 10 µg Pb in 10, 50, 100, 600, 1000, 2000, 2500, and 3000 ml solutions under optimal experimental conditions were passed through the disks, the Pb(II) was quantitatively retained in all cases. Thus, the breakthrough volume for the method must be greater than 2500 ml, providing a concentration factor of > 600. The limit of detection (LOD) of the method for the determination of Pb(II) was studied under the optimal experimental conditions. The LOD based on 3σ of the blank (5 ml of methanol) is 5 ng per 1000 ml.

The capacity of the modified disks (5 mg TPP-NHCO-SPFGraphene) was determined by passing 50 ml portions of sample solutions containing 8 mg of Pb and 0.1 M sodium acceate-acetic acid buffer with pH 4.0 to 4.5, followed by the determination of the retained metal ions in the eluting solution using AAS. The maximal capacity of the disk obtained from three replicate measurements was 398 ± 3 µg of Pb^{2+} on the disk.

In order to investigate the selective separation and determination of Pb^{2+} ions from its binary mixtures with various metal ions, an aliquot of aqueous solutions (50 ml) containing 10 µg Pb^{2+} and mg amounts of other cations was taken and the recommended procedure was followed. The results are summarized in Table 5. The results show that the Pb(II) ions in binary mixtures are retained almost completely by the modified disk, even in the presence of up to about 100 mg of various ions. Meanwhile, retention of other cations by the disk is very low and they can be separated effectively from the Pb^{2+} ion. Its is interesting to note that, in other experiments, we found that in the presence of high enough concentrations, $NH_2OH.HCl$ is a suitable reducing agent (> 0.5 M) (Moghimi, 2007; Tuzen et al., 2005).

Analysis of water samples

To assess the applicability of the method to real samples, it was applied to the extraction and determination of Pb from different water samples. Tap water (Tehran, taken after 10 min operation of the tap), rain water (Tehran, 20 January, 2012), snow water (Tehran, 16 February, 2012) and Sea water (taken from Caspian Sea, near the Mahmoud-Abad shore) samples were analyzed (Table 6). As can be seen from Table 3, the added Pb ions can be quantitatively recovered from the water samples used. As it's seen, the recovered Pb ion revealed that the results are quite reliable and are in satisfactory agreement with those obtained by ICP-AES.

Table 5. Separation of Pb from binary mixtures[a].

Diverse ion	Amount taken (mg)	Found (%)	Recovery of Pb^{2+} ion (%)
Na^+	92.5[s]	1.15 (2.4)[b]	97.5 (2.7)
K^+	92.4	1.32 (2.3)	98.7 (2.2)
Mg^{2+}	24.23	0.70 (2.2)	97.5 (1.7)
Ca^{2+}	26.4	2.25 (3.0)	98.7 (1.8)
Sr^{2+}	2.45	2.85 (2.1)	98.7 (2.0)
Ba^{2+}	3.66	3.16 (2.1)	97.7 (2.3)
Mn^{2+}	2.64	1.75 (2.2)	96.3 (2.8)
Co^{2+}	2.13	6.40 (1.8)	95.0 (1.9)
Ni^{2+}	1.64	2.23 (2.4)	99.5 (2.4)
Zn^{2+}	2.76	4.97 (2.1)	98.7 (2.6)
Cd^{2+}	2.57	2.90 (2.0)	98.2 (2.5)
Hg^{2+}	1.65	2.81 (2.1)	97.7 (2.3)
Ag^+	2.34	3.45 (2.9)	98.6 (2.5)
UO^{2+}	2.85	2.80 (2.1)	97.3 (2.6)

[a]Initial samples contained 10µg Pb^{2+} and different amounts of various ions in 100 ml water(0.1 M acetate ion); b Values in parentheses are RSDs based on five individual replicate analysis.

Table 6. Recovery of Pb added to 1000 ml of different water samples (containing 0.1 M acetate at pH = 4.0-4.5).

Sample	Pb^{2+} added (µg)	Pb^{2+}determined ($ng.ml^{-1}$)	ICP-AES
Tap water	0.0	1.65 (2.8)[a]	ND
	10.0	11.90 (2.5)	11.8
Snow water	0.0	4.84 (2.3)	ND
	10.0	14.96 (2.3)	14.5
Rain water	0.0	2.85 (2.5)	ND
	10.0	12.82 (2.4)	12.9
Sea water	0.0	12.43 (2.3)	12.5
	10.0	22.70 (2.2)	23.6

[a]Values in parentheses are RSDs based on five individual replicate analysis; [b] Not detected.

Conclusion

Results presented in this work demonstrate very well the tremendous possibilities offered by the solid phase extraction of trace amounts of Pb(II) in water samples using Octadecyl Silica membrane disks modified by (TPP-NHCO-SPFGraphene) and its determination by FAAS. The method developed was simple, reliable, and precise for determining Pb in water. Also, the proposed method was free of interference compared to conventional procedures to determine Pb. (Moghimi, 2007; Shamsipur et al., 1999; Soylak et al., 2002; Saracoglu et al., 2001; Melek et al., 2006). The method can be successfully applied to the separation and determination of Pb in binary mixtures.

ACKNOWLEDGEMENTS

The author wish to thank the Chemistry Department of Varamin (pishva) branch Islamic Azad University and East Tehran branch Islamic Azad University for their financial support.

REFERENCES

Alexandrova A, Arpadjan S (1993). Elimination of sulfide interference by sodium hypochlorite solution in the cold vapor atomic absorption spectrometric determination of mercury using tin(II) reduction in alkaline medium. Analyst 118:1309-1313.

Arpadjan S, Vuchkova L, Kostadinova E (1997). Study of the adsorption

behavior of heavy metal ions on nanometer-size titanium dioxide with ICP-AES. Analyst 122:243-250.
Becerril HA, Mao J, Liu ZF, Stoltenberg RM, Bao ZN, Chen YS (2008). Evaluation of solution-processed reduced graphene oxide films as transparent conductors ACS Nano. 2:463-469.
Boll I, Kramer R, Brunner J, Mokhir A (2005). Oligonucleotide-Templated Reactions for Sensing Nucleic Acids J. Am. Chem. Soc. 27:7849-7855.
Boudreau SP, Cooper WT (1989). Analysis of thermally and chemically modified silica gels by heterogeneous gas-solid chromatography and infrared spectroscopy. Anal. Chem. 61:41-48.
Bruening ML, Mitchell DM, Bradshaw JS, Izatt RM, Bruening RL (1991). Removal of cesium from alkaline waste solution: Part II – Column ion exchange study. Anal. Chem. 63:21-27.
Brunner J, Mokhir A, Kramer R (2003). Copper(II)-Quenched Oligonucleotide Probes for Fluorescent DNA Sensing. J. Am. Chem. Soc. 125:12410-12415.
Campderros ME, Acosta A, Marchese J (1998). Selective separation of copper with Lix 864 in a hollow fiber module. Talanta 47:19-23.
Caroli C, Alimanti A, Petrucci F, Horvath Z (1991). Selective pre-concentration and solid phase extraction of mercury(II) from natural water by silica gel-loaded dithizone phases. Anal. Chim. Acta 248:241-247.
Choi YS, Choi HS (2003). Studies on Solvent Sublation of Trace Heavy Metals by Continuous Flow System as Ternary Complexes of 1,10-Phenanthroline and Thiocyanate Ion. Bull. Korean Chem. Soc. 24:222-228.
Cuculic V, Mlakar M, Branica M (1997). Synergetic adsorption of copper(II) mixed ligand complexes onto the SEP-PAK Cl8 column. Anal. Chim. Acta 339:181-189.
Dadler V, Lindoy LF, Sallin D, Schlaepfer CW (1987). Selective pre-concentration and solid phase extraction of mercury(II) from natural water by silica gel-loaded dithizone phases. Aust. J. Chem. 40:1557-1563.
Gennaro MC, Baiocchi C, Campi E, Mentasti E, Aruga R (1983). Undesirable and harmful metals in wines-Determination and removal. Anal. Chim. Acta 151:339-344.
Hagen DF, Markell CG, Schmitt GA (1990). Enhanced biosensor performance for on-site field analysis of explosives in water using solid-phase extraction membranes. Anal. Chem. 67:3363-3369.
Hummers WS, Jr. Offeman RE (1958). Preparation of graphitic oxide. J. Am. Chem. Soc. 80:1339-1339.
Izatt RM, Bradshaw JS, Bruening RL (1996). Non-destructive separation of metal ions from wastewater containing excess aminopolycarboxylate chelant in solution with an ion-selective immobilized macrocyclic. Pure Appl. Chem. 68:1237-1242.
Jones JS, Harrington DE, Leone BA, Bramdstedt WR (1983). Determination of trace elements in analytical-reagent grade sodium salts by atomic absorption spectrometry and inductively coupled plasma atomic emission spectrometry after preconcentration by column solid phase extraction. Atom. Spectrosc. 4:49-57.
Krueger CJ, Fild JA (1995). Method for the analysis of triadimefon and ethofumesate from dislodgeable foliar residues on turfgrass by solid-phase extraction and in-vial elution. Anal. Chem. 67:3363-3369.
Kvitek RJ, Evans JF, Carr PW (1982). Denaturation of purple membranes at the air/water interface studied by SEM. Anal. Chim. Acta 144:93-98.
Leyden DE, Luttrell GH (1975). Comparison of chelating agents immobilized on glass with chelex 100 for removal and preconcentration of trace copper(II). Anal. Chim. 47:1612-1616.
Leyden DE, Luttrell GH, Nonidez WK, Werho DB (1976). Adsorption of Co(II) and Cu(II) on silica gel surface modified with pyridinium ion from acetone and ethanol solutions. Anal. Chem. 48:67-72.
Leyden DE, Luttrell GH, Sloan AE, DeAngelis NJ (1976). Automated separation and preconcentration of copper(II) from natural waters using a column treatment. Anal. Chim. Acta 84: 97-102.
Liu ZF, Liu Q, Zhang XY, Huang Y, Ma YF, Yin SG, Chen YS (2008). Substrate effect on the electronic structures of CuPc/graphene interfaces. Adv. Mater. 20:3924-3928.
Mahmoud ME (1997a). Silica gel-immobilized Eriochrome black-T as a potential solid phase extractor for zinc (II) and magnesium (II) from calcium (II). Talanta 45:309-314.
Mahmoud ME (1997b). Silica-immobilized formylsalicylic acid as a selective phase for the extraction of iron(III). Talanta 44:15-21.
Mahmoud ME (1998). Proceeding of the 25th FACSS Conference, Austin, TX, USA, 11–15 October. pp. 297-303.
Mahmoud ME (1999). Selective solid phase extraction of mercury(II) by silica gel-immobilized-dithiocarbamate derivatives. Anal. Chim. Acta 398:297-302.
Mahmoud ME, Soliman EM (1997). Study of the selective extraction of iron (III) by silica-immobilized 5-formyl-3-arylazo-salicylic acid derivatives. Talanta 44:1063-1069.
Mazlum AM, Ensafi AA, Naeimi H, Dastanpour A, Shamelli A (2003). Synthesis and analytical application of a novel tetradentate N_2O_2 Schiff base as a chromogenic reagent for determination of nickel in some natural food samples. Russ. J. Electrochem. 39:269.
Melek E, Tuzen M, Soylak M (2006). Flame Atomic Absorption Spectrometric Determination of Cadmium(II) and Lead(II) after their Solid Phase Extraction as Dibenzyldithiocarbamate Chelates on Dowex Optipore V-493. Anal. Chim. Acta 578:213-219.
Moghimi A (2006). Solid phase extraction of lead on octadecyl bonded silica membrane disk modified with Isopropyl 2- [(isopropoxy carbothioyl)disulfanyl]ethane thioate and determination by flame atomic absorption spectrometry. Oriental J. Chem. 22(3):527-532.
Moghimi A (2007a). Solid Phase Extraction of Trace Copper(II) Using Octadecyl Silica Membrane Disks Modified with *N,N'*-Disalicylideneethylenediamine. Chin. J. Chem. 25(10):1536-1542.
Moghimi A (2007b). Preconcentration and Determination of Trace Amounts of Heavy Metals in Water Samples Using Membrane Disk and Flame Atomic Absorption Spectrometry. Chin. J. Chem. 25(10):640-645.
Moghimi A (2008). Preconcentration of Copper(II) Using Mesoporous Organo-Silicas and Determination by Flame Atomic Absorption Spectrometry. Chin. J. Chem. J. Korean Chem. Soc. 52(2):155-163.
Nambiar DC, Patil NN, Shinde VM (1998). Effects of Mercuric Chloride on the Immunological, Hematological, Biochemical Parameters and Diseases Resistance of Nile Tilapia Challenged with Aeromnas hydrophila. Fresenius J. Anal. Chem. 360:205-212.
Narin I, Soylak M, Elic L, Dogan M (2000). Solid Phase Extraction of Copper(II) from Aqueous Solutions by Adsorption of its 2-propylpiperidine-1-carbodithioate Complex on Alumina Column. Talanta 52:1041-1047.
Nayebi P, Moghimi A (2006). Preconcentration of mercury by adsorption of its diphenylthiocarbazone complex on an silica gel-immobilized Schiff base diphenylthiocarbazone (DPC) column and spectrophotometric determination. Oriental J. Chem. 22(3):507-512.
Ohta K, Tanahasi H, Suzuki T, Kaneco S (2001). Degradation of Reactive Yellow 86 with photo-Fenton process driven by solar light. Talanta 53:715-722.
Pawliszyn J (1997). Solid-Phase Microextraction, Theory and Practice, Wiley-VCH, New York. pp. 233-254.
Qiaosheng P, Qiaoyu S, Zhide H, Zhixing S (1998). Application of 2-mercaptobenzothiazole self-assembled monolayer on polycrystalline gold electrode as a nanosensor for determination of Ag(I). Anal. 123:239-243.
Saber TM, Rastegar F, Parchehbaf A, Rezvani Z (2005). Preconcentration and Determination of Copper(II)AND zinc(II) Using Octadecyl Silica Membrane Disks Modified by 1,5-Diphenylcarhazide and Flame Atomic Absorption Spectrometry. Chin. J. Chem. 23:1437-1442.
Saracoglu S, Soylak M, Elci L (2001). Enrichment and Separation of Traces of Cadmium, Chromium, Lead and Manganese Ions in Urine by Using Magnesium Hydroxide Coprecipitation Method. Trace Elem. Electrolytes 18:129-133.
Shamsipur M, Ghiasvand AR, Sharghi H (2001). Electrochemical oxidation of Catechol derivatives in presence of 4-Hydroxycoumarin Int. J. Environ. Anal. Chem. 82:23-29.
Shamsipur M, Ghiasvand AR, Yamini Y (1999). Preconcentration, Determination and Speciation of Chromium(III) Using Solid Phase Extraction and Flame Atomic Absorption Spectrometry. Anal. Chem. 71:4892-4897.
Shin DH, Ko YG, Choi US, Kim WN (2005). Synthesis and Properties of some Transition Metal Complexes with Water Soluble Hydroxy Functionalized Polyacrylonitrile. Ind. Eng. Res. 43:2060.

Soylak M, Uzun A, Elci L (2002). Sorbent Extraction of Copper, Lead, Nickel and Cadmium Ions in Dialysis Concentrates before their Atomic Absorption. Spectrometric Determinations, Trace Elem. Electrolytes 19:15-19.

Tabarzadi M, Abdouss M, Hasani SA, Shoushtary AM (2010). Mat.-wiss. u.Werkstofftech 41:4-221.

Tahaei P, Abdouss M, Edrissi M, Shoushtari AM, Zargaran M (2008). Mat.-wiss. u. Werkstofftech. 39:839.

Tajodini N, Moghimi A (2010). Preconcentration of Mercury by Adsorption from Natural Water by Silica Gel -Loaded Schiff's Base Diphenylthiocarbazone (DPC) phases. Asian J. Chem. 22(7):4994-5000.

Taylor KZ, Waddell DS. Reiner EJ (1995). Direct Elution of Solid Phase Extraction Disks for the Determination of Polychlorinated Dibenzo-p-dioxins and Polychlorinated Dibenzofurans in Effluent Samples. Anal. Chem. 67:1186-1192.

Tehrani MS, Moghimi A, Waqif HS (2005). Preconcentration and determination of trace amounts of heavy metals in water samples using membrane disk and flame atomic absorption spectrometry. Mater. Sci. Res. India 3(2):135-142.

Thurman EM, Mills MS (1998). Solid-Phase Extraction, Principles and Practice, Wiley, New York. pp. 344-352.

Tong A, Akama Y, Tanaka S (1990). Selective preconcentration of Au(III), Pt(IV) and Pd(II) on silica gel modified with γ-aminopropyltriethoxysilane). Anal. Chim. Acta 230:179-185.

Tuzen M, Soylak M, Parlar K (2005). Cadmium and Lead Contamination in Tap Water Samples from Tokat, Turkey. Bull. Environ. Contam. Toxicol. 75:284-289.

Unger K, Porous S (1979). Comparison of an ordered mesoporous aluminosilicate, silica, alumina, titania and zirconia in normal-phase high-performance liquid chromatography. Elsevier, Amsterdam. pp. 156-166.

Yamini Y, Ashraf-Khorassani M (1994). Solid-Phase Extraction and Determination of Trace Amounts of Lead(II) Using Octadecyl Silica Membrane Disks Modified with a Recently Synthesized Anthraquinone Derivative and Atomic Absorption Spectrometry. High Resolut. J. Chromatogr. 17:634-639.

Zelder FH, Brunner J, Kramer (2005). A fluorogenic and chromogenic probe that detects the esterase activity of trace copper(II)s). R. Chem. Commun. pp. 902-911.

6

Physicochemical properties and rheological behaviour of *Ficus glumosa* gum in aqueous solution

Paul Ocheje Ameh

Department of Chemistry, Ahmadu Bello University, Zaria, Kaduna State, Nigeria. E-mail: nocaseoche@yahoo.com.

The physicochemical properties, cationic composition and rheological behaviour of exudates from *Ficus glumosa* were investigated. The gum was found to be mild acidic, ionic and soluble in water, sparingly soluble in ethanoland insoluble in acetone and chloroform. The trend for the decreasing concentration of elements determined in the gum sample is Mg > Ca > Mn > Zn > Fe > Ni > Zn > Cu > Cd > Pb. Rheological properties of the gums have been adequately modeled using Huggins, Kraemer, Arrhenius, Tanglertpaibul and Rao theories. The results indicated that the viscosity of the gum increases with increasing pH and concentration but decreases with increase in temperature. Values of intrinsic viscosity obtained from Huggins and Kraemer plots were similar (11.197). The value of activation energy (E_{vis}) of the gum studied was found to be 1.915 kJ/mol. The low E_{vis} values obtained suggests the existence of few inter and intra-interactions between the molecules of the gums within the investigated temperature. *F. glumosa* gum exudate hence, possesses the potentials to be used as an additive in food and other industries.

Key words: *Ficus glumosa*, gum, heavy metal, characterization, rheological modeling.

INTRODUCTION

Gums are considered to be pathological products formed following injury to the plant or owing to unfavourable conditions, such as drought, by a breakdown of cell walls (extra cellular formation; gummosis) (Jani et al., 2009). They are complex carbohydrate derivatives of a polysaccharide nature and are either soluble in water as in the case of gum *Arabic* or form mucilages by the absorption of large amounts of water (gum tragacanth). Their principal use is in foodstuffs by nature of their ability to impart desired qualities to foods by influencing their viscosity, body and texture; most frequently in confectionery food, flavouring and soft drinks. They also have pharmaceutical and industrial applications as demulcents, adhesives in pill manufacture, lithography, paints, inks, corrosion inhibitors and as emulsifying agents. These gums found wider application because of their physical, rheological and chemical properties (such properties include solubility, water sorption, swelling capacity, pH, effect of temperature, and viscosity among others) (De Paula et al., 2001).

There is increasing in demands for gums globally because of its vast application, which causes the increase in prices of the existing gums in the local and international market. For the price to be stable and less expensive there is need to look for alternative or discoveries of more suitable natural gums which will be of the same quality or even surpass the existing ones. *Ficus glumosa* is an example of lesser known plant producing gum. The plant commonly called African rock fig tree in English and ‘kawuri’ in Hausa, belongs to the Family, Moraceae. *F. glumosa,* a tall plant of about 24 m in height and about 3 m in diameter, occurs on rocky outcrops, where it splits rocks; along dry watercourses or in open country; frequently in valleys, where it reaches its greatest size (Keay, 1989). The species also occurs in fringe forest in savannah areas, especially in swampy ground, and in swamp forest in coastal areas (Arbonnier, 2004). The bark also contains abundant sticky white latex which is used in Northern Nigeria like bird-lime to trap crickets (Ameh et al., 2012). The exudate is also chewed as chewing-gum and used for fastening arrowheads to their shaft.

Studies on the physicochemical and rheological properties of some gum exudates have been carried out.

Gum exudates from *Khaya senegalensis* (Family, Meliaceae) plants grown in Northern Nigeria were investigated for its physicochemical properties such as pH, water sorption, swelling capacity and viscosities at different temperatures using standard methods by Mahmud et al. (2008). It was found that *Khaya* gum appeared to be colourless to reddish brown translucent tears. 5%w/v mucilage has pH of 4.2 at 28°C. The gum is slightly soluble in water and practically insoluble in organic solvents. Water sorption studies revealed that it absorbs water readily and is easily dehydrated in the presence of desiccants. A 5%w/v mucilage concentration gave a viscosity value which was unaffected at temperature ranges (28 to 40°C). The results indicated that the swelling ability of *K. senegalensis* gum may provide potentials for its use as a disintegrant in tablet formulation, as a hydro gel in modified release dosage forms and the rheological flow properties may also provide potentials for its use as suspending and emulsifying agents owing to its pseudo plastic and thixotropic flow patterns.

Mhinzi (2002) analyzed gum samples from three selected *Albizia* species from Tanzania and determined their commercial potential by comparing their properties with those of *Albizia zygia* and *Acacia* gums. The properties of the gum exudates from *Albizia amara, Albizia pertesiana* and *Albizia harveyi* were found to be similar to those of *A. zygia* gum except that their aqueous solutions possess slightly lower viscosity and higher levels of tannin. The *Albizia* gums were much less soluble in water than *Acacia* gums; however, their methoxyl contents and acid equivalent weights (AEW) were similar to those of some *Acacia* gums.

Rheological properties of Xanthan and locust beans gums have been studied by Higiro et al. (2006) and the gums were found to obey Huggins and power law models and from the rheological modeling of the gums, the conformation of the gums were established. The aim of this work therefore is, to establish some of the physicochemical and rheological properties of *F. glumosa* gum, hence providing the information about the nutritional values, the metal composition and the functional properties of the gum.

MATERIALS AND METHODS

Collection of samples

Crude *F. glumosa* gum was obtained as dried exudates from their parent trees grown at Adepe - Otukpain Ogbadibo LGA of Benue State, Nigeria. The plant material had earlier been identified and authenticated in the herbarium Department of Biological Sciences of Ahmadu Bello University Zaria. The gum were collected from the plant species by tapping during the mid of July and in the day time (Smith and Montgomery, 1959).

Purification of the gum

The crude sample of *F. glumosa* gum consists of mixture of large and small modules admixed with bark and organic debris. These were hand sorted to remove fragments of bark and other visible impurities and then the gum particles were spread out in the sun to dry for 1 to 2 weeks.

The crude gum was dissolved in cold distilled water and the solution was strained through muslin and was then centrifuged, depositing a small quantity of dense gel. The straw coloured supernatant liquor was separated and acidified to pH of 2 with dilute hydrochloric acid. Ethanol was then added until it was 80%. The gum that precipitated out was removed by centrifugation at 2000 r/min, washed with alcohol followed by ether and was finally dried in a desiccator. The dried flakes were pulverized using a blender and stored in an air tight container.

Physiochemical analysis

In order to characterize the gums, it was subjected to the following physiochemical tests:

Determination of percentage yield of the purified gum

The dried, precipitated and purified gum obtained from the crude dried exudate was weighed and the percentage yield was expressed in percentage using the weight of the crude gum, as the denominator.

Determination of solubility

The solubility of the gum was determined in cold and hot distilled water, acetone, chloroform and ethanol. 1.0 g sample of the gum was added to 50 ml of each of the above mentioned solvents and left overnight. 25 ml of the clear supernatants were taken in small pre-weighted evaporating dishes and heated to dryness over a digital thermostatic water bath. The weights of the residue with reference to the volume of the solutions were determined using a digital top loading balance (Model.XP-3000) and expressed as the percentage solubility of the gums in the solvents (Carter, 2005).

Determination of concentration of metals

Concentrations of Zn, Mg, Ca, Mg, Mn, Fe, Cu, Cd and Pb were determined using perkin Elmer atomic absorption spectrophotometer. Calibration curve for each metal was prepared and the concentration of the metal was in the analyte and was estimated by extrapolation.

Determination of nitrogen and protein content

Crude protein content of the gum was determined using the Kjeldahl method with the nitrogen content being multiplied by a factor of 6.25 (Rodriguez et al., 2004).

Determination of pH

The sample powder was thoroughly mixed and 1 g and was dissolved in 100 ml of hot distilled water. The mixture was allowed to stand for 5 min at room temperature before the pH and temperature was recorded using a pre-calibrated pH meter (Oaklon pH meter, Model 1100).

Viscosity measurements

The apparent viscosity of the mucilage was measured using a digital Brookfield DV I prime viscometer. The intrinsic viscosity of the gum samples was determined in distilled water. The gum

Table 1. Physiochemical parameters of *F. glumosa* gum.

Parameter	*F. glumosa* gum
Colour	Pale – yellow
Odour	Honey
Taste	Bland
pH	4.3
Percentage yield (% w/w)	68%
Solubility (% w/v)	
Cold water	10.20
Hot water	11.10
Acetone	0.0
Chloroform	0.0
Ethanol	0.4
Nitrogen (%)	0.41
Protein (%)	2.56

Table 2. Cationic composition of *F. glumosa* gum.

Element	*F. glumosa* gum
Mg (%w/w)	1.28
Ca (%w/w)	0.60
Zn (ppm)	18.27
Mn (ppm)	105.53
Fe (ppm)	14.57
Ni (ppm)	10.12
Cu (ppm)	7.32
Cd (ppm)	1.69
Pb (ppm)	1.20

solutions were prepared by dispersing 50 mg of each gum sample (db, dry basis) separately in 100 ml of the distilled water at room temperature and mixed using a magnetic stirring overnight. The solution (2 ml) was transferred into a Cannon Ubbelohde capillary viscometer (Cannon Instruments, model I-71) which was immersed in a precision water bath to maintain the temperature at 25.0 ± 0.1°C and after equilibration for 10 min, the flow time was determined between the two etched marks. Serial dilution was performed *in situ* and three readings were taken for each dilution and the average obtained. The relative viscosity (η_{rel}) was calculated using Equation 1:

$$\eta_{rel} = \frac{T - T_o}{T_o} \quad (1)$$

WhereT is the flow time of gum solution in seconds, T_o is the flow time of solvent (water) in seconds. Microsoft Excel 2010 (Microsoft Corporation, Seattle, WA) was used to plot viscosities against concentrations, as well as to obtain linear regression lines with the corresponding equations and correlation coefficients (R^2) in order to assess the best model.

RESULTS AND DISCUSSION

Physiochemical parameters of *F. glumosa* gum

Table 1 presents physiochemical parameters of *F. glumosa* gum. The colour of crude *F. glumosa* gum is pale yellow. After purification, the yield of gum obtained was 68%w/w. The purified gum had bland taste and is odourless. The measured pH of the gum is 4.3 indicating that the gum is mild acidic. This compared favourably with that of gum *Arabic* which is reported to be between 4.0 and 4.8 (Belitz et al., 2004). The solubility of *F. glumosa* in cold water and hot water are 10.20 and 11.10%, respectively indicating that the solubility of the gum is temperature dependent. However, since solubility is expected to increase with increase in temperature, the solubility of the gum in hot water is higher than the corresponding solubility in cold water. On the other hand, the samples were not soluble in acetone and chloroform but were sparingly soluble in ethanol. These indicate that the studied gum is ionic in character, unlike the covalent counterpart, that is soluble in organic solvents (Sarah et al., 1998). The sparingly solubility of the gum observed in ethanol, is due to the fact that it can ionize to produce hydroxyl ion (OH^-) and the value of its dielectric constant is higher than those of acetone and chloroform. Consequently, ethanol has some polar character over acetone and chloroform which are characterized by low values of dielectric constant.

The nitrogen content of *F. glumosa* gum was found to be 0.41% which is higher than *Karaya* - 0.20% but lower than that of kondagogu gum 1.0% (Janaki and Sashidhar, 1998) and *Acacia* gum - 0.44% (Mhinzi and Mrosso, 1995). Protein content of *F. glumosa* gum was found to be 2.56%. This indicates a possibility of its usage as a supplement to other protein sources in the food.

Cationic composition of *F. glumosa* gum

Table 2 presents concentration of elements identified in the gum sample. Cationic composition of *F. glumosa* gum exudate demonstrated that calcium and magnesium are the main metals present among the metals determined. These major metals are also present in *Acacia senegal* gum (Anderson and Douglas, 1988). The high level of the magnesium and calcium shows that the exudates could provide an alternative source of magnesium and calcium in diets. Magnesium is required in large quantities by the body for the activation of enzymes involved in protein synthesis. Possible outcomes of deficiency are growth failure, behavioral disturbances, weakness and spasms (Murray et al., 2000).

The trend for the decreasing concentration of other elements determined in the gum sample is Mn > Zn > Fe > Ni > Zn > Cu > Cd > Pb. The body requires in a minute quantity of Mn for the activation of various enzymes, Co is nutritionally Co^{2+} and can function as a replacement

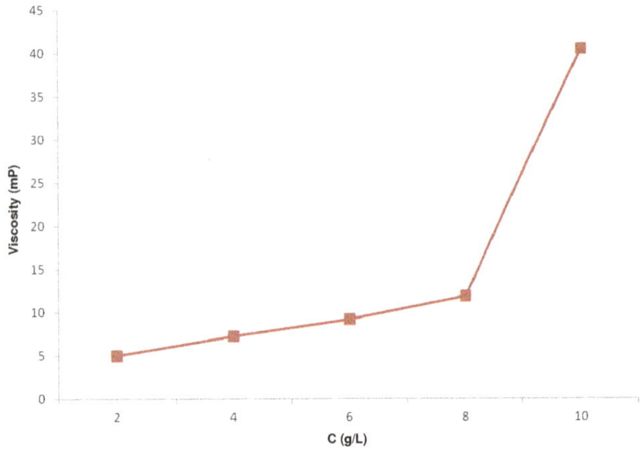

Figure 1. Variation of viscosity with concentration of *F. glumosa*, gums.

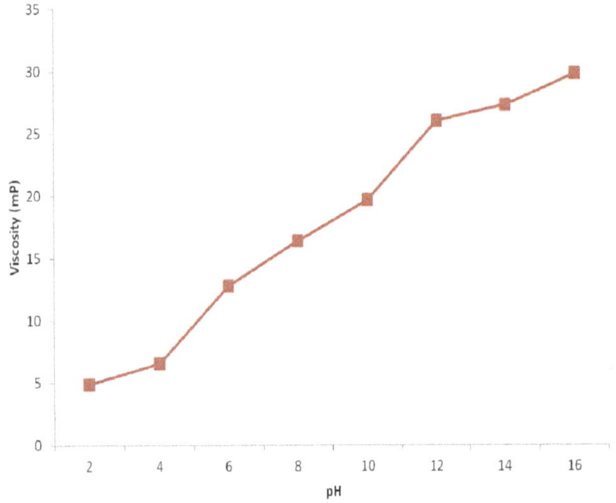

Figure 2. Variation of viscosity with pH for *F. glumosa* gum.

in vitro for other divalent cations in human (Underwood, 1977). Fe is needed in trace amount by the body as it is a constituent of hemoglobin and enzymes involved in energy metabolism in iron transport (Malhotra, 1998). The concentration of Pb detected in the gum was found to be within the permissible range of 2 ppm for substances used as food additives (FAO, 1999).

Rheological study

Rheological measurements provide useful behavioural and predictive information for various products in addition to knowledge of the effects of processing, formulation changes, and aging phenomena (Ebewele, 2000). Rheology deals with those properties of materials that determine their response to mechanical force. For solids, rheological studies involve elasticity and plasticity. For

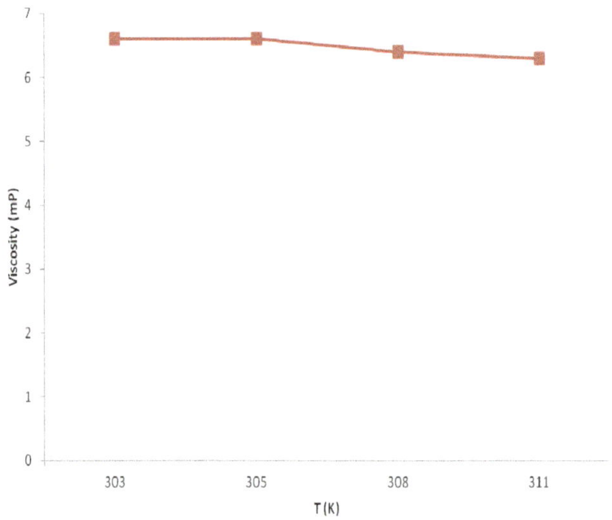

Figure 3. Variation of viscosity with temperature for *F. glumosa* gum.

fluids, on the other hand, this involves viscosity measurements, viscosity being a measure of the internal friction of a fluid.

Anderson and Dea (1967) stated that viscosity is one of the most important analytical and commercial parameters in polymers because it is affected by the size and shape of macro-molecules.

Figure 1 shows plots for the variation of viscosity with concentration of *F. glumosa* gum. From Figure 1, it is evident that the viscosity of *F. glumosa* gum tends to increase with increase in concentration. Jefferies et al. (1977) and Ashton et al. (1975) have reported that the viscosity of gum *Ghatti* and *Albizia* gums increases with increasing proportion of dispersible gel. This rheological behaviour is also exhibited by gum exudate studied. The high viscosity experience at higher concentrations may be due to increase in the strength of molecules-molecules interaction and the corresponding reduction in molecule-solvent interaction.

Figure 2 presents plots for the variation of viscosities of the gums with pH. Figure 2 reveals that the viscosity of the gum tend to increase with increase in pH indicating that the emulsifying properties of the gum is pH-dependent (Calvo et al., 1998).

In Figure 3, plot for the variation of viscosity with temperature for the studied gum is presented. In order to verify the onset of degradation or conformational transitions during heating, the viscosity of the gums was again measured as cooling proceeded. The tests did not indicate any difference between the two set of data suggesting that there was no degradation or conformational transition when the gums were heated (De Paula et al., 2001). Figure 3 revealed that the viscosity of the gum decreases with increasing temperature. This trend occurs because the increased kinetic motion at higher temperatures promotes the

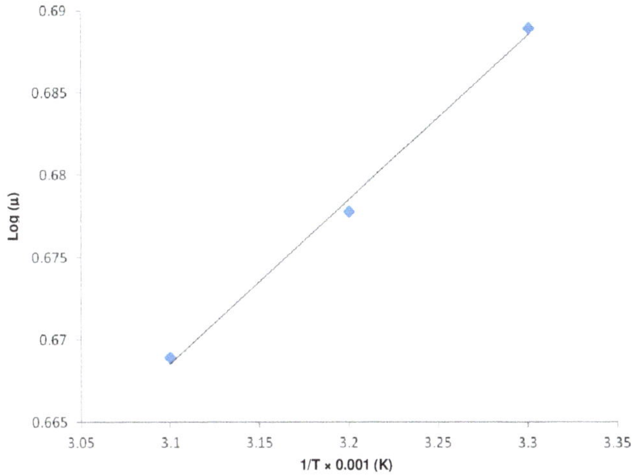

Figure 4. Arrhenius molecular kinetic plots of *F. glumosa* gum.

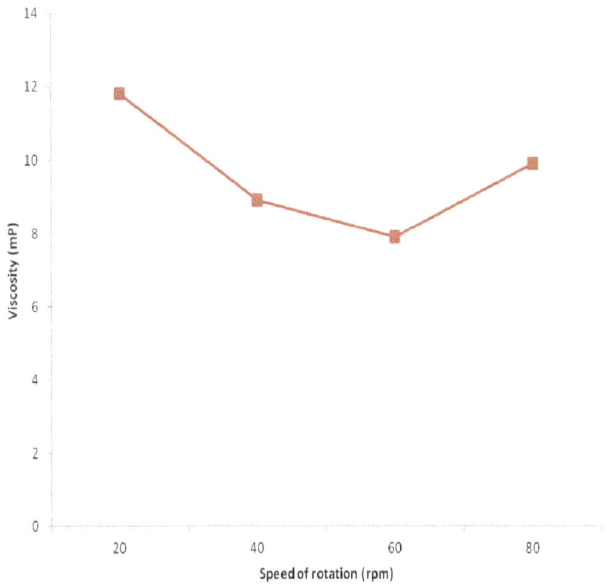

Figure 5. Variation of viscosity with rotation speed of *F. glumosa* gum.

breaking of intermolecular bonds between adjacent layers. From the viscosity results, the activation energy of viscous flow, E_{vis}, of the gum exudate was determined using the Arrhenius equation given by Equation 2:

$$[\eta] = Ae - E_{vis} / RT \qquad (2)$$

Where *T* is temperature, A is the Arrhenius coefficient and *R* is the universal gas constant. A plot of log [*η*] against *l*/*T* was made. The results obtained indicated a linear (R^2 = 0.9959) relationship between log(η) and 1/T. The slope of the plot (Figure 4) is equal to $-E_{vis} / 2.303R$ from which E_{vis} was evaluated. The value of E_{vis} of the gum studied was found to be 1.915 kJ/mol. The calculated activation energy of flow is relatively low when compared with data obtained for some plant gums such as *Arabic* gum (15 kJ/mol) (Chikamai and Banks, 1993) and *Albizia lebbeck* gum (15.9 kJ/mol) (De Paula et al., 2001). The low E_{vis} values obtained suggests the existence of few inter and intra-interactions between the molecules of the gums within the investigated temperature.

In order to study the effect of centrifugal forces on the viscosity of the gum, the viscosity of the gum were measured at various speeds of rotation. Figure 5 shows the plot of viscosity versus the rotation speed of a centrifuge containing similar concentration of the gum. From Figure 5, it can be seen that the viscosity of the gum tend to decrease with increasing speed of rotation until a critical value is reached after which the viscosity increases with increase in speed of rotation. Several explanations may be ascribed to the observed behaviour of the gum. One of such explanations is the increase in the extent of shaking as a result of increasing speed of rotation. Gundlah and Kapur (2003) found that the viscosity behaviour of polystyrene fractions in benzene and methyl ethyl ketone is markedly dependent on the extent of shaking of the solutions and concluded that the anomalous viscosity behaviour of polymers at high dilutions results from configurational changes of the macromolecules. Therefore, alteration in the composition of the various gums as a result of rotation may be responsible for the observed changes in viscosity of the gums with rotation speed of the centrifuge.

Intrinsic viscosity [η], represents the volume occupied by the unit mass of the polymer molecule. It is not actually a viscosity. This parameter is generally measured in order to obtain information about the molecular weight and conformation of a polymer (Hill et al., 1998; Ross-Murphy, 1995). Intrinsic viscosity is determined experimentally by measurements of the viscosity of solutions of very low concentrations (c<c*). Denoting solution and solvent viscosity as, η and η_s, respectively, it is defined formally by the following standard relationships:

$$\text{Relative viscosity: } \eta_{rel} = \eta / \eta_s \qquad (3)$$

$$\text{Specific viscosity: } \eta_{sp} = (\eta - \eta_s) / \eta_s = \eta_{rel} - 1 \qquad (4)$$

$$\text{Intrinsic viscosity: } [\eta] = \lim_{c \to 0} \frac{\eta_{sp}}{C} \qquad (5)$$

The intrinsic viscosity [η] was determined using Huggins, Kraemer Tanglertpaibul and Rao plots as proposed by Higiro et al. (2006).

The Huggins equation is given as Equation 6:

$$\eta_{sp} / c = [\eta] + K' [\eta]^2 C \qquad (6)$$

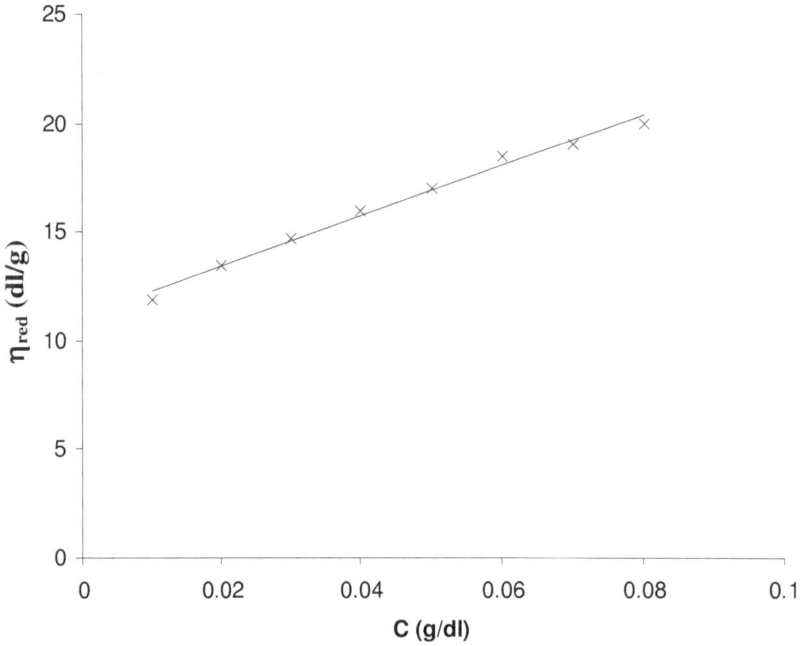

Figure 6. Variation of reduced viscosity with concentration (Huggins plot) for *F. glumosa* gum.

Table 3. Huggins and Kraemer parameters for *F. glumosa* gum.

Plot	Slope	[η]	$[\eta]^2$	k	R^2
Huggins	114.32	11.197	120.734	0.955	0.9874
Kraemer	8.300	11.397	136.306	0.059	0.9544

It can be deduced from Equation 6 that a plot of η_{red} versus C should be linear with slope equal to $[\eta]^2$ and intercept equal to [η]. Figure 6 shows Huggins plot for the gum exudate. Values of Huggins constant deduced from the plot are presented in Table 3. Based on experimental observations, the physical meaning of the dimensionless Huggins coefficient k can be summarized as follows (Sakai, 1968):

1) A polymer exhibits a higher value of k in a poor solvent than in a good solvent, that is, when the polymer-polymer interactions become favourable over polymer-solvent interactions,
2) It has a value of 0.5 to 0.7 in a theta solvent,
3) k is very sensitive to the formation of molecular aggregates.

The value of k obtained for the studied sample is 0.955. Therefore, the degree of polymer-polymer interaction in *F. glumosa* gum is low.
The intrinsic viscosity was also analyzed using the Kraemer equation, which is given by Equation 7 (Higiro et al., 2006) as follows:

$$\ln(\eta_{rel}/C) = [\eta] + k\,[\eta]^2C \qquad (7)$$

Where k″ is the Kraemer constant. From Equation 7, a plot of $\ln(\eta_{rel}/C)$ versus C should be a straight line with intercept and slope equal to [η] and k″[η], respectively. Figure 7 shows Kraemer plots for the studied gum. Values of Kraemer parameters deduced from the plot are recorded in Table 3. From the results obtained, k″values obtained from Kraemer plot is comparable to k' values obtained from Huggins plot. However, value of [η] deduced from Huggins plots are relatively higher than those obtained from Kraemer plot.

Tanglertpaibul and Rao (1987) obtained three equations that can also be used in the determination of the intrinsic viscosity of a polymer. These are Equations 8, 9 and 10.

$$\eta_{rel} = 1 + [\eta]C \qquad (8)$$

$$\eta_{rel} = \exp([\eta]C) \qquad (9)$$

$$\eta_{rel} = 1/(1-[\eta]C \qquad (10)$$

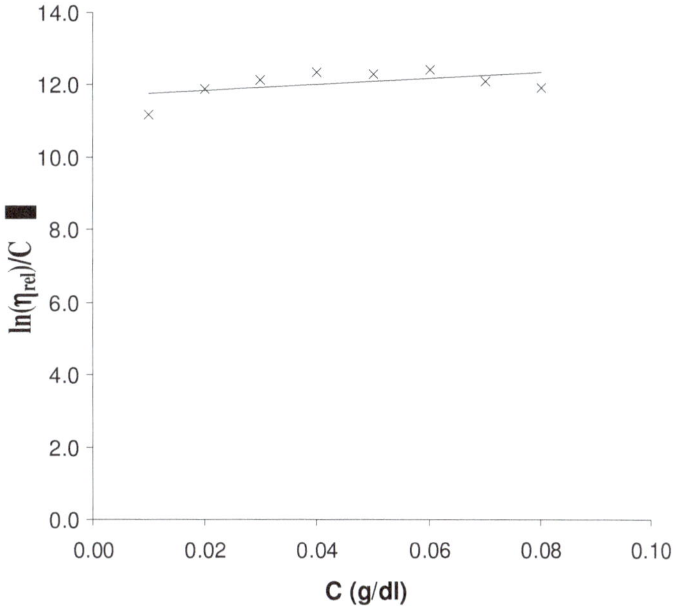

Figure 7. Variation of ln(relative viscosity) with concentration (Kraemer plot) of *F. glumosa* gum.

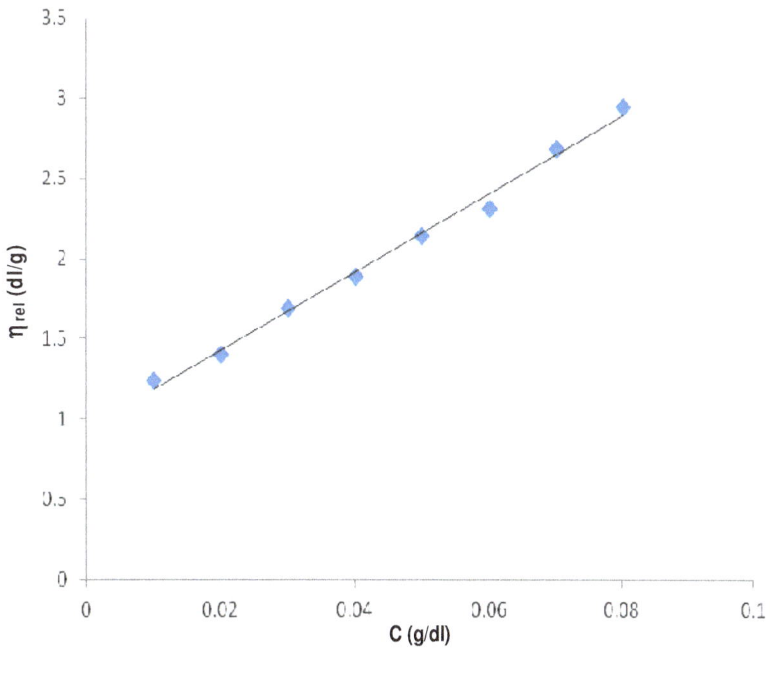

Figure 8. Variation of η_{rel} with concentration of *F. glumosa* gum.

From Equation 8, [η] is the slope obtained by plotting η_{rel} versus C. Also, the implication of Equation 9 is that, [η] is the slope obtained from the plot of $\ln\eta_{rel}$ versus C and from Equation 10, [η] is a slope obtained by plotting $(1-1/\eta_{rel})$ versus C. Tanglertpaibul and Rao plots for the studied gum (from Equations 8 to 10) are presented in Figures 8 to 10, respectively. Values of Tanglertpaibul and Rao parameters deduced from the plots are presented in Table 4. The excellent correlation between the values of [η] computed from the various methods

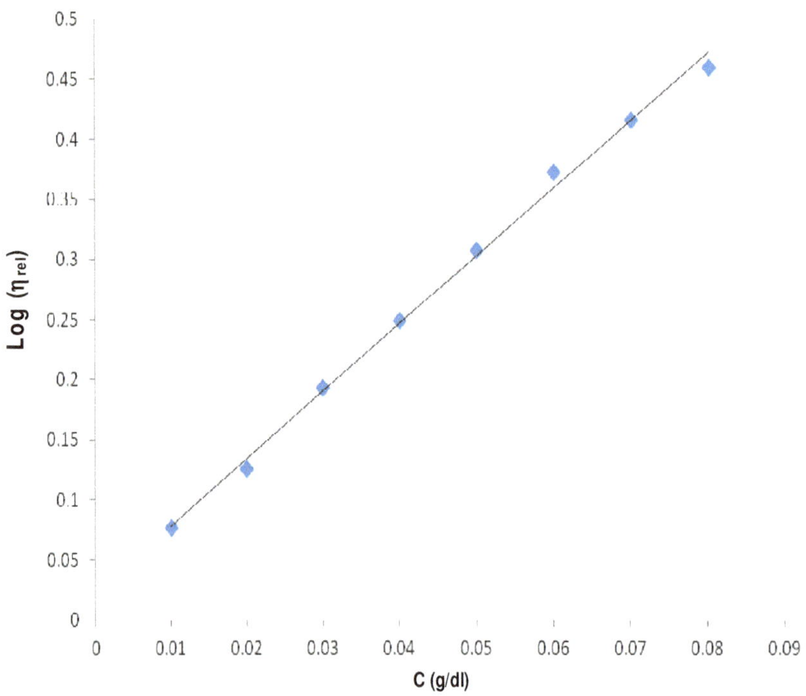

Figure 9. Variation of Log(η_{rel}) with concentration of *F. glumosa* gum.

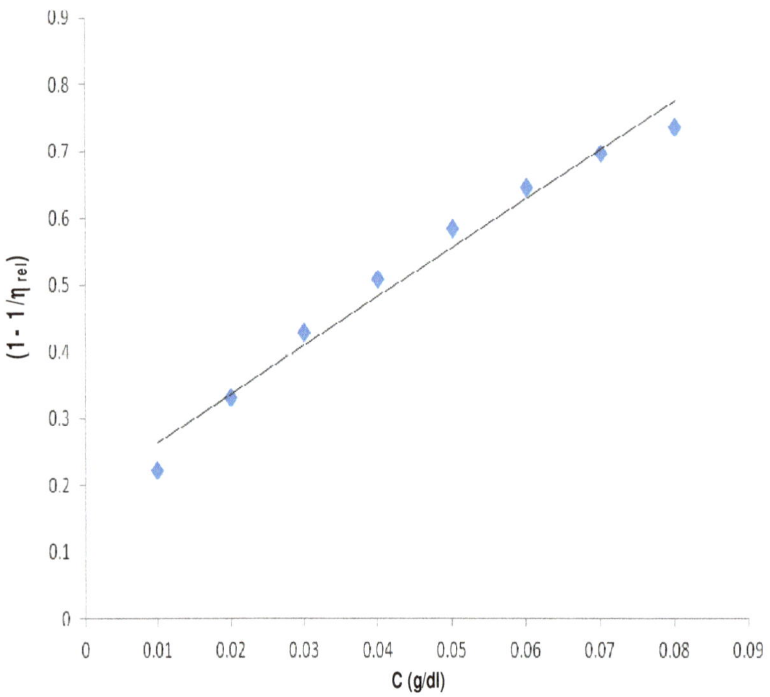

Figure 10. Variation of (1- 1/η_{rel}) with concentration of *F. glumosa* gum.

used in evaluating intrinsic viscosity indicates the consistency of the methods for the study of intrinsic viscosity of the system.

Conclusions

From the results obtained from the present study, the

Table 4. Tanglertpaibul and Rao parameters for *F. glumosa* gum.

Gum	Equation 8		Equation 9		Equation 10	
	[η]	R^2	[η]	R^2	[η]	R^2
F. glumosa	24.40	0.9928	5.6254	0.9967	7.3244	0.9756

following conclusions are drawn:

1) *F. glumosa* gum occurs as pale yellow, translucent tears. The gum mucilage has a bland taste, odourless, and mild acidic with a pH of 4.3 at 303 K. It is soluble in water, sparingly soluble in ethanol and practically insoluble acetone and chloroform.
2) The mineral elements compositions of the gum are within the range specified by World Health Organization (WHO) and Food and Agriculture Organization (FAO). Hence, *F. glumosa* gum exudate can be used as an additive in food and other industries.
3) *F. glumosa* gum is a macromolecule whose viscosity depends on temperature, concentration, speed of rotation and pH.
4) The rheological properties of the gum can be adequately modeled using the Arrhenius model, Huggins model, Kraemer model, Tanglertpaibul and Rao models. The excellent correlation between the values of [η] computed from the various methods used in evaluating intrinsic viscosity indicates the consistency of the methods for the study of intrinsic viscosity of the system.

REFERENCES

Ameh PO, Ladan M, Salihu T (2012). Corrosion inhibition and adsorption behaviour for mild steel by *Ficus glumosa*gum in H_2SO_4 solution. Afr. J. Pure Appl. Chem. 6(7):100-106.

Anderson DMW, Dea ICM (1967). The composition of the Gum from *Acacia drepanolobium*. Carbohydr. Res. 5(4):461-469.

Anderson DMW, Douglas DM (1988). The composition of the proteinaceous gums exuded by some *Leucaena*species, subspecies and hybrids. Food Hydrocoll. 2(3):247-253.

Arbonnier M (2004). Trees, shrubs and lianas of West African dry zones. CIRAD, Margraf Publishers Gmbh, MNHN, Paris, France. pp. 10-12.

Ashton WA, Jefferies M, Morley RG, Pass G, Phillips GO, Power DMJ (1975). Physical properties and applications of aqueous solutions of *Albizia zygia* gum. J. Sci. Food Agric. 26:697-704.

Belitz HD, Grosch W, Schieberle P (2004). Food Chemistry 3rd edition, Springer, Berlin. P. 3.

Calvo C, Martinez-Checa F, Mota A, Quesada E (1998). Effect of cations, pH and sulfate content on the viscosity and emulsifying activity of the Halomonaseurihalina Exopolysaccharide. J. Ind. Microbiol. Biotechnol. 20:205-209.

Carter SJ (2005). Tutorial Pharmacy: Solution. Great Britain: Pitman Press. pp. 15-18.

Chikamai BN, Banks WB (1993). Gum arabic from *Acacia senegal* (L.) wild in Kenya. Food Hydrocoll. 7(6):521-527.

De Paula RCM, Santana SA, Rodrigues JF (2001). Composition and rheological properties of *Albizia lebbeck* gum exudates. Carbohydr. Polym. 44:133-139.

Ebewele RO (2000). Polymer Science and Technology, CRC Press, New York.

FAO (1999). Compendium of food additive specifications addendum 7. Food and nutrition paper, No. 52. Add. 7. Joint FAO/WHO ExpertCommittee on Food Additives 53rd Session Held in Rome, 1–10 June 1999. Rome: Food and Agriculture Organization of the United Nations.

Gundlah S, Kapur SL (2003). Polymer solutions, IV. Anomalous viscosity behaviour of polystyrene and polymethyl acrylate solutions at high dilutions. Macromol. Chem. Phys. 42(1):127-139.

Higiro J, Herald TJ, Alavi S (2006). Rheological study of xanthan and locust bean gum in dilute solution. Food Res. Int. 39(2):165-175.

Hill SE, Ledward DA, Mitchell JR (1998). Functional Properties of Food Macromolecules (2nd edition). Gaithersburg, Maryland. 1-3

Janaki B, Sashidhar RB (1998). Physicochemical analysis of *gum kondagogu:* A potential food additive. Food Chem. 61(1-2):231-236.

Jani GK, Shah DP, Vipul D, Prajapati VD, Vineet C, Jain VC (2009). Gums and mucilages: Versatile excipients for pharmaceutical formulations. Asian J. Pharm. Sci. 4(5):309-323.

Jefferies M, Pass G, Phillips GO (1977). Viscosity of aqueoussolutions of *gum ghatti*. J. Sci. Food Agric. 28:173-179.

Keay RW (1989). Trees of Nigeria. Claredon Press Oxford. pp. 18-19.

Mahmud HS, Oyi AR, Allagh TS (2008). Studies on some physicochemical properties of *Khaya senegalensis* gum Nig. J. Pharm. Sci. 7(1):146-152.

Malhotra VK (1998). Biochemistry for Students.10th Edition. Jaypee Brothers Medical Publishers Ltd, New Delhi, India. pp. 68-69.

Mhinzi GS (2002). Properties of gum exudates from selected *Albizia* species from Tanzania. Food Chem. 77(3):301-304.

Mhinzi GS, Mrosso DS (1995). Chemotaxonomic aspects of gum exudates from some acacia species. Food Chem. 54(3):261-265.

Murray RK, Granner DK, Mayes PA, Rodwell VW (2000). Harper's Biochemistry, 25th Edition, McGraw-Hill, Health Profession Division, U.S.A. pp. 36-38.

Rodriguez GO, De Ferrier BS, Ferrier A, Rodriguez B (2004). Characterization of honey produced in Venezuela. Food Chem. 4:499-502.

Ross-Murphy SB (1995). Structure-property relationships in food biopolymer gels and solutions. J. Rheol. 39(6):1451-1463.

Sakai T (1968). Huggins constant k' for flexible chain polymers. J. Polym. Sci. A-2, 6:1535-1549.

Sarah LV, Singer BW, Hictchen SM, Townsend JH (1998). The Development and initial application of a gas chromatographic method for the characterization of gum media. J. Am. Inst. Conversation 37:295-311.

Smith F, Montgomery R (1959). The Chemistry of plant gums and mucilages. Reinhold publishing corporation New York. pp. 23-25.

Tanglertpaibul T, Rao MA (1987). Intrinsic viscosity of tomato serum as affected by methods of determination and methods of processing concentrates. J. Food Sci. 52(6):1642-1688.

Underwood EJ (1977). Trace Elements in Human and Animal Nutrition. 4th ed. Academic Press Inc. New York.

Kinetic, mechanism and spectral studies of bis (dibenzoyl-methane) copper (II) complex and its adducts

Mary Adelaide Oladipo[1], Olugbenga Solomon Bello[1]* and Adewumi Idowu Adeagbo[2]

[1]Department of Pure and Applied Chemistry, Ladoke Akintola University of Technology, P. M. B 4000, Ogbomoso, Oyo State, Nigeria.
[2]Department of Chemistry, Emmanuel Alayande College of Education Oyo, Oyo State, Nigeria.

The copper (II) complex of 1, 3-diphenyl-1, 3-propanedione (dbm) and its adducts with 2, 2-bipyridine (bipy) and 1, 10-phenantroline (phen) were synthesized. They were characterized by micro analytical data, infrared and ultraviolet (UV)-visible spectra and magnetic susceptibility data. The kinetics of the formation of the complex follows a rapid pre-equilibrium in which the rate is independent of the metal concentration, suggesting an interchange associative mechanism.

Key words: Copper (II)-adducts, kinetics, mechanism of formation, interchange associative mechanism.

INTRODUCTION

β-dicarbonyl compounds are very versatile ligands. They exhibit various bonding modes to metal ions (Stephen, 1989). Copper (II) complexes and adducts play important roles in both synthetic and structural research because of their preparative accessibility, structural diversity (Ahmed et al., 2007; Aliyu and Sanni, 2008) and biological activities (Patel et al., 2008; Shalin and Saxena, 2009). Various complexes involving copper II ions and β-dicarbornyl compounds have been synthesized and characterized (Ahmed et al., 2007; Patel, 2009; Sylvester, 2006). Studies show that the commonest types of diketonate complexes are those with stoichiometries: [M -(β-dike)$_3$] and [M- (β-dike)$_2$] (Sylvester, 2006). Bis-(β-diketonato) metal (II) complex behaves as Lewis acid and forms five or six coordinate adducts with neutral donor molecules (Harding et al., 2010). The kinetics and mechanism of the reaction of Cu (II) with pentane-2,4-dione and 4, 4, 4.-trifluoro-1-(2-thienyl) butane-1, 3-dione have been investigated by different authors. It was reported that the keto form of pentane-2, 4-dione reacts with Cu (II) at an abnormally slow rate. Complex formation was proposed to occur via a precursor complex in which the metal ion is loosely bound to the protonated keto tautomer.

The rate of the reaction was 2 x $10^4 dm^3 mol^{-1} s^{-1}$. As a comparison, the solvent exchange rate for Cu (II) amounts to $1x10^9 dm^3 mol^{-1} s^{-1}$ (Hugee et al., 1993). In another study, it was further asserted that rapid reactions such as those involving the formation of complexes of Cu (II) are likely to be affected by steric factors (Micheal and Brigid, 1979). Sterically, square planar complexes are less favored, but with large, bulky ligands, steric hindrance may prevent additional ligands from co-ordinating to the metal. Dibenzoylmethane (1,3-diphenyl-1,3-propanedione,dbm) has been used as chelate "blocking" ligand due to the bulky phenyl groups which reduce packing efficiency, providing opportunities for solvated structures. In [Cu (dbm)$_2$], the Cu (II) atom is located on a symmetry centre and is surrounded by four O atoms from two dbm molecules forming a plane-square coordination environment. The aims of this work are (i) to

*Corresponding author E-mail: osbello@yahoo.com.

Table 1. Analytical data of the [Cu $(dbm)_2(H_2O)_2$] complex and its –bipy and phen adducts.

Compound	Molecular formula	Formula mass	Colour	Yield (%)	Melting point (°C)	Analysis; found (Calc.) (%)		
						Cu	C	H
Cu $(dbm)_2$. $(H_2O)_2$	$Cu[C_6H_5COCH_2COC_6H_5]_2$ $2H_2O$	546.092	Green	60	270	11.54 (11.64)	65.46 65.98)	4.61 (4.29)
Cu $(dbm)_2$ phen	$Cu[C_6H_5COCH_2COC_6H_5]_2$ $C_{12}H_8N_2)$	710.22	Green	46	220	6. 17 (6.05)	76.01 (75.45)	4.48 (4.41)
Cu$(dbm)_2$ bipy	$Cu[C_6H_5COCHCOC_6H_5]_2$ $(C_{10}H_8N_2)$	666.242	Green	50	230	9.87 (9.54)	72.20 (72.11)	4.54 (4.53)

Table 2. Infrared bands for the ligand, complex and adducts (cm^{-1}).

Compound	Y_{c-o}	$Y_{C=O+C=C}$	Y_{Cu-N}	Y_{Cu-o}
dbm	1228	1596	-	-
[Cu$(dbm)_2(H_2O)_2$]	1218	1542	-	548
[Cu$(dbm)_2$phen]	1231	1593	570	548
[Cu$(dbm)_2$bipy]	1231	1593	570	548

prepare [Cu (dbm) $_2(H_2O)_2$], (ii) to react it with 2, 2-bipyridine (bipy) and 1, 10-phenantroline (phen) to form adducts,(iii) to determine the kinetic of formation of [Cu$(dbm)_2(H_2O)_2$] in view of exploring the mechanism of reactions and (iv) to use the proposed mechanism alongside spectroscopic data to suggest the formula of the probable intermediates.

MATERIALS AND METHODS

Synthesis

Preparation of the Cu(II) complex with dbm

2.99 g (0.015 moles) of Cu (II) acetate monohydrate in 100 cm^3 of aqueous methanol was reacted with 6.72 g (0.03 moles) of 1,3-diphenyl-1,3, propanedione in 20 cm^3 of methanol on a magnetic stirrer. Precipitation occurred immediately, and the resulting compound was washed with water and aqueous methanol before drying in a desicator over anhydrous calcium chloride

Preparation of the bipy and phen adducts

0.77 g (0.0040 moles) of the solid complex was added gradually to 1.00 g (0.0066 moles) of bipy in 25 cm^3 of chloroform on a magnetic hot plate stirrer at 70°C. This was stirred for about 1 h before filtering by suction; the collected adduct was then dried in a desiccator over silica gel. The phen adduct was similarly prepared.

Electronic absorption spectra

The electronic spectra of the complex and adducts were measured in aqueous methanol on a Lambda 3B ultraviolet (UV)/Visible spectrophotometer at 298 K.

Infra red spectra

The infrared spectra of the complex and adducts were recorded between 400 to 4000 cm^{-1} on a Nicolet Aviatar 330 Fourier Transform InfraRed (FT-IR) spectrophotometer using KBr discs.

Kinetic measurements

Kinetic measurements were carried out in thermostated water bath at 298K. Solutions of aqueous methanol in large excess of metal (10^{-3}M) to ligand (10^{-5}M) were mixed together to ensure pseudo-first order kinetic runs at various concentrations. This was done on a Lambda 3B UV/Visible spectrophotometer at 259 nm.

RESULTS AND DISCUSSION

Elemental analysis

The results of the C, H and metal analysis show that the copper-dbm complex was formed with 1:2 metal: ligand ratio as expected for divalent metal complexes and from the crystal structure (Table 1).

Infrared spectra

Table 2 reports the IR bands for the ligand, complex and adducts. The cis-enol form of β-diketones are characterized by strong intramolecular hydrogen bond (Tayyari et al., 2007).

Thus, a weak but broad band at 3442 cm^{-1} is attributed to enolic OH stretch. On chelation, the only band in this

Table 3. Observed rate const *(k_{obs})*, at varied concentration of Cu (II) at 298 K.

	$[Cu^{2+} \times 10^{-3} M]$			
	1.0	**1.5**	**2.0**	**2.5**
[DBM] x 10^{-6}M	k_{obs}	k_{obs}	k_{obs}	k_{obs}
0.625	0.0122 ±0.004	0.0322±0.001	0.0420±0.002	0.0408±0.004
1.250	0.0219±0.003	0.0361±0.002	0.0447±0.004	0.0480±0.003
2.500	0.0322±0.001	0.0407±0.003	0.0465±0.001	0.0520±0.002
5.000	0.0376±0.002	0.0420±0.002	0.0479±0.003	0.0550±0.001
10.000	0.0509±0.001	0.0450±0.002	0.0502±0.001	0.0575±0.002

region is strong and sharp at 3021cm^{-1}which could be attributed to H_2O. There was virtually no absorption in this region in the adducts. The characteristic $Y_{c=o+}$ $Y_{-c=c}$ bands of dbm appear at 1596 cm^{-1},insertion of Cu (II) lowers it to 1542 cm^{-1} in the bis complex, while it appears at 1593 cm-^{1} in the phen and bipy adducts, respectively.

Agwara et al. (2010) ascribe such bands to ring vibrations of uncoordinated bipy and phen at 1588 cm^{-1}. Similar band observed in the spectra of dibenzoylmethane and benzoylacetone at 1600 cm^{-1} was assigned to aromatic stretching vibration (Aliyu and Mustapha, 2009). The bands at 570 cm^{-1} and 548 cm^{-1} are assigned to Cu-N and Cu-O, respectively; similar assignments were made by Rahmam et al. (2010) and Agwara et al. (2010).

Electronic spectra

Dibenzoylmethane in methanolic media showed three intense bands at 205, 262 and 349 nm, which are assigned to π-π* and n-π* transitions (Prasad et al., 2007). The copper (II) ion gave a single broad band at 973 nm. On complexation, peaks due to the ligand appeared as a single sharp band at 241 nm. This confirms the involvement of diketonate oxygen atoms in bonding (Sakar et al., 2008) the band resulting from d-d transition shifted to 976 nm, in the dibenzoylmethane complex, sustaining a bathochromic shift on adduct formation.

Magnetic susceptibility measurements

The moment usually observed for magnetically dilute Cu (II) complexes is between 1.7-2.2 BM with compounds whose geometry approaches octahedral geometry having moments at the lower end while those approaching tetrahedral geometry are at the higher end (Hugee et al., 1993).

The magnetic moment value of $[Cu(dbm)_2(H_2O)_2]$ complex is 1.80 B.M which suggests a distorted octahedral geometry around the metal ion.

Kinetics of the reaction

Plots of $-\ln(A_\alpha\text{-}A_t)$ versus time resulted in straight line graphs. The slopes of the graphs were the observed pseudo first order rate constant, $k_{obs.}$ Table 3 reports the k_{obs} values against the various concentrations of the ligand (dbm).

For each metal concentration, a graph of kobs values against the various concentrations of the ligand serves as a basis for suggesting the rate law that governs the reaction between Cu (II) and dbm. The nature of the profiles shown in Figure 1 (a to d) suggests a rapid pre-equilibrium mechanism (Odiaka, 2004) involving the formation of an intermediate prior to the formation of the product. The intermediate is formed instantaneously followed by a slow rate determining step (k_2) leading to the formation of the product. The proposed mechanism for the reaction of Cu (II) and dbm methane is shown as follows:

$$Cu^{2+} + DBM \longrightarrow Int \longrightarrow product(p). \quad (1)$$

Since the rate of reaction is governed by the slow step, then the rate for Equation (1) is given by

$$\text{Rate} = \frac{dp}{dt} = k_2 [DBM][Int] \quad (2)$$

The pre – equilibrium constant for intermediate formation is given by Equation (3)

$$k_1 = \frac{[Int]}{[Cu^{2+}][DBM]} \quad (3)$$

The [Int] is given by

$$[Int] = k_1 [Cu^{2+}] [DBM] \quad (4)$$

Substituting (4) into (2)

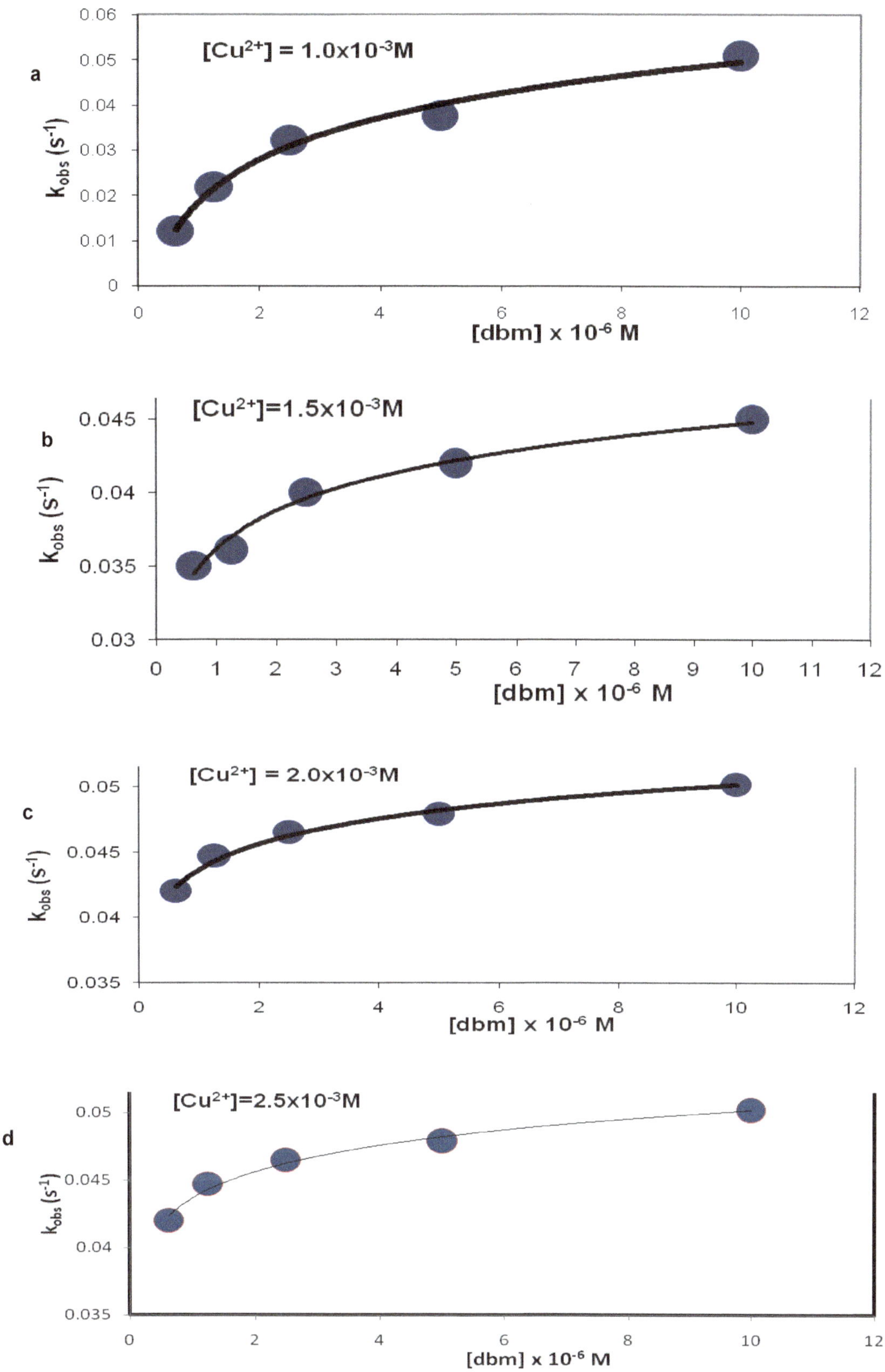

Figure 1. a. The plot of pseudo first order rate constant, (k_{obs}) against the concentration of dbm, [dbm] at 298 K for Cu^{2+} = 1.0 x10^{-3}M; b. The plot of pseudo first order rate constant, (k_{obs}) against the concentration of dbm, [dbm] at 298 K for Cu^{2+} = 1.5 x10^{-3}M; c. The plot of pseudo first order rate constant, (k_{obs}) against the concentration of dbm, [dbm] at 298 K for Cu^{2+} = 2.0 x10^{-3}M; d. The plot of pseudo first order rate constant, (k_{obs}) against the concentration of dbm, [dbm] at 298 K for Cu^{2+} = 2.5 x10^{-3}M.

Table 4. Rate constants for the formation of intermediate and product.

$[Cu^{2+}] \times 10^{-3}$ M	$k_1 \times 10^5 (s^{-1})$	$k_2 (s^{-1})$ (T)	$k_2 (s^{-1})$ (I)
1.0	0.49 ±0.02	0.051±0.03	0.052±0.02
1.5	3.7±0.01	0.045±0.01	0.043±0.03
2.0	10.0±0.03	0.050±0.02	0.049±0.01
2.5	17.2±0.02	0.058±0.01	0.058±0.02

T= Tangent of k_{obs} vs [DBM] graphs; I= Inverse of the intercept of $\frac{1}{k_{obs}}$ vs $\frac{1}{[DBM]}$ graphs.

$$\text{Rate} = \frac{dp}{dt} = k_2\, k_1\, [Cu^{2+}][DBM]^2 \qquad (5)$$

Experimentally, we followed the disappearance of the [Int] and $[Cu^{2+}]$. Hence

$$\text{Rate} = \frac{dp}{dt} = k_{obs}\ \{[Cu^{2+}] + [Int]\} \qquad (6)$$

$$\text{Rate} = \frac{dp}{dt} = k_{obs}\ \{[Cu^{2+}] + k_1 [Cu^{2+}][DBM]\} \qquad (7)$$

Substituting (4) into (6) we have

$$\text{Rate} = \frac{dp}{dt} = k_{obs}\ [Cu^{2+}]\{1 + k_1 [DBM]\} \qquad (8)$$

From Equation (5 and 8). Rate = $k_2\, k_1\, [Cu^{2+}]\, [DBM]^2$

$$= k_{obs}\, [Cu^{2+}]\{1 + k_1 [DBM]\} \qquad (9)$$

$$\therefore k_{obs} = \frac{k_2\, k_1 [DBM]^2}{1 + k_1 [DBM]} \qquad (10)$$

Equation (10) is the general rate expression for a pre-equilibrium mechanism. The intermediate in Equation (1) was not attacked by a nucleoplile that is DBM. It only rearranges in a rate determining manner to form the product.

Then Equation 10 reduces to Equation (11)

$$\therefore k_{obs} = \frac{k_2\, k_1 [DBM]}{1 + K_1 [DBM]} \qquad (11)$$

The intermediate in Equation (1) was formed rapidly.

Then k_1 in Equation (11) must be very large such that k_1 [DBM] is so large that we can reasonably ignore 1 in

Equation 11 which now collapses to Equation (12)

$$k_{obs} = k_2 - \qquad (12)$$

Equation (12) explains the extrapolation of the flattened portion of the graph to the k_{obs} axis which results in getting k_2 (tabulated in Table 4). Inversion of equation 11 yields.

$$\frac{1}{k_{obs}} = \frac{1}{k_2 k_1 [DBM]} + \frac{1}{k_2}. \qquad (13)$$

Therefore plots of $\frac{1}{k_{obs}}$ versus $\frac{1}{[DBM]}$ results in straight lines (Figure 2 a to d) whose slope = $\frac{1}{k_2 k_1}$

From where k_1 values are obtained and intercept on the y-axis = $\frac{1}{k_2}$

Conclusion

The reaction between Cu (II) and dbm follows a rapid pre-equilibrium mechanism. Although, the rate of complex formation (k_2) is constant for various metal concentration (Table 4) and is independent of the metal concentration. This suggests an interchange associative (I_a) mechanism, implying that bond making is important than bond breaking.

The k_1 value obtained is lower than the expected value for water exchange rate ($10^5 \leq 10^8\ M^{-1}s^{-1}$) hence, it is suspected that the formation of the intermediate is not strictly the loss of water from the Cu(II) ions, but some metal – ligand bond formation takes place before the Cu-OH_2 bond is completely broken.

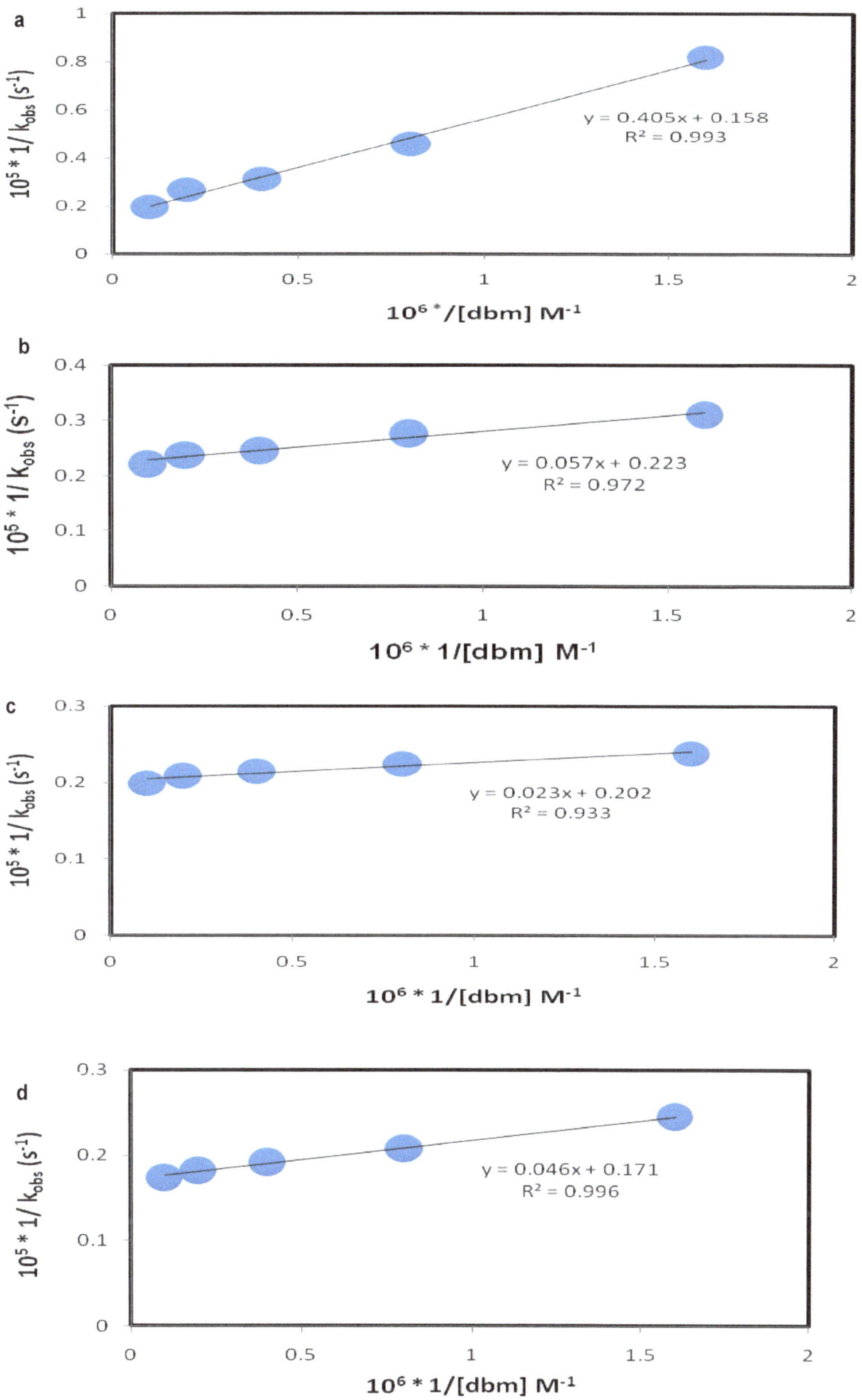

Figure 2 a. Plot of inverse of pseudo first order rate constant, ($1/k_{obs}$) against inverse of [dbm], (1/[dbm]) at 298K for [Cu^{2+}] = 1x 10^{-3} M; b. Plot of inverse of pseudo first order rate constant, ($1/k_{obs}$) against inverse of [dbm], (1/[dbm]) at 298K for [Cu^{2+}] = 1.5x 10^{-3} M; c. Plot of inverse of pseudo first order rate constant, ($1/k_{obs}$) against inverse of [dbm], (1/[dbm]) at 298K for [Cu^{2+}] = 2.0x 10^{-3} M; d. Plot of inverse of pseudo first order rate constant, ($1/k_{obs}$) against inverse of [dbm], (1/[dbm]) at 298 K for [Cu^{2+}] = 2.5 x 10^{-3} M.

REFERENCES

Agwara MO, Ndifon PT, Ndosiri NB, Paboudam AG, Yufanyi DM Mohamadou A (2010).Synthesis, characterisation and antimicrobial activities of Cobalt (II) Copper(II), and Zinc(II) Mixed ligand complexes containing 1,10-phenantroline and 2,2-bipyridine.Bull. Chem. Soc. Ethiop., 24: 383-389.

Ahmed AA, Benguzzi SA, El-Hadi AA (2007). Synthesis and Characterization of divalent transition metals complexes of Schif Bases derived from O-phenyldiamine and Benzoylacetone and Related Species. J. Sci. Appl., 1: 79-90.

Aliyu HN, Mustapha A (2009). Synthesis and characterization of oxovanadium (iv) β-diketonate complexes. Afr. Sci., 10: 124-126

Aliyu HN, Sani U. (2008). Potentiometric and Spectrophotometric Studies of Bis (Acetyl Acetonato) Ethylene Diamine Cobalt (II) Complex. Int. J. P/A Sci., 2: 42-46.

Harding P, Harding DJ, Soponrat N, Tinpun K, Samuadnuan S, Adams H (2010) Synthesis and elecrochemical studies of Nickel β-diketonate complexes incorporating asymmetrical dimine ligands. Aust. J. Chem., 63: 75-82.

Hugeey JE, Keiter EA, Keiter RL (1993). Inorganic Chemistry. principles of structure and reactivity. Harper Collins College Publishers.

Michael JH, Brigid DO (1979). Kinetic and mechanisms of the reaction of Nickel (II) and pentane -2, 4-dione. J. Chem. Soc., Dalton Trans., 1: 162-166.

Odiaka TI (2004). Modern organometallic chemistry, Ibadan, University Press.

Patel RN, Gundla VLN, Patel DK (2008). Synthesis, characterization and properties of some ternary copper (II) complexes containing NOS donor schiff base and NN donor bidentate ligands. Indian J. Chem., 47: 353-360.

Patel RN (2009). Synthesis, characterization and superoxide dismutase activity of a series of copper (II) complexes containing NNO donor schiff base and benzimidazole. ligands. Indian J. Chem., 18A: 173-180.

Prasad RN, Sharma KM, Agrawal A (2007). Mixed ligand complexes of Co(II) containing 5-bromosalicylaldehyde and β-diketones, hydroxyaryl aldehydes or ketones. Indian J. Chem., 46 A: 600-604.

Rahman AML, Hossain MB, Halim MB, Chowdhury DA, and Salam MA (2010).Studies on some diazo coupled products of chromium (III) chelates of acetylacetone and benzoylacetone. Afr. J. P/A Chem., 4 (10): 216-220.

Sakar B, Drew MGB, Estrader MDC, Ghosh A (2008). Cu^{II} acetate complexes involving NNO donor ligands: mono-atomic oxygen bridged dimers and $Cu_2(OAc)_4$ Polyhedron, 27: 2625-2633.

Shalin K, Durga ND. Saxena PN (2009). Applications of metal complexes of schiff bases – A review. J. Sci. Ind. Res., 68: 181-187.

Slyvester B (2006). Synthesis, Characterization and properties of copper (II) βeta-Diketonate Macrocycles PhD Thesis of Lowstana State University.

Stephen RCBS (1989). The synthesis and Spectroscopic Characterization of Bis (acetylacetonenate) Palladium (II) Phosphine Complexes. An M.Sc Thesis of Texas Tech. University.

Tayyari SF, Rahemi, H, Nekoei, AR, Zahedi M, Wang YA (2007). Vibrational assignment and structure of dibenzoylmethane. A density functional theoretical study. Spec. Chem. Acta. Part A., 66: 394-404.

Study of physical chemistry on biosorption of zinc by using *Chlorella pyrenoidosa*

Hassan Rezaei*, Satish D. Kulkarni and Praveen G. Saptarshi

Environmental Sciences Department, University of Pune – Pune, India.

Discharge of heavy metals from metal processing industries is known to have adverse effects on the environment. Biosorption of heavy metals by metabolically inactive biomass of microbial organisms is an innovative and alternative technology for removal of these pollutants from aqueous solution. Presence of heavy metals in the aquatic system is posing serious problems and zinc has been used in many industrials and removal of Zn ions from waste waters is significant. Biosorption is one of the economic methods that is used for removal of heavy metals. In the present study, the biomass generated from the dried *Chlorella pyronoidsa* was used for evaluating the biosorption characteristics of Zn ions in aqueous solutions. Batch adsorption experiments were performed on these leaves and it was found that the amount of metal ions adsorbed increased with the increase in the initial metal ion concentration. In this study effect of agitation time, initial metal ion concentration, temperature, pH and biomass dosage were studied. Maximum metal uptake was observed at pH = 5. Maximum metal uptake (q_{max}) was 101.11 mg/g. The biosorption followed both Langmuir and Freundlich isotherm model. The adsorption equilibrium was reached in about 1 h. The kinetic of biosorption followed the second – order rate. The biomass could be regenerated using 0.1 M HNO_3. A positive value of $\Delta H°$ indicated the endothermic nature of the process. A negative value of the free energy (ΔG°) indicated the spontaneous nature of the adsorption process. A positive value of $\Delta S°$ showed increased randomness at solid–liquid interface during the adsorption of heavy metals, it also suggests some structural changes in the adsorbate and the adsorbent. Fourier transform infrared (FTIR) spectrums of *C. pyrenoidosa* revealed the presence of hydroxyl, amino, carboxylic and carbonyl groups. The scanning electron micrograph (SEM) clearly revealed the surface texture and morphology of the biosorbent.

Key words: Biosorption, *Chlorella pyronoidosa*, Zn, isotherm models, kinetic.

INTRODUCTION

Heavy metal pollution has posed a serious threat to the aquatic environment. A thigh concentrations, metals are toxic to animals and plants alike, as they could be dispersed in water and consequently in human beings through food chain biomagnifications that could cause serious health hazards. In view of the human health impacts, each metal imparts different effects and symptoms. For instance, in the case of minor zinc exposure, irritability, muscular stiffness, loss of appetite and nausea are common (Bhattacharya et al., 2006). Zinc is an essential element as enzyme activators in humans, but it is equally toxic at levels of 100.500 mg/day and it is a known carcinogen (Volesky and Holan, 1995). However, these methods were found to be either inefficient or expensive when metal ions exist in low concentrations (<100 mg/L) and may also be associated with the generation of secondary environmental problems from waste disposal (Ahalya et al., 2003). Biosorption is the binding and concentration of heavy metals from aqueous solutions (even very dilute ones) by certain types of inactive, dead, microbial biomass (Macaskie et al., 1992). Some of the advantages of biosorption include competitive performance, heavy metal selectivity, cost-effectiveness, regenerative and no sludge generation. Sources of biomass include seaweeds, microorganisms

*Corresponding author. E-mail: Hassanrezaei1979@gmail.com.

Abbreviations: FTIR, Fourier transform infrared; **SEM,** scanning electron micrograph; **BOD,** biochemical oxygen demand; **COD,** chemical oxygen demand; **NCIM,** national collection of industrial microorganisms; **EDTA,** ethylene diamine tetraacetic acid; **HCL,** hydrogen chloride.

(bacteria, fungi, yeast and molds), activated sludge and fermentation waste. Studies using biosorbents have shown that both living and dead microbial cell are able to uptake metal ions and offer potential inexpensive alternative to conventional absorbents (Khoo and Ting, 2001; Knorr, 1991).

However, living cell is subject to toxic effect of the heavy metals, resulting in cell death. Moreover, living cell often require the addition of nutrients and hence increase the Biochemical oxygen demand (BOD) and Chemical Oxygen Demand (COD) in the effluent. For these reasons, the use of non-living biomaterials or dead cells as metal binding compounds has been gaining advantage because toxic ions do not affect them. In addition, dead require less care and maintenance and cheaper (Mofa, 1995). Furthermore, dead biomass could be easily regenerated and reused. The capability of some living microorganisms to accumulate metallic elements has been observed at first from toxicological point of view (Volesky, 1990a,b,c) However, further researches have revealed that inactive/dead microbial biomass can passively bind metal ions via various physicochemical mechanisms. Therefore researches on biosorption have become an active field for the removal of metal ions or organic compounds. Biosorbent behavior for metallic ions is a function of the chemical make-up of the microbial cells of which it consists (Volesky and Holan, 1995). Mechanisms responsible for biosorption, although understood to a limited extent, may be one or combination of ion exchange, complexation, coordination, adsorption, electrostatic interaction, chelation and micro precipitation (Veglio and Beolchini, 1997; Vijayaraghavan and Yun, 2008; Wang and Chen, 2006).

MATERIALS AND METHODS

Biomass and culture medium

In this study *Chlorella pyrenoidosa* (2738) obtained from National Collection of Industrial Microorganisms (NCIM) from PUNE - India, which was isolated and thoroughly pure. The *C. pyrenoidosa* was maintained in modified Fog's Media at 28°C with using 3000 lx light intensity. After 21 days cultivation period cells were harvested by centrifugation and were washed several times with deionised water in order to remove culture media and was kept on a filter paper to reduce the water content. The biomass was dried at 60°C in an oven for 24 h and milled to a gritty consistency. The biomass was sieve for particle size smaller than 1 mm and stored in dark bottle and keeps in a dry cabinet for experiments. All of the media are sterilized by autoclaving at 121°C for 20 min.

Preparation of synthetic sample

A stock solution of 1000 mg/l of Zn was obtained by dissolving Zinc chloride (Merck Company) in distilled water. The test solutions of various concentrations range from 10 mg/L to 100 mg/L were prepared from the stock solution. The solution pH was adjust using 0.1 M HNO_3 and 0.1 M NaOH at the beginning of the experiment and not controlled afterwards. The conical flasks (250 ml) were shaken at 120 rpm in a temperature controlled rotatory shaker.

Analysis of zinc ions

Zinc ions were determined spectrophotometrically by atomic adsorption spectrophotometer (UNICAM, model 929, United Kingdom).

Batch biosorption studies

Batch mode adsorption studies for individual metal compounds were carried out to investigate the effect of different parameters such as adsorbate concentration, adsorbent dose, agitation time and pH. Solution containing adsorbate and adsorbent was taken in 250 ml capacity conical flask and agitated at 120 rpm in a shaker at predetermined time intervals. The adsorbate was decanted and separated from the adsorbent using whatman No. 41 filter paper.

Effect of agitation time and initial concentration

For the determination of rate of metal biosorption by biomasses from 100 ml (at 10, 20, 50 and 100 mg/L) on conical flask 250 mL, the supernatant was analyzed for residual metal at different time intervals. The pH and the adsorbent dosage was kept constant at pH = 5±0.01, which varied according to the adsorbent and adsorbate under consideration. Amount of biomass dosage was 0.1±0.001 g for biomass (*C. pyronoidosa*) and temperature was 25±1°C and agitation speed of shaker was 120 rpm.

Effect of adsorbent dosage and initial concentration

The effect of adsorbent dosage that is, the amount of the biomasses on the adsorption of metals was studied at different dosages ranging from 0.1 to 3 g with varied metal concentrations of 10, 20 and 50 mg/L. The equilibrium time and the pH were kept constant depending on the metal under consideration. The pH and the adsorbent dosage was kept constant at pH = 5±0.01, which varied according to the adsorbent and adsorbate under consideration. Agitation time was 120 min for biomass (*C. Pyronoidosa*) and temperature was 25±1°C and agitation speed of shaker was 120 rpm.

Effect of pH and initial concentration

To determine the effect of pH on the adsorption of metal solutions (100 ml) of different concentration ranges (10, 20 and 50 mg/L) at conical flask 250 ml were adjusted to desired pH values and mixed with constant amount of adsorbent and agitated at preset equilibrium time. The equilibrium time and adsorbent dosage varied with the metal and adsorbent under consideration. Amount of biomass dosage was 0.1±0.001 g for e biomass (*C. pyronoidosa*) and temperature was 25±1°C and agitation speed of shaker was 120 rpm and contact time was 120 min.

Effect of temperature

Optimum biomass concentration with optimum pH was used to monitor the temperature effect on biosorption. Experiments were carried out at different temperatures from 10 to 40°C for each culture and kept on rotary shaker at 120 rpm. The samples were allowed to attain equilibrium. To determine the effect of temperature on the adsorption of metal solutions (100 ml) of concentration 50 mg/L at conical flask 250 ml were adjusted to desired pH values and mixed with constant amount of adsorbent and agitated at preset equilibrium time. The equilibrium time and adsorbent dosage

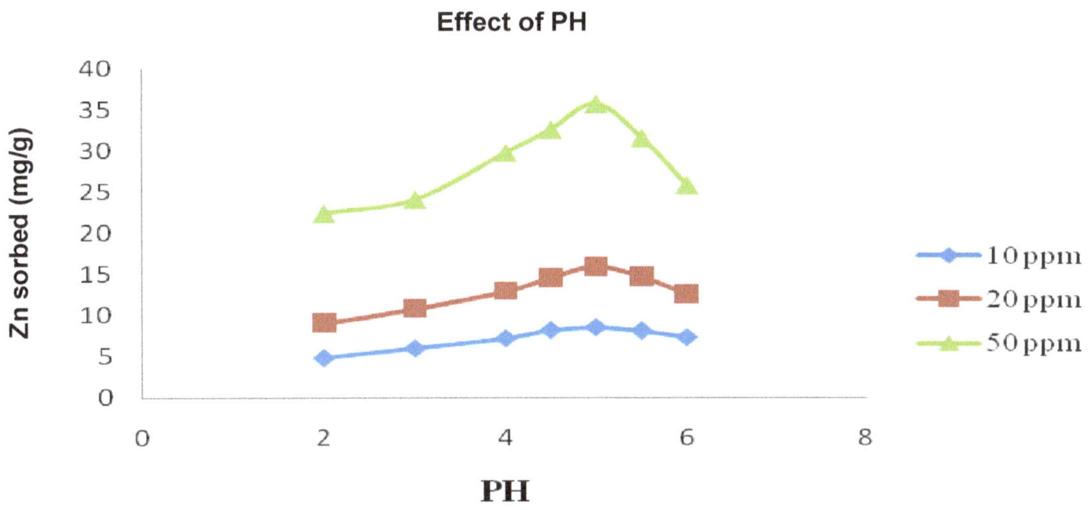

Figure 1. Effect of pH on biosorption of Zinc by *C. pyrenoidosa* (biomass dose = 0.1 g, initial Zinc ion concentration = 10, 20 and 50 mg L^{-1}; temperature = 25°C; agitation speed = 120 rpm; contact time = 120 min).

varied with the metal and adsorbent under consideration. Amount of biomass dosage was 0.1±0.001 g for biomass (*Chlorella*) and pH was 5±0.001 and agitation speed of shaker was 120 rpm and contact time was 120 min.

Desorption studies

After adsorption, the adsorbates loaded adsorbent were separated from the solution by centrifugation and the supernatant was drained out. The adsorbent was gently washed with water to remove any unadsorbed adsorbate. Regeneration of adsorbate from the adsorbate laden adsorbent was carried out using the desorbing media distilled water at pH ranges using dilute solutions of ethylene diamine tetraacetic acid (EDTA), Hydrogen chloride (HCl) and HNO_3 (Stirred at 120 rpm for 120 min at 25°C). Then they were agitated for the equilibrium time of respective adsorbate. The desorbed adsorbate in the solution was separated and analyzed for the residual heavy metals.

FT-IR spectroscopy (Fourier transform infrared)

In order to determine the functional groups responsible for Zn biosorption, infrared (IR) spectroscopy was used that about 0.1 g biomass was mixed with KBr for FT-IR spectra analysis (Shimadzu model 8400).

SEM (scanning electron microscopy)

The SEM was used to investigate the morphology of the biosorbent. We used samples with pH=5 and C_0 = 10 mg.L^{-1}. SEM (SEM, JEOL, JSM-6360A) used for this study.

RESULTS

A result on the effect of pH of Zn at different initial metal ion concentrations by *C. pyrenoidosa* is presented in Figure 1. Maximum percentage of biosorption accursed at the initial concentration of 10 mg/L at the time of 120 min at pH= 5 for Zn was 85.55% and metal ion uptake capacity was 8.55 mg/g and when initial concentration of Zn increased to 50 mg/L, percentage of biosortion of Zn was 71.33% and uptake capacity was 35.66 mg/g for *C. pyrenoidosa.* With increase the initial concentration percentage of biosorption decreased and metal ion uptake capacity was increased. A result on the contact time of zinc at different initial metal ion concentrations by *C. pyrenoidosa* is presented in the Figure 2. The time required to reach equilibrium for Zn adsorption by *C. pyrenoidosa* was 60 min for all initial metal ion concentrations. In the initial concentration of 10 mg/L at the time of 60 min percentage of remove of Zn was 82.27% and metal ion uptake capacity was 8.22 mg/g and when initial concentration of Zn increased to 100 mg/L, percentage of remove of Zn was 64.61% and uptake capacity was 64.61 mg/g for *C. pyrenoidosa*. The time taken for Zn adsorption by *C. pyrenoidosa* was dependent on initial metal ion concentration and increased with increase in concentration of Zn. With increase the initial concentration percentage of biosorption decreased and metal ion uptake capacity was increased.

A result on the effect of biomass dosage of Zn at different initial metal ion concentrations by *C. pyrenoidosa* is presented in the Figure 3. percentage of biosorption accursed at the initial concentration of 10 mg/L at the time of 120 min at pH = 5 and 0.1± 0.001 g of biomass for Zn was 85.55% and metal ion uptake capacity was 8.56 mg/g and when initial concentration of Zn increased to 50 mg/L, percentage of biosortion of Zn was 71.33% and uptake capacity was 35.66 mg/g for *C. pyrenoidosa.* When amount of biomass increased from 0.1±0.001 g to 3±0.001 g, percentage of biosorption accursed at the initial concentration of 10 mg/L at the

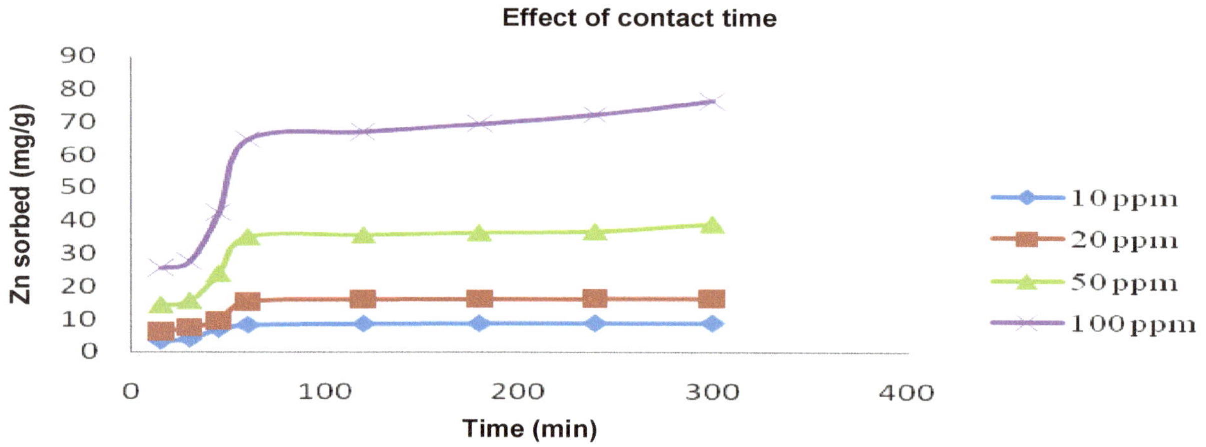

Figure 2. Effect of Contact Time on biosorption of Zinc by *C. pyrenoidosa* (biomass dose=0.1 g, pH=5, initial Zinc ion concentration = 10, 20 50 and 100 mg L^{-1}; temperature = 25°C; agitation speed = 120 rpm).

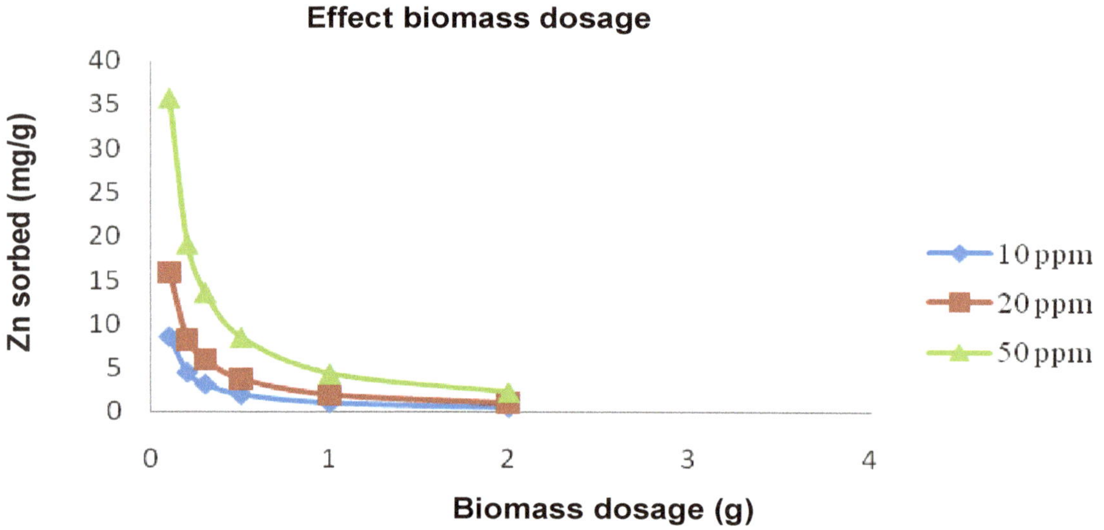

Figure 3. Effect of biomass dosage on biosorption of zinc by *C. pyrenoidosa* (pH = 5, initial Zinc ion concentration = 10, 20 and 50 mg l^{-1}; temperature = 25°C; agitation speed = 120 rpm; contact time = 120 min).

time of 120 min at pH= 5 for Zn was 97.45% and metal ion uptake capacity was 0.325 mg/g and when initial concentration of Zn increased to 50 mg/L, percentage of biosortion of Zn was 91.55% and uptake capacity was 1.52 mg/g for *C. pyrenoidosa.* With increase the initial concentration percentage of biosorption decreased and metal ion uptake capacity was increased. With increased the amount of biomass observed that percentage of biosorption increased and metal ion uptake capacity was decreased. A result on the effect of temperature of Zn at initial metal ion concentration of 50 mg/L by *C. pyrenoidosa* is presented in the Figure 4. Maximum percentage of biosorption accursed at the initial concentration of 50 mg/L at the time of 120 min at pH= 5 for Zn was 75.44% and metal ion uptake capacity was 37.72 mg/g at the temperature of 40°C for *C. pyrenoidosa.* The findings of *C. pyrenoidosa* indicate that the sorption percentage increased with increase in temperature up to 40°C.

Equilibrium isotherms

The isotherm studies were performed in the solution with the initial concentrations ranging from 10 to 100 mg/L at optimum pH values for ions (pH=4.5 or pH = 5). After shaking the flask containing the mixture of biomass (120 rpm, 25°C) and ions for 120 min, the amount of residual ions in the filtrated solution was analyzed. The biosorption equilibrium uptake capacity for each sample

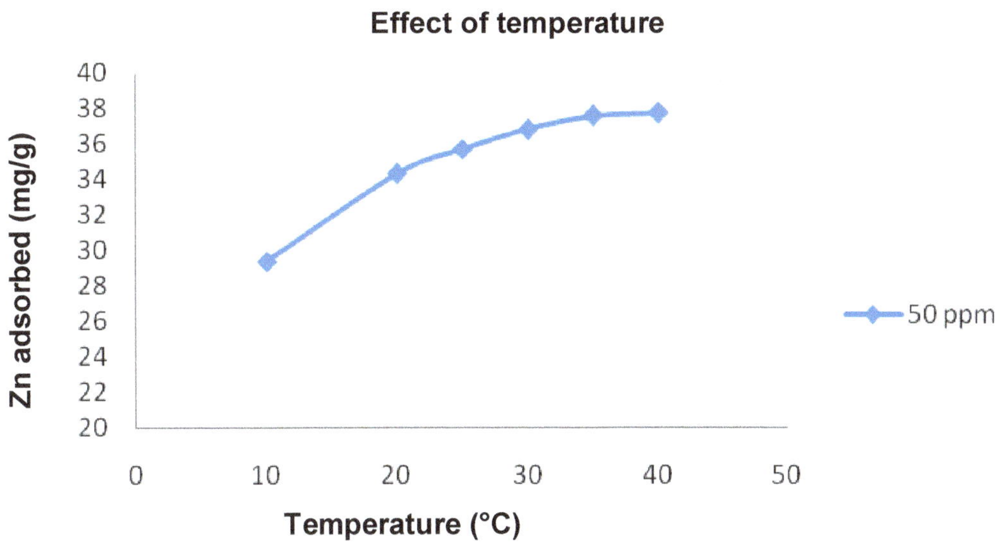

Figure 4. Effect of temperature on biosorption of zinc by *C. pyrenoidosa* (biomass dose = 0.1 g, pH=5, initial Zinc ion concentration = 50 mg l^{-1}; temperature = 25°C; agitation speed = 120 rpm; contact time = 120 min).

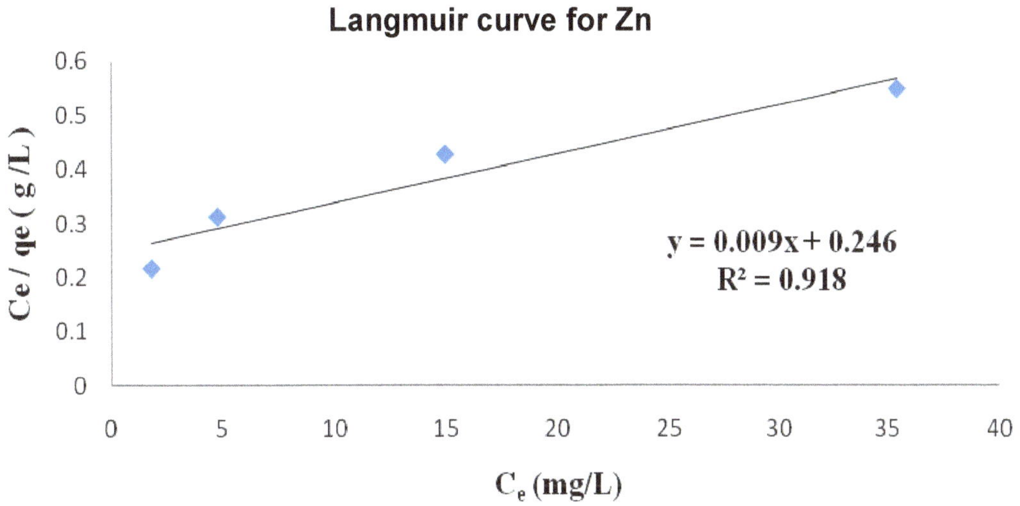

Figure 5. Langmuir adsorption isotherm for Zn by *C. pyrenoidosa* (q_{max} = 111.11 mg/g, b = 0.036 L/mg).

was calculated according to mass balance on the ions expressed in this equation: $q_e = \frac{(C_0 - C_e)}{M} \times V$

where V is the sample volume (L), C_0 is the initial ion concentration (mg/L), C_e is the equilibrium or final ion concentration (mg/L) , M is the biomass dry weight (g), and q_e is the biomass biosorption equilibrium ions uptake capacity (mg/g). Langmuir and Freundlich isotherms, the two classical adsorption models, were used to describe the equilibrium between adsorbed ions on the biomass cell (q_e ,q) and ions in the solution (C_e, q) in this study.

Langmuir isotherm model: $q_e = \frac{q_{\max} C_e b}{1 + C_e b}$

That after arrange we have; $\frac{C_e}{q_e} = \frac{1}{q_{\max} b} + \frac{C_e}{q_{\max}}$

These values q_{max} and b (where b, is the adsorption equilibrium constant) can be obtained from the slopes and the intercepts of the linear plots respectively, where experimental data of Ce/qe as the function of C_e as shown in Figure 5. The empirical Freundlich equation

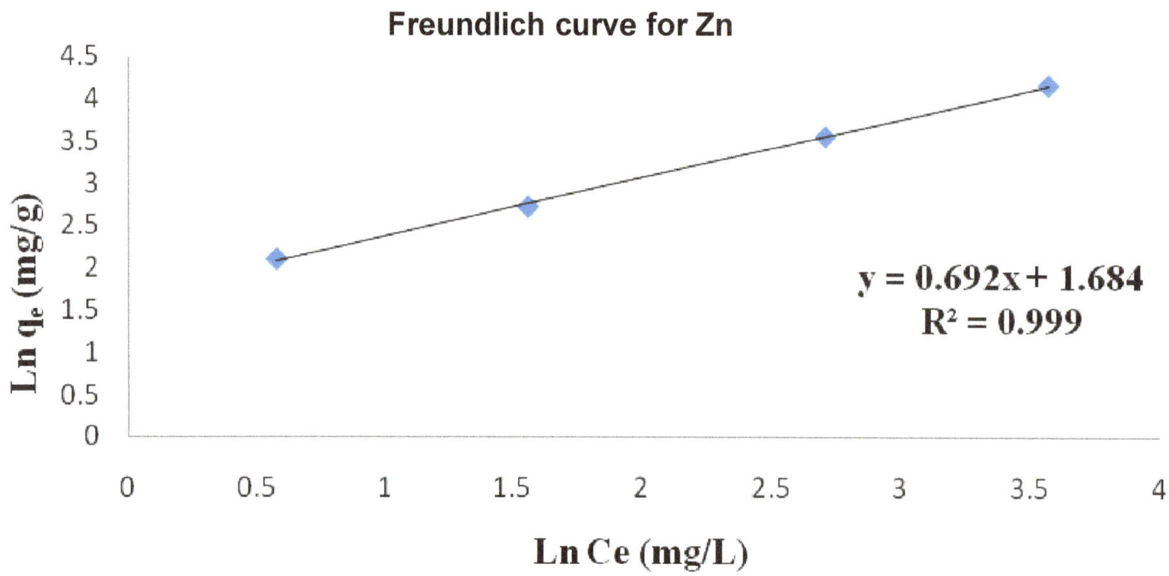

Figure 6. Freundlich adsorption isotherm for Zn by *C. pyrenoidosa* (n = 1.44, K = 5.38).

Table 1. Parameters of isotherm models for chromium.

Biomass	Langmuir Parameters			Freundlich parameters		
	q_{max} (mg/g)	b(L/mg)	R^2	K_f	n	R^2
Chlorella pyrenoidosa	101.11	0.036	0.918	5.38	1.44	0.999

based on sorption on a heterogeneous surface, on the other hand, is as follows:

$q_e = K_f (C_e)^n$

K and n: An experimental constant, K is an indication of the adsorption capacity of the adsorbent; n indicates the effect of concentration on the adsorption capacity and represents adsorption intensity. The equation can be linearized in the following logarithmic form:

$$\ln q_e = \ln k_f + \frac{1}{n} \ln C_e$$

These values n and K_f can be obtained from the slopes and the intercepts of the linear plots, respectively, where experimental data of Ln q_e as the function of Ln C_e. .The equilibrium experimental results of Zinc ions have been fitted in the Langmuir and Freundlich models. For biosorption of Zinc using *C. pyrenoidosa* the coefficient of determination (R^2) of both models was mostly close to 1 as shown in Figure 6. This indicates that both models adequately describe the experimental data of the biosorption of Zinc. In the biosorption of Zinc by *Chlorella*, most of the metal ions were sequestered very fast from the solutions in the first phase of contact time 60 min and almost no increase in the level of bound metal have been occurred after this time interval. Biosorption equilibrium isotherms were plotted for metal uptake q against the residual metal concentration in the solution. The q verses C_f sorption isotherm relationship was mathematically expressed by Langmuir and Freundlich models. The higher the values of k and n; lower the value of b, the higher the affinity of the biomass. Table 1 describes summaries of linear regression data for Langmuir and Freundlich isotherms for Zinc biosorption using *C. pyrenoidosa* biomass. Langmuir and Freundlich constant k were obtained from the linear equations of both models. As indicated in Table 1, the coefficients of determination (R^2) of both models are close to 1. In the Table 1 the values of K_f, n, q_{max} and b were given.

Kinetic modeling

Figure 7 shows the experimental break through curves for the effects of contact time on a bound rate of Zn. It can be observed that the adsorption of Zinc ions quickly increased at the beginning of biosorption, but after 15 min, the adsorption slowed down. The result indicated that the maximum adsorbed amount of the Zinc ions was achieved within 60 min and then followed by a longer equilibrium period. After this equilibrium period, the

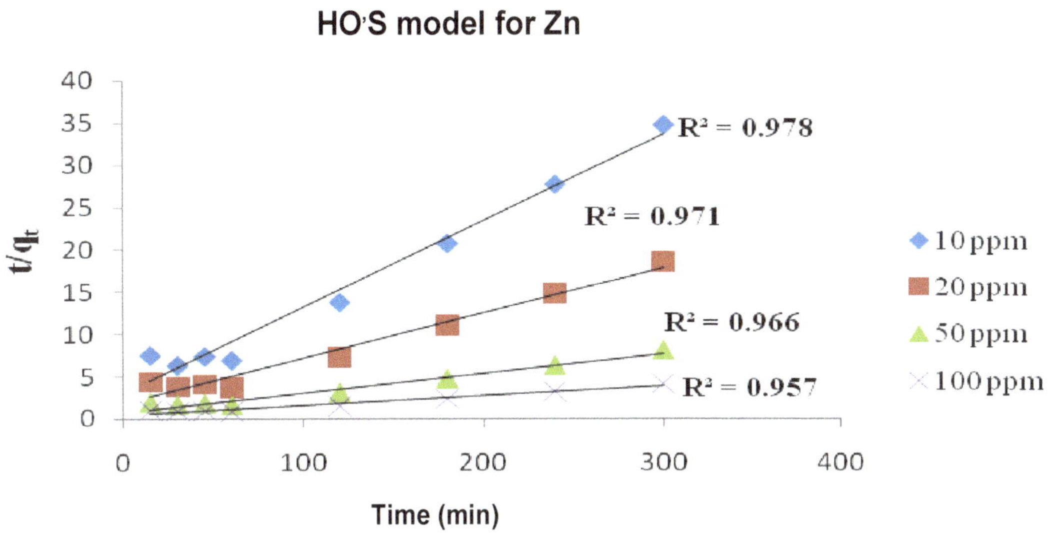

Figure 7. Plot of HO'S Model (pseudo second order rate) for Zinc by *C. pyrenoidosa*.

Table 2. Type of isotherm for various R_L.

R_L	$R_L>1$	$R_L=1$	$0< R_L<1$	$R_L = 0$
Type of isotherm	Un favorable	Linear	Favorable	Irreversible

Table 3. Parameter of kinetic model for chromium of *C. pyrenoidosa*.

Concentration of Cr	Equation	R^2	K_2
10 mg/L	y = 0.103x + 2.468	0.991	4.30×10^{-3}
20 mg/L	y = 0.055x + 1.577	0.989	1.92×10^{-3}
50 mg/L	y = 0.023x + 0.762	0.988	9.94×10^{-4}
100 mg/L	y = 0.011x + 0.479	0.986	2.53×10^{-4}

amount of adsorbed ions did not significantly change with the adsorption time. Therefore, for the following experiments, the contact time was maintained for 60 min to ensure that equilibrium was fully achieved. The pseudo-second-order equation is also based on the sorption capacity, which is expressed as:

$$\frac{t}{q_t} = \frac{1}{\left(K_2 q^2{}_e\right)} + \frac{t}{q_e}$$

Where K_2 is the rate constant of pseudo-second-order sorption (g· mg^{-1}·min^{-1}). K_2qe^2 is the initial rate constant (represented by h, mg·g^{-1}·min^{-1}). Plotting t/q_t versus t will give a straight line. The values of qe and K_2 can be determined from the slope and intercept of the plot, respectively. The results showed that the pseudo-second-order model fitted the simulation curve much better than the pseudo-first-order model for Zn. The results of pseudo-second-order model are shown on Table 2. The coefficient of determination (R^2) and K_2 HO'S Model for the different metal ion concentration under study has been established as: 10 > 20 > 50 > and 100 ppm. With increase the initial concentration coefficient of determination (R^2) and K_2 decreased.

Equilibrium parameter R_L

The essential characteristics of a Langmuir isotherm can be expressed in terms of a dimensionless constant separation factor or equilibrium parameter R_L, which is defined by the equation:

$R_L = 1/1+bC_o$

Where C_o is the initial adsorbate concentration (mg/L) and b is the Langmuir constant (L/mg). The parameter indicates the shape of the isotherm as follows (Table 3). The R_L values at different initial adsorbate

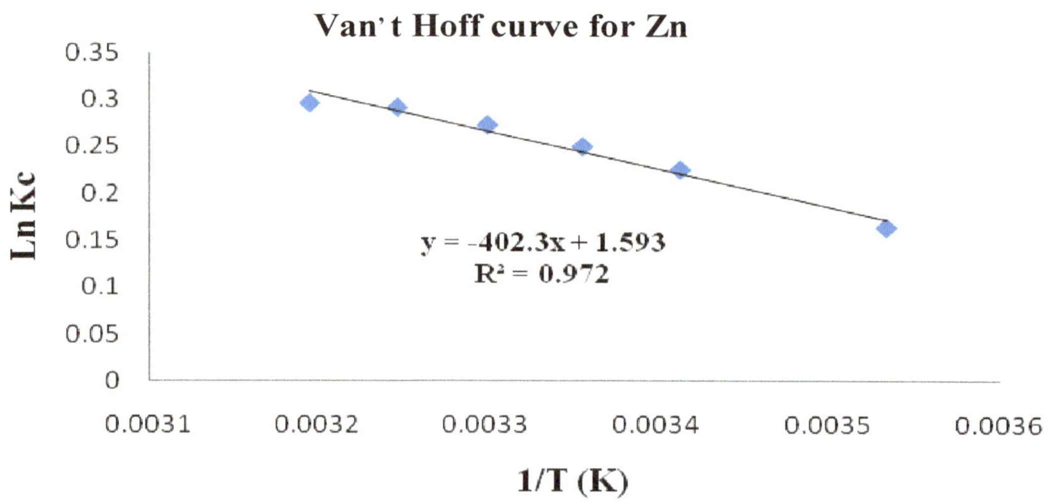

Figure 8. Plot of Van' t Hoff equation for the estimation of thermodynamic parameters for biosorption of Zinc by *C. pyrenoidosa*.

Table 4. Thermodynamic parameters for the adsorption of heavy metals by *C. pyrenoidosa*.

Heavy metal	ΔH° (J mol^{-1}K^{-1})	ΔS° (J K^{-1}mol^{-1})	-Δ°G (J.mol^{-1}K^{-1})					
			283 K	293 K	298 K	303 K	308 K	313 K
Zn	3344.72	13.24	13.61	37.39	51.84	68.01	84.74	98.31

concentrations indicate favorable adsorption for all the adsorbents and adsorbates studied.

Thermodynamic studies

Adsorption, studies in the temperature range 283 to 313 K were conducted to determine thermodynamics constants such as Gibbs free energy change (ΔG°), enthalpy change (ΔH°) and entropy change (ΔS°) for the system and to ascertain the sorption mechanism. For this study, adsorbent dosage selected was 0.1 gr. and Zinc concentration was 50 mg L^{-1} with pH = 5 in a conical flask and allowed to equilibrate for 1 h at the different temperatures ranging from 283 to 313K. In order to determine thermodynamic parameters, experiments were carried out at different temperatures in the range of 283 to 313K for heavy metals adsorption. The thermodynamic parameters such as standard Gibbs free energy change (ΔG°), enthalpy change (ΔH°) and entropy change (ΔS°) were estimated to evaluate the feasibility and nature of the adsorption process. The Gibbs free energy change, of the process is related to equilibrium constant by the equation:

$$\Delta G^\circ = -RT \ln K_C$$

Where, T is temperature in K, R is ideal gas constant having value as 8.314, J mol^{-1} K^{-1} and K_C is thermodynamic equilibrium constant. The thermodynamic equilibrium constant, was determined as:

$$K_C = \frac{C_a}{C_e}$$

Where, C_a is mg of adsorbent adsorbed per liter and C_e is the equilibrium concentration of solution, mg/L. According to thermodynamics, the Gibbs free energy change is also related to the enthalpy change (ΔH°) and entropy change (ΔS°) at constant temperature by the following Van' t Hoff equation:

$$\ln K_C = \frac{-\Delta H^\circ}{R}\frac{1}{T} + \frac{\Delta S^\circ}{R}$$

The values of enthalpy change (ΔH°) and entropy change (ΔS°) were calculated from the slope and intercept of the plot lnKc versus, 1/T. Results on the plots of van' t Hoff equation for the estimation of thermodynamic parameters by *C. pyrenoidosa* is presented in Figure 8. The calculated values of thermodynamic parameters are reported in Table 4. A positive value of ΔH° indicated the endothermic nature of the process. A negative value of the free energy (ΔG°) indicated the spontaneous nature of the adsorption process. It was also noted that the

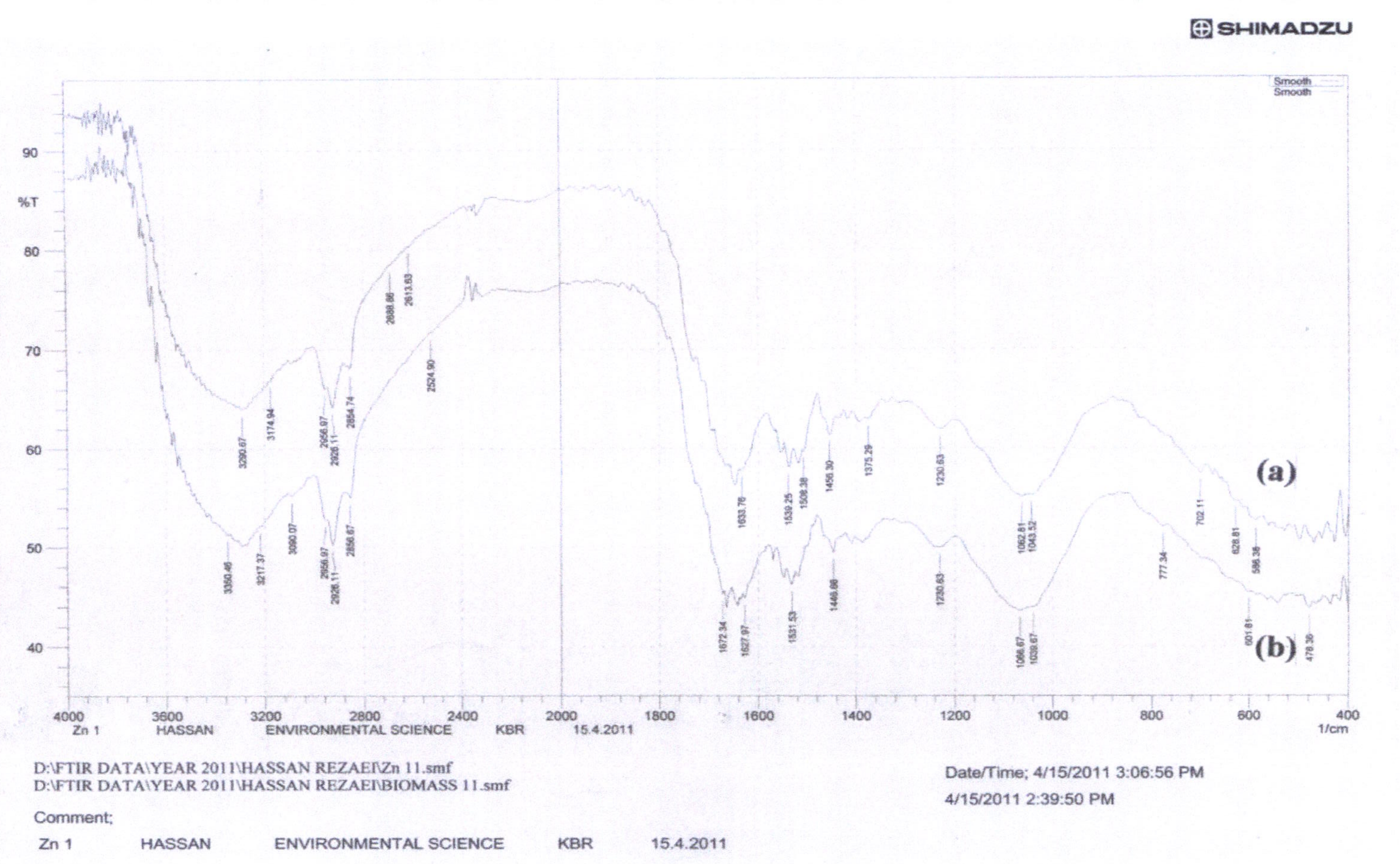

Figure 9. FTIR Spectrum of the *C. pyrenoidosa* (before (a) and after (b) biosorption of Zinc).

change in free energy, increases with increase in temperature, which exhibits an increase in adsorption with rise in temperature. This could be possibly because of activation of more sites on the surface of biomasses with increase in temperature. A positive value of $\Delta S°$ showed increased randomness at solid–liquid interface during the adsorption of heavy metals, it also suggests some structural changes in the adsorbate and the adsorbent.

FTIR spectroscopy of *C. pyrenoidosa*

Spectrums of *C. pyrenoidosa* are presented in Figure 9 that revealed the presence of hydroxyl, amino, carboxylic and carbonyl groups. The

Table 5. Assignment of bands to functional group on the surface of *C. pyrenoidosa* as observed from FTIR spectroscopy.

Wave numbers (Cm^{-1})		Assignment
Before	After	
3433.41	3373.61	O-H stretching /
3325.39	3174.94	N-H stretching
3255.95	3091.99	
3213.51		
3174.94		
2924.18	2924.18	C-H stretching aliphatic
2854.74	2854.74	C-H stretching
2746.73	2744.80	
1737.92	1732.13	C=O stretching vibration
1639.55	1645.33	C=O Stretching vibration /
	1618.33	N-H Stretching vibration
1450.52	1454.38	CH_2Bonding vibration
1317.43	1315.50	CH_3 Bonding vibration
1033.88	1031.95	C-O Stretching
756.12		C-X (X = F, Cl, Br and I)

presence of OH group along with carbonyl group confirmed the presence of carboxylic acid groups in the biosorbent. The presence of NH group and OH group along with carbonyl group might be attributed the presence of amino acid groups in the biosorbent. All of them are summarized in Table 5.

Scanning electron microscopy (SEM)

SEM of *C. pyrenoidosa* at before and after biosorption of Zinc is presented in Figures 10 and 11, respectively. The SEM clearly revealed the surface texture and morphology of the biosorbent at different magnifications. The SEM analysis revealed important information on surface morphology. In these micrographs structures with large surface area were evident.

Desorption studies

Desorption and regeneration studies of the adsorbates showed that regeneration and recovery of the adsorbates is possible. Chemisorption/ion exchange was the main mechanism by which the adsorbates (metals) were attached to the adsorbents. Physical adsorption played a minimal role in the process .The result of desorption studies of Zn in a batch system showed that HNO_3 (0.1 M) was more efficient in Zn desorption, which remove 95% Zinc ions (Table 6).

Conclusions

The batch experiment conducted with the biosorption demonstrated that biomass of *C. pyrenoidosa* exhibited the potential for Zn removal from aqueous solution. Optimum pH and temperature for biosorption in this study were 5 and 25°C, respectively. The time taken for Zn adsorption by *C. pyrenoidosa* was dependent on initial metal ion concentration and increased with increase in concentration of Zn. With increase the initial concentration percentage of biosorption decreased and metal ion uptake capacity was increased. Which increased the amount of biomass observed that percentage of biosorption increased and metal ion uptake capacity was decreased. The findings of *C. pyrenoidosa* indicate that the sorption percentage increased with increase in temperature up to 30°C and there was a decrease in sorption percentage with further increase in temperature. The removal of Zn increase with increase in biosorbent. The biosorption process was followed both Langmuir and Freundlich isotherm model. The pseudo second-order kinetics described the experimental data well. The equilibrium time was 60 min.

The R_L values at different initial adsorbate concentrations indicate favorable ($0<R_L<1$) adsorption for all the adsorbents and adsorbates studied.HNO_3 (0.1M) had higher efficiency of Zn desorption than EDTA (0.1M) and HCL(0.1M) with 95% efficiency desorption. A positive value of $\Delta H°$ indicated the endothermic nature of the process. A negative value of the free energy ($\Delta G°$) indicated the spontaneous nature of the adsorption process. A positive value of $\Delta S°$ showed increased randomness at solid–liquid interface during the adsorption of heavy metals, it also suggests some structural changes in the adsorbate and the adsorbent. FTIR spectrums of *C. pyrenoidosa* revealed the presence of hydroxyl, amino, carboxylic and carbonyl groups. The

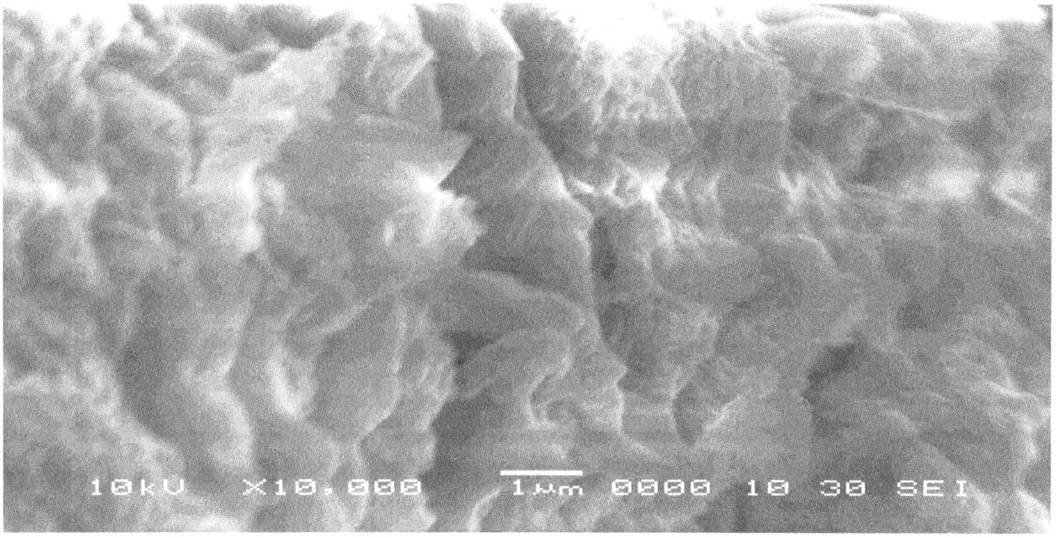

Figure 10. Scanning electron microscopy (SEM) micrographs of the *C. pyrenoidosa* before biosorption.

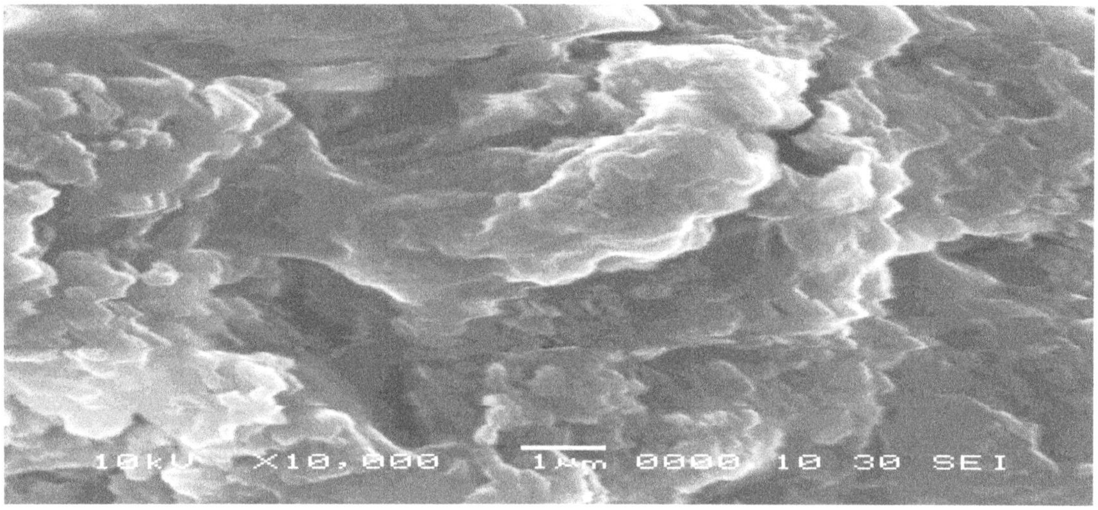

Figure 11. Scanning electron microscopy (SEM) micrographs of the *C. pyrenoidosa* after biosorption of zinc.

Table 6. The desorption efficiency of different desorbent.

Desorbent	EDTA (0.1M)	HCL (0.1M)	HNO_3 (0.1M)
% Desorption of Zn	76.14±4.12	91.36±4.43	94.84±4.01

SEM clearly revealed the surface texture and morphology of the biosorbent.

REFERENCES

Ahalya N, Ramachandra TV, Kanamadi RD (2003). Biosorption of heavy metals. J. Chem. Environ., 7(4): 71-79.

Bhattacharya AK, Mandal SN, Das SK (2006). Adsorption of Zn(II) from aqueous solution by using different adsorbents. Chem. Eng. J., 123: 43-51.

Khoo KM, Ting YP (2001). Biosorption of gold by immobilized fungal biomass. Biochem. Eng. J., 8: 51-59.

Knorr D (1991). Recovery and utilization of chitin and chitosan in food processing waste management. Food Technol., 45: 114-122.

Macaskie LE, Empson RM, Cheetham AK, Grey CP, Skarnulis AJ (1992). Uranium bioaccumulation by a Citrobacter sp. as a result of enzymically mediated growth of polycrystalline HUO2PO4. Sci., 257: 782-784.

Mofa AS (1995). Plants proving their worth in toxic metal cleanup. Sci., 269: 302-305.

Veglio F, Beolchini F (1997). Removal of metals by biosorption: A review. Hydrometallurgy, 44: 301-16.
Vijayaraghavan K, Yun YS (2008). Bacterial Biosorbents and biosorption. Biotechnol. Adv., 26: 266-291.
Volesky B, Holan ZR (1995). Biosorption of heavy metals. Biotechnol. Prog., 11: 235-50.
Volesky B (1990a). Biosorption by fungal biomass. Florida: CRC press, pp. 139-171.
Volesky B (1990b). Introduction in Biosorption of heavy metals. Boca Raton: CRC press, pp. 3-5.
Volesky B (1990c). Removal and recovery of heavy metals by biosorption. Florida: CRC press, pp. 8-43.
Wang JL, Chen C (2006). Biosorption of heavy metals by Saccharomyces cerevisiae: a review. Biotechnol. Adv., 24: 427-451.

Simple and simultaneous spectrophotometric determination of Pd (II) in alloys using hydrazone compounds

Malathi Challa[1]* and T. Srinivasulu Reddy[2]

[1]Department of Chemistry, MSRIT, Bangalore-54, Karnataka, India.
[2]Department of Chemistry, S. K. University, Anantapur - 515003, Andra Pradesh, India.

A highly selective, reproducible and rapid spectrophotometric method for the determination of palladium (II) is developed based on the reaction of Pd (II) with 2,4-dihydroxyl acetophenone-p-hydroxylbenzoylhydrazone [RPPHBH] in the pH range 5 to 6 which forms a stable pale green complex. The complex shows absorbance maximum at 375 nm at pH of 5.5. Beer's law is obeyed in the range 1.06 to 8.51 μg/ml of Pd (II). The molar absorptivity and Sandell's sensitivity of the complex are 0.85 × 10^4 L $mol^{-1}cm^{-1}$ and 0.0126 μg cm^{-2} respectively. Au (III) also reacts with RPPHBH at pH 5.5 to form a violet coloured complex. This complex shows an absorbance maximum at 590 nm. Thus, Pd (II) and Au (III) are simultaneously determined at pH of 5.5. The molar absorptivity of Pd (II) and Au (III) complexes of RPPHBH at 375 nm are 0.85 × 10^4 L mol^{-1} cm^{-1} and 0.42 × 10^4 L mol^{-1} cm^{-1} respectively. Pd (II): RPPHBH complex has no absorbance at 590 nm. The molar absorptivity of the gold complex at 590 nm is 0.5 × 10^4 L mol^{-1} cm^{-1}. This fact is successfully exploited for the simultaneous determination of palladium (II) and Au (III) in alloys.

Key words: 2, 4–dihydroxylacetophenone–p-hydroxylbenzoylhydrazone (RPPHBH), palladium (II), Au (III), alloy samples, UV- Vis, spectrophotometry.

INTRODUCTION

Palladium is used in industry as a catalyst. It is also used in cosmetic and jewelry industries. With the increasing use of palladium in jewelry and cosmetic industry, a need arose for the development of rapid and effective methods for the determination of palladium (II) in real samples. Pd (II) had strong corrosion resistance in the presence of acids (except nitric acid) at ordinary temperatures. It always occurred naturally with platinum group metals and was found in native metallic state. It had great affinity to absorb hydrogen on its surface. For higher density, it is alloyed with Ag, Au and Cu and used in the manufacture of watch bearings, springs and balance wheels and also for mirrors in scientific instruments. Palladium salts are widely employed in making special photographic printing paper.

Palladium is one of the strongest allergens from a health hazard view point. Hence, a rapid and accurate method for the determination of palladium is necessary. For the determination of palladium at micro levels, there are several analytical techniques such as X-ray fluorescence spectroscopy, atomic absorption spectrophotometry, spectrofluorimetry and other techniques. Among these, UV-Visible spectrophotometric methods are preferred because they are cheaper, easy to handle and highly sensitive. Though numerous spectrophotometric methods are available for the determination of palladium (II), the literature survey revealed that they are not sensitive enough (Gaurav et al., 2010; Janardhana et al., 2008; Lakshmi et al., 2007; Nain et al., 2006; Pourreza and Elhami, 2006; Roa et al., 1992; Rajeshet al., 1999; Sanke et al., 1981; Sarma et al., 2002; SreeVani et al., 2007).

This method is found to be rapid, sensitive, selective, reproducible and free from interference of a large number of common metal ions. Very few reports on direct spectrophotometric method for the simultaneous determination of Pd (II) and Au (III) are there in the literature (Donglan et al., 2010). However, a third

*Corresponding author. E-mail: maalathichalla@gmail.com.

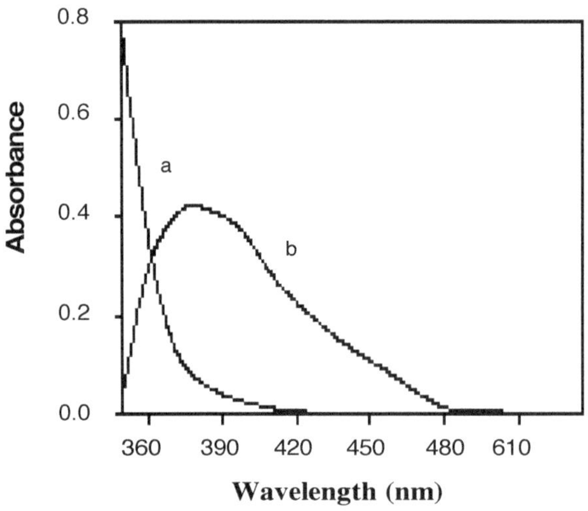

Figure 1. Absorption spectra of (a) RPPHBH Vs buffer blank; (b) [Pb (II)]: RPPHBH Vs reagent blank. Pd (II): 2, 4-dihydroxylacetophenone–p-hydroxylbenzoylhydrazone a pale green colored highly water soluble complex is formed within 10 min and has λ_{max} = 375 nm (Figure 1), where reagent has low absorbance. It is stable for 3 h.

derivative spectrophotometric method was reported for the simultaneous determination of Pd (II) and Au (III) in the synthetic mixtures but was not applied for real samples like alloys (Vojkovic et al., 2003).

EXPERIMENTAL

Schimadzo (UV–160 A) spectrophotometer, (Koyoto in Japan) is used for absorbance measurements. pH meter (ELICO) was used for measuring pH of buffer solutions. All chemicals and solvents used were of analytical grade. RPPHBH is synthesized by refluxing equimolar quantities of 2, 4-dihydroxylacetophenone and p-hydroxylbenzoylhydrazine in aqueous alcoholic medium in presence of small amount of CH_3COONa. The pale yellow coloured product obtained was recrystalised from hot ethanol in presence of charcoal. Its melting point is 248 to 252℃. The structure of RPPHBH was determined from IR, H^1NMR, Mass spectral studies and the purity was checked by HPLC. 1 × 10^{-2} M solution of RPPHBH was prepared by dissolving 0.777 g of RPPHBH in dimethylformamide (DMF) in a 25 ml volumetric flask and diluted up to the mark with DMF. Fresh reagent solution is prepared every day before use.

Stock solution of $PdCl_2$

One gram of palladium chloride was dissolved in distilled water after adding a few drops dilute HCl and diluted up to the mark in 100 ml volumetric flask. It was standardized.

Stock solution of chloroauric acid

One gram of $AuCl_3$ was dissolved in distilled water after adding a few drops of dilute HCl and made up to the mark with distilled water in a 100 ml volumetric flask and standardized. All the stock solutions were suitably diluted whenever necessary.

Method for simultaneous determination

One milliliter of 4 × 10^{-4} M solutions of each of Pd (II) and Au (III) were mixed in a 10 ml volumetric flask containing 5 ml buffer solution of pH 5.5. 0.6 ml of RPPHBH solution of 4 × 10^{-2} M was added and made up to the mark with distilled water. The absorbance of the solution was measured at 375 nm and 590 nm against reagent blank (5 ml of buffer solution and 0.6 ml of RPPHBH solution diluted up to the mark with distilled water).

Preparation of alloy sample

0.1 to 0.5 g of the sample was dissolved in a mixture of 2 ml HCl and 10 ml $HN0_3$. The resulting solution is evaporated to a small volume and mixed with 5 ml of 1:1 H_2O: H_2SO_4 mixture, and evaporated to dryness. The residue was dissolved in 15 ml of distilled water and filtered through Whatman filter paper No.41. The filtrate was collected in a 100 ml volumetric flask and made up to the mark with distilled water. The solution was further diluted as required.

RESULTS AND DISCUSSION

[Pd (II): RPPHBH] complex

0.5 milliliter of 1 × 10^{-4} M palladium solution and 1.5 ml of 1 × 10^{-2} M reagent solution along with 5 ml of pH 5.5, buffer solution are taken in 10 ml volumetric flask and diluted up to the mark with distilled water. The spectrum of [Pd (II) - RPPHBH], a pale green coloured complex, was recorded over the range 300 to 600 nm against a reagent blank. It shows maximum absorbance at 375 nm and is shown in Figure 1. The effect of pH on the absorbance of the complex is shown in Figure 2. The

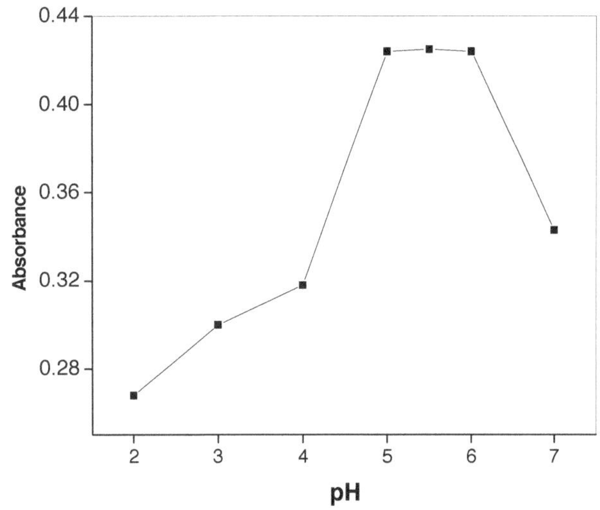

Figure 2. Effect of pH on absorptance of Pb (II). RPPHBH system. [Pb (II)] = 5.0 × 10^{-5} M: RPPHBH = 1.5 × 10^{-2} M. Wave length = 375 nm.

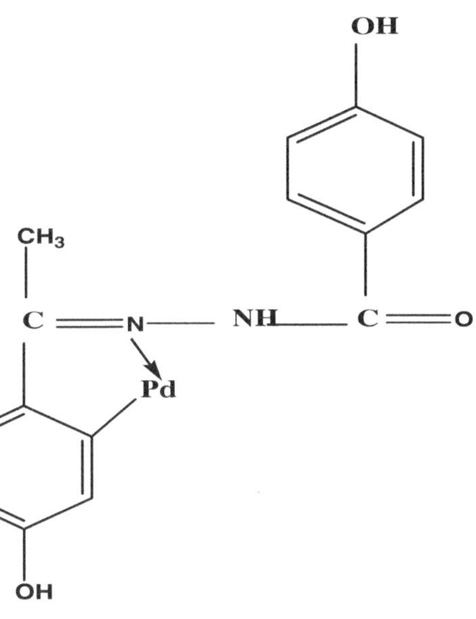

Figure 3. Structure of Pb (II): RPPHBH complex.

absorbance is maximum and constant in the range of pH 5 to 6, therefore pH 5.5 is selected. Fifteen fold excess of the reagent is necessary to get maximum absorbance. Beer's law was obeyed in the range 1.06 to 8.51 µg/ml of Pd (II).

Analysis of the results of Pd (II) – RPPHBH gave linear plots corresponding to A = 0.0783C + 0.0061. The optimal concentration range of determination of Pd (II) is found to be 2.12 to 7.45 μg/ml from Ringbom's plot. The molar absorptivity and Sandell's sensitivity are found to be 0.85 × 10^4 L mol^{-1} cm^{-1} and 0.0126 μg cm^{-2} respectively. The standard deviation in the determination of 5.32 μg/ml of Pd (II) was 0.0069. The composition of the complex was established as 1:1 [Pd (II): RPPHBH] from Job's continuous variation method and molar ratio method. The stability constant of the complex from Job's method was determined as 3.10 × 10^6. The possible structure of the complex is shown in Figure 3.

Au (III): RPPHBH complex

Au (III) reacts with RPPHBH solution to form a violet colored water soluble complex. It has maximum intensity

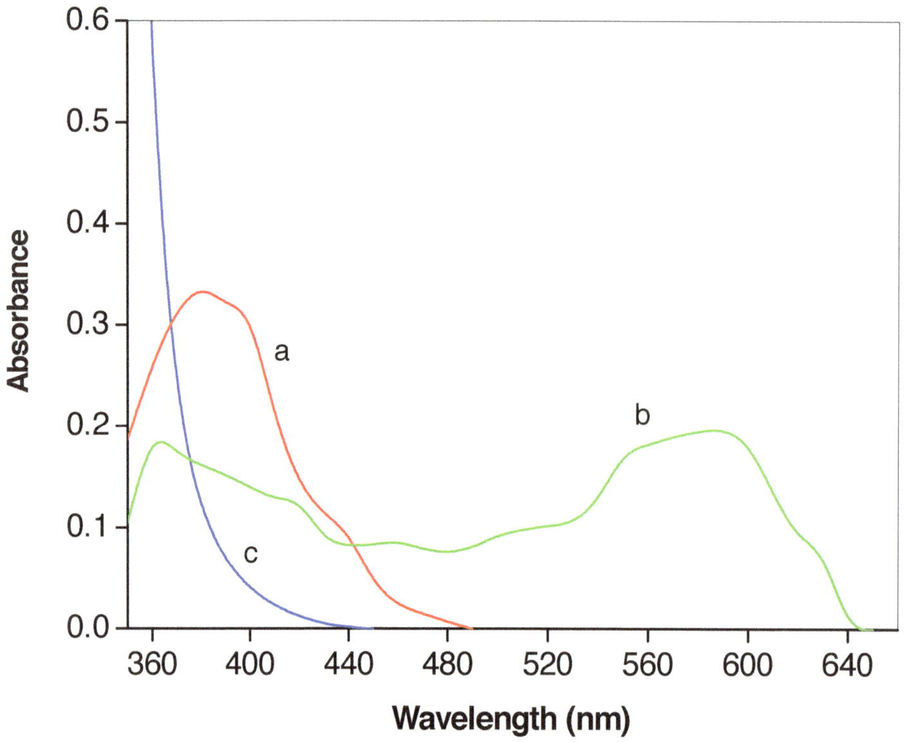

Figure 4. (a) Absorptance spectrum of Pb (II): RPPHBH Vs reagent blank. (b) Absorptance spectrum of Au (II): RPPHBH Vs reagent blank. (c) Absorptance spectrum of RPPHBH Vs buffer solution. Simultaneous spectra of Pd (II): RPPHBH and Au (III): RPPHBH complexes.

in the pH range 4.0 to 7.0. The complex shows maximum absorbance at 590 nm. A 10 fold excess of RPPHBH solution is necessary to attain maximum absorbance.

Simultaneous determination of Pd (II) and Au (III)

Pd (II) and Au (III) complexes of RPPHBH have maximum absorbances at pH 5.5. The advantage of this system is that Pd (II): RPPHBH complex shows maximum absorbance at 375 nm and it had practically no absorbance after 480 nm (Figure 4). It is observed that the total absorbance of the solutions containing a mixture of Au (III) and Pd (II) complex species at 375 nm is equal to the sum of the absorbance values due to individual Pd (II) and Au (III) species. Thus, it is possible to determine simultaneously Au (III) and Pd (II) using RPPHBH at pH 5.5. The absorbance of the Au (III) and Pd (II) complexes in the sample are measured at 590 nm and 375 nm.

Analytical features

Au (III): RPPHBH complex obeys Beer's law in the range 3.94 to 27.50 µg/ml at 375 nm and in the range 1.96 to 19.70 µg/ml at 590 nm. Pd (II) system obeys Beer's law in the range 1.06 to 8.15 µg/ml of Pd (II) at 375 nm. The molar absorptivities for Au (III) and Pd (II) species at 375 nm are 4.20×10^3 L mol^{-1} cm^{-1} and 0.85×10^4 L mol^{-1} cm^{-1}, respectively. The molar absorptivity of Au (III) at 590 nm was 0.5×10^4 L mol^{-1} cm^{-1}. The absorbance of Au (III) species at 375 nm is found to be 0.83 times its absorbance at 590 nm. The absorbance due to Pd (II) at 375 nm is obtained by subtracting the absorbance contribution due to Au (III) at this wavelength which is 0.83 times its absorbance at 590 nm, where Au (III) alone absorbs. It was calculated by using the equation.

$$\text{Pd(II)} = \frac{A_{375} - (A_{590} \times 0.83)}{\varepsilon_{375}^{Pd(II)}} \times 106.42 \times 10^3$$

Where at A_{375} and A_{590} nm were the absorbance values of the solution mixture at 375 nm and 590 nm respectively. $\varepsilon_{375}^{Pd(II)}$ is the molar absorptivity of Pd (II) at 375 nm. 106.42×10^3 is the factor to convert the concentration of Pd (II) into amount.

Effect of foreign ions

The interference of other associated ions in the determination of Pd (II) was studied using 3.192 µg/ml of palladium. Most of the ions do not interfere except

Table 1. Tolerance limit of foreign ions.

Foreign ion	Tolerance limit (μg/ml)
Ti(IV)	Interferes
Mn(II)	44
Se (IV)	40
Zr(II)	66
Ca (II)	407
Au (III)	20
Mg(II)	243
Pb (II)	84
Cd(II)	68
Co(II)	25
Zn(II)	320
Mo(VI)	29
V (V)	Interferes
Ru(III)	10
Cu(II)	Interferes
Cr(VI)	60
Th (IV)	60
Fe(III)[a]	220
Al (III)	42
Ce(IV)	84

Most of the anions do not interfere except thiocyante, thiourea, EDTA and thiosulphate anions. These interfering ions used as a masking agent.

Table 2. Simultaneous determination of Au (III) and Pd (II) in synthetic mixture.

Amount taken (μg/ml)		Amount found (μg/ml)		Relative error (%)	
Au (III)	**Pd (II)**	**Au (III)**	**Pd (II)**	**Au (III)**	**Pd (II)**
1.97	4.26	1.97	4.24	-0.50	-0.47
3.94	4.26	3.92	4.24	0.50	-0.47
7.88	4.26	7.85	4.29	-0.38	0.70
9.85	4.26	9.89	4.25	0.41	-0.23
11.80	4.26	11.75	4.23	-0.42	-0.70
7.88	2.12	7.85	2.11	-0.38	-0.85
7.88	3.19	7.93	3.16	0.63	-0.94
7.88	4.26	7.83	4.29	-0.63	0.70
7.88	5.35	7.92	5.33	0.51	-0.37
7.88	6.38	7.81	6.39	-0.89	0.16

*Average of seven determinations.

copper, titanium, vanadium and iron. These ions were masked with suitable anions to avoid their interference during the determination of palladium (Table 1).

APPLICATION

The method is applied for the determination of palladium in certain synthetic mixtures corresponding to its alloys and also for real alloy samples. The results are presented in Tables 2 and 3.

Conclusion

The proposed spectrophotometric method for the determination of Pd (II) using RPPHBH as a spectrophotometric reagent is sensitive, rapid and selective at pH 5.5. A direct spectrophotometric method

Table 3. Analysis of alloy samples.

Sample	Found (%)	
	Pd (II)	Au (III)
P 30; B 50TF198, [30.5 Pd, 56.55 Ni, 10.5 Cr, 2.45 B]	30.2	-
Pd 72%, Ag 26%, and Ni 2%.	70.9	-
Pd 60%, Au 40%.	60.3	39.7
Pd 34 %, Au 10%, Co 22%, Ni 34%.	34 .2	9.8
Au 75%, Pd 10%, Ni 10%, Zn 5%.	10.2	74.9

The results in the table were average of seven determinations.

for the simultaneous determination of Pd (II) and Au (III) in alloys is developed for the first time.

REFERENCES

Donglan MA, Guosheng D, Jinye W (2010). Simutaneous spectrophotometric determination of Pd(II) and Au(III) with N-phenyl-N' – [sodium-p-amino benzene sulfonate] thourea. Analy. Bioanalyti. Chem., 372(5-6): 740-743.

Gaurav BP, Subash, Bhadange G, Anad SA (2010). Extractive spectrophotometric determination of Pd(II) with 2-hydroxy-5-methylacetophenoneisonicotinoylhydrazone. Pelagia Research Library. Adv. Appl. Sci. Res., 1(2): 58-64.

Janardhana RK, Rajesh KJ, Ramachandraiah C, Adinarayana RS, Varadha RA (2008). Selective and sensitive extractive spectrophotometric determination of Pd(II) with N-ethyl-3-carbazolecarbaxaldehydethiosemocarbazone. Environ. Monit. Assess., 136: 337-346.

Lakshmi NS, Janardhana RK, Somala AR, Jyothi RK, VardhaReddy AR (2007). Selective, sensitive and extractive spectro photometric determination of Pd (II) in water samples and synthetic mixtures. J. Chin. Chem. Soc., 54: 1233-1241.

Nain S, Agnihotri N, Kakkar LR (2006). Micro determination of Pd(II) using benzyldioxime and thicyanate. J. Ind. Chem. Soc., 83(7): 707-708.

Pourreza N, Elhami (2006). Simutaneous solid phase extraction of Au (III) and Pd (II) as chloride complexes on naphthalene – methylstrioctyl ammonium chloride sorbent. Canad. J. Anal. Sci. Spectroscopy, 51(5): 260-266.

Roa DM, Reddy KH, Reddy DV (1992).Spectrophotometric determination of Pd(II) with 5,6-di mthyl-1,3-indanedione-2-oxime. Talanta, 39(9): 1037-1040.

Rajesh S, Sondhi SM, Bina G (1999). Extraction spectrophtometric determination of Pd(II) with 3,4,4a,5-tetrahydro-3,3,4a-trimethyl-7-[substituted]-pyrimido[1,6-a]-benzimidazole-1-thiol.Talanta, 42(3): 410-415.

Sanke GH, Padmaji KA, Thimmaiah KN (1981). Simutaneous spectrophotometric determination of Pd(II) and Au(III) with methiomeprazine hydrochloride analysis of alloys and minerals. Anal., 10: 198-205.

Sarma LS, Kumar JR, Reddy KJ, Reddy AV (2002).A rapid and sensitive extractive spectrophotometric determination of Pd(II) in synthetic mixtures and hydrogenation catalysis using pyridoxal -4-phenyl -3- thio asemi carbazone. Anal. Sci., 18(11): 1257-1261.

SreeVani I, Praveen KA, Raveendra RP, Mohan R, Krishna R (2007). Spectrophotometric determination of Pd(II). Anali. de.Chimica., 32: 1237-1245.

Vojkovic V, Druskovic V (2003). Simutaneous spectrophotometric determination of Pd(II) and Au(III) in mixtures by third derivative spectrophotometry using 3- hydroxyl-2-methyl-1-phenyl-4-pyridone ligand. Crotica Chem. Acta., 76(1): 87-92.

Yoyseff A, Sayed E, Fathi A (1998). Spectrophotometric and derivative spectrophoto-metric determination of Pd (II) with pyridopyridazinedithione in the presence of non ionic surfactant. Microchem. Acta., 129: 225-231.

Syntheses and crystal structures of two o-phthalate based transition metal complexes

Xuehui Zhao[1], Hongmei peng[2] and Liangcai Yu[2*]

[1]Key Laboratory of Green Packaging and Application Biological Nanotechnology of Hunan Province, Hunan University of Technology, Zhuzhou 412008, P. R. China.
[2]Key Laboratory of Theoretical Chemistry and Molecular Simulation of Ministry of Education, Hunan Provincial University Key Laboratory of QSAR/QSPR, School of Chemistry and Chemical Engineering, Hunan University of Science and Technology, Xiangtan Hunan, 411201, P. R. China.

We report the synthesis and the crystal structures of two complexes. The two metal complexes; (1) $[Zn(bipy)_2(HPht)](HPht)\cdot H_2O$ and (2) $[Cu_2(Pht)_2(Bipy)_2]\cdot 4H_2O$ (where Pht^{2-} = dianion of o-phthalic acid, $HPht^-$ = anion of o-phthalic acid, Bipy = 2, 2′-bipyridine) have been synthesized and their crystal structures were determined by elemental analyses, IR and single crystal diffraction analyses. X-ray diffraction studies show that the complex 1 crystallizes in triclinic system, space group $P\bar{1}$, with lattice parameters a = 10.0134(4), b = 11.7090(4), c = 14.0979(5)Å, α = 111.0560(10), β = 105.9690(10), γ = 91.028°(2), V = 1617.27(10)Å^3 and Z = 2; Complex 2 crystallizes in monoclinic system, space group P2(1)/n, with lattice parameters a = 8.8148(6), b = 12.9931(9), c = 15.7454(11)Å, β = 100.131°(2), V = 2954.24 Å^3 and Z = 4.

Key words: 2, 2′-bipyridine, zinc, cupric, crystal structure, potassium biphthalate.

INTRODUCTION

The construction of metal–organic hybrid compounds are of considerable interest and importance in recent years, owing to not only their intriguing structural motifs but also their potential applications in catalysis, medicine, host–guest chemistry and the promising photomaterials, electromaterials and magnetic materials (Leininger et al., 2000; Hagrman et al., 1999; Zhang et al., 2002; Feng and Xu, 2001). O-Phthalic acid is a versatile ligand for coordinating to metal ions in various modes, such as monodentate bonding through a carboxylic O atom, bidentate fashion *via* two carboxylic O atoms and triple coordination through its three carboxylic O atoms, and bridging mode with two or four O atoms of its carboxylate groups. The coordination properties thus allow for the preparation of complexes with a large variety of architectures, forming isolated mono-, di- and polynuclear, as well as one-, two- and three-dimensional networks. Mixed ligand complexes with o-Phthalic acid and 2, 2′-bipyridine are investigated extensively recently and some crystal structures have been determined for Cu^{2+} (Zhang et al., 2000; Hao et al., 2004), Co^{2+} (Baca et al., 2005; Hu et al., 2004), Mn^{2+} (Ma et al., 2004; Yang et al., 2005), Ni^{2+} (Poleti et al., 1900; Xue et al., 1993) and Pb^{2+} (Yu et al., 2008). In this paper, we report the hydrothermal synthesis and structural characterization of two complexes (1) $[Zn(bipy)_2(HPht)](HPht)\cdot H_2O$ and (2) $[Cu_2(Pht)_2(bipy)_2]\cdot 4H_2O$.

MATERIALS AND METHODS

Materials and physical measurements

All reagents were commercially available and used without further purification. C, H and N data were obtained using American PE 2400 II CHNS/O elemental analyzer. Infrared spectra were measured from KBr pellets using a Nicolet 5DXB system.

Preparation of complexes 1-2

Synthesis of [Zn(bipy)$_2$(Pht)](HPht) ·H$_2$O(1)

$Zn(SO_4)_2\cdot 7H_2O$ (0.2870 g, 1 mmol), KHphth (Potassium

*Corresponding author: E-mail:ylcphd@gmail.com.

Table 1. Crystal data and structure refinement for 1 and 2.

Parameter	1	2
Empirical formula	$C_{36}H_{28}N_4O_9Zn$	$C_{36}H_{32}Cu_2N_4O_{12}$
Formula weight	725.99	839.74
Temperature(K)	296(2)	296(2)
Wavelength(Å)	0.71073	0.71073
Crystal system	Triclinic	Monoclinic
Space group	$P\bar{1}$	P2(1)/n
a (Å)	11.0134(4)	8.8148(6)
b (Å)	11.7090(4)	12.9931(9)
c (Å)	14.0979(5)	15.7454(11)
α(°)	111.0560(10)	90
β (°)	105.9690(10)	100.131(2)
γ (°)	91.028(2)	90
V ($Å^3$)	1617.27(10)	1775.2(2)
Z	2	2
Dc (mg/m^3)	1.491	1.571
F(000)	748	860
h	-13 ≤ h ≤ 13	-10 ≤ h ≤ 10
k	-14 ≤k ≤ 9	-14 ≤ k ≤ 16
l	-13 ≤ l≤ 17	-19 ≤ l ≤ 15
Completeness	94.3	99.9
Final *R* indices[*I>2sigma(I)*]	R_1=0.0645, wR_2=0.1873	R_1=0.0402, wR_2=0.1220
R indices(all data)	R_1=0.0725, wR_2=0.1970	R_1=0.0461, wR_2=0.1277
Goodness-of-fit on F^2	1.092	1.112

biphthalate) (0.2040 g, 1 mmol) and 2, 2'-bipyridine (0.1560 g, 1 mmol) were mixed in 10 mL deionized water and 10 mL C_2H_5OH. After stirring half an hour, the mixture was placed in 25 mL Teflon-lined reactor and heated at 130℃ in an oven for 3 da ys, cooled slowly to room temperature and the pale yellow block crystals of the title complex 1 suitable to X-ray diffraction analysis were obtained. $C_{36}H_{28}N_4O_9Zn$ (725.99) Anal. Found: C 59.50, H 3.86, and N 6.61. Calcd:C 59.56, H 3.56, N 6.58. IR data: (KBr pellet, cm^{-1})，3414(m), 2607(w), 1910(w), 1716(s), 1579(s), 1543(s), 1478(s), 1404(s), 1315(m), 1252(m), 1157(m), 1022(s), 864(m), 773(s), 779(w), 650(m), 575(m).

Synthesis of [$Cu_2(Pht)_2(bipy)_2$] ·$4H_2O$ (2)

The procedure is similar to that of 1, except that $Cu(SO_4)_2 \cdot 5H_2O$ was used. $C_{36}H_{32}Cu_2N_4O_{12}$ (839.74) Anal. Found: C 51.46, H 3.84, and N 6.63. Calcd: C 51.44, H 3.81, N 6.67. IR data: (KBr pellet, cm^{-1}), 3448 (m), 3064 (m), 2324 (w), 1604 (s), 1472 (m), 1443 (m), 1376 (s), 1152 (m), 1107 (m), 850 (s), 807 (m), 770 (s),730 (m), 651(w), 522 (w).

Crystal structure determination

The X-ray single crystal data collections for complex 1 and 2 were performed on a Bruker Smart CCD diffractometer equipped with a graphite monochromator Mo-Kα radiation (*λ*=0.71073 Å). Multi-scan absorption correction was applied using the SADABS program (Sheldrick, 1996). The structures were solved by direct method using the SHELXS-97 program (Sheldrick, 1997). Refinements on F^2 were performed using SHELXL-97 (Sheldrick, 1997) by full-matrix least-squares with anisotropic parameters for all non-hydrogen atoms. The hydrogen atoms of Pht^{2-}, $HPht^-$ and 2, 2'-bipyridine were generated geometrically, while the H atoms of lattice water molecules were located from Fourier difference map and refined subject to an O-H=0.85(1)Å. Part H atoms of lattice water molecule in 2 were not located for high values of temperature factor of O atoms. The crystal data is given in Table 1.

Supplementary material

Crystallographic data for title compounds 1 and 2 were deposited to the Cambridge crystallographic data center with deposition numbers CCDC**,** 701976 and 615071, respectively.

RESULTS AND DISCUSSION

Spectroscopic properties

The IR spectrum exhibit a medium-intensity and broad band with maxima at 3414 and 3448 $cm^{-1}$for 1 and 2 corresponding to the stretching vibration of lattice water molecules. The strong peaks at 1579 to 1543 and 1478 to 1404 cm^{-1} regions for 1, 1604 and 1472 to 1443 cm^{-1} for 2, can be attributed to the asymmetric and symmetric stretching vibrations of the coordinated carboxylate

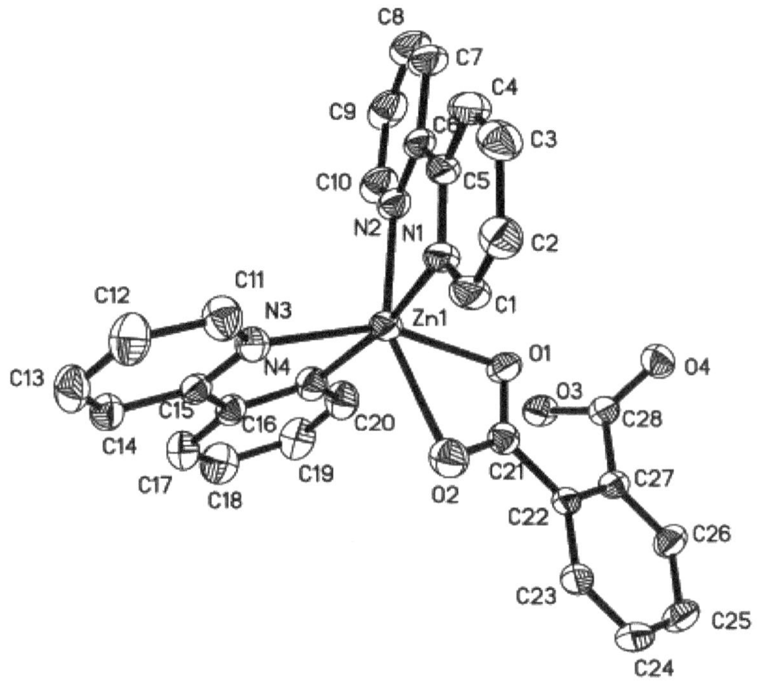

Figure 1. The coordination environment of the Zn(II) ion in 1. All hydrogen atoms are omitted for clarity.

groups of the phthalate ligands, respectively. Absorption band at 1716 cm^{-1} indicates the presence of uncoordinated o-phthalic acid in the complex 1.

Crystal structures

The structure of [Zn(bipy)$_2$(HPht)](HPht) ·H_2O(1)

X-ray diffraction analysis revealed that 1 consists of a mononuclear [Zn(bipy)$_2$(HPht)]$^+$ cation, $HPht^-$ anion and one solvate H_2O molecule, which is different from the structure of a dimer [(bpy)$_2$Zn(Pht)H(Pht)Zn(bpy)$_2$](HPht)(H_2Pht).2H_2O (Svetlana et al., 2003), as shown in Figure 1. Zn atom in the mononuclear unit has a distorted octahedral geometry with four nitrogen atoms from two chelating bpy ligands (Zn–N bond distances range from 2.104(3) to 2.149(3) Å) and two oxygen atoms from one chelating carboxylate group of $HPht^-$ ligand (Zn–O 2.094(3) and 2.336(3) Å) (Table 2). Bond lengths of C(28)–O(3) (1.287(5) Å) are distinctly longer than those of C(28)–O(4) (1.216(5) Å), indicating that the uncoordinated carboxylate groups have not been ionized.

Nevertheless, the bond lengths of C(21)–O(1) and C(21)–O(2) (1.255(5) and 1.246(5) Å) are very close, which shows that the chelating carboxylate group are deprotonated. The observed distortion from octahedral geometry can be caused by small bite angle of the chelating carboxylate group [O(1)-Zn(1)-O(2) bond angle is 58.66(10)°]. The other two bite angles N(2)-Zn(1)-N(1), N(4)-Zn(1)-N(3) (77.71(13) and 76.54(13)°) deviate normal angle for octahedral geometry. The bicyclical rings A: N1/C1/C2/C3/C4/C5 and B:N2/C6/C7/C8/C9/C10, C: N3/C11/C12/C13/C14/C15 and D:N4/C16/C17/C18/C19/C20, are essentially coplanar with corresponding dihedral angles of 8.3° and 1.6°, respectively. The coordinated carboxylate group (C21/O1/O2) is coplanar with the base plane E:C22/C23/C24/C25/C26/C27 with the dihedral angle 8.5°, while the uncoordinated carboxylate group (C28/O3/O4) is non-coplanar with base plane E:C22/C23/C24/C25/C26/C27 with dihedral angle 76.3°. The lattice water and uncoordinated carboxylate O atoms participate to form hydrogen bonds, thus forming three-dimensional network (Figure 2 and Table 3).

The structure of [Cu$_2$(Pht)$_2$(bipy)$_2$] ·4H_2O (2)

X-ray diffraction analysis revealed that the title complex 2 consists of isolated dicarboxylato-bridged dinuclear [Cu$_2$(bipy)$_2$(pht)$_2$] unit and lattice water molecules, which is different from the structure of a reported o-phthalate-bridged copper(II) chain complex of {[Cu(bipy)(m-phthalate)H_2O]·3.5H_2O}*n* (Zhang et al., 2000). The molecular structure of the title complex 2 is shown in Figure 3, with all hydrogen atoms and water molecules omitted for clarity. The coordination environment around central copper (II) ion in the structure is slightly distorted

Table 2. Selected bond distances (Å) and bond angles (°) for compounds 1 and 2.

Bond	***d*, Å**	**Bond**	***d*, Å**
Zn(1)-O(1)	2.094(3)	Zn(1)-N(1)	2.124(3)
Zn(1)-O(2)	2.336(3)	Zn(1)-N(2)	2.104(3)
Zn(1)-N(3)	2.149(3)	Zn(1)-N(4)	2.134(4)
O(3)-C(28)	1.287(5)	O(4)-C(28)	1.216(5)
O(1)-C(21)	1.255(5)	O(2)-C(21)	1.246(5)
Angle	ω, deg	Angle	ω, deg
O(1)-Zn(1)-N(2)	96.53(12)	O(1)-Zn(1)-N(1)	98.66(12)
N(2)-Zn(1)-N(1)	77.71(13)	O(1)-Zn(1)-N(4)	91.92(12)
N(1)-Zn(1)-N(3)	93.43(13)	N(4)-Zn(1)-N(3)	76.54(13)
O(1)-Zn(1)-O(2)	58.66(10)	N(1)-Zn(1)-O(2)	94.66(12)
N(4)-Zn(1)-O(2)	90.09(13)	N(3)-Zn(1)-O(2)	97.28(12)
Bond	*d*, Å	Bond	*d*, Å
Cu(1)-O(1)	2.4080(19)	Cu(1)-O(1)#1	1.9446(18)
Cu(1)-O(4)#1	1.903(2)	Cu(1)-N(1)	1.993(2)
Cu(1)-N(2)	1.980(2)		
Angle	ω, deg	Angle	ω, deg
O(4)#1-Cu(1)-O(1)#1	91.63(9)	O(1)#1-Cu(1)-N(2)	94.63(9)
O(4)#1-Cu(1)-N(1)	92.27(10)	N(2)-Cu(1)-N(1)	81.38(10)
O(4)#1-Cu(1)-O(1)	94.51(9)	O(1)#1-Cu(1)-O(1)	76.56(8)
N(2)-Cu(1)-O(1)	89.63(9)		

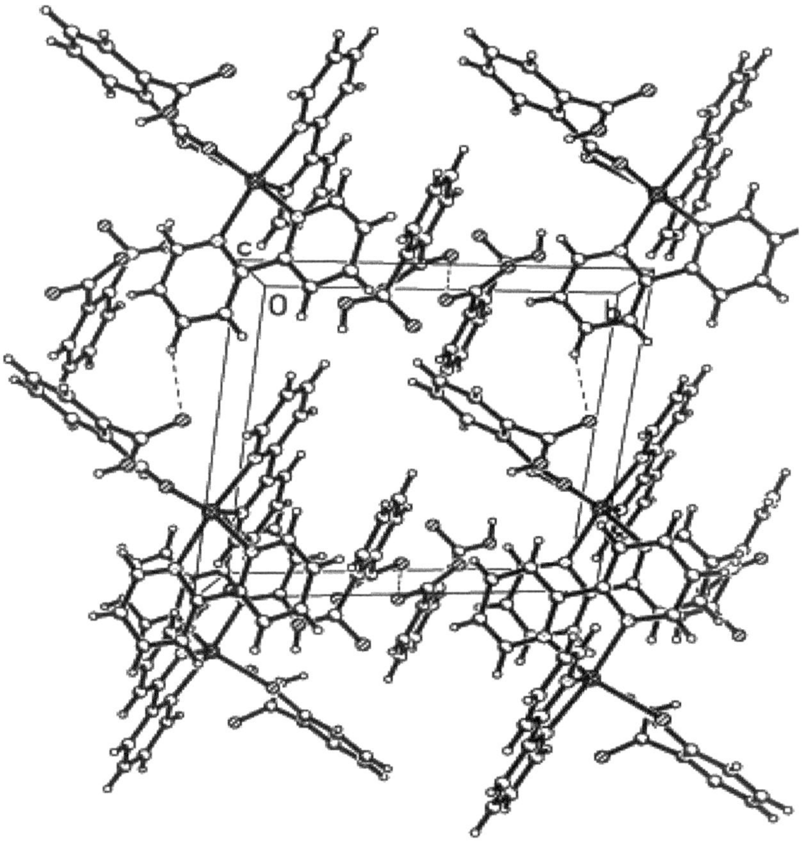

Figure 2. Crystal packing view of 1 along the c axis.

Table 3. Hydrogen bonds for 1.

D-H...A	d(D-H)	d(H...A)	d(D...A)	<(DHA)
O(4)-H(4)...O(9)#1	0.82	2.27	2.838(6)	126.8
O(9)-H(9B)...O(7)#2	0.852(10)	1.86(5)	2.572(7)	140(8)
O(9)-H(9B)...O(8)#2	0.852(10)	2.58(3)	3.362(11)	153(5)

Symmetry transformations used to generate equivalent atoms: #1 x, y, z-1, #2 -x,-y+1,-z+1.

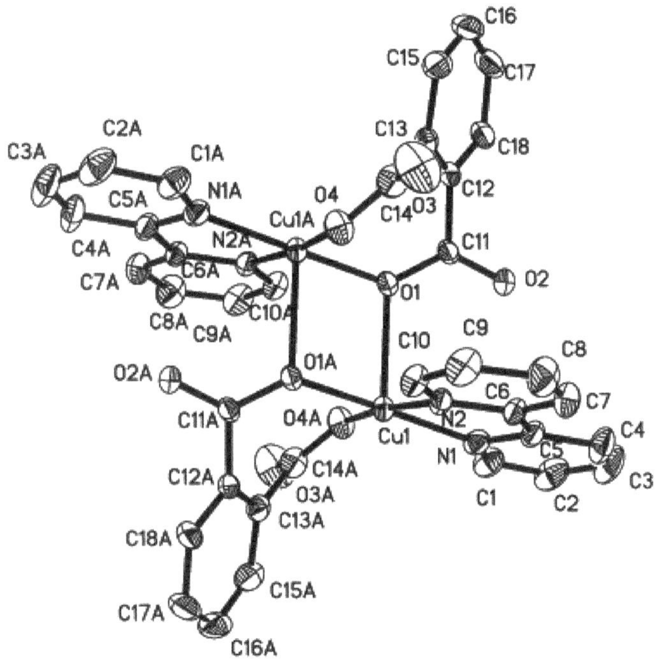

Figure 3. The coordination environment of the Cu(II) ion in 2. All hydrogen atoms are omitted for clarity.

square pyramid. The Cu^{2+} is located 0.0314 Å above the average plane defined by oxygens N1, N2, O1A and O4A, which deviate from such average plane 0.0016 Å above (N1), 0.0131 Å above (N2), 0.0031 Å above (O1A) and 0.0137 Å above (O4A). In the equatorial plane, the Cu^{2+} is coordinated by two nitrogen atoms[N(1) and N(2)) from one chelate 2, 2′-bipyridine molecule with the bond distances of [Cu(1)-N(1) = 1.993(2)Å, Cu(1)-N(2) = 1.980(2)Å] and two oxygen atoms[O(1A) and O(4A)] from one bridging carboxylate group of Pht^{2-} with the bond distances of [Cu(1)-O(1A) = 1.9446(18)Å, Cu(1)-O(4A) = 1.903(2)Å].

The apical water molecule is coordinated at a longer distance [Cu(1)-O(1) = 2.4080(19)Å] for the Jahn-Teller effect (Figure 3 and Table 2). Herein the two Pht^{2-} anion acted as a tridentate ligand bridging two Cu atoms and forms an isolated dinuclear unit. The observed distortion from square pyramid geometry can be caused by small bite angle of the chelating 2, 2′-bipyridine ligand [N(1)-Cu(1)-N(2) bond angle is 81.38(10)°] and other bond angles O–Cu(1)–N [vary from 89.63(9) to 94.63(9)°]. The bicyclical rings A: N1/C1/C2/C3/C4/C5 and B:N2/C6/C7/C8/C9/C10 of bipy ligand are essentially coplanar with corresponding dihedral angles of 2.1°. The coordinated carboxylate groups (C11/O1/O2) and (C14/O3/O4) is non-coplanar with the base plane C:C12/C13/C14/C15/C16/C17 with the dihedral angle 59.2 and 48.4°, respectively, while the two coordin ated carboxylate groups are almost orthogonal with the dihedral angle 84.3°. The lattice water and carboxy late O atoms participate to form hydrogen bonds, thus forming three-dimensional network (Figure 4).

Conclusion

In summary, assembling hybrid ligands of Phth, Bipy and H_2O with Cu^{2+} and Zn^{2+}, the complexes 1 and 2 have been synthesized by hydrothermal process. The complex 1 is a mononuclear molecule, while the 2 is a dimmer with the two Pht^{2-} anion acted as a tridentate ligand bridging two Cu atoms to form an isolated binuclear unit.

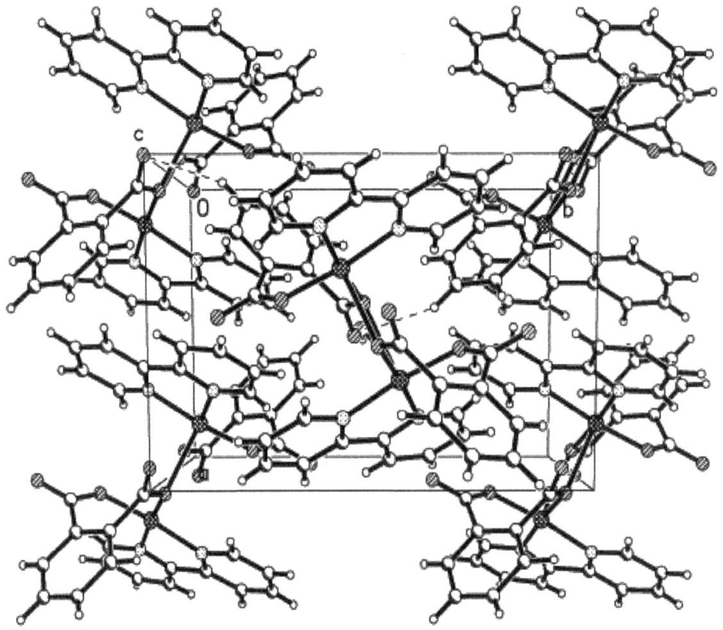

Figure 4. Crystal packing view of 2 along the c axis.

ACKNOWLEDGEMENTS

The author acknowledges the financial supports from National Natural Scientific Foundation of China (No. 50874041) and from Natural Science Foundation of Hunan (No. 08JJ5036).

REFERENCES

Baca SG, Filippova IG, Ambrus C, Gdaniec M, Simonov YA, Gerbeleu N, Gherco OA, Decu-rtins S (2005). From Mono- and Dinuclear to Polynuclear Cobalt(II) and Cobalt(III) Coordination Compounds Based on o-Phthalic Acid and 2,2'-Bipyridine: Synthesis, Crystal Structures and Properties. Eur. J. Inorg. Chem., 15: 3118-3130.

Feng SH, Xu RR (2001). New Materials in Hydrothermal Synthesis. Acc. Chem. Res., 34(3): 239-247.

Hagrman PJ, Hagrman D, Zubieta J (1999). Organic–Inorganic Hybrid Materials: From “Simple” Coordination Polymers to Organodiamine Templated Molybdenum Oxides. Angew. Chem. Int. Ed., 38(18): 2638-2684.

Hao N, Li YG, Wang EB, Shen EH, Hu CW, Xu L (2004). Hydrothermal synthesis and crystal structure of an infinite 1D ladderlike metal-organic compound: $[Cu_2(btec)(2,2'\text{-bipy})_2]_\infty$ (btec=1,2,4,5-benzenetetracarboxylate). J. Mol. Struct., 697(1-3): 1-8.

Hu ML, Ye MD, Xiao HP, Yuan JX (2004). Crystal structure of bis[triaqua(2,2-bipyridine)cob- alt(II)] 1,2,4,5-benzenetetracarboxylate dihydrate, $[Co_2(C_{10}H_8N_2)_2(H_2O)_6(C_{10}H_2O_8)] \cdot 2H_2O$. Z. Kristallogr., 219: 19-20.

Leininger S, Olenyuk B, Stang PJ (2000). Self-Assembly of Discrete Cyclic Nanostructures Mediated by Transition Metals. Chem. Rev., 100(3): 853-908.

Ma CB, Wang WG, Zhang XF, Chen CN, Liu QT, Zhu HP, Liao DZ, Li LC (2004). Molecular, One and Two-Dimensional Systems Built from Manganese(II) and Phthalate/Diimine Ligands: Syntheses, Crystal Structures and Magnetic Properties. Eur. J. Inorg. Chem., 17: 3522-3532.

Poleti D, Karanovic L, Prelesnik BV(1900). Structure of triaqua[1,2-benzenedicarboxylato(2-(2,2'-bipyridine)nickel(II) monohydrate. Acta. Cryst. Sect. C(Cr.Str.Comm.), 46: 2465-2467.

Sheldrick GM (1996). SADABS, Program for Empirical Absorption Correction of Area Detector Data. University of Göttingen, Göttingen.

Sheldrick GM (1997). SHELXL 97, Program for Crystal Structure Refinement. University of Göttingen, Göttingen.

Sheldrick GM (1997). SHELXS 97, Program for Crystal Structure Solution. University of Göttingen, Göttingen.

Svetlana GB, Yurii AS, Maria G, Nicolae G, Irina GF, Grigore T (2003). Synthesis and structural characterisation of unprecedented dinuclear zinc(II) complex with H-bonded bridging phthalate ions. Inorg. Chem. Comm., 6: 685-689.

Xue FC, Jiang ZH, Liao DZ, Ma SL, Yan SP, Wang GL, Yao XK, Wang RJ (1993). Novel structure of nickel(II) complex $Ni(phth)(bpy)(H_2O)_3$ ($phth^{2-}$ = the dianion of phthalic acid; BPY = 2,2′-bipyridine). Polyhedron., 12(23): 2787-2789.

Yang JM, Zhou ZH, Zhang H, Wan HL, Lu SJ (2005). Temperature effect on the conversions of phthalato and maleato manganese(II) complexes with diamine ligands. Inorg. Chim. Acta., 358(6): 1841-1849.

Yu LC, Liu SL, Liang EX, Wen CL (2008). A 1D coordination polymer$\{Pb[C_6H_4(COO)_2]$- $[phen]\}_n$ with strong blue fluorescent emission. J. Coord. Chem., 61(5): 810-815.

Zhang XM, Tong ML, Chen XM (2002). Hydroxylation of N-Heterocycle Ligands Observed in Two Unusual Mixed-Valence Cu^I/Cu^{II} Complexes. Angew. Chem. Int. d., 41(6): 1029-1031.

Zhang Y, Li J, Su Q, Wang Q, Wu X (2000). Synthesis, structure and spectroscopic properties of an o-phthalate-bridged copper (II) chain complex. J. Mol. Struct., 516(2-3): 231-236.

Zhang Y, Li J, Su Q, Wang Q, Wu X (2000). Synthesis, structure and spectroscopic properties of an *o*-phthalate-bridged copper (II) chain complex. J. Mol. Struct., 516: 231-236.

Anthocyanidin 3-galactosides from flowers of *Abrus canescens* Bak

Robert Byamukama[1], George Ogweng[2], Monica Jordheim[3], Øyvind M. Andersen[3] and Bernard T. Kiremire[1*]

[1]Chemistry Department, Makerere University, P. O. Box 7062, Kampala, Uganda.
[2]Chemistry Department, Kyambogo University, P. O. Box 1, Kyambogo- Kampala, Uganda.
[3]Department of Chemistry, University of Bergen, Allegt.41, 5007 Bergen, Norway.

Three anthocyanins, the 3-galactoside of delphinidin (1) (8.5%), petunidin (2) (9.2%) and malvidin (3) (73.8%) were isolated from the deep red-purple flowers of *Abrus canescens* by a combination of chromatographic techniques. Their structures were elucidated by on-line diode array detection chromatography and homo- and hetero-nuclear magnetic resonance (NMR) techniques.

Key words: *Abrus canescens*, Fabaceae, anthocyanin, malvidin 3-galactoside, petunidin 3-galactoside, delphinidin 3-galactoside, taxonomy.

INTRODUCTION

The genus *Abrus* belongs to the family Fabaceae (Leguminosae), and consists of about 13 species distributed in the tropic regions worldwide (Hutchinson and Dalziel, 1958; Verdcout, 1971; Labat, 1991; Thulin, 1994). In Africa, *Abrus gawenensis* Thulin., *Abrus canescens* Bak., *Abrus pulchellus* Thw. and *Abrus precatorius* L. are the most common species (Agbagwa, 2007; Thulin, 1994; Hutchinson and Dalziel, 1958). *Abrus canescens* is a climbing shrub that grows to a height of 0.5 to 3 m at the edges of rivers or swamps in the altitude range of 600 to 1500 m above sea level (Lock, 1989). Root extracts from *A. canescens* have been used as a genital stimulant (Burkill, 1995) and roots macerated in palm wine are considered as an aphrodisiac (Bouquet, 1969).

During our bioprospecting survey of anthocyanins in the East-African flora, we came across anthocyanidin 3-galactosides to be responsible for the colour of *A. canescens* flowers. The taxonomical potential of the glycosidic pattern of the anthocyanins in *A. canescens* is discussed.

*Corresponding author. E-mail: kiremire@chemistry.mak.ac.ug.

MATERIALS AND METHODS

Plant material

Flowers of *Abrus canescens* were collected from the outskirts of Budongo National Forest Reserve in Western Uganda in September 2008. The flowers were stored in a deep freezer before extraction. The identification of the plant was carried out in the Botany Department at Makerere University and voucher specimen (No.GO22/2008) has been deposited in the herbarium of that Department.

Isolation of anthocyanins

The flowers (80 g) were weighed and extracted with 600 mL of methanol containing 0.5% (v/v) trifluoroacetic acid (TFA) for 8 h. The filtered extract was concentrated under reduced pressure at 30°C, purified by partition (several times) against ethyl acetate before application to an Amberlite XAD-7 column. After washing the column with H_2O, the anthocyanins adsorbed to the column were eluted with methanol containing 0.5% (v/v) TFA. The concentrated anthocyanin extract was further purified by Sephadex LH-20 chromatography using H_2O-MeOH-TFA (80:19.5:0.5, v/v/v) as eluent. The individual anthocyanins were separated using preparative HPLC (Gilson 305/306 pump equipped with a UV 6000LP detector) equipped with an ODS Hypersil column (25 × 2.2 cm; i.d.; 5 mm). Two solvents were used for elution: A = formic acid-water (1:9, v/v) and B = formic acid- water- methanol (1:4:5; v/v/v). The elution profile consisted of a linear gradient from 10 to

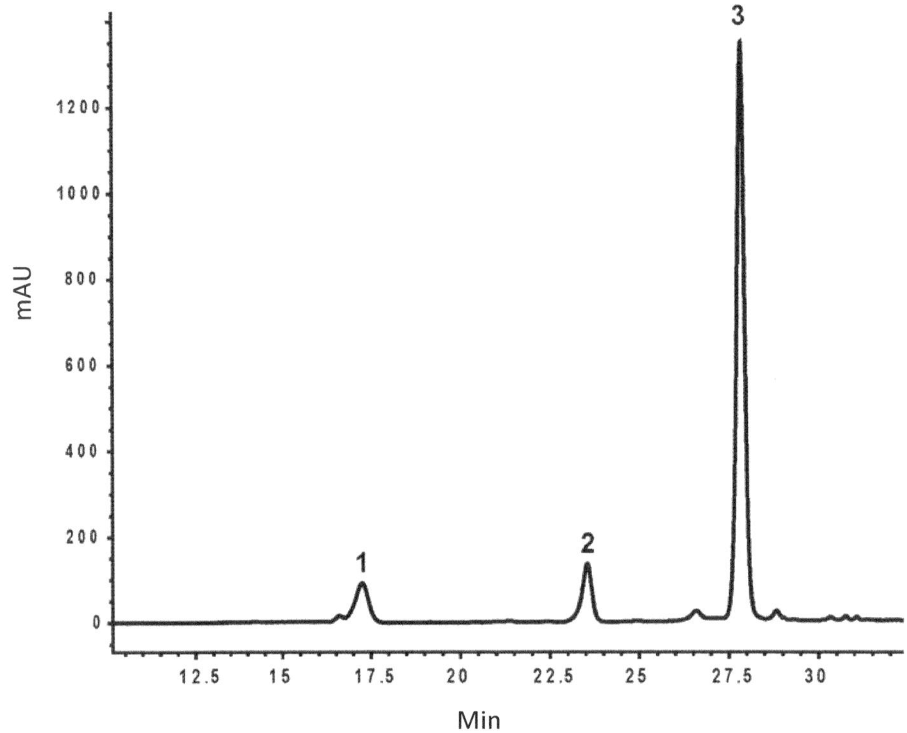

Figure 1. HPLC chromatogram of anthocyanin **1-3** in an acidified extract of flowers of *Abrus canescens* Bak. Detected at 520 ± 20 nm.

100% B for 30 min, isocratic elution (100% B) for the next 10 min, followed by a linear gradient from 100 to 10% B for 1 min. The flow rate was 14 ml/min for 41 min and aliquots of 500 µl were injected.

Co-chromatography (TLC, On-line HPLC)

Co-chromatographic methods included TLC and on-line HPLC. The TLC was carried out on microcrystalline cellulose (F 5556, Merck) with the solvent HCO_2H-HCl-H_2O (1:1:2, v/v/v). The Agilent 1100 HPLC system was equipped with a HP 1050 diode-array detector and a 200 × 4.6 mm i.d., 5 µm ODS Hypersil column, (Supelco, Bellefonte, USA). Two solvents; C, water (0.5% TFA) and D, acetonitrile (0.5% TFA) were used for elution. The elution profile for HPLC consisted of initial conditions with 90 C and 10% D followed by linear gradient elution for the next 10 min to 14% D, isocratic elution (10 to 14 min) and the subsequent linear gradient conditions; 14 to 18 (to 16% D), 18 to 22 (to 18% D), 22 to 26 (to 23% D), 26 to 31 (to 28% D) and 31 to 32 min (to 40% D), isocratic elution 32 to 40 min (40% D) and final linear gradient elution 43 to 46 min (to 10% B). The flow-rate was 1.0 ml/min and aliquots of 15 µL were injected with an Agilent 1100 Series, Micro Autosampler. Before injection, all samples were filtered through a 0.45 mm Millipore membrane filter. Anthocyanins from crowberries, *Empetrum nigrum* Coll. (Käppä et al., 1984; Ogawa et al., 2008) were used as references. These TLC and HPLC methods have been earlier described by Byamukama et al. (2006).

Spectroscopy

The UV–visible absorption spectra were recorded on-line during HPLC analysis and the spectral measurements were made over the wavelength range 200 to 600 nm in steps of 2 nm. The NMR experiments were obtained at 600.13 and 150.92 MHz for ^{1}H and ^{13}C, respectively, on a Bruker Biospin AV-600 MHz instrument equipped with a cryogenic probe at 298 K. The deuteriomethyl ^{13}C signal and the residual ^{1}H signal of the solvent, CF_3COOD–CD_3OD (95:5; v/v), were used as secondary references (δ 49.0 and δ 3.4 ppm from TMS for ^{1}H and ^{13}C, respectively) (Andersen and Fossen, 2003). The NMR experiments ^{1}H, ^{1}H-^{13}C HSQC, ^{1}H-^{13}C HMBC, ^{1}H-^{1}H DQF-COSY and ^{1}H-^{1}H TOCSY were recorded.

RESULTS AND DISCUSSION

The HPLC chromatogram (Figure 1) of a weakly acidified methanol extract of the flowers of *Abrus canescens* revealed the presence of three major (**1-3**) anthocyanins, 8.5, 9.2 and 73.8%, respectively. The anthocyanins in the extracts were purified by partition against ethyl acetate followed by Amberlite XAD-7 column chromatography and separated by Sephadex LH-20 chromatography and preparative HPLC. The isolated anthocyanins were checked for homogeneity by analytical HPLC. The UV-visible spectra of **1-3** recorded on-line during HPLC analysis showed visible maxima around 532 nm, and $A_{440}/_{Avis\text{-}max}$ of about 29% (Table 1), in accordance with 3-glycosides of either delphinidin, petunidin or malvidin (Andersen, 1987; Guisti and Wrolstad, 2001).

The aromatic region of the ^{1}H NMR spectrum of **3** showed a singlet at δ 9.16 (br, H-4), a 2H singlet at δ

Table 1. Chromatographic and on-line spectral data of the anthocyanidin 3-galactosides (**1-3**) isolated from flowers of *Abrus canescens* Bak.

Compound	Relative amount (%)	On-line HPLC			R_f (TLC)
		Vis_{max} (nm)	$A_{440}/A_{Vis\text{-}max}$ (%)	t_R (min)	
1	8.5	530	29	16.32	0.11
2	9.2	530	29	22.93	0.19
3	73.8	532	29	27.60	0.27

Table 2. ^{1}H and ^{13}C NMR spectral data for malvidin 3-*O*-β-galactopyranoside (**3**) isolated from flowers of *Abrus canescens* Bak.

	^{1}H δ (ppm), J (Hz)	^{13}C δ (ppm)
2		164.0
3		145.6
4	9.15 d 0.6	137.3
5		159.1
6	6.75 d 1.9	103.4
7		170.2
8	7.05 dd 0.6, 1.9	96.9
9		157.8
10		113.3
1'		119.7
2'	8.12 s	110.6
3'		149.7
4'		147.5
5'		149.7
6'	8.12 s	110.6
OCH_3	4.10 s (∫6 H)	57.2
3-O-β-galactopyranoside		
1"	5.39 d 7.8	104.6
2"	4.08 dd 7.8, 9.8	72.2
3"	3.77 dd 3.3, 9.8	75.0
4"	4.03 dd 1.1, 3.3	70.2
5"	3.89 m	77.8
6A"	3.85 m	62.3
6B"	3.85 m	

8.12 (H-2'/6') and an AX system at δ 7.04 (dd, J = 0.9, 1.9 Hz, H-8) and δ 6.76 (d, J = 1.9 Hz, H-6) indicating an anthocyanidin with a symmetrically substituted B-ring. The chemical shifts of the corresponding carbons of the aglycone were assigned from the HSQC experiment while the remaining quaternary carbon atoms were assigned from the HMBC resonances (Table 2). The singlet at δ 4.10 (2 × OCH_3) and the crosspeak at δ 4.10/149.9 (OCH_3/ C-3', 5') in the HMBC spectrum showed that the aglycone was malvidin. The sugar region of the 1H NMR spectrum indicated one anomeric signal at δ 5.40 with a coupling constant of 7.7 Hz indicating presence of one monosaccharide with a β-configuration. Starting from the anomeric proton at δ 5.40 (d, 7.7 Hz, H-1"), the observed cross peak at δ 4.09 in the DQF-COSY spectrum permitted the assignment of H-2". The chain of sequentially coupled protons, H-3", H-4", H-5" and H-6A"/H-6B", were thereafter assigned from other cross peaks in the COSY spectrum (Table 2). The carbon atoms of the sugar of Pigment 3 were assigned from the HSQC spectrum. The chemical shifts, coupling constants and crosspeaks of the sugar moiety (Table 2) were in accordance with a β-galactopyranosyl. The cross peak at δ 5.40/ 146.67 (H-1"/C-3) in the HMBC spectrum confirmed the linkage of the β-galactopyranose to the 3-position of the aglycone. The major pigment 3 was therefore confirmed as malvidin 3-O-β-galactopyranoside (Figure 2).

Pigments **1-3** co-chromatographed (TLC and HPLC) with the 3-galactoside of delphinidin, petunidin and malvidin, respectively, from crowberry, *Empetrum nigrum* Coll. (Käppä et al., 1984; Ogawa et al., 2008). Pigments

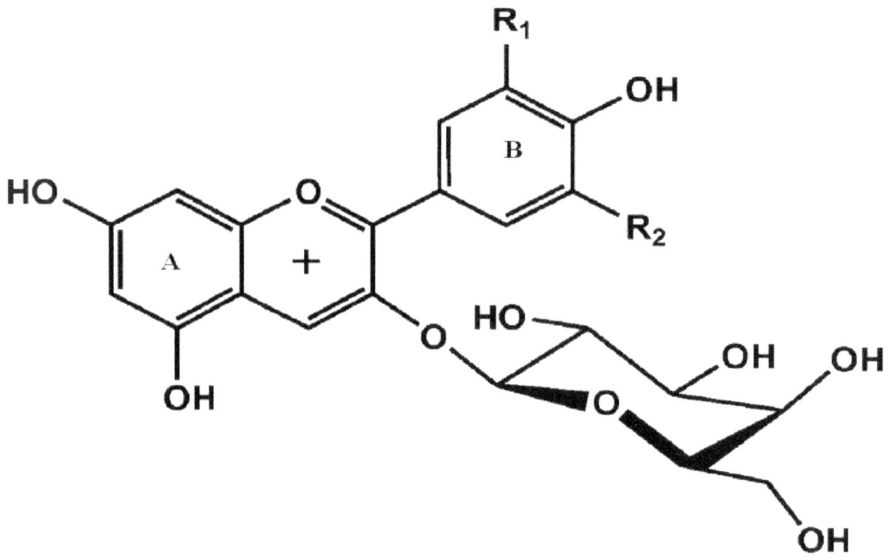

Figure 2. The structure of the three anthocyanidin 3-galactosides isolated from flowers of *Abrus canescens* Bak.: $R_1 = R_2 = OH$ (**1**), delphinidin 3-galactoside; $R_1 = OCH_3$, $R_2 = OH$ (**2**), petunidin 3-galactoside and $R_1 = R_2 = OCH_3$ (**3**), malvidin 3-galactoside.

1 and **2** were hence identified as delphinidin 3-galactoside and petunidin 3-galactoside, respectively, using chromatographic methods (Table 1).

All the reported anthocyanins (**1-3**) isolated from flowers of *A. canescens* have 3',4',5'-tri-O-substitutions on their anthocyanidin B-rings. This pattern is similar to the pattern found for most other species in the subfamily Faboideae; with malvidin glycosides as main pigments with petunidin glycosides and delphinidin glycosides and other type of anthocyanins present in decreasing order (Harborne, 1967; Ishikura, 1978). This observation may be the prototype of a situation in the subfamily Faboideae, where some anthocyanidins (malvidin and petunidin) are formed as a result of modifications of other anthocyanidins (delphinidin) and not by the activity of anthocyanidin synthase on flavan-3,4-cis-diol during the biosynthetic process. In *Petunia hybrida*, for example, peonidin, petunidin and malvidin have been reported to be formed from modification of their respective cyanidin and delphinidin B-rings by methyltransferases (Brugliera et al., 1994; Jonsson et al., 1984). However, although the aglycone pattern seem to be consistent within the family Fabaceae, the glycosidic patterns varies a lot with the main being the 3-rhamnoside, 3-rhamnoside-5-glucoside, 3-glucoside, 3-(2-xylosylgalactoside) and 3-glucoside-5-glucoside (Harborne, 1967; Ishikura, 1978). In this context the 3-galactosides of the anthocyanins (**1-3**) of *A. canescens* in the genus *Abrus* may have taxonomical potential.

ACKNOWLEDGEMENTS

We acknowledge Carnegie cooperation of New York for financial support. GO gratefully acknowledges Kyambogo University for tuition fees.

REFERENCES

Agbagwa IO (2007). Evaluation of Diagnostic Vegetative and Reproductive Characters among Abrus species in Nigeria. Aust. J. Basic Appl. Sci., 1(4): 841-852.

Andersen ØM (1987). Anthocyanins in fruits of *Vaccinium uliginosum* L. (Bog whortleberry). J. Food Sci., 52: 665-666 and 680.

Andersen ØM, Fossen T (2003). Characterisation of anthocyanins by NMR, In: Current protocols in Food analytical Chemistry, Wrolstad, R. E., ed. John Wiley, New York Chapter 1.

Bouquet AJ (1969). Natural products as an alternative remedy. . Royal Botanic Gardens. Kew 2(4): 166-179.

Brugliera F, Holton TA, Stevenson TW, Farcy E, Lu CY and Cornish EC (1994). Isolation and characterisation of cDNA clone corresponding to the Rt locus of *Petunia hybrida*. Plant J., 5: 81-92.

Burkill HM (1995). The Useful Plants of West Tropical Africa. 3. 2nd Ed. Families J-L. Roy. Bot. Gardens Kew, 3(2): 610.

Byamukama R, Jordheim M, Kiremire B, Namukobe J, Andersen ØM (2006). Anthocyanins from flowers of *Hippeastrum* cultivars. Sci. Hort., 109: 262-266.

Guisti MM, Wrolstad RE (2001). Characterization and measurement of anthocyanins by UV-visible spectroscopy, in Current protocols in Food Analytical Chemistry, Wrolstad RE and Schwartz RE, Eds., John Wily, New York.

Harborne J (1967). Comparative Biochemistry of the Flavonoids. Academic Press, London Chapter 5.

Hutchinson J, Dalziel JM (1958). Flora of West Tropical Africa, Part 2. Crown Agents for Oversea Governments and Administrations, London.

Ishikura N, Ito S, Shibata M (1978). Paper chromatographic survey of anthocyanins in Leguminosae. III. Identification and distribution pattern of anthocyanins in twenty-two legumes. Bot. Mag. Tokyo, 91: 25-30.

Jonsson LMV, Aarsman MEG, Poulton JE, Schram AW (1984). Properties and genetic control of four methyltransferases involved in methylation of anthocyanins in flowers of *Petunia hybrida*. Planta, 160: 174-179.

Käppä J, Kallio H, Peltonen I, Link R (1984). Anthocyanins of Crowberry, *Empetrum nigrum* Coll. J. Food Sci., 49: 634-636.

Labat JN (1991). *Abrus longibracteatus*, a new species of Leguminosae Papilionoidae from Laos and Vietnam. Bull. Mus. Nation. Hist. Natl. Sec. B., Adansonia, 13: 167-171.

Lock JM (1989). Legumes of Africa: A checklist. Royal Botanic Gardens, Kew No. 357.

Ogawa K, Sakakibara H, Iwata R, Ishii T, Sato T, Goda T, Shimoi K, Kumazawa S (2008). Anthocyanin composition and antioxidant activity of the crowberry (*Empetrum nigrum*) and other berries. J. Agric. Food Chem., 56: 4457-4462.

Thulin M (1994). Two new species of *Abrus* (Leguminosae – Papilionoideae) from Somalia. Nord. J. Bot., 14: 55-58.

Verdcourt B (1971). *Abrus*. – In: Milne – Redhead E and Polhill RM (Eds), Flora of tropical East Africa, Leguminosae (part 3) subfamily Papilionoideae (1). Crown agents for oversea governments and administrations. London, pp. 113-118.

Spectrophotometric determination of cobalt (ii) in low concentrations using hydroxytriazene as selective chelating agents

Ombaka O.* and Gichumbi J. M.

[1]Department of Chemistry, Chuka University College, Box 109, Chuka, Kenya.

The present work describes a selective, rapid and simple spectrophotometric method for the determination of cobalt (II) at trace levels using some selected hydroxytriazene. The method is based on the reaction of hydroxytriazene at pH 5.5 to 6.2, 6.0 to 6.8, 7.1 to 7.35, 7.3 to 7.9, 6.75 to 7.25 and 7.2 to 7.75 for hydroxytriazenes no (i), (ii), (iii), (iv), (v) and (vi) respectively and ethanol media with cobalt (II) to produce a highly absorbent yellow coloured chelate product with an absorption maximum at 393,397,417,406,410 and 412 nm for hydroxytriazenes no. (i), (ii), (iii), (iv), (v) and (vi), respectively. The reaction is instantaneous and the absorption remains stable for 24 Hrs. The average molar absorption coefficient were 1.1 x 10^4, 1.2 x 10^4,1.8 x 10^4, 2.5 x 10^4, 2.1 x 10^4, 1.9 x 10^4 mol^{-1} dm^3 cm^{-1} for hydroxytriazene no (i), (ii), (iii), (iv), (v), (vi) respectively and Sandell's sensitivity in ng/cm^2 were 5.4, 4.9, 3.3, 2.4, 2.9 and 3.1 for hydroxytriazene no. (i), (ii), (iii), (iv), (v) and (vi), respectively. The stoichiometric composition of the chelate is 1:3 for hydroxytriazene no (i), (iv), (v), (vi), 1:2 for (ii) and 1:1 for (iii). A large excess of over 30 cations, anions and complexing agent do not interfere in the determination at equivalent amounts. The method was successfully used in *Mussaenda arcu*. The method has high precision and accuracy.

Key words: Spectrophotometry, *Mussaenda arcu,* hydroxytriazene, cobalt, interference.

INTRODUCTION

Essential elements are of important in many biological systems. Metals can be found incorporated or naturally complexed in proteins, acting like the well known metalloenzymes and cofactors. The metal of interest in the study (Co) has a major biochemical role in biological systems especially in vitamin B12, a coenzyme in some biochemical processes (Sharga and Soomro, 2004).

Element complexing is an important chemical process in chemical analysis and separation of compounds, besides other uses. Spectrophotometric techniques offer considerable advantages in trace analysis and therefore the selection of more specific methods for the wide variety of environmental and biological sample analysis is very important especially where simpler procedures, low cost and high quality of analytical techniques is guaranteed (Pearson and Seim, 1977).

Several spectrophotometric chelating reagents are available for the determination of the cobalt in trace amounts. For example β-nitroso-α-naphthol forms pink cobaltic complex in citrate medium at pH 2.5 to 5, but its disadvantage is having less sensitivity. Sodium1-nitroso-2-hydroxynaphthalene-3, 6-disulphonate which is preferred due to its sensitivity has disadvantages due to interferences from some of the other heavy metals and also its complex is usually formed in hot acetate-acetic acid medium. EDTA is a non selecting chelating reagent and also forms colourless solution which makes its use in spectrophotometric method limited.

Exposure to cobalt can occur through inhalation, oral or dermal (skin) routes (Jefferson et al., 2002). Human are exposed to natural sources of cobalt in their food, water and air (Agnihotri et al., 1999; Amin et al., 2001; Agwara et al., 2010; Birsen et al., 2000). In addition to naturally occurring forms in the environment cobalt substances

*Corresponding author. E-mail: Ombaka2020@yahoo.com.

may also be present in certain occupational settings and in some consumer products (Jamaluddin and Jakir, 2008). Cobalt in the form of vitamin B12 (hydroxocyanocobalamin) is essential in humans; vitamin B12 supports important synthetic reactions in metabolic processes and is essential for the production of red blood cells (Linna et al., 2004). In the past, cobalt was recommended for treating anemia, nephriis and infection in addition to the usual hemopoietic agents (Goizman et al., 2005; Jamaluddin and Jakir, 2008).

In clinical situations where decreased numbers of red blood cells were associated with anaemia, ingestion of inorganic cobalt {Co^{2+} ion in the form of cobalt chloride} stimulated an increase in the production of red blood cells{polycythemia or erythroytosis}(Jefferson et al., 2002). Having high amounts of cobalt for long periods of time could affect the heart and might decrease fertility in men. An excessively high intake of cobalt may damage the heart muscles and may cause a over-production of red blood cells or damage to the thyroid glands (Kaul et al., 2001). A deficiency of cobalt leads to anaemia, a lethal disorder. Increased consumption of vitamin B12 via food /or supplements would alleviate any deficiency symptoms (Jefferson et al., 2002; Linna et al., 2004). Therefore, the accurate determination of cobalt at trace levels using simple, rapid and selective methods is paramount importance.

The synthesized hydroxytriazenes were used to determine the amount of cobalt in *Mussaenda arcu* Poir (Local name: Adieremo), which is a shrub of about 2 to 4.4 m high. It has elliptic leaves which are cuneate to rounded, apex acuminate, 4 to 14 by 2 to 6 cm, glabrous o sparsely pubescent. The shrub has yellow flowers with orange hairy centre and it produces blue fruits which are subglobose or ellipsoid. It is found in wooded bush land forests margins. The reason it was chosen for this study is because it is used for local treatment of anemic patients (Kokwaro and Johns, 1972).

Spectrophotometry is essentially a trace analysis technique and is one of the most powerful tools in chemical analysis. The six reagents have never been used as a spectrophotometric method for the trace determination of cobalt. The objective of this work is therefore to come up with a spectrophotometric method which is selective, sensitive, rapid, involving simple procedures and providing high quality results with less costs for cobalt determination in trace amounts.

MATERIALS AND METHODS

Experimental apparatus

Systronics double beam UV-V is recording spectrophotometer 108 (Brand: Systronics, manufacturer: Systronics (India) Limited) and a Systronics pH meter 324 with a combination of electrode were used for the measurements of absorbance and pH, respectively. Atomic absorption spectrophotometer (model CTA 2000 AAS) was used for Comparing the results. The melting point and composition verification of hydroxytriazenes were carried out using melting point apparatus {Model Kruess M500} and elemental analyzer {Perkin Elmer 2400 Series II CHN Analyzer}, respectively.

Reagents and solutions

All the chemicals used were of analytical reagent grade. Doubly distilled deionized water, which is non-absorbent under visible radiation, was used throughout. Glass vessels were cleaned by soaking in acidic solutions of K_2Cr_2O7, followed by washing with concentrated HNO_3 and rinsed several times with deionized water. Stock solutions of 1000 ml were kept in polyethylene bottles containing 1 ml of concentrated HNO_3.

Synthesis and characterization of hydroxytriazenes

The six hydroxytriazenes were synthesized according to the recommended method (Purohit et al., 2003; Sharma et al., 2006), by coupling of arylhydroxylamine in sodium acetate medium of pH 5 to 6 and temperature range of 0 to 5°C. The yellow precipitate formed was filtered off on cooling, washed with ethanol and recrystallized from ethanol and dried under vacuum over silica gel. The prepared hydroxytriazenes were characterized by elemental analyses and melting point determination apparatus.

Solutions for spectrophotometric determination

Standard stock solution of cobalt (II)

A 1.0 x 10^{-2} M stock solution of cobalt (II) was prepared by dissolving required quantity of A.R grade cobalt nitrate hexahydrate in double distilled deionized water. To prevent hydrolysis a few drops of concentrated nitric acid were added to the solution. These solution was then standardized using 1.0 x 10 $^{-2}$ M EDTA solution and xylenol orange as indicator.

Reagent solution

Stock solution of each hydroxytriazene (1.0 x 10^{-2} M) was prepared by dissolving the required quantity of it in ethanol.

Solutions for pH adjustments

1.0% tris-buffer (1.0 gm of tris-buffer was dissolved in minimum quantity of water and then diluted to 100 ml with alcohol) and 1% perchloric acid solution (1.0 ml of perchloric acid was diluted to 100 ml with alcohol) were used to adjust the desired pH.

Solutions for atomic absorption spectrophotometric determination

Standard solutions

Standard solutions for cobalt were prepared from 1000 ppm spectro econ stock solution by serial dilution to the range of standards as recommended by the manufacturers of AAS CTA 2000 model.

Sample collection and treatment

Preparation of sample solutions

The samples were collected from different villages of Mfangano Island in Nyanza province Kenya. Samples were dried in air, grounded into fine powder and mixed by coning and quartering to obtain an analytical sample. 100.0 gms sample was taken into a 500 ml Kjeldahl flask.10 ml of 60% $HClO_4$ was then added .This was followed by swirling gently and digesting slowly at moderate heat. The heating was increased later. Digestion continued until white fumes appeared. The solution was then cooled and filtered into 50 ml volumetric flask through Whatmann Filter paper No.44.The solution was thereafter made to the mark using doubled distilled water. The blank determination was also prepared in similar manner (Stewart, 2003).

Determination of working wavelength

1 ml of 5 x 10^{-4} M Co^{2+} was taken into a 10 ml volumetric flask. Then 5 ml of 5x 10^{-4} M hydroxytriazenes was added. The pH of this solution was adjusted to pH of 6.15 to 6.5 using tris-buffer and perchloric acid .The solution was topped to the mark using double distilled deionized water. The absorbance was then determined in entire visible region against reagent blank. The spectrum of reagent was also measured in the same wavelength region against ethanol. The plot of absorbance against wavelength was then done on the same graph. The working wavelength was chosen where there was maximum difference between the absorbance due to complex and the reagent.

Determination of pH range

A set of solutions having concentrations of Co^{2+} and hydroxytriazenes were prepared in such a manner that, the ratio of [M]:[R] was 1:5 .The pH of these solutions was adjusted in such a manner that, each solution was having the pH different from others. Then the absorbance of each solution was measured against a reagent blank. The absorbance obtained was then plotted against the corresponding pH.

Molar composition of Cobalt (II) complex with each of the hydroxytriazenes

Three different methods, namely Job's method, mole ratio method and slope ratio method of Harves and Manning were used for the determination of molar composition of metal: reagent.

Validity of Lambert-beer's law

Under corresponding optimum conditions of pH, solvent and cobalt (II) to reagent ratio, the validity of Beer –Lambert's Law was studied. The results of absorbance obtained were plotted against the corresponding concentration of cobalt (II). The calibration curve obtained was used to determine the concentration of cobalt in sample solution.

Precision studies

This was carried out by measuring the absorbance of ten solutions. Each solution contained the same concentration of cobalt (II) and the reagent. The absorbance was measured against reagent blank under optimum conditions of pH, solvent and cobalt (II) to reagent ratio. On the basis of their absorbance values, standard deviation was determined.

Interference studies

Cobalt (II) was determined with each reagent under its optimum conditions of pH, solvent and cobalt (II) to reagent ratio, in the presence of equivalent amounts, five and ten fold excess of 35 diverse ions. From group VIII elements only Ni^{2+} was checked for interference, this was due to time and financial limitations.

Procedure for atomic absorption spectrophotometric analysis

The instrument was put on and left to warm up for about 0.5 h. The optimum operating conditions were set before any measurement as recommended by the manufacturer. Thereafter, aspiration of the sample, series of standard containing the sample having the same concentration of cobalt (II) were aspirated hence standard addition method was used.

Hydroxytriazene used as a complexing reagent in spectrophotometric determination of Cobalt (II)

(i). 3-hydroxy-3-0-tolyl-1-p-sulphonato (sodium salt) phenyltriazene
(ii). 3-hydroxy-3-0-tolyl-1-o-carboxyphenyltriazene
(iii). 3-hydroxyl-3-m-tolyl-1-o-carboxyphentyltriazxene
(iv). 3-hydroxy-3-m-tolyl-1-p-methoxyphenyltriazene
(v). 3-hydroxy-3-p-tolyl-1-m-hydroxyphenyltriazene
(vi). 3-hydroxy-3-m-tolyl-1-m-hydroxyphenyltriazene.

Quality control

The quality of the method was compared with the known reagent for cobalt spectrophotometric analysis, nitroso-R-salt [sodium-1-nitroso-2-hydroxyl naphthalene-3, 6-disulphonate). It was found out that the hydroxytriazenes synthesized were better reagents, with higher sensitivity and selectivity and also time saving. The purity of standard hydroxytriazenes and other reagents were checked for the presence of cobalt (II) using AAS.

RESULTS AND DISCUSSION

Table 1 consists of the results of the physical characteristics and elemental analysis. The hydroxytriazenes prepared were either light yellow coloured shining needles or orange–yellow coloured needles. The melting point of these compounds ranged from 112 to 178°C. Ethanol was used to crystallize these compounds. The results of C, H, and N indicate that the theoretical values and experimental values are identical.

The spectrum of the complex formed with each reagent ([M]:[R]=1:5) was obtained in wavelength region 375 to 500 nm against a reagent blank. Further, spectrum of reagent was also measured in the same wavelength region against ethanol. The working wavelength was chosen such that there was maximum difference between the absorbance due to complex and the reagent. The result in Table 2 indicates that, the working wavelengths are very close to λ max. This ensures that sensitivity is high, molar absorptivity remains constant and Lambert-beers law is obeyed.

The absorbance of a series of solutions containing cobalt (II) and reagent in the molar ratio 1:5 ([M]:[R]) was measured against reagent blank at corresponding working wavelength at various pH values to determine

Table 1. Physical characteristic and elemental analysis of hydroxytriazenes.

Reagent No.	Synthesized hydroxytriazene	Physical characteristics		M.P	Elemental analysis						Molecular formula
		Colour and shape of crystals	Crystallized from		C (%)		H (%)		N (%)		
					Th	Exp	Th	Exp	Th	Exp	
i	3-Hydroxy-3-o-tolyl-1-p-sulphanato (sodium salt)phenyltriazene)	Light yellow shining needles	ethanol	128	47.39	47.26	3.67	3.65	12.76	12.37	$C_{13}H_{12}N_3O_4SNa$
ii	3-Hydroxy-3-o-tolyl-1-o-carboxyphenyltriazene	light yellow shining needles	ethanol	138	61.96	60.48	4.83	4.63	15.49	14.69	$C_{14}H_{13}N_3O_3$
iii	3-Hydroxy-3-m-tolyl-1-o-carboxyphenyltriazene	Light yellow shining needles	ethanol	178	61.96	61.44	4.83	4.69	15.49	14.66	$C_{14}H_{13}N_3O_3$
iv	3-Hydroxy-3-m-tolyl-1-p-methoxyphenyltriazene	Light yellow shining needles	ethanol	112	65.33	64.56	5.88	5.77	16.34	16.87	$C_{14}H_{15}N_3O_2$
v	3-Hydroxy-3-p-tolyl-1-m-hydroxyphenyltriazene	Orange- yellow needles	ethanol	124	64.16	63.40	5.38	5.03	17.28	17.89	$C_{13}H_{13}N_3O_2$
vi	3-Hydroxy-3-m-tolyl-1-m-hydroxyphenyltriazene	Orange- yellow needles	ethanol	122	64.16	63.21	5.38	4.87	17.28	17.63	$C_{13}H_{13}N_3O_2$

Table 2. Spectrophotometric determination of cobalt (ii) with hydroxytriazenes.

Reagent No.	Composition of the complex Co:R	Colour of the complex	λmax in nm	Working wavelength nm	Optimum pH range	Beer's law rangex 10^{-5}M	Molar absorptivity in $mol^{-1}dm^3cm^{-1}$	Sandell's sensitivity ng/cm2	Stability constant for complex	Free energy of formation of the complex at 25°C at Kcal/mol	Standard deviation in ppm
(i)	1:3	Yellow	387	393	5.5-6.20	0.4-2.4	11,000	5.357	5.24×10^{14}	-19.383	0.021
(ii)	1:2	Yellow	392	397	6.0-6.8	0.3-1.8	11,944	4.934	3.100×10^{11}	-15.4012	0.012
(iii)	1:1	Yellow	409	417	7.1-7.35	0.2-1.2	18,000	3.274	1.609×10^{7}	-9.336	0.010
(iv)	1:3	Yellow	399	406	7.3-7.9	0.15-0.9	24,556	2.400	5.519×10^{15}	-21.733	0.007
(v)	1:3	Yellow	403	410	6.75-7.25	0.2-1.2	20,167	2.922	1.733×10^{16}	-22.012	0.009
(vi)	1:3	Yellow	406	412	7.2-7.75	0.2-1.2	19,333	3.048	1.719×10^{15}	-20.433	0.009

pH range of constant and maximum absorbance. The pH of the solution was adjusted using tris-buffer and perchloric acid solutions. The colour developed in pH range of constant absorbance was yellow for each reagent. The colour development in each case was

Table 3. Determination of cobalt (II) in prescence of interfering species at equivalent amount.

Reagent no.	Working wavelength (nm)	pH range	$[Co^{2+}] \times 10^{-5}$ M	Absorbance in absence of interfering species	$[Co^{2+}]:[R]$	Absorbance when interfering species was present						
						Cl^-	CH_3COO^-	CO_3^{2-}	Br^-	PO_4^{3-}	SO_4^{2-}	$C_2O_4^{2-}$
(i)	393	5.5-6.2	3	0.327	1:6	0.329	0.321	0.326	0.322	0.219	0.325	0.310
(ii)	397	6.0-6.8	2	0.242	1:6	0.241	0.240	0.240	0.237	0.209	0.242	0.230
(iii)	417	7.1-7.35	1.5	0.270	1:6	0.270	0.261	0.267	0.269	0.160	0.268	0.181
(iv)	406	7.3-7.9	1.5	0.374	1:6	0.367	0.367	0.376	0.377	0.344	0.371	0.365
(v)	410	6.75-7.25	2	0.412	1:6	0.422	0.406	0.406	0.402	0.216	0.402	0.380
(vi)	412	6.75-7.25	1.5	0.290	1:6	0.290	0.294	0.297	0.285	0.123	0.290	0.278

Reagent no.	Absorbance when interfering anions were present									
	I^-	$S_2O_3^{2-}$	NO^{2-}	SO_3^{2-}	F^-	S^{2-}	HPO_4^{2-}	NO_3^-	$Mo_7O_{24}^{6-}$	WO_4^{2-}
(i)	0.327	0.320	0.328	0.327	0.327	0.329	0.301	0.325	0.321	0.326
(ii)	0.242	0.242	0.238	0.240	0.244	0.245	0.210	0.242	0.240	0.239
(iii)	0.276	0.270	0.265	0.268	0.270	0.267	0.193	0.269	0.262	0.188
(iv)	0.369	0.367	0.376	0.365	0.368	0.377	0.297	0.374	0.370	0.370
(v)	0.403	0.416	0.412	0.418	0.403	0.422	0.320	0.410	0.412	0.416
(vi)	0.286	0.289	0.286	0.297	0.293	0.293	0.167	0.290	0.290	0.283

Reagent no.	Absorbance when interfering cations were present									
	K^+	UO^{2+}	Mn^{2+}	Ba^{2+}	Pb^{2+}	Hg^{2+}	Sn^{2+}	Th^{4+}	Cd^{2+}	Mg^{2+}
(i)	0.320	0.334	0.330	0.323	0.327	0.321	0.253	0.321	0.327	0.324
(ii)	0.242	0.242	0.237	0.242	0.265	0.244	0.240	0.242	0.236	0.236
(iii)	0.268	0.275	0.272	0.271	0.270	0.269	0.273	0.267	0.270	0.263
(iv)	0.368	0.366	0.367	0.371	0.369	0.374	0.375	0.369	0.374	0.369
(v)	0.408	0.420	0.408	0.414	0.420	0.420	0.280	0.404	0.410	0.410
(vi)	0.296	0.288	0.287	0.290	0.291	0.293	0.108	0.290	0.289	0.289

Reagent no.	Absorbance when interfering cations were present						
	Ca^{2+}	NH_4^+	ZrO^{2+}	Na^+	Ni^{2+}	Cu^{2+}	Zn^{2+}
(i)	0.326	0.320	0.323	0.329	0.293	0.320	0.325
(ii)	0.239	0.241	0.240	0.239	0.557	0.323	0.350
(iii)	0.273	0.271	0.274	0.270	0.534	0.320	0.359
(iv)	0.369	0.371	0.369	0.369	0.538	0.453	0.484
(v)	0.420	0.410	0.402	0.408	0.572	0.416	0.416
(vi)	0.285	0.291	0.284	0.291	0.406	0.290	0.297

Table 4. Determination of cobalt (II) in prescence of interference species at five fold excess.

Reagent no.	Working wavelength (nm)	pH range	$[Co^{2+}]$ x 10^{-5} M	Absorbance in absence of interfering species	$[Co^{2+}]$:[R]	Absorbance when interfering species was present						
						Cl^-	CH_3COO^-	CO_3^{2-}	Br^-	PO_4^{3-}	SO_4^{2-}	$C_2O_4^{2-}$
(i)	393	5.5-6.2	3	0.327	1:6	0.329	0.327	0.329	0.320	-	0.326	-
(ii)	397	6.0-6.8	2	0.242	1:6	0.242	0.240	0.245	0.238	-	0.237	0.173
(iii)	417	7.1-7.35	1.5	0.270	1:6	0.274	0.265	0.265	0.267	-	0.261	-
(iv)	406	7.3-7.9	1.5	0.374	1:6	0.367	0.372	0.377	0.376	-	0.365	0.218
(v)	410	6.75-7.25	2	0.412	1:6	0.416	0.402	0.410	0.414	-	0.402	-
(vi)	412	6.75-7.25	1.5	0.290	1:6	0.294	0.290	0.294	0.287	-	0.287	-

Reagent no.	Absorbance when interfering anions were present									
	I^-	$S_2O_3^{2-}$	NO^{2-}	SO_3^{2-}	F^-	S^{2-}	HPO_4^{2-}	NO_3^-	$Mo_7O_{24}^{6-}$	WO_4^{2-}
(i)	0.321	0.294	0.299	0.364	0.320	0.323	-	0.328	0.319	0.319
(ii)	0.237	0.244	0.242	0.312	0.234	0.242	-	0.238	0.214	0.224
(iii)	0.264	0.189	0.263	0.153	0.263	0.270	-	0.271	0.263	-
(iv)	0.367	0.365	0.369	0.357	0.354	0.376	-	0.379	0.369	0.379
(v)	0.402	0.409	0.405	0.416	0.406	0.405	-	0.402	0.380	0.407
(vi)	0.283	0.283	0.282	0.298	0.285	0.296	-	0.287	0.277	0.281

Reagent no.	Absorbance when interfering cations were present									
	K^+	UO^{2+}	Mn^{2+}	Ba^{2+}	Pb^{2+}	Hg^{2+}	Sn^{2+}	Th^{4+}	Cd^{2+}	Mg^{2+}
(i)	0.319	0.326	0.314	0.327	0.319	0.327	-	0.299	0.325	0.321
(ii)	0.245	0.249	0.243	0.243	-	0.249	0.229	0.273	0.229	0.239
(iii)	0.265	0.280	0.263	0.274	0.263	0.263	0.143	0.349	0.173	0.265
(iv)	0.367	0.368	0.380	0.373	0.390	0.383	0.312	0.328	0.365	0.367
(v)	0.410	0.416	0.418	0.408	0.402	0.406	-	0.220	0.407	0.421
(vi)	0.293	0.284	0.291	0.283	0.286	0.283	-	0.200	0.281	0.293

Reagent no.	Absorbance when interfering cations were present						
	Ca^{2+}	NH_4^+	ZrO^{2+}	Na^+	Ni^{2+}	Cu^{2+}	Zn^{2+}
(i)	0.326	0.325	0.319	0.325	-	0.267	0.267
(ii)	0.238	0.248	0.246	0.238	-	-	-
(iii)	0.277	0.269	0.279	0.270	-	-	-
(iv)	0.365	0.369	0.364	0.370	-	-	-
(v)	0.417	0.408	0.412	0.402	-	0.544	0.404
(vi)	0.295	0.290	0.289	0.289	-	0.297	0.395

- Indicates that interfered at lower concentration and hence interference studies at next concentration.

Table 5. Determination of cobalt (Ii) In prescence OF interference species at ten fold excess.

Reagent no.	Working wavelength (nm)	pH range	$[Co^{2+}]$ x 10^{-5} M	Absorbance in absence of interfering species	$[Co^{2+}]$:[R]	Absorbance when interfering species was present						
						Cl^-	CH_3COO^-	CO_3^{2-}	Br^-	PO_4^{3-}	SO_4^{2-}	$C_2O_4^{2-}$
(i)	393	5.5-6.2	3	0.327	1:6	0.329	0.320	0.323	0.328	-	0.318	-
(ii)	397	6.0-6.8	2	0.242	1:6	0.242	0.240	0.244	0.236	-	0.225	-
(iii)	417	7.1-7.35	1.5	0.270	1:6	0.271	0.269	0.263	0.264	-	0.177	-
(iv)	406	7.3-7.9	1.5	0.374	1:6	0.367	0.370	0.373	0.375	-	0.341	-
(v)	410	6.75-7.25	2	0.412	1:6	0.418	0.407	0.402	0.418	-	0.260	-
(vi)	412	6.75-7.25	1.5	0.290	1:6	0.293	0.293	0.295	0.289	-	0.122	-

Reagent no.	Absorbance when interfering anions were present									
	I^-	$S_2O_3^{2-}$	NO^{2-}	SO_3^{2-}	F^-	S^{2-}	HPO_4^{2-}	NO_3^-	$Mo_7O_{24}^{6-}$	WO_4^{2-}
(i)	0.278	-	-	-	0.263	0.284	-	0.320	0.250	0.289
(ii)	0.230	0.222	0.220	-	0.214	0.246	-	0.241	-	-
(iii)	0.174	-	0.197	-	0.179	0.263	-	0.267	0.188	-
(iv)	0.341	0.370	0.328	-	0.347	0.374	-	0.368	0.371	0.354
(v)	0.234	0.254	0.399	0.262	0.254	0.388	-	0.406	-	0.254
(vi)	0.111	0.240	0.131	0.133	0.151	0.281	-	0.293	-	0.259

Reagent no.	Absorbance when interfering cations were present									
	K^+	UO^{2+}	Mn^{2+}	Ba^{2+}	Pb^{2+}	Hg^{2+}	Sn^{2+}	Th^{4+}	Cd^{2+}	Mg^{2+}
(i)	0.325	0.229	-	0.316	0.300	0.319	-	-	0.258	0.319
(ii)	0.241	0.253	0.249	0.245	-	0.242	-	-	-	0.242
(iii)	0.271	0.287	0.282	0.279	0.290	0.261	-	-	-	0.269
(iv)	0.371	0.369	0.387	0.375	-	0.395	-	-	0.339	0.365
(v)	0.414	0.404	0.406	0.388	0.398	0.326	-	-	0.246	0.422
(vi)	0.295	0.257	0.297	0.289	0.306	0.278	-	-	0.191	0.297

Reagent no.	Absorbance when interfering cations were present						
	Ca^{2+}	NH_4^+	ZrO^{2+}	Na^+	Ni^{2+}	Cu^{2+}	Zn^{2+}
(i)	0.328	0.327	0.228	0.325	-	-	-
(ii)	0.239	0.239	0.223	0.240	-	-	-
(iii)	0.280	0.263	0.285	0.270	-	-	-
(iv)	0.366	0.373	0.302	0.374	-	-	-
(v)	0.426	0.410	0.421	0.412	-	-	0.382
(vi)	0.299	0.290	0.247	0.289	-	0.210	-

- Indicates that interfered at lower concentration and hence interference studies at next concentration not done.

Table 6. Determination of cobalt in BAS-5g Brass, {Cu=67.4%, Sn=1.09%, Pb=2.23%, Zn=28.6%, Ni=0.33%, P=0.01%) a certified reference material.

Reagent no.	Cobalt		
	Certified value in mg/L	Found (n=5) in mg/l	RSD (%)
(i)	1.0	1.053	1.69
(ii)	1.0	1.048	1.48
(iii)	1.0	1.041	1.56
(iv)	1.0	1.017	1.32
(v)	1.0	1.028	1.45
(vi)	1.0	1.031	1.37

Table 7. Estimation of cobalt (II) in *Mussaenda arcu* using uv-visible spectrophotometric and atomic absorption spectrophotometric methods.

Reagent no.	Sampling points/concentrations (Mg/Kg)						Mean	% Error	F-test at the 95% confidencE level
	SENA	NYAKWERI	UOZI	UNGINA	SOKLO	GULWE			
(iv)	0.551	0.530	0.562	0.522	0.489	0.511	0.527	+4.98	1.998
AAS	0.526	0.510	0.512	0.500	0.471	0.495	0.502	-	-

instantaneous, stable for about 24 h. The data for pH range and type of colour developed have been incorporated in Table 2. The molar composition of cobalt (II) complex with each of the hydroxytriazenes has been found using three different methods, namely: Job's method, mole ratio method and slope ratio method of Harvey and Manning, under corresponding optimum conditions of pH, solvent and cobalt (II) to reagent ratio. The results have been included in Table 2. The ratio of [M]:[R], which was employed during verification of Labert-beer's law was obtained from mole ratio curve and was such that, absorbance remained constant as the concentration of reagent was being increased and this contributed to Lambert-beer's law being obeyed.

The stability constants of cobalt (II) complexes were determined using Harvey and Manning method. The results from Table 2 reveals that, the complex formed by reagent (iv) and (v) are more stable while that formed by reagent (iii) is the least stable. The free energy of formation were calculated and the results indicates that during the formation of the complex between Co^{2+} and reagent (iv) and (v) high amount of energy is given out while using reagent (iii) the energy produced was least.

Sandell's spectrophotometric sensitivity was calculated from the values of molar absorptivity. This value indicates that reagent (iv) is the most sensitive while reagent (i) is the least sensitive.

The precision studies were carried out in order to check the reliability of the results of cobalt (II) determination with each of the hydroxytriazenes and the results have been given in Table 2. A close examination of Table 2 shows a fair replication of the results with the reagent studied.

Cobalt (II) has been determined spectrophotometrically in the presence of 35 diverse ions at equivalent amounts; five fold excess with each of the hydroxytriazenes under their respective optimum conditions. The results obtained from these studies have been given in Tables 3 to 5. From these results Cl^-,CH_3COO-,CO_3^{2-},Br^-,SO_4^{2-},I^-,NO_3^-, K^+,Ba^{2+},Hg^{2+},Mg^{2+},Ca^{2+},NH_4^+,Na^+, do not interfere with the determination of cobalt (II) even if these are present in ten fold excess with all hydroxytriazenes used. PO_4^3, $C_2O_4^{2-}$,HPO_4^{2-} Sn^{2+},Ni^{2+},Cu^{2+},Zn^{2+}interfered with cobalt (II) determination with all hydroxytriazenes even at equivalent amounts. These tables of interferences reveals that, reagent nos (vi), (v) and (iv) are the most selective reagents while reagent nos (i), (ii) and (iii) are the least selective reagent.

The validity of the method was tested by analyzing a standard reference material Table 6 and also comparing the results with convectional analysis (AAS) (Table 7).

The estimation of cobalt (II) in *M. arcu* samples were done spectrophotometrically using the most sensitive, selective hydroxytriazenes and the results obtained were compared with those of atomic absorption spectrometer. The results obtained have been summarized in Table 7. The concentration of cobalt in the sample was ranging from 0.489 to 0.562 mg/kg, 0.47 to 0.526 mg/kg for reagent no (iv) and AAS, respectively. The percentage error of reagent no (iv) in comparison with AAS was +4.98%. The slight differences in values using hydroxytriazenes and AAS might have been due to the presence of foreign species which interfered with the determination of cobalt (II).

The tabulated F-value for V1=5 and V2=5 is 5.05.Since the calculated value is less than this, hence there is no significant difference in the precision of the two methods. The detection limit (3S of the blank) was found to be 2.0 ng/ml which is comparable with that of AAS used (1.0

ng/ml). Thus this method is suitable for the determination of cobalt in a wide range of application in environmental and biological samples.

Conclusion

Six hydroxytriazenes are introduced as new reagents for spectrophotometric determination of cobalt (II). One of these is very selective, sensitive and can be used for the determination of cobalt in the presence of a large number of cations and anions and the results are comparable with those of atomic absorption spectrophotometer.

This hydroxytriazenes can be used to determine cobalt content in *M. arcu* samples. However, the differences obtained in the present method and AAS might be due to the following cations and anions which might have interfered with the determination: PO_4^{3-}, HPO_4^{2-}, and Zn^{2+}.

The ease of preparation over many reported reagents, wide pH range, easy methods (direct method) for spectrophotometric determination of cobalt (II), better yield are further advantages of hydroxytriazenes.

ACKNOWLEDGEMENTS

Authors are grateful to Prof. Njoka the principal of Chuka University College for providing a good environment for research. The authors are also grateful to Mr. Samuel Ombaka of KCB for sponsoring the research. We are grateful to the Chairman of Chemistry Department, Egerton University, Kenya for allowing us to use laboratory facilities.

REFERENCES

Agnihotri NK, Bhajan S, Vinay, KS (1999).Non-extractive derivative spectrophotometric determination of cobalt in neutral micellar medium. Talanta., 48(3): 623-631.

Amin AS, Ahmed IS, Moustafa ME (2001).Selective spectrophotometric determination of cobalt (II) in synthetic mixtures and pharmaceutical formulates with 5-[0-carboxyphenylazo] 2, 4-dihydroxybenzoic acid. Anal. Lett., 34(5): 749-759.

Agwara MO, Ndifon PT, Ndosiri NB, Paboudam AG, Yufanyi B, ,Mohamadou A (2010). Synthesis, characterization and antimicrobial activities of cobalt(II) and Zinc(II) mixed ligand complexes containing 1,10-phenanthroline and 2,2'-bipyridine. Bull. Chem. Soc. Ethiop., 24(3): 383-389.

Birsen DO, Hayati F, Esma T, Resat A (2000). Simultaneous derivative spectrophotometric determination of cobalt (II) and nickel (II) by dithizone without extraction. Talanta., 53(1): 263-269.

Goizman MS, Tikhomrova GB, Kalidichenko VN, Kochenova EM (2005). Quantitative determination of cobalt containing pthalocyanine fragments mantitumor drugs based on chemically modified terapthal. Pharm. Chem. J., 39(10): 551-555.

Jamaluddin A and Jakir KH (2008). A rapid spectrophotometric method for the determination of cobalt in industrial, Environmental, Biological, Pharmaceutical and soil samples using Bis(5-bromosalicylaldehyde) or thophenylenediamine. J. Ir. Chem. Soc., 5(4): 677-688.

Kaul KN, Lark BS, Malik AK, Faubel W, Rao AL (2001). Spectrophotometric determination of cobalt, nickel, palladium, copper, ruthenium and molybdenum using sodium isoamylxanthalate in presence of surfactants. Turk. J. Chem., 99: 105.

Kokwaro JA, Johns T (1972). Luo Biological Dictionary. East African Publishers. Nairobi, Kenya.

Jefferson J. Escudero E, Hurtado M, Pando J, Tapia R, Swenson E, Pich IJ, Schriener G, Schoene R, Hurtado A, Johnson R(2002).Excessive Erythrocytosis chronic mountain sickness and serum cobalt levels. The Lanc., 359(9304): 407-408.

Linna A, Oksa P, Groundstroem K, Halkosaari M, Palnroos P, Huikka S, Uitti J (2004). Exposure to cobalt in the production of cobalt and cobalt compounds and its effect on the heart. Occp. Environ. Med., 61: 877-885.

Pearson RM, Seim HJ (1977). Spectrophotometric determination of cobalt complexed with the p-nitophenyl hydrazone of diacetylmonoxime and ethylene diamine. Anal. Chem., 49(14): 580-582.

Purohit DN, Kumar S, Goswami AK (2003). A review of hydroxytriazenes. Rev. Anal. Chem. (Israel)., 22(1): 73-80.

Shar GA, Soomro GA (2004). Spectrophotometric determination of cobalt (II), Nickel (II) and Copper (II) with 1-(2pyridylazo)-2-naphlol in micellar medium. The Nucle., 41(1-4): 77-82.

Sharma JC, Upadhyav M, Jain MC, Dashora R (2006). Complexometric determination of copper in pharmaceutical sample using hydroxytriazenes. Asian J. Chem., 18(4): 3144-3146.

Stewart EA (2003). Chemical Analysis of Ecological materials 3rd ed., Blackwell Scientific Publications.

13

Structural study of pyridine-2,6-dicarbohydrazide nitrate salt

Mohammad Hakimi[1]*, Keyvan Moeini[1], Zahra Mardani[1] and Sanaz Khorasani[2]

[1]Chemistry Department, Payame Noor University, 19395-4697 Tehran, Iran.
[2]Molecular Sciences Institute, School of Chemistry, University of the Witwatersrand, Johannesburg, South Africa.

Reaction between dimethyl-2,6-pyridinedicarboxylate and anhydrous hydrazine by microwave synthesis under solvent free conditions gave pyridine-2,6-dicarbohydrazide (pdch). This compound was recrystallized in methanol at low pH condition and gave the nitrate salt of pdch. In the crystal structure of this compound, protonated pdch cations and nitrate anions are linked via N–H···$N_{nitrate}$ and N–H···$O_{nitrate}$ hydrogen bonds into a three-dimensional network. The structure features extensive hydrogen bonding, with 13 distinct interactions. There are also weak intermolecular interactions including O···*π*, O···O and N–H···*π*.

Key words: Pyridine-2,6-dicarbohydrazide, spectral characterization, X-ray crystal structure.

INTRODUCTION

Synthesis of compound based on 2,6-disubstituted pyridine has attracted a great deal of attention, with a broad array of both acyclic and macrocyclic ligands reported by Wilkinson et al. (1987). In particular, diamides incorporating a -NH-CO-pyridine-CO-NH- core based on pyridine-2,6-dicarboxylic acid (H_2dipic) have been examined, with interest in these compounds extending to potential applications as diverse as azo dyes (Liu et al., 2002) and telomerase inhibitors (Bouchard and Hittinger, 2002). Our research interests lie in the study of multi N-donor compounds which can be coordinated to metals and in the past few years, we have published a series of paper in this field (Hakimi et al., 2011b, 2012b). In this paper, preparation of pyridine-2,6-dicarbohydrazide nitrate salt (L, Scheme 1) was described and elemental analysis, spectroscopic characteristic (IR), and crystal structure of this compound were determined.

MATERIALS AND METHODS

Instrumentation

All starting chemicals and solvents were reagent or analytical grade and used as received. The infrared spectrum of a KBr pellet in the range of 4000 to 400 cm^{-1} were recorded with a FT-IR 8400-Shimadzu spectrometer. The carbon, hydrogen and nitrogen contents were determined in a Thermo Finnigan Flash Elemental Analyzer 1112 EA. The melting point was determined with a Barnsted Electrothermal 9200 electrically heated apparatus.

Synthesis of L

1 cm^3 of hydrazine hydrate and 1.95 g (10 mmol) of dimethyl pyridine-2,6-dicarboxylate were irradiated using a microwave oven (MicroSYNTH, Milestone s.r.l) for 10 min with a power up to 1 kW. After microwave irradiation, the white crystalline solid was obtained and washed with dichloroethane. The residue was recrystallized from methanol in pH adjusted to 3 by addition of HNO_3. After a day colorless crystals of L suitable for X-ray diffraction studies were obtained by slow evaporation of a solution and filtered. Yield 2.63 g (91%). m. p. 188°C. Elemental analysis: calculated for $C_{14}H_{21}N_{13}O_{13}$: C 29.02, H 3.65, N 31.43%; found: C 29.21, H 3.65, N 31.76%. IR (cm^{-1}, KBr): 3340 and 3215 m (ν NH_3^+), 3160 m (ν NH), 3078 m (ν CH), 1682 s (ν C=O), 1610 m (δ NH_2), 1527 s (ν C=N and/or C=C), 1389 vs (ν_4 NO_3), 1333 m (ν_1 NO_3), 1203 m (ν CN), 1011 m (ν_2 NO_3), 856 w (ν_6 NO_3).

Single crystal structure determination

Intensity data were collected on a Bruker APEX II CCD area detector diffractometer with graphite monochromated Mo K_α radiation (50 kV, 30 mA) using the APEX 2 (Bruker, 2005a) data collection software. The collection method involved ω-scans of width 0.5° and 512 × 512 bit data frames. Data reduction was

*Corresponding author. E-mail: mohakimi@yahoo.com.

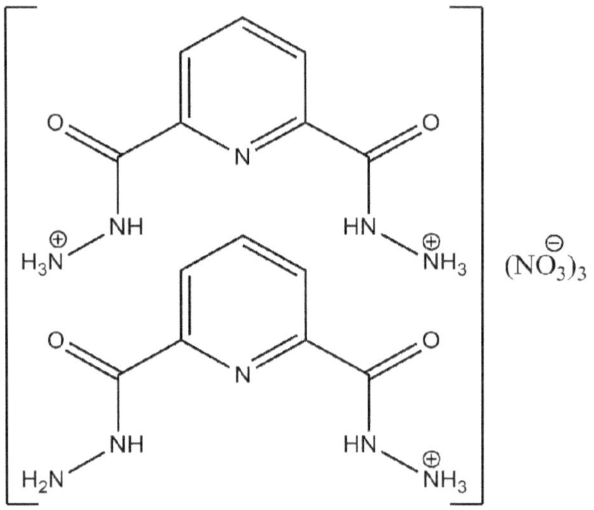

Scheme 1. Structure of the L.

Table 1. Crystal data and structure refinement for L.

Empirical formula	$C_{14}H_{21}N_{13}O_{13}$
Formula weight (g mol^{-1})	579.44
Temperature (K)	173(2)
Crystal system	Monoclinic
Space group	*C2/c*
Unit cell dimensions (Å, °)	
a	13.0563(5)
b	5.5522(2)
c	31.3941(12)
β	91.658(2)
Volume (Å^3), *Z*	2274.84(15), 4
Calculated density (mg m^{-3})	1.692
Absorption coefficient (mm^{-1})	0.150
F(000)	1200
Crystal size (mm^3)	0.30 × 0.18 × 0.10
θ range for data collection (°)	1.30–27.98
h, *k*, *l* ranges	–17:17, –7:7, –41:41
Reflections collected	8853
Independent reflections	2749
R_{int}	0.0504
Data / restraints / parameters	2749 / 1 / 191
Goodness-of-fit on F^2	0.878
Final *R* indexes [$I > 2\sigma(I)$]	R_1 = 0.0417, wR_2 = 0.0889
R indexes (all data)	R_1 = 0.0780, wR_2 = 0.0993
Largest diff. peak and hole (e.Å^{-3})	0.732 and –0.206

carried out using the program *SAINT+* (Bruker, 2005b).

The crystal structure was solved by direct methods using *SHELXTL* (Bruker, 1999). Non-hydrogen atoms were first refined isotropically followed by anisotropic refinement by full matrix least-squares calculations based on F^2 using *SHELXTL*. Hydrogen atoms were first located in the difference map then positioned geometrically and allowed to ride on their respective parent atoms. Diagrams and publication material were generated using SHELXTL, PLATON (Spek, 2003), ORTEP-3 (Farrugia, 1997), and Mercury (Macrae et al., 2008). Crystallographic data and details of the data collection and structure refinement are listed in Table 1. Bond lengths (Å) and angles (°) are reported in Table 2.

Table 2. Selected bond length (Å) and angles (deg) for L with estimated standard deviations in parentheses.

Distances		Angles	
C1–N1	1.344(2)	C5–C7–O2	123.2(2)
C5–N1	1.337(2)	C5–C7–N4	114.8(2)
C6–O1	1.226(3)	C7–N4–N5	116.7(2)
C7–O2	1.230(2)	O3–N6–O4	119.3(2)
N2–N3	1.413(2)	O4–N6–O5	119.7(2)
N4–N5	1.423(2)	O5–N6–O3	121.0(2)
N6–O3	1.245(2)		
N6–O4	1.257(2)		
N6–O5	1.238(2)		

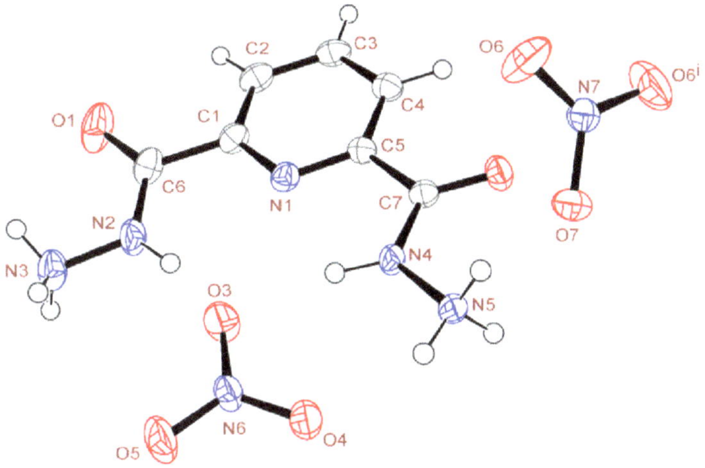

Figure 1. View of the L with the atom numbering scheme. The displacement ellipsoids are drawn at the 50% probability level.

RESULTS

In the IR spectrum of L, the absorptions at 3340 and 3215 cm^{-1} were assigned to the asymmetric and symmetric vibrations of NH_3^+. Other band at 3160 cm^{-1} was confirmed the presence of NH. Four bands in the IR spectrum near 1389, 1333, 1011, 856 cm^{-1} for L can be assigned to vibrations of the nitrate group (respectively the vibrations ν_4, ν_1, ν_2 and ν_6).

The molecular structure of the L (Figure 1) reveals presence of two protonated portions and three nitrate ions. The plane of pyridine ring with planes of two acetohydrazide moiety including α (N3/N2/C6/O1) and β (N5/N4/C7/O2) forms the dihedral angles of 12.14° and 1.86°, respectively, which means that the L is not planar (r.m.s. deviation 0.271 Å for O1 atom). In the crystal structure of L, one of the nitrate ions (containing N6 atom) lies closely to plane of cationic portion and makes five interactions with it. The dihedral angle between plane of this nitrate ion (O4/O5/N6/O3) and plane of cationic portion is 10.47°. In the crystal, the molecules are stabilized due to N–H···O and N–H···N hydrogen bonds (Table 3 and Figure 2). This 3D supramolecular structure is further stabilized by the O···π, O···O and N–H···π interactions (Figure 2).

DISCUSSION

Free nitrate ion has D_{3h} symmetry and three infrared active vibrations but it was lowered to C_{2v} and C_s upon interaction of two and one oxygen atoms in complex, respectively (Drago, 1992; Nakamoto, 2009) with six infrared active vibrations. In L, nitrate groups did not coordinate to metal atom and must have D_{3h} symmetry but the interactions such as hydrogen bonds and others influenced the nitrates geometry and lowered it. In title compound the bond lengths and angles are normal and are comparable to the values observed in similar compound (Jia et al., 2006).

Table 3. Hydrogen-bond geometry (Ǻ, °).

D–H···A	***d*(D–H)**	***d*(H···A)**	***d*(D···A)**	**<(DHA)**
N(3)–H(3A)···O(5)#1	0.91	2.01	2.883(2)	160
N(3)–H(3B)···O(1)#2	0.91	2.29	2.990(2)	134
N(3)–H(3B)···O(1)#3	0.91	2.52	3.313(2)	147
N(3)–H(3C)···N(3)#1	0.91	1.83	2.741(3)	175
N(3)–H(3C)···N(2)#1	0.91	2.65	3.469(2)	150
N(5)–H(5A)···O(2)#4	0.91	1.88	2.785(2)	170
N(5)–H(5B)···O(4)	0.91	2.00	2.796(2)	145
N(5)–H(5B)···O(6)#3	0.91	2.32	2.987(2)	130
N(5)–H(5C)···O(7)	0.91	1.89	2.7746(18)	163
N(4)–H(4H)···O(3)	0.917	1.919	2.820(2)	167
N(4)–H(4H)···O(4)	0.917	2.57	3.035(2)	111.7
N(4)–H(4H)···N(6)	0.917	2.60	3.352(2)	140
N(2)–H(2H)···O(3)	0.915	2.010	2.888(2)	160

Symmetry transformations used to generate equivalent atoms: #1 −x+1.5, −y+1.5, −z+1; #2 −x+1, −y+1, −z+1; #3 x, y+1, z; #4 −x+1.5, y+0.5, −z+1.5.

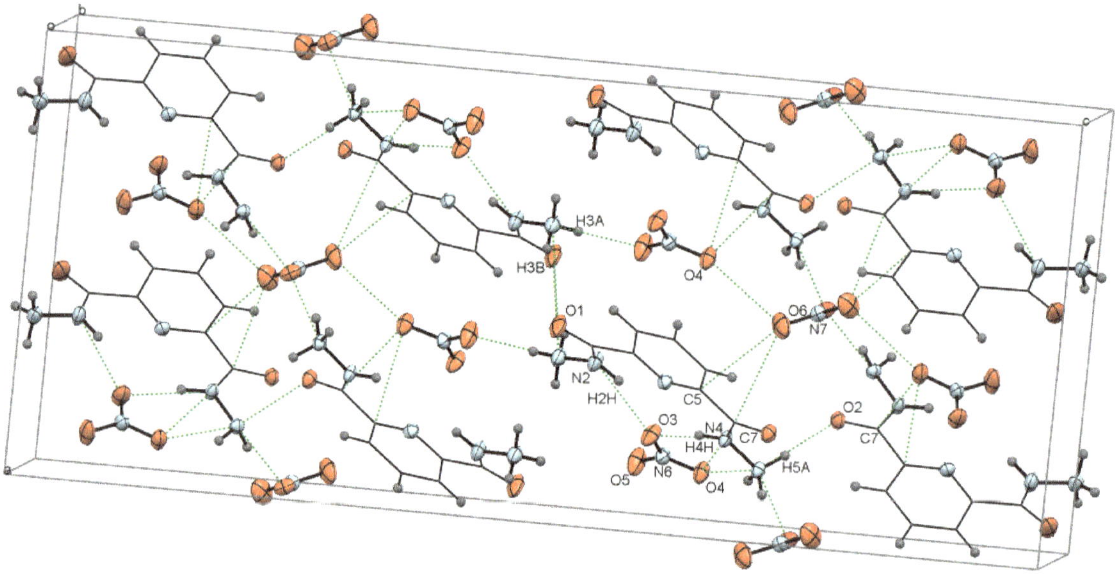

Figure 2. Packing of crystal, showing the hydrogen bonded chain formation and short contacts in the *ac* plane.

Conclusion

In this paper, preparation of the pyridine-2,6-dicarbohydrazide nitrate salt, L, was presented and its spectral and structural properties were described. The crystal structure determination of L revealed that this crystal consist of two protonated pyridine-2,6-dicarbohydrazide and three nitrate ions. In the crystal of L, hydrogen bonds and weak interactions such as $N–H\cdots N_{nitrate}$, $N–H\cdots O_{nitrate}$, $O\cdots\pi$, $O\cdots O$ and $N–H\cdots\pi$ gave rise to a supramolecular network. The synthesis of complexes with the pyridine-2,6-dicarbohydrazide is continued by our research group.

CCDC 833546 contains the supplementary crystallographic data for this paper. These data can be obtained free of charge from The Cambridge Crystallographic Data Centre *via* www.ccdc.cam.ac.uk/data_request/cif.

ACKNOWLEDGEMENT

We are grateful to Payame Noor University of Iran for financial support.

REFERENCES

Bouchard H, Hittinger A (2002). PCT Int. Appl. WO2002096903.

Bruker (1999). SHELXTL. Version 5.1. (includes XS, XL, XP, XSHELL) Bruker AXS Inc., Madison, Wisconsin, USA.

Bruker (2005a). APEX2. Version 2.0-1. Bruker AXS Inc., Madison, Wisconsin, USA.

Bruker (2005b). SAINT-NT. Version 6.0. (includes XPREP and SADABS) Bruker AXS Inc., Madison, Wisconsin, USA.

Drago RS (1992). Physical methods for chemists, 2rd Edn. Ft. Worth: Sainders College Poblishing, pp. 199.

Farrugia LJ (1997). ORTEP-3 for Windows - a version of ORTEP-III with a Graphical User Interface (GUI). J. Appl. Cryst. 30: 565–566.

Hakimi M, Maeder M, Lawrance GA (2011b). Synthesis, characterization, and crystal structure of a tricadmium complex of 3,4-diamino-5-methyl-1,2,4-triazole. J. Coord. Chem. 64: 105-114.

Hakimi M, Mardani Z, Moeini K, Fernandes MA (2012b). Coordination geometries and crystal structures of cadmium(II) complexes with a new amino alcohol (NN'O) ligand. J. Coord. Chem. 65: 2221-2233.

Jia B, Hu ZQ, Deng XT, Cheng CX, Shi SM (2006). Acta Cryst. Pyridine-2,6-dicarbohydrazide. E62: o4902–o4903.

Liu H, Wang L, Xiong J, Zhang J, Ding J (2002). Faming Zhuanli shenqing Gongkai shumingshu, CN1335303.

Macrae CF, Bruno IJ, Chisholm JA, Edgington PR, McCabe P, Pidcock E, Rodriguez-Monge L, Taylor R, van de Streek J, Wood PA (2008). Mercury CSD 2.0 - new features for the visualization and investigation of crystal structures. J. Appl. Crystallogr. 41: 466–470.

Nakamoto K (2009). Infrared and Raman Spectra of Inorganic and Coordination Compounds, 6rd Edn, Hoboken: John Wiley and Sons, p. 251.

Spek AL (2003). Single-crystal structure validation with the program PLATON. J. Appl. Cryst. 36: 7-13.

Wilkinson G, Gillard RD, McCleverty GA (1987). Comprehensive Coordination Chemistry, Oxford: Pergamon, P. 585.

Potential of nano crystalline hydroxyapatite for lead (II) removal from aqueous solutions: Thermodynamic and Adsorption isotherm study

I. Mobasherpour[1]*, E. Salahi[1] and M. Pazouki[2]

[1]Ceramics Department, Materials and Energy Research Center, P. O. Box 31787/316, Karaj, Iran.
[2]Energy Department, Materials and Energy Research Center, P. O. Box 31787/316, Karaj, Iran.

The potential of the synthesized nano hydroxyapatite to remove Pb (II) from aqueous solutions was investigated in batch reactor under different experimental conditions. The study also investigates the effects of process parameters such as initial concentration of Pb (II) ion, temperature and adsorbent mass. Various thermodynamic parameters, such as $\Delta G°$, $\Delta H°$ and $\Delta S°$ have been calculated. The thermodynamics of Pb (II) ion onto nano HAp system indicates spontaneous and endothermic nature of the process. Lead uptake was quantitatively evaluated using the Langmuir, Freundlich and Dubinin–Kaganer–Radushkevich (DKR) model. The adsorption data follow the Langmuir model better than the Freundlich and DKR model and the adsorption equilibrium was described well by the Langmuir isotherm model with maximum adsorption capacity of 714.286 mg/g of Pb (II) ions on nano HAp.

Key word: Adsorption, nano crystalline hydroxyapatite, Pb (II), thermodynamic parameters, adsorption isotherms.

INTRODUCTION

The presence of heavy metals in aqueous waste streams has become a problem due to its harmful effects on human health and to the fauna and flora of receiving water bodies. It is known that legal standards on environment control are becoming strict and as a result, the discharge of heavy metals into aquatic bodies and sources of potable water is being rigorously controlled. The US EPA allows solutions containing heavy metals to be discharged if the concentration is usually less than 5.0 mg L^{-1} (Sahin et al., 2005).

Lead is one of the most ubiquitous contaminants in the soil and aqueous environments. A severe environmental Pb contamination can often be found at shooting ranges where the soil Pb concentration sometimes exceeds 10000 mg kg^{-1} because of spent lead bullets. In Iran, many shooting ranges are generally located in mountainous regions and suffer from the degradation of natural vegetation due to Pb toxicity, which may have the potential to augment the Pb contamination via soil erosion. Therefore, development of cost-effective technologies is necessary to reduce the mobility and bioavailability of Pb in soil and water environments (Vangronsveld et al., 1995).

Calcium hydroxyapatite (HAp), $Ca_{10}(PO_4)_6(OH)_2$, is used for the removal of heavy metals from contaminated soils, wastewater and fly ashes (Chen et al., 1997; Laperche et al., 1996; Ma et al., 1993,1994; Mavropoulos et al., 2002; Nzihou et al., 2002; Takeuchi et al., 1990). Calcium hydroxyapatite (CaHAp) is a principal component of animal hard tissues and has been of interest in industry and medical fields. Its synthetic particles find many applications in bioceramics, chromatographic adsorbents to separate protein and enzyme, catalysts for dehydration and dehydrogenation of alcohols, methane oxidation, and powders for artificial teeth and bones paste germicides (Elliott, 1994). These properties relate to various surface characteristics of HAp, e.g., surface functional groups, acidity and basicity, surface charge, hydrophilicity and porosity. It has been found that CaHAp surface possesses 2.6 groups nm^{-2} of P–OH groups acting as sorption sites (Tanaka et al., 2005). The sorption properties of HAp are of great importance for both environmental processes and industrial purposes

*Corresponding author. E-mail: I_Mobasherpour@merc.ac.ir, Iman.Mobasherpour@gmail.com.

Hydroxyapatite is an ideal material for long-term containment of contaminants because of its high sorption capacity for actinides and heavy metals, low water solubility, high stability under reducing and oxidizing conditions, availability, and low cost (Helferrich, 1962). It was conducted in stabilization of a wide variety of metals (e.g., Cr, Co, Cu, Cd, Zn, Ni, Pu, Pb, As, Sb, U and V) by many investigators (Krestou et al., 2004; Czerniczyniec et al., 2003; Vega et al., 1999; Reichert et al., 1996; Leyva et al., 2001; Fuller et al., 2002; McGrellis et al., 2001). They reported that the sorption take place through ionic exchange reaction, surface complexation with phosphate, calcium and hydroxyl groups and/or co-precipitation of new partially soluble phases.

The objective of this study was to investigate the possible use of nano crystalline hydroxyapatite as an alternative adsorbent material for removal of Pb (II) ions from aqueous solutions. The Langmuir, Freundlich and D-K-R models were used to fit the equilibrium isotherm. The dynamic behavior of the adsorption was investigated on the effect of initial metal ion concentration, temperature, adsorbent mass and pH of solution. The thermodynamic parameters were also evaluated from the adsorption measurements.

MATERIALS AND METHODS

Preparation of nano crystalline hydroxyapatite sorbents

All chemicals used in this work were of analytical grade and the aqueous solutions were prepared using double distilled water. Nanocrystalline hydroxyapatite compounds were prepared by a solution-precipitation method (Mobasherpour et al., 2007) using $Ca(NO_3)_2.4H_2O$ (Analar No.10305) and $(NH_4)_2HPO_4$ (Merck No. 1205) as starting materials and ammonia solution as agents for pH adjustment. A suspension of $Ca(NO_3)_2.4H_2O$ was vigorously stirred and its temperature was maintained at 25°C. A solution of $(NH_4)_2HPO_4$ was slowly added dropwise to the $Ca(NO_3)_2.4H_2O$ solution. In all experiments the pH of $Ca(NO_3)_2.4H_2O$ solution by ammonia solution was 11. The precipitin HAp was removed from solution by the centrifuge method at a rotation speed of 3000 rpm. The resulting powder was dried at 100°C.

Transmission electron microscopy (TEM) was used to characterize the particles of HAp. For this purpose, particles were deposited onto Cu grids, which support a "holey" carbon film. The particles were deposited onto the support grids by deposition from a dilute suspension in acetone or ethanol. The crystalline shapes and sizes were characterized by diffraction (amplitude) contrast and for crystalline materials, by high resolution (phase contrast) imaging. The specific surface area was determined from N_2 adsorption isotherm by the BET method using a micromeritics surface area analyzer model ASAP 2010.The particles thus synthesized were characterized by the following methods. The crystal phase was identified by powder X-ray diffraction (XRD) using Siemens (30 kV and 25 mA) X-ray diffractometer with Cu $K\alpha$ radiation ($\lambda = 1.5404$A°) and XPERT software.

Sorption study

Aqueous solutions containing Pb (II) ions of concentration 100, 200, 300 and 400 mg/L were prepared from Lead nitrate ($Pb(NO_3)_2$, Merck No.7397). 0.1 g of nano HAp was introduced in a stirred tank reactor containing 500 ml of the prepared solution. The stirring speed of the agitator was 300 rpm. The temperature of the suspension was maintained at 20±1°C. The initial pH of the solution was adjusted to the value 5.5 by adding NH_3 and HCl.

Samples were taken after mixing the adsorbent and Pb (II) ion bearing solution at pre determined time intervals (5, 10, 20, 30, 60, and 120 min) for the measurement of residual metal ion concentration in the solution and to ensure equilibrium was reached. After specified time the sorbents were separated from the solution by centrifuge and filtration through the filter paper (Whatman grade 6). The exact concentration of metal ions was determined by AAS (GBC 932 Plus atomic absorption spectrophotometer). All experiments were carried out twice. The mass balance of lead is given by:

$$mq = V\,(C_0-C) \tag{1}$$

Where *m* is the weight of nano-HAp (g), *q* the amount of lead removed by unit of weight of HAp (Uptake capacity: mg Pb/g HAp), *V* the volume of lead solution (L), C_0 the initial lead concentration of solution (mg Pb/L) and *C* is the concentration of lead at the time *t* of adsorption (mg Pb/L). After a long time (120 min), *C* and *q* will reach equilibrium value *Ce* and *qe*. The percent removal (%) and distribution ratio (K_d) were calculated using the equations.

$$\%\,\mathrm{Re}\,moval = \frac{(c_i - c_f)}{c_i} \times 100 \tag{2}$$

Where C_i and C_f are the concentrations of the metal ion in initial and final solutions (after 120 min) respectively, and

$$K_d = \frac{amount\ of\ metal\ in\ adsorbent}{amount\ of\ metal\ in\ solution} \times \frac{V}{m} \tag{3}$$

Where *V* is the volume of the solution (mL) and m is the weight of the adsorbent (g).

RESULTS AND DISCUSSION

Characteristics of adsorbent

TEM micrograph of the HAp powders after drying is seen in Figure 1. The microstructure of the HAp crystalline after drying is observed be almost like irregular circular plates, with size in the range 8 to 10 nm. The crystal structure analysis of HAp particles was performed, using X-ray diffraction and the obtained diffractograms are represented in Figure 2. The produced reflection patterns match the ICDD standards (JCPDS) for HAp.

The patterns show only the peaks characteristic to the synthesized HAp with no obvious evidences on the presence of other additional phases. The broadened peaks indicate that the crystallites are very tiny in size. XRD grain size has been calculated by scherrer method. The apparent size of hydroxyapatite crystallites obtained from XRD profile analysis by Scherrer method is 7.75 nm

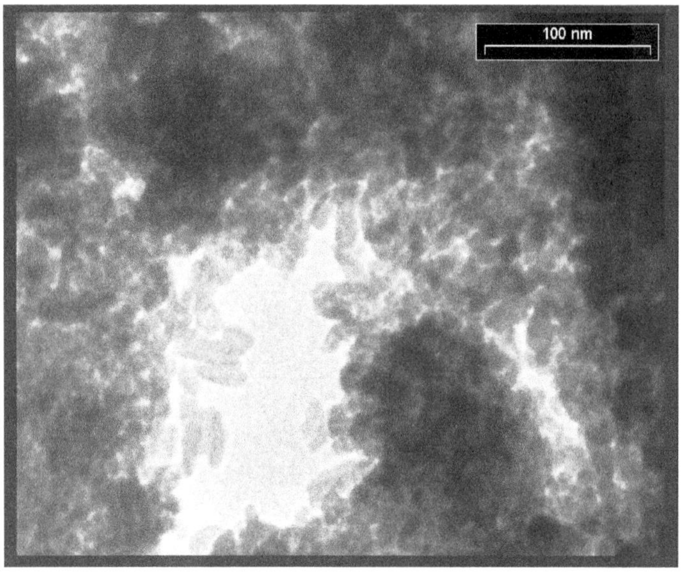

Figure 1. TEM micrograph of the nanocrystalline hydroxyapatite after drying at 100°C.

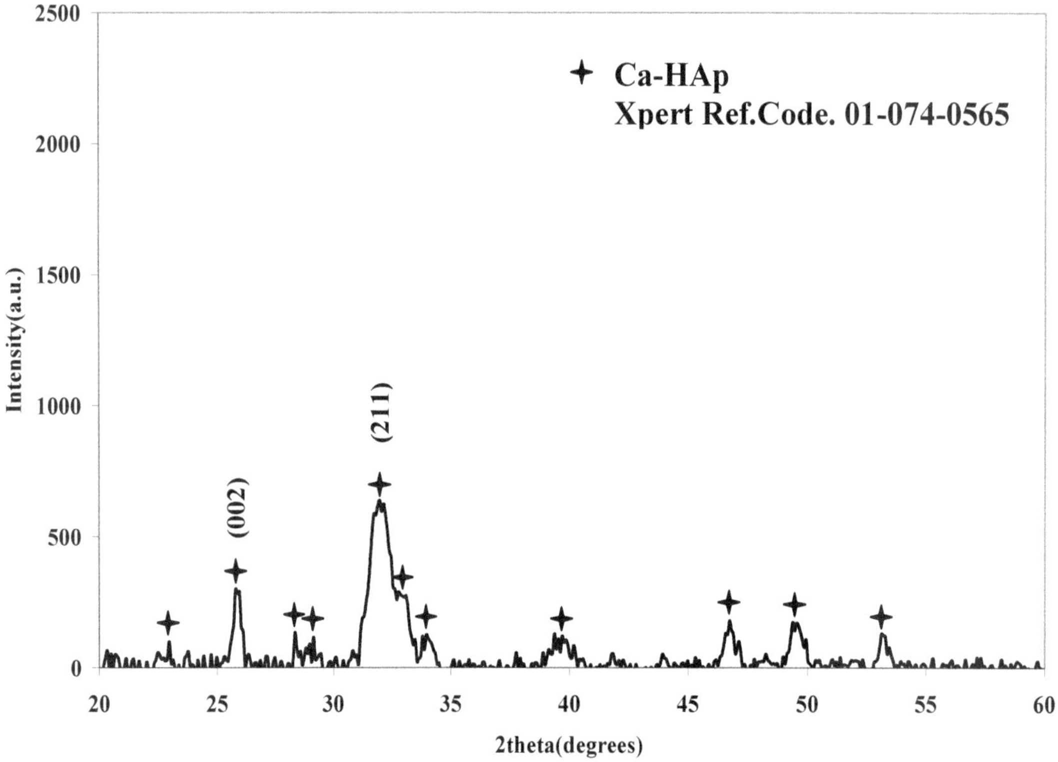

Figure 2. XRD pattern of calcium nanocrystalline hydroxyapatite powders after drying at 100°C.

(Mobasherpour et al., 2007). This TEM observation on the shape of particles agrees with our earlier XRD studies. The analysis of the HAp sample has confirmed a low-crystalline product, with the specific surface area 94.9 m^2/g.

Effect of adsorbent dosage and initial Pb (II) concentration

The sorption of Pb (II) ions was carried out at different initial lead ion concentrations ranging from 200 to 400

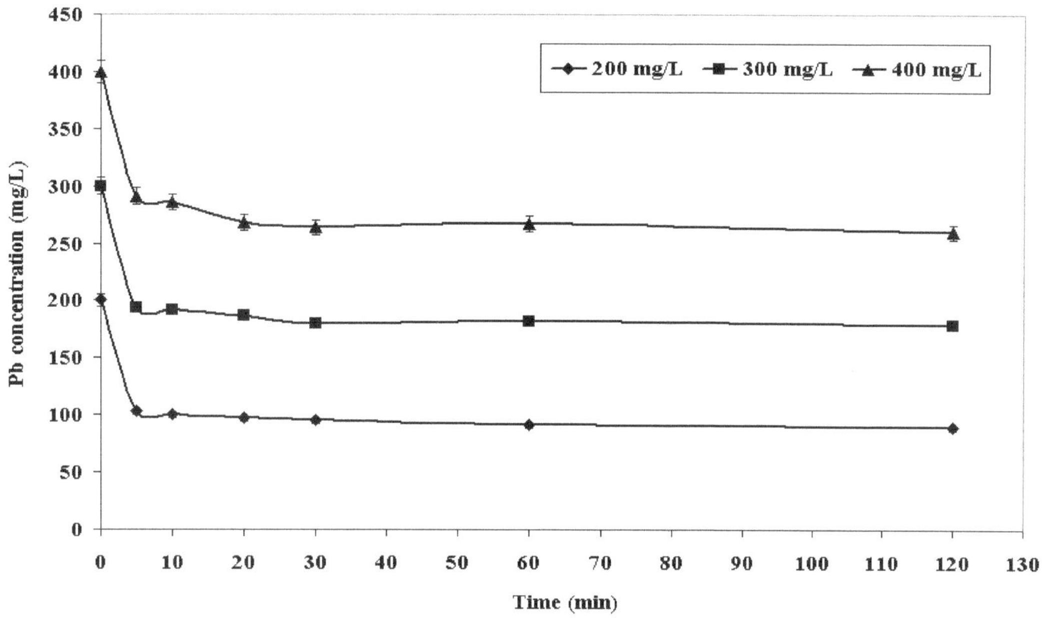

Figure 3. Time dependent concentration of aqueous Pb (II) by nano hydroxyapatite sorbents (pH 5.5, adsorbent dosage=0.2 gr/L, 300 rpm agitating rate).

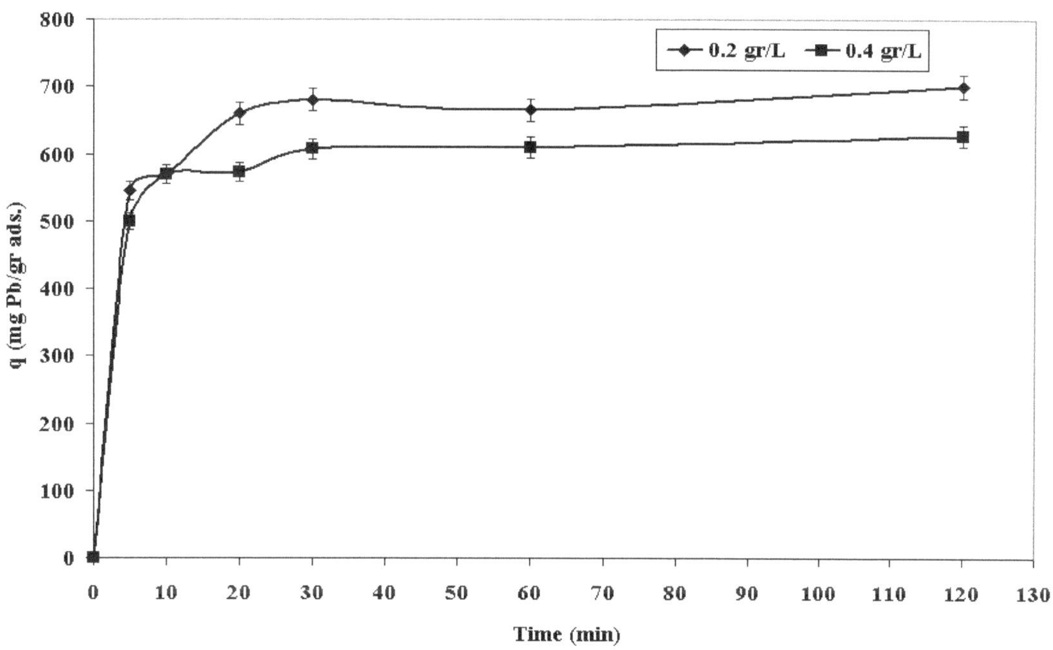

Figure 4. Effect of adsorbent dosage on removal of Pb (II) by nano hydroxyapatite (pH 5.5, initial metal concentration = 400 mg/L, 300 rpm agitating rate).

mg/L, at pH 5.5, at 300 rpm with 120 min of contact time using nano-HAp. A rapid kinetic reaction of Pb removal by sorbent occurred within the first 5 min (Figure 3.). The aqueous Pb concentration at 5 min decreased to 103, 194 and 291 mg/L by nano-HAp for 200, 300 and 400 mg/L initial concentration, respectively. Our findings on rapid kinetic reaction of synthesized nano-HAp agreed with those described elsewhere (Aklil et al., 2004; Chaturvedi et al., 2006; Prasad et al., 2004). The effect of nano-HAp dosage is presented in Figure 4. It is evident that adsorption increases with the increase in the mass of sorbent and the uptake capacity of Pb (II) decreased from

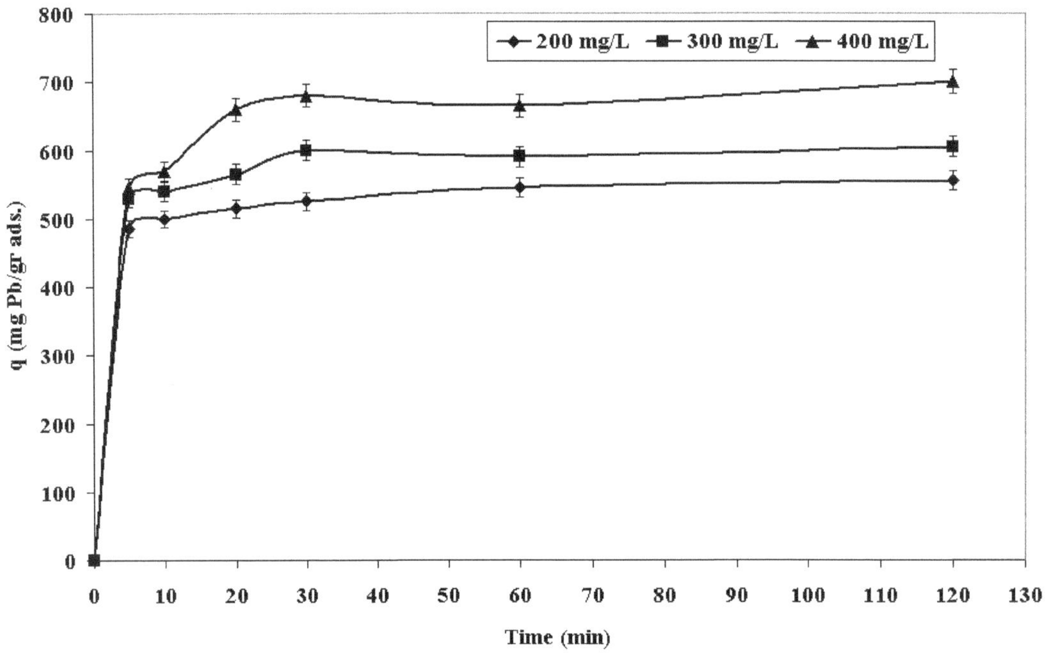

Figure 5. Effect of initial concentration on removal of Pb (II) by nano hydroxyapatite (pH 5.5, adsorbent dosage=0.2 gr/L, 300 rpm agitating rate).

700 (35% removal) to 627.5 mg/g (62.75% removal) with the increasing nano-HAp concentration from 0.2 to 0.4 g/L. This is because at the higher dosage of sorbent due to increased surface area, more adsorption sites are available causing higher removal of Pb (II). A higher initial concentration provides an important driving force to overcome all mass transfer resistances of the pollutant between the aqueous and solid phases, thus increases the uptake (Aksu et al., 2005). Uptake of the Pb (II) also increased with increasing the initial metal concentration tending to saturation at higher metal concentrations. As shown in Figure 5, when the initial Pb (II) concentration increased from 200 to 400 mg/L, the uptake capacity of nano HAp increased from 555 to 700 mg/g.

Effect of temperature

To study the effect of temperature on the uptake of Pb (II) ions by nano-HAp, we selected the following temperatures: 20, 50 and 70°C. Figure 6 illustrates the relationship between temperature and the amount of Pb (II) ions adsorbed onto nano-HAp at equilibrium time (120 min). As seen in Figure 6, the adsorption of Pb (II) on nano-HAp increased from 700 (35% removal) to 1050 mg/g (52.5% removal) when temperature was increased from 20 to 70°C at an initial concentration of 400 mg/L. The increase in the equilibrium sorption of Pb (II) with temperature indicates that Pb (II) ions removal by adsorption on nano-HAp favors a high temperature. This may be a result of increase in the mobility of the Pb (II) ion with temperature. An increasing number of molecules may also acquire sufficient energy to undergo an interaction with active sites at the surface. Furthermore, increasing temperature may produce a swelling effect within the internal structure of the nano-HAp enabling large metal ions to penetrate further (Do˘gan et al., 2003).

Adsorption isotherms

Analysis of the equilibrium data is important to develop an equation which accurately represents the results and which could be used for design purposes (Aksu, 2002). Several isotherm equations have been used for the equilibrium modeling of adsorption systems.

The sorption data have been subjected to different sorption isotherms, namely, Langmuir, Freundlich and Dubinin–Kaganer–Radushkevich (DKR). The equilibrium data for metal cations over the concentration range from 100 to 400 mg/L at 20°C have been correlated with the Langmuir isotherm (Langmuir, 1918):

$$\frac{C_e}{q_e} = \frac{1}{Q_0 K} + \frac{C_e}{Q_0} \tag{4}$$

Where C_e is the equilibrium concentration of metal in solution (mg/L), q_e is the amount absorbed at equilibrium onto nano-HAp (mg/g), Q_0 and K are Langmuir constants

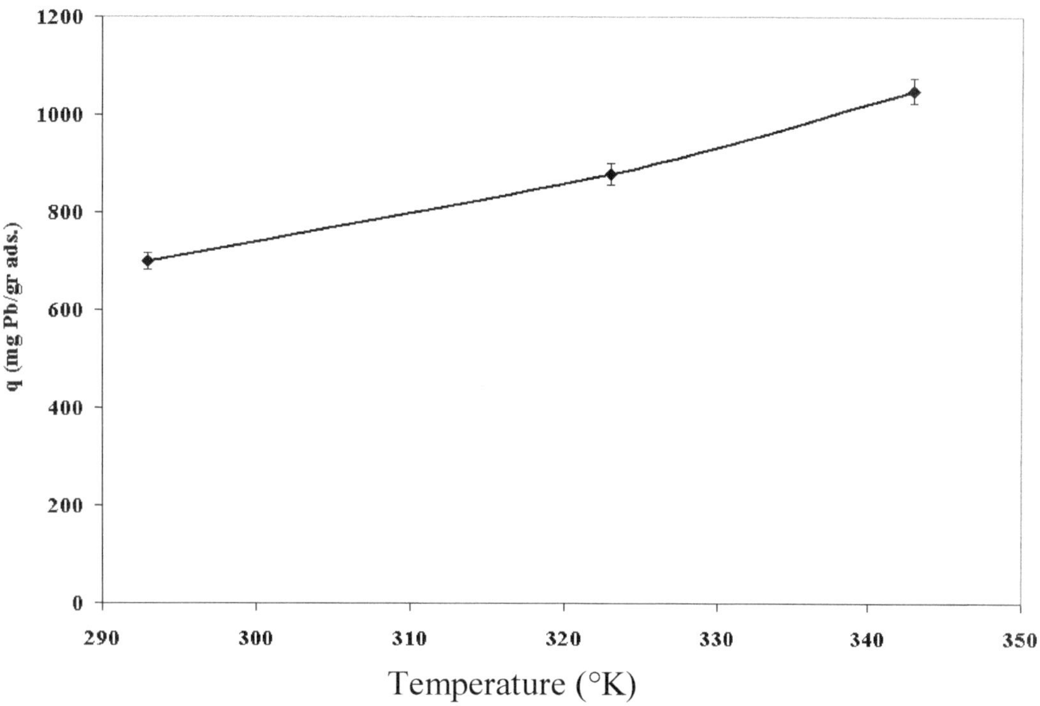

Figure 6. The uptake capacity of Pb (II) ions at different temperature (pH 5.5, initial metal concentration = 400 mg/L, adsorbent dosage=0.2 gr/L, 300 rpm agitating rate).

related to sorption capacity and sorption energy, respectively. Maximum sorption capacity (Q_0) represents monolayer coverage of sorbent with sorbate and K represents enthalpy of sorption and should vary with temperature. A linear plot is obtained when C_e/q_e is plotted against C_e over the entire concentration range of metal ions investigated.

The linearized Langmuir adsorption isotherms of Pb (II) ions are given in Figure 7 a. An adsorption isotherm is characterized by certain constants that express the surface properties and affinity of the sorbent and can also be used to find the sorption capacity of sorbent.

The Freundlich sorption isotherm, one of the most widely used mathematical descriptions, usually fits the experimental data over a wide range of concentrations. This isotherm gives an expression encompassing the surface heterogeneity and the exponential distribution of active sites and their energies. The Freundlich adsorption isotherms were also applied to the removal of Pb (II) on nano-HAp (Figure 7 b).

$$Lnq_e = Lnk_f + \frac{1}{n} Lnc_e \quad (5)$$

where q_e is the amount of metal ion sorbed at equilibrium per gram of adsorbent (mg/g), C_e the equilibrium concentration of metal ion in the solution (mg/L), k_f, n the Freundlich model constants (Malkoc et al., 2003; Kadirvelu et al., 2001). Freundlich parameters, k_f and n, were determined by plotting lnq_e versus lnC_e. The numerical value of $1/n < 1$ indicates that adsorption capacity is only slightly suppressed at lower equilibrium concentrations. This isotherm does not predict any saturation of the sorbent by the sorbate; thus infinite surface coverage is predicted mathematically, indicating multilayer adsorption on the surface (Hasany et al., 2002).

The Dubinin–Kaganer–Radushkevich (DKR) equation has been used to describe the sorption of metal ions on clays. The DKR equation has the form,

$$\text{Ln } C_{ads} = \text{Ln } X_m - \beta\varepsilon^2 \quad (6)$$

Where C_{ads} is the number of metal ions adsorbed per unit weight of adsorbent (mol/g), X_m (mol/g) is the maximum sorption capacity, β (mol^2/J^2) is the activity coefficient related to mean sorption energy and ε is the Polanyi potential, which is equal to

$$\varepsilon = RT\, ln(1 + 1/Ce) \quad (7)$$

Where R is the gas constant (kJ/mol.K) and T is the temperature (K). The saturation limit X_m may represent the total specific micropore volume of the sorbent. The sorption potential is independent of the temperature but varies according to the nature of sorbent and sorbate (Khan et al., 1995). The slope of the plot of ln C_{ads} versus ε gives β (mol^2/J^2) and the intercept yields the sorption capacity, X_m (mol/g). The sorption space in the vicinity of

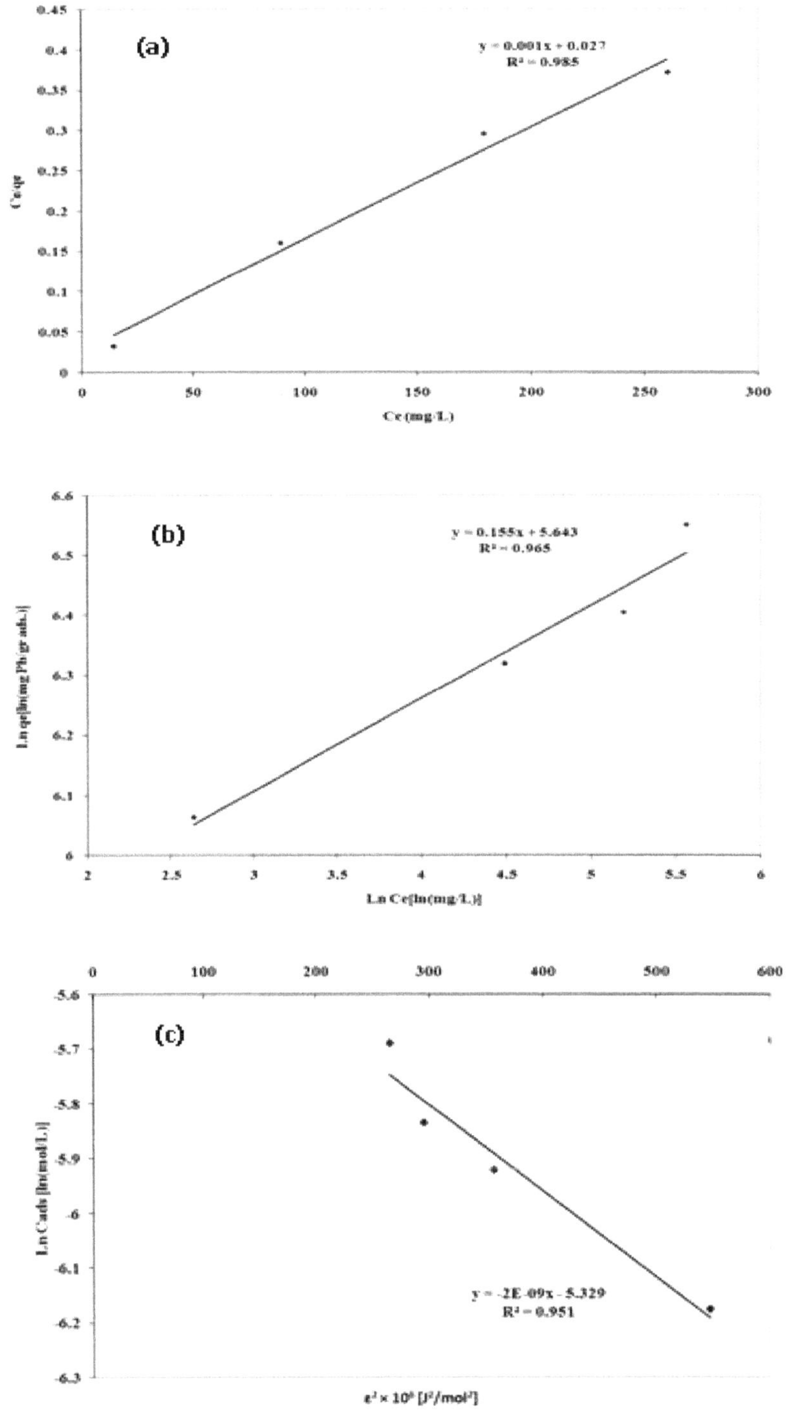

Figure 7. Linear fits of experimental data obtained using Langmuir (a), Freundlich (b) and DKR (c) sorption isotherms for the adsorption of Pb (II) onto nano Hydroxyapatite.

a solid surface is characterized by a series of equipotential surfaces having the same sorption potential. This sorption potential is independent of the temperature but varies according to the nature of sorbent and sorbate. The sorption energy can also be worked out using the following relationship:

$$E = 1/\sqrt{-2\beta}\,. \qquad (8)$$

The plot of Ln C_{ads} against ε^2 for metal ion sorption on nano-HAp is shown in Figure 7c. The Langmuir, Freundlich and DKR adsorption constants from the

Table 1. Langmuir, Freundlish and DKR constants for adsorption of Pb (II) onto nano Hydroxyapatite.

Langmuir adsorption isotherms constants		
Q_0 (mg/g)	K (L/mg) 0.052	R^2
714.286		0.985
Freundlich adsorption isotherms constants		
k_f (mg/g)	n	R^2
282.421	6.464	0.965
DKR adsorption isotherms constants		
X_m (mg/g)	β (mol^2/J^2)	R^2
1003.525	-2×10^{-9}	0.951

isotherms and their correlation coefficients are also presented in Table 1.

The correlation factors R (0.985, 0.965 and 0.951 for Langmuir, Freundlich and DKR model, respectively) confirm good agreement between both theoretical models and our experimental results. The maximum sorption capacity, Q_0, calculated from Langmuir equation is 714.286 mg/g, while Langmuir constant K is 0.052 L/mg. The values obtained for Pb (II) from the Freundlich model showed a maximum adsorption capacity (K_f) of 282.421 mg/g with an affinity value (n) equal to 6.464. The values of sorption constants, derived from DKR model are: 1003.525 mg/g (4.848 mmol/g) for X_m, -2×10^{-9} mol^2/J^2 for β and 15.811 kJ/mol for E.

The values indicate that the adsorption pattern for Pb (II) on nano-HAp followed third the DKR isotherm (R^2 > 0.951), the Freundlich isotherm (R^2 > 0.965) and the Langmuir isotherm (R^2 > 0.985) at all experimental situations. It is clear that the Langmuir isotherm has best fitted for the sorption of Pb (II) on nano-HAp. When the system is in a state of equilibrium, the distribution of Pb (II) between the nano HAp and the Pb (II) solution is of fundamental importance in determining the maximum sorption capacity of nano HAp for the Lead ion from the isotherm. The E values are 15.811 kJ for Pb (II) on the nano HAp. It is the orders of an ion-exchange mechanism, in which the sorption energy lies within 8 to 16 kJ/mol (Helferrich, 1962; Rieman et al., 1970). Generally, HAp selectivity towards divalent metal cations is a result of the ion-exchange process with Ca^{2+} ions (MonteilRivera et al., 2002). Ionic radius of Lead (1.19A°) slightly differ from that of Ca^{2+} (0.99A°), and it can substitute Ca^{2+} in the HAp crystal lattice. Figure 8 presents the XRD patterns of Pb^{2+}-loaded sample.

No structural changes of nano HAp were detected by the powder X-ray diffraction analysis of the solid residue with maximum amount of uptake capacity of Pb (II), obtained after interaction 0.2 g/L of nano HAp with 400 mg/L Pb (II) solution, at pH 5.5, at 300 rpm with 120 min of contact time. The sample was indexed in the hexagonal system with space group $P6_3/m$ and Calcium Lead Hydroxide Phosphate ICSD name (Ref.Code.01-084-0814 in XPERT software). The unit cell parameters of the starting nano Ca-HAp (Figure 2) were a = b = 0.9424 nm and c = 0.6879 nm, while values of a = b = 0.9880 nm and c = 0.7417 nm were calculated for the Pb-exchanged sample. These shifts are indicative to the increase in unit cell dimensions which is due to the replacement of Ca^{2+} by Pb (II) (ionic radius 1.19 Å), which is larger than Ca^{2+} (ionic radius 0.99 Å), into the crystal lattice of apatite molecules. These data strongly support the ion-exchange mechanism for Pb (II) sorption by nano HAp. The reaction mechanism corresponds to equimolar exchange of lead and calcium yielding $Ca_{10-x}Pb_x(PO_4)_6(OH)_2$, where x can vary from 0 to 10 depending on the reaction time and experimental conditions. Our results of synthesized nano-HAp agreed with those described elsewhere that the proposed mechanism for lead removal by HAp comprises two steps: firstly, rapid surface complexation of the lead on the ≡POH sites of the HAp which causes the decrease of the pH (from pH = 5.5 to pH = 4.5 for initial metal concentration = 400 mg/L, dosage = 0.2 g/L, 300 rpm agitating rate) and secondly, partial dissolution of calcium followed by the precipitation of an apatite with formula: $Pb_xCa_{10-x}(PO_4)_6(OH)_2$ (Baillez et al., 2007).

In which lead ions are first adsorbed on the nano HAp surface and substitution with Ca occurs as described by the following equation:

$$Ca_{10}(PO_4)_6(OH)_2 + xPb^{2+} \rightarrow Ca_{10-x}Pb_x(PO_4)_6(OH)_2 + xCa^{2+} \quad (9)$$

Determination of thermodynamic parameters

Thermodynamic parameters such as free energy change ($\Delta G°$), enthalpy change ($\Delta H°$) and entropy change ($\Delta S°$) can be estimated using equilibrium constants changing with temperature. The free energy change of the sorption reaction is given by the following equation.

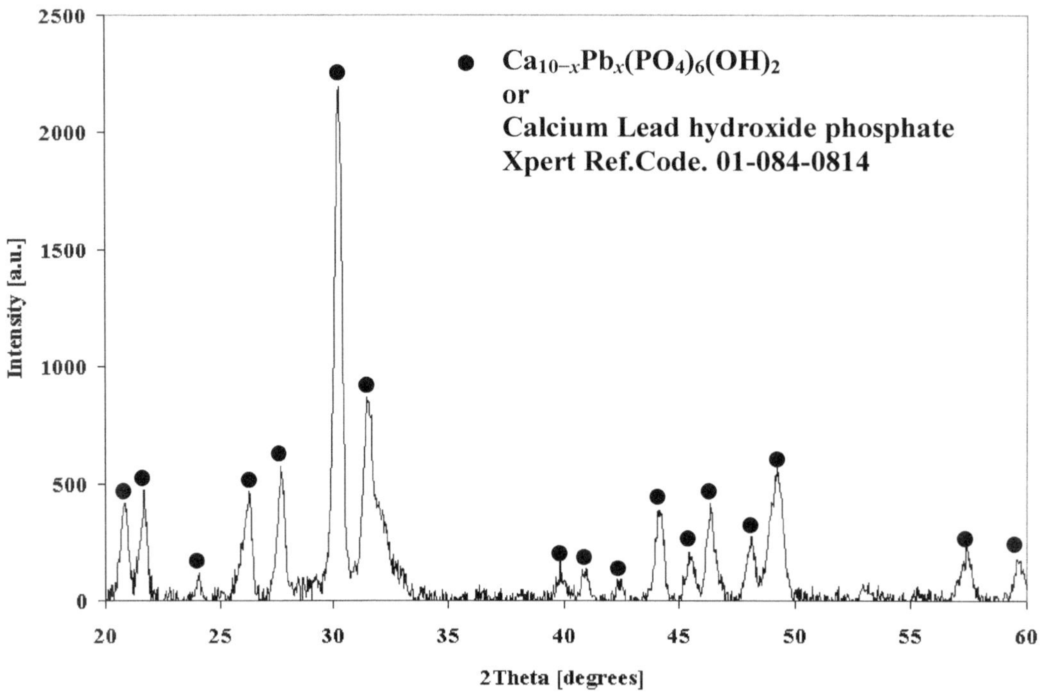

Figure 8. XRD pattern of the solid residue with maximum amount of uptake capacity of Pb (II).

Table 2. Thermodynamic parameters for the adsorption of Pb (II) onto nano Hydroxyapatite.

T °K	K_d	ΔG^0 (J/mol)	ΔH^0 (J/mol)	ΔS^0 (J/mol.°K)
293	1346.15	-17551.40		
323	1954.77	-20350.20	11901	100.32
343	2763.20	-22597.30		

$$\Delta G° = -RT\ Ln\ K_d \qquad (10)$$

Where $\Delta G°$the standard free energy change (J), R is is is the universal gas constant, 8.314 J/mol.K and T the absolute temperature (K).

$$\Delta G° = \Delta H° - T\ \Delta S° \qquad (11)$$

The distribution ratio (K_d) values increased with temperature, indicating the endothermic nature of adsorption. A plot of Gibbs free energy changes, $\Delta G°$ versus temperature, T(K); was found to be linear. The values of $\Delta H°$ and $\Delta S°$ were determined from the slope and intercept of the plots. The thermodynamic parameters Gibbs free energy change, $\Delta G°$, are shown in Table 2.

The enthalpy change, $\Delta H°$ and the entropy change, $\Delta S°$, for the sorption processes are calculated to be 11.90 and 100.32 kJ/mol, respectively. The negative values of $\Delta G°$ at various temperatures indicate the spontaneous nature of the adsorption process. The positive value of $\Delta S°$ indicates that there is an increase in the randomness in the system solid/solution interface during the adsorption process. In addition, the positive value of $\Delta H°$ indicates that the adsorption is endothermic. The positive value of $\Delta S°$ reflects the affinity of the nano-HAp for Pb (II) ions and suggests some structural changes in Lead and nano-HAp (Ho, 2003).

Conclusions

Nano HAp exhibited high adsorption capacity. The present investigation shows that the nano HAp is an effective adsorbent for the removal or Pb (II) from aqueous solutions. The aqueous Pb concentration (400 mg/L) at 5 min was reduced to 291 mg/L by nano HAp. The adsorption process is a function of the adsorbent dosage, initial Pb (II) concentration, temperature. The efficiency of Pb (II) adsorption increased with an increase in the adsorbent dosage. Isotherm studies indicate that the Langmuir model fits the experimental data better than Freundlich and DKR model. The adsorption equilibrium was described well by the Langmuir isothermal model with maximum adsorption capacity of 714.286 mg/g of Pb (II) on nano crystallite HAp particles. The results of XRD analysis strongly support the ion-exchange as a main mechanism for Pb (II) removal by nano HAp. The results show that the lead uptake by nano hydroxyapatite proceeds with a rapid surface complexation of the lead on the ≡POH site before the formation of a pyromorphite compound of formula $Ca_{10-x}Pb_x(PO_4)_6(OH)_2$. Thermodynamic calculations showed that the Lead sorption process of nano HAp has endothermic and spontaneous nature. Thermodynamic calculations showed that the lead sorption process of nano crystallite HAp particles had endothermic and spontaneous nature.

ACKNOWLEDGMENT

This research was supported by the Research Project Unit at the Materials and Energy Research Center, Karaj, Iran under the project No. 388873.

REFERENCES

Aklil A, Mouflih M, Sebti S (2004). "Removal of heavy metal ions from water by using calcined phosphate as a new adsorbent." J. Hazard. Mater., 112: 183-190.

Aksu Z (2002). "Determination of the equilibrium, kinetic and thermodynamic parameters of the batch biosorption of nickel(II) ions onto *Chlorella vulgaris*." Process Biochem., 38: 89-99.

Aksu Z, Tezer S (2005). "Biosorption of reactive dyes on the green alga *Chlorella vulgaris*." Process Biochem., 40: 1347-1361.

Baillez S, Nzihou A, Bernache-Assolant D, Champion, E.,Sharrock P (2007). "Removal of aqueous lead ions by hydroxyapatites: Equilibria and kinetic processes." J. Haz. Mater., A139: 443-446.

Chaturvedi PK, Seth CS, Misra V (2006). "Sorption kinetics and leachability of heavy metal from the contaminated soil amended with immobilizing agent (humus soil and hydroxyapatite)." Chemosphere, 64: 1109-1114.

Chen X, Wright JV, Conca JL, Peurrung LM (1997). "Effects of pH on heavy metal sorption on mineral apatite." Environ. Sci. Technol., 31: 624-631.

Czerniczyniec M, Farias S, Magallanes J, Cicerone D (2003). Arsenic adsorption on biogenic HAP: Solution composition effects. 11th International Conference on Surface and Colloid Science, Foz do Iguazu, Brazil, p. 269.

Do˘gan M, Alkan M (2003). "Adsorption kinetics of methyl violet onto perlite." Chemosphere, 50: 517-528.

Elliott JC (1994). Structure and Chemistry of the Apatites and Other Calcium Orthophosphates. Amsterdam: Elsevier.

Fuller C, Bargar J, Davis J, Piana M (2002). "Mechanisms of uranium interactions with hydroxyapatite: Implications for groundwater remediation." Environ. Sci. Technol., 36: 158-165.

Hasany SM, Saeed MM, Ahmed M, Radioanal J (2002). "Sorption and thermodynamic behavior of zinc(II)-thiocyanate complexes onto polyurethane foam from acidic solutions." Nucl. Chem., 252: 477-484.

Helferrich F (1962). Ion Exchange. New York : McGraw–Hill.

Ho YS (2003). "Removal of copper ions from aqueous solution by tree fern." Water Res., 37: 2323-2330.

Kadirvelu K, Thamaraiselvi K, Namasivayam C (2001). "Adsorption of nickel (II) from aqueous solution onto activated carbon prepared from coirpith." Sep. Purif. Technol., 24: 497-505.

Khan SA, Rehman UR, Khan MA (1995). "Adsorpton of chromium (III), chromium (VI) and silver (I) on bentonite." Waste Manage., 15: 271-282.

Krestou A, Xenidis A, Panias D (2004). "Mechanism of aqueous uranium (VI) uptake by a natural zeolitic tuff." Miner. Eng., 16: 1363-1370.

Langmuir I (1918). "The adsorption of gases on plane surfaces of glass, mica and platinum." J. Am. Chem. Soc., 40: 1361-1403.

Laperche V, Traina SJ, Gaddam P, Logan TJ, Ryan JA (1996). "Chemical and mineralogical characterizations of Pb in a contaminated soil: Reactions with synthetic apatite." Environ. Sci. Technol., 30: 3321-3326.

Leyva A, Marrero J, Smichowski P, Cicerone D (2001). "Sorption of antimony onto hydroxyapatite." Environ. Sci. Technol., 35: 3669-3675.

Ma QY, Traina SJ, Logan TJ, Ryan JA (1993). "*In situ* lead immobilization by apatite." Environ. Sci. Technol., 27: 1803-1810.

Ma QY (1994). "Effects of Aqueous Al, Cd, Cu, Fe(II), Ni, and Zn on Pb immobilization by hydroxyapatite." Environ. Sci. Technol., 28: 1219-1228.

Malkoc E, Nuho˘glu Y (2003). "The removal of chromium(VI) from synthetic wastewater by Ulothrix zonata." Fres. Environ. Bull., 12: 376-381.

Mavropoulos E, Rossi AM, Costa AM, Perez CA, Moreira JC, Saldanha M (2002). "Studies on the mechanisms of lead immobilization by hydroxyapatite." Environ. Sci. Technol., 36: 1625-1629.

McGrellis S, Serafini J, Jean J, Pastol J, Fedoroff M (2001). "Influence of the sorption protocol on the uptake of Cd ions in calcium hydroxyapatite." Sep. Purif. Technol., 24: 129-138.

Mobasherpour I, Heshajin MS, Kazemzadeh A, Zakeri M (2007). "Synthesis of nanocrystalline hydroxyapatite by using precipitation method." J. Alloys Comp., 430: 330-333.

MonteilRivera F, Fedoroff M (2002). Sorption of Inorganic Species on Apatites from Aqueous Solutions In: Encyclopedia of Surface and Colloid Science. New York: Marcel Dekker. Inc.

Nzihou A, Sharrock P (2002). "Calcium phosphate stabilization of fly ash with chloride extraction." Waste Manage., 2002: 235-239.

Prasad M, Saxena S (2004). "Sorption mechanism of some divalent metal ions onto low-cost mineral adsorbent." Ind. Eng. Chem. Res., 43: 1512-1522.

Reichert J, Binner J (1996). "An evaluation of hydroxyapatite-based filters for removal of heavy metal ions from aqueous solutions." J. Mater. Sci., 31: 1231-1241.

Rieman W, Walton H (1970). Ion Exchange in Analytical Chemistry. Oxford: Pergamon.

Sahin Y, Ozturk A (2005). "Biosorption of chromium(VI) ions from aqueous solution by the bacterium Bacillus thuringiensis." Process Biochem., 40: 1895-1901.

Takeuchi Y, Arai H (1990). Removal of coexisting Pb2+, Cu2+, Cd2+ ions from water by addition of hydroxyapatite powder. J. Chem. Eng. Jpn., 23: 75-80.

Tanaka H, Futaoka M, Hino R, Kandori K, Ishikawa T (2005). "Structure of synthetic calcium hydroxyapatite particles modified with pyrophosphoric acid" J. Colloid Interface Sci., 283: 609-612.

Vangronsveld J, Colpaert J, Tichelen KV (1995). "Reclamation of a bare industrial area contaminated by non-ferrous metals: physicochemical and biological evaluation of the durability of soil treatment and revegetation." Environ. Pollut., 94: 131-140.

Vega ED, Pedregosa JC, Narda GE (1999). "Interaction of oxovanadium(IV) with crystalline calcium hydroxyapatite: surface mechanism with no structural modification." J. Phys. Chem. Solids, 60: 759-766.

Sodium nano-particle effects on montmorillinite structure

A. Bahari*, M. Asgharzadeh and M. Najafikhah

Department of Physics, University of Mazandaran, Iran.

The main aim of the work is to study the expansion of the inter-layers clay to see what will happen to the mechanical properties of clays. The expansion of clay layers with sodium nano- particles makes for the exfoliation and dispersion of the crystal in fine particles in the clay positions. Patterns obtained by X-ray diffraction (XRD) of nano montmorillinite (MMT) illustrate that its shape is irregular and the range of sizes is very wide as found by X- powder method. XRD patterns represent significant improvement in acid treatment relative to direct or other methods. Mg^{-2}, Ca^{-2} and ammonium are the dominant exchangeable cations which show relatively low dispersion degree and large size of the particle. However, in the present work, Na nano - particles have been inserted in the clays and the extension of the destruction clay layers is revealed.

Key words: Nanostructures, nanoclay, montmorillinite, sol – gel method.

INTRODUCTION

Recently, many researchers (Mani et al., 2005; Quang et al., 2007; Sánchez, 2009; Bahari, 2006; Morgen, 2006; Bahari et al., 2005; Bahari and Pashayan, 2011) have studied polymer layered silicon nano – composites as suitable material both in industry area and in academia, due to their potential for enhanced physical, chemical and mechanical properties compared to conventially filled composites. Indeed, two categories of clay minerals are: first type known 1:1 clays which is one tetrahedral sheet presented for one octahedral sheet; second type known 2:1 clays with two tetrahedral sheets presented for one octahedral sheet. They have been studied both experimentally and theoretically. The 2:1 clay minerals are further subdivided into smectites, micas, vermiculites, and pyrophyllites. We have studied MMT which is 3:4 clay mineral and classified as a smectite, which is widely used in the preparation of polymer nanocomposites.

MMT is a nano clay and seems something like polymer – based nano compositions because of their suitable charge layer density, 2:1 tetrahedral – octahedral layer structure. From Mani et al. (2005) and Quang et al. (2007), we found that the structure of MMT is derived from clay by the substitution of certain atoms for other atoms. By looking at their structure, it is clear that the tetrahedral sheet is tetravalent Si with little trivalent Al atoms. But, the octahedral sheet includes divalent Mg which replaces trivalent Al without filling the third vacant octahedral position. In fact, the pointed mineral contains an atom of lower positive valence which is replaced with one of higher valence atom. This resulted in a deficit of positive charge (that is, excess negative charge), excess negative charge as compensated by the adsorption of cations. They are too large to be accommodated in the interior of the crystal. In the presence of water, the compensating cations on the layer surfaces may be easily exchanged by other cations, as expected by looking at the structures using crystal phases of clays and X- powder method.

*Corresponding author. E-mail: a.bahari@umz.ac.ir, ali.panbechoole@newcastle.edu.au.

METHODOLOGY

The sol-gel process is a wet-chemical technique (chemical solution deposition) and/ or a bottom – up procedure, widely used recently in the fields of materials science and ceramic engineering. Such methods are used primarily for the fabrication of materials starting from a chemical solution which acts as the precursor for an integrated network (or gel) of either discrete particles or network nano composites. Typical precursors are acetic, nitric, chloride, formic and sulphuric acids, which undergo various forms of reactions.

Thus, the sol evolves towards the formation of a gel-like diphase system containing both a liquid phase and solid phase whose

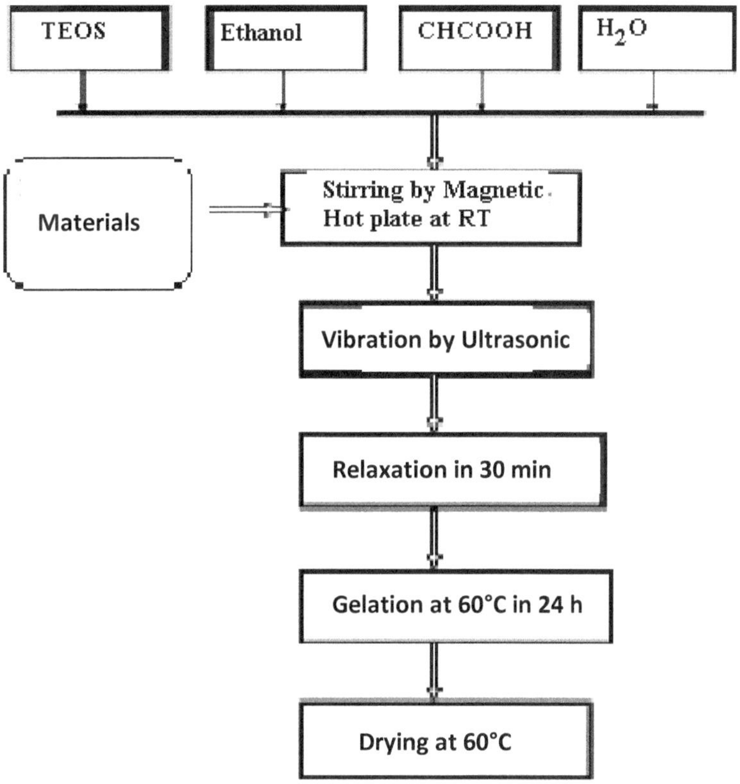

Figure 1. Flowchart of synthesis stainless steel nano- particles.

Table 1. The procedure of experiment in sol – gel method with H_2O_2, acetic chloride and sodium treatments.

Specimen	Composites and procedures	MMT $d_{(001)}$ (nm)
MMT -0	Natural clay	6.65
MMT -1	MMT -0 with H_2O_2	7.85
MMT -2	MMT -1 with acetic chloride	5.36
MMT -3	MMT -1 with Na	8

morphologies range from discrete particles to continuous networks. In this process, sodium clay powders are dissolved in de-ionized water to which is added citric acid. The mixture is then stirred carefully using a magnetic stirrer while ammonium hydroxide is added to obtain a pH of 7. Initially a zero gel and finally a powder are obtained. MMT powder is synthesized via simple sol – gel method as summarized in Figure 1.

RESULTS AND DISCUSSION

XRD technique is used for crystal phase identification and estimation of the crystallite size. XRD patterns are measured on a (GBC-MMA 007 (2000)) X-ray diffractometer. The diffractograms are recorded with 0.02° step size in which the speed is 10°/min radiations over a 2θ range of 10 to 80° at a sampling width of 0.2° and a scanning speed of 10°/ min. The XRD patterns of MMT are composed of calcined at different treatments. The crystalline peaks corresponded to MMT in the form of either highly pure mixed MMT - quartz, except the as – prepared nanostructure showing amorphous phase. This point indicates that increase in MMT - quartz transformation of clay peak place with some materials is given in Table 1.

Crystallite size, lattice micro strain and lattice parameter of each nano clay powder have been calculated from the XRD peak broadening. Following Scherrer equation, the crystallite sizes have been determined for MMT and clays powders (Bertoluzza et al., 1982; Ramez and Occurrence, 1996; Bahari et al., 2008; Hernandez-Torres and Mendoza-Galvan, 2005).

$$\Gamma = \frac{K\lambda}{\beta \cos\theta}$$

Where Γ, K, λ, β and θ is mean crystallite dimension, X-

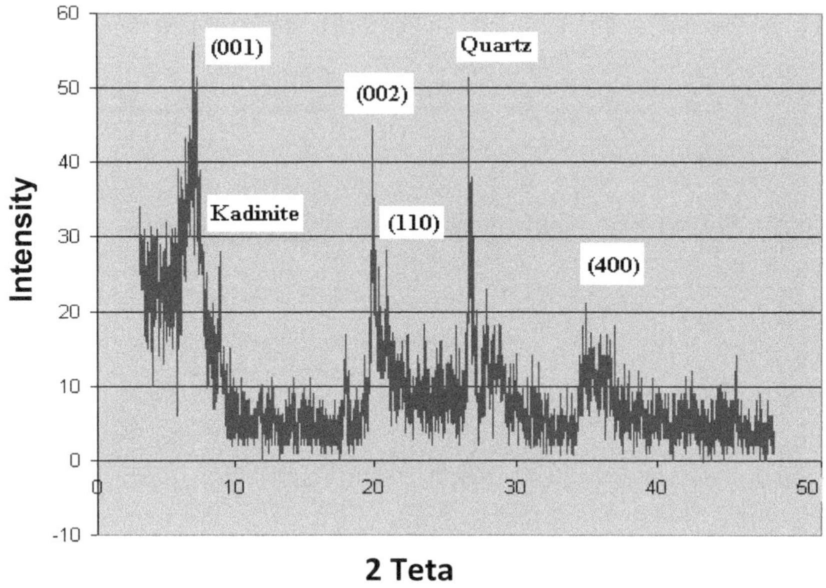

Figure 2. XRD pattern of nano clay before adding Na nano-particles corresponding to Table 1.

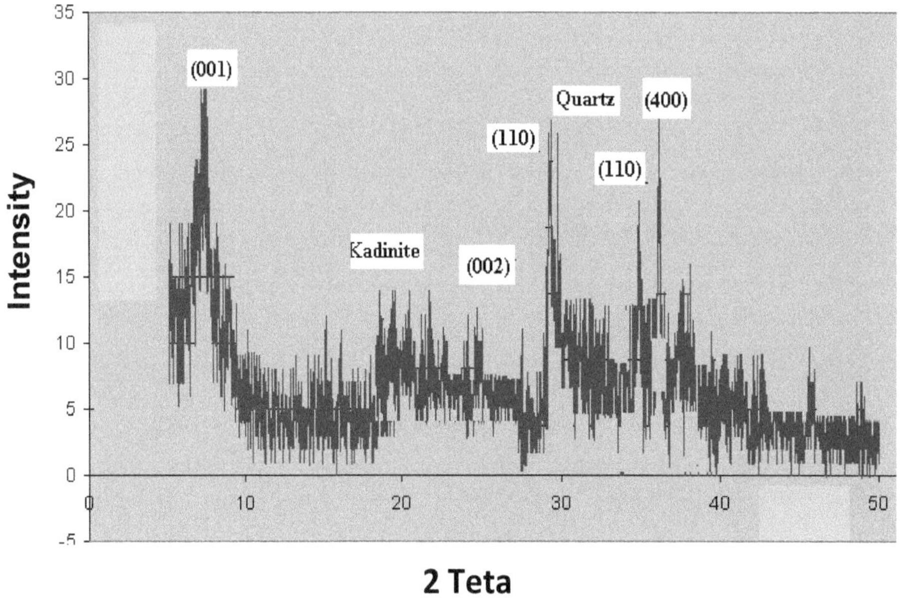

Figure 3. XRD pattern of nano clay after adding Na nano-particles corresponding to Table 1. The shifted peak around 20° Indicates the extend spacing between layer.

ray wavelength, FWHM (in radians) and Bragg angle, respectively.

Based on the diffraction peak of MMT in XRD patterns (Figures 2 to 5), under different treatments , their X-ray peaks are calculated to obtain the interlayer distance change as shown in Table 1. The characteristic peak for the pristine clay is that of the diffraction at MMT (100); such peak interdistance is expressed by d_{100}.

Because the Scherrer relation cannot determine the accurate size of nano - particles, we have thus used X-powder method (Figures 6 to 9) and found the accurate crystalline size.

From the diffraction patterns in Figures 2 to 5, we can see that the peak, MMT (001) is shifted to a lower angle compared to the natural MMT (001) peak. This shifted peak indicates some expansion of the clay gallery with

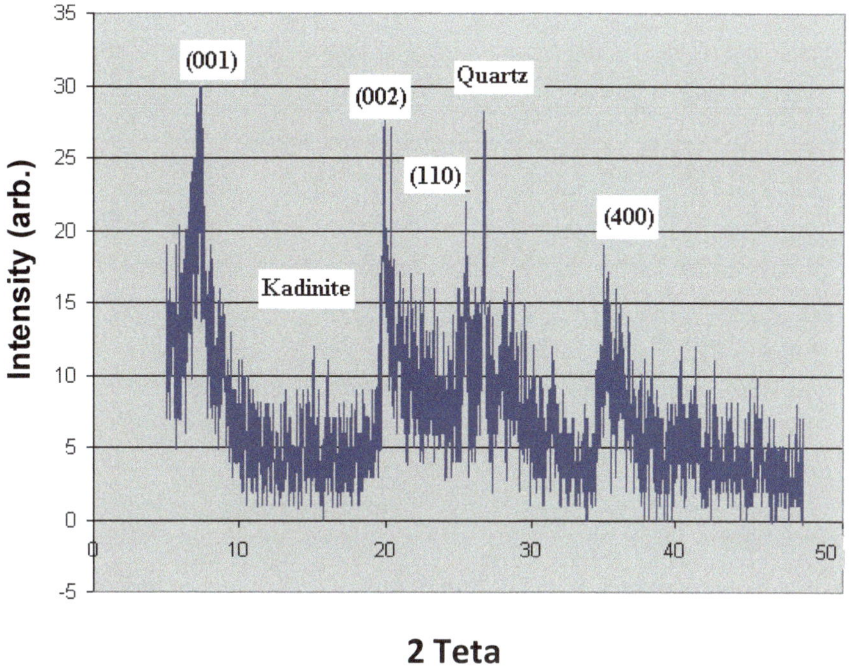

Figure 4. XRD pattern of nano clay which shows lower intensity of MMT (001) due to fifty – fifty Na nano-particles- acids in Table 1.

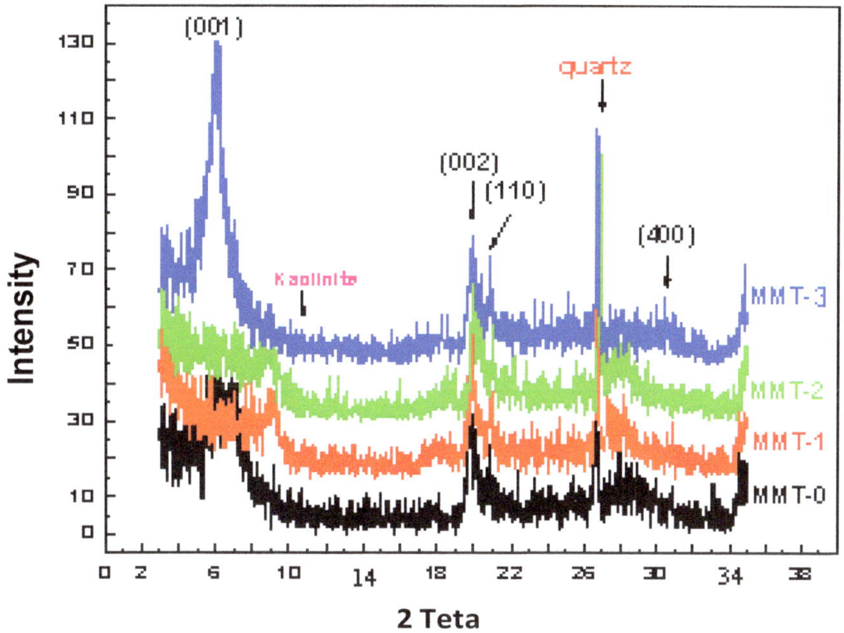

Figure 5. XRD pattern of nano clay corresponding to Table 1.

the diffusion of Na- nano- particles chains into the clay galleries. Comparing the diffraction patterns of the Figures 2 and 3, we see that the peak of sample prepared is shifted, which leads us to believe that the expansion at the clays is due to Na nano- particles. On the other hand, XRD patterns of extruded samples (Figures 2, 4 and 5) prepared via Na cations show no significant shifting issues peaks, which may indicate fully exfoliated nanostructures. From this observation, we can conclude that the exfoliation of the layered silicate that resulted from Na - nano- particles is not caused by the foaming of the Na - Si bonds when exiting the extruder.

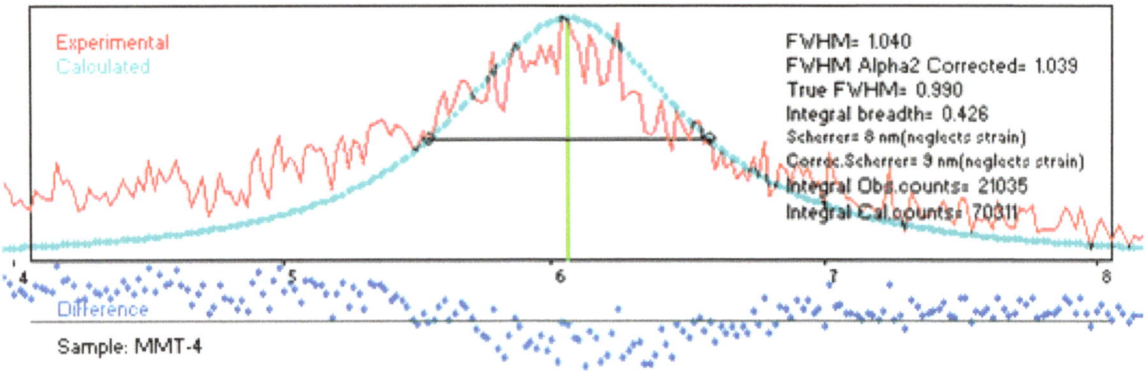

Figure 6. The size of nano - particle is 8 nm as determined with using X- powder method.

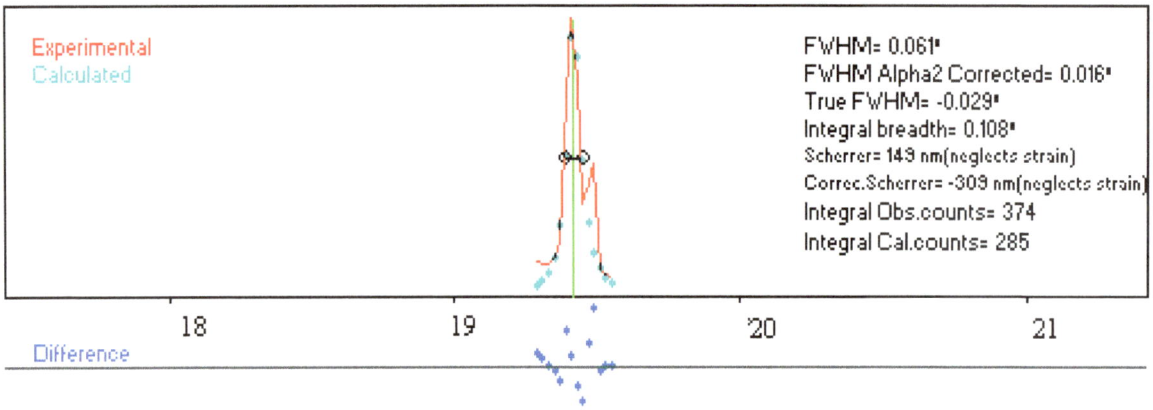

Figure 7. The size of nano- particle is 14.3 nm as determined with using X- powder method.

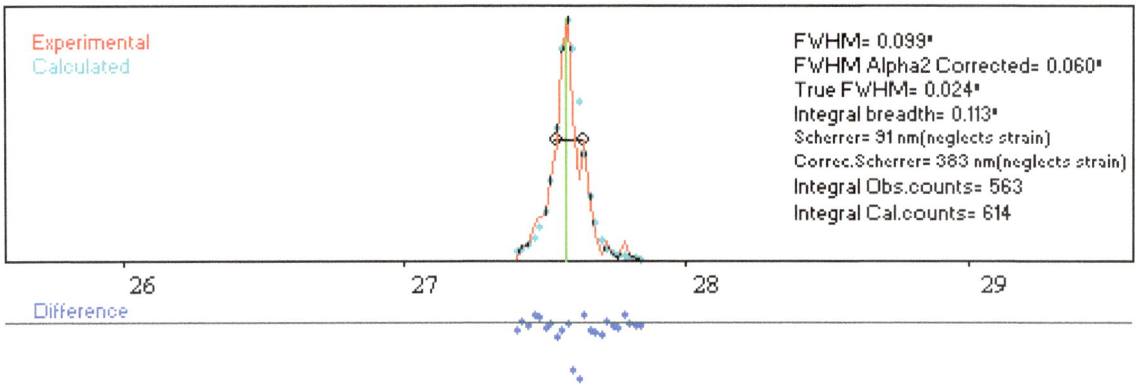

Figure 8. The size of nano- particle is 31 nm as determined with using X- powder method.

Due to the limitation of the current technique, that is, chamber design, we are not able to get exfoliated structure for clay concentration XRD pattern for the nano-composite as this concentration clearly indicates expansion of the d-spacing as given in Table 1.

The crystallite size estimated from the XRD analysis of the powder is close to 8 to 31 nm. The MMT (001) peak intensity in Figure 5 increased by adding Na content as shown with broaden peak. It can be observed that after adding Na nano-particles, kaolinite peak appeared which

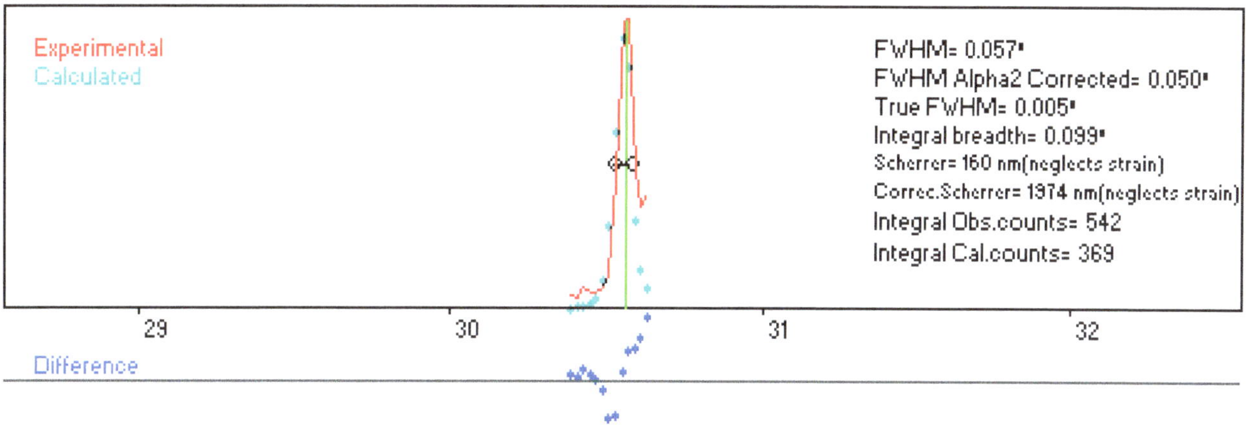

Figure 9. The size of nano- particle is 16 nm as determined with using X- powder method.

demonstrates the presence of the other crystal phase besides the other major crystal phases in the diffraction patterns and suggests only complete solubility. As shown in X - powder peaks (Figures 6 to 9), the significant broadening of the MMT peaks may be due to the reduction and refinement of the grain size and the increase in the internal strain induced from the repeated fracturing and welding process during mechanical alloying. This suggests that major structural changes and dissolution of the clay elements were almost shifted, due to the entrance of Na atoms into the lattice of the clay which changes distortion structure to expanded structure in it.

The hump peaks MMT (001), MMT (004) and quartz originate due to nonreversible relaxation by partial chain movement and clay - sodium nano - particles interaction and also for polar – modified samples due to entire movement of MMT (004) peak in the clay layers. The sample is pillared which is proven by the change in d_{100} distance from 8 nm for MMT (001) to 31 nm for MMT (002). The rearranged structure is only partly accessible to Na atoms or molecules.

Indeed, network porosity may have a key role in the nano clustering size, because they are in the same time an efficient surface and interface for absorbing water and/or the other unwanted materials (hydropholic materials), and also a suitable morphological unit for the nucleation of nano clay charges via trapping some cations to replace Mg^{2+} (Mani et al., 2005; Quang et al., 2007; Kuznetsov et al., 2010; Zhongzhong, 2008).

The difference in the miller index and the shifted crystal phase can be attributed to polarity of the structure, which causes difference in internal phase separation from clay layers. Based on the diffraction peak of MMT, the inter-layer distance change is increased as calculated with X-powder method as shown in Table 1. Nano- structural analysis of MMT reveals the different forms of MMT orientations. The characteristic peak for the MMT clay is that of the diffraction at MMT (100); such peak inter-distance is expressed by d_{100}.

Conclusion

The dispersion of the nanoclay- MMT particles is studied with XRD technique. An increased d-spacing value of the clay particles in the nanocomposites is observed with Na-nano-particles treatment.

The obtained results with XRD patterns indicate that number of intercalated layers in a single clay crystallite is increased to three nano- crystal phases of clays. Each phase can be used for two distinct areas of materials science: MMTs and ceramics due to their nano- structural properties.

REFERENCES

Bahari A, Morgen P, Li ZS , Pederson K (2006). Growth of a Stacked Silicon Nitride/Silicon Oxide Dielectric on Si (100), J. Vacuum Sci. Technol., B, 24: 2119-2123.

Bahari A, Morgen P, Li ZS (2008). Ultra thin silicon nitride films on Si 100) studied with core level photoemission, Surface Sci., 602: 2315-2324.

Bahari A, Pashayan M (2011). Nano structural properties of stainless steel for Ultra high vacuum chambers, Chem. Tech., 3: 403-407.

Bahari A, Robenhagen U , Morgen P (2005). Grown of Ultra thin Silicon Nitride on Si (111) at low temperatures, physical Review *B.*, 72: 205323-205329.

Bertoluzza A, Fagnano C, Morelli MA, Gotfardi V , Guglielmi M (1982). Raman and Infra-Red spectra on silica gel evolving towards glass, J. Non Crystalline Solids, 48: 1-17.

Hernandez – JT, Mendoza – AG (2005). Formation of NiO – SiO_2 nano composite thin films by the sol – gel method, J. Nan crystalline solids, 351: 2029-2035.

Kuznetsov DV, Lysov DV, Levina VV, Kondrat rva MN, Byzhonkov DI, Kaloshkin SD (2010). Structural special features in nano dispersed Ni – SiO_2 composite materials produces by method of chemical dispersion, Inorg. Mater. Appl. Res., 1: 57-63.

Mani GQ, Fan S, Ugbolue C, Yang Y (2005). Morphological Studies of Polypropylene–Nanoclay Composites, J. Appl. Polymer Sci., 97: 218-226.

Morgen P, Bahari A, Pederson K (2006). Functional properties of Nanostructured Material, Springer, 223: 29-257.

Quang T, Nguyen T, Baird DG (2007). An improved technique for exfoliating and dispersing nanoclay particles into polymer matrices using supercritical carbon dioxide, Polymer, 48: 6923-6933.

Ramez A, Occurrence M (1996). Mineralogy of a deposit of

Shampoo-clay in southern Iran, Appl. Clay Sci., 11: 43-45.
Sánchez LC, Beltran JJ, Osorio J, Calle AM, Barrero CA (2009). Fe-doped SnO_2 nanopowders obtained by sol–gel and mechanochemical alloying with and without thermal treatment , Phys. Astron., 185-189: 1-29.
Zhongzhong Q (2008). Preparation and xharacterzation of MMT-sillica nanocomposite: A sol-gel approach modifying clay surface, Physic *B*, 403: 3231-3238.

16

Chemoselective 3-*O*-alkylation of L-ascorbic acid under phase transfer catalysis

Shankar Ramchandra Thopate* and Santosh Rajaram Kote

Department of Chemistry, Dr. John Barnabas Post Graduate School for Biological Studies, Ahmednagar College, Ahmednagar, Maharashtra. 414001. India.

A detailed investigation for synthesis of chemo-selective 3-*O*-alkylation of L-ascorbic acid under phase transfer catalysis has been carried out. All the compounds were fully characterized by infra red (IR), NMR and microanalysis. Reaction of 5,6-*O*-isopropylidene-L-ascrobic acid with one equivalent of corresponding alkylating agent and tetrabutyl ammonium iodide as phase transfer catalyst in biphasic system (water:ethyl acetate, 1:1) afforded the 3-*O*-alkyl-5,6-*O*-isopropylidene-L-ascorbic acid derivatives (2a-f) in moderate to good yield as a sole product. Under this reaction condition formation of 2-*C*-alkylated and 2,3-di-*O*-alkylated product was not observed. Furthermore, present protocol allowed preparation of 2,3-di-*O*-alkyl and differentially protected 2,3-di-*O*-alkyl derivatives under mild reaction conditions at room temperature in a one-pot reaction.

Key words: 3-*O*-Alkylation, L-ascorbic acid, phase transfer catalysis, chemoselective alkylation.

INTRODUCTION

L-ascorbic acid, commonly known as Vitamin C, is widely distributed in aerobic organisms and is involved in several biological processes. The biological and pharmacological activity, as well as the therapeutic potential of L-ascorbic acid and its derivatives has been studied extensively (Tripathi et al., 2009). 2-*O* and 3-*O*-alkylated derivatives are known to protect against peroxidation of lipids of the bio-membrane (Kato et al., 1988; Nihro et al., 1991). In addition these derivatives have also been used as an inexpensive chiral synthon for the synthesis of a variety of natural products and pharmacologically active agents (Tripathi et al., 2009).

Many a times, while undertaking a natural product synthesis from L-ascorbic acid, selective preparation of its 2-*O* and 3-*O*-alkyl derivatives is a prerequisite and is a complex issue. The C3 OH is more reactive towards electrophiles under mild basic conditions as compaired to C2 OH group. However, attending the reaction conditions to selectively alkylate the C3 OH group is quite difficult, as the reaction is fairly sensitive to reaction conditions leading to mixture of 3-*O*-alkyl, 2-*O*-alkyl, 2,3-di-*O*-alkyl and/or 2-*C*-alkyl derivatives. This is quite evident from reports in the literature; in solvents of high dielectric constant (Jackson and Jones, 1965; Fodor et al., 1984; Poss and Belter, 1988), for example, in water 2-*C*-alkylation is favored. Similarly 2-*C*-alkylation was also observed in Michael addition (Fodor et al., 1983; Arnold et al., 1987; Sussangkarn et al., 1988; Poss and Smyth, 1987), of L-ascorbic acid to conjugated carbonyls in highly polar solvents. So far only one method is reported for 2-*O*-alkylation of L-ascorbic acid using *t*-BuOk in DMSO/THF (Olabisi and Wimalasena, 2004). There exist few reports in the literature for selective 3-*O*-alkylation of L-ascorbic acid or 5,6-*O*-isopropylidene derivative (Wimalasena and Mahindaratne, 1994; Kulkarni and Thopate, 1996; Tahir and Hindsgaul, 2000; Beifuss et al., 2000; Kulkarni and Kate, 2004; Thierry et al., 2009), using various bases and solvents. All previous methods for selective 3-*O*-alkylation of L-ascorbic acid require anhydrous reaction conditions and gives mixture of products while in highly polar solvents like water 2-*C*-alkylation is favored. Phase transfer catalysis had been used for *O* and *C*-alkylations in literature (Kim et al., 2006; Yadav and Tekale, 2010; Bender et al., 2010). As a part of ongoing program to explore the utility of L-ascorbic

*Corresponding author. E-mail: srthopate@gmail.com.

acid in organic synthesis, we required to synthesize 3-*O*-alkyl derivatives of L-ascorbic acid. It has been reported that alkylation of L-ascorbic acid in aqueous medium affords mainly 2-*C*-alkylated compound. So it is interesting to study the alkylation of L-ascorbic acid in biphasic system (for example, water: ethyl acetate, 1:1) under phase transfer conditions.

MATERIALS AND METHODS

All solvents and reagents were used as obtained from commercial source. Melting points (M.P) (uncorrected) were determined in open capillary tubes using paraffin oil bath. Infra red (IR) spectra were recorded on Perkin Elmer Model 1600 series Fourier Transform (FT) instrument. Optical rotations were recorded on JASCO DIP Digital 181 Polarimeter. ^{1}H NMR and ^{13}C NMR were recorded on Varian Mercury 300 and 75 MHz respectively in $CDCl_3$ solution and tetramethylsilane (TMS) as internal reference (□δ scale). The elemental analysis was obtained on Perkin-Elmer Series II 2400 CHNS/O elemental analyzer.

General procedure for 3-*O*-alkylation of 5,6-*O*-isopropylidene-L-ascorbic acid under phase transfer catalysis

To a magnetically stirred solution of 5,6-*O*-isopropylidene-L-ascorbic acid (1.00 g, 4.6 mmol) and tetrabutyl ammonium iodide (0.169 g, 0.4 mmol, 10 mol %) in water (5.00 ml), ethyl acetate (5.00 ml), potassium hydroxide (0.259 g, 4.6 mmol) was added (pH = 7.5 to 8.0) followed by addition of corresponding alkylating agent (4.6 mmol). Stirring was continued at room temperature for 17 to 24 h. Product was extracted in ethyl acetate (3 × 10.00 ml). Combined ethyl acetate extracts were dried over anhydrous sodium sulphate. Ethyl acetate was removed under reduced pressure and the crude product was purified on silica gel column chromatography using hexane-ethyl acetate solvent system.

General procedure for 2,3-di-*O*-alkylation of 5,6-*O*-isopropylidene-L-ascorbic acid under phase transfer catalysis

Method A

To a magnetically stirred solution of 5,6-*O*-isopropylidene-L-ascorbic acid 1 (1.00 g, 4.6 mmol) and tetrabutyl ammonium iodide (0.169 g, 0.46 mmol 10 mol %) in water (5.00 ml), ethyl acetate (5.00 ml), potassium hydroxide (0.544 g, 9.7 mmol) was added (pH = 7.5 to 8.0) followed by addition of corresponding alkylating agent (4.6 mmol). Stirring was continued at room temperature for 17 to 24 h. Product was extracted in ethyl acetate (3 × 10.00 ml). Combined ethyl acetate extracts were dried over anhydrous sodium sulphate. Ethyl acetate was removed under reduced pressure and the crude product was purified on silica gel column chromatography using hexane-ethyl acetate solvent system.

General procedure for differentially protected 2,3-di-*O*-alkyl derivatives in a one pot reaction

Method B

To a magnetically stirred solution of 5,6-*O*-isopropylidene-L-ascorbic acid 1 (1.00 g, 4.6 mmol) and tetrabutyl ammonium iodide (0.169 g, 0.4 mmol, 10 mol %) in water:ethyl acetate 1:1 (5.00 ml each), potassium hydroxide (0.259 g, 4.6 mmol) was added (pH = 7.5 to 8.0) followed by addition of corresponding alkylating agent (4.6 mmol). Stirring was continued at room temperature for 17 to 24 h (TLC check). Then potassium hydroxide (0.259 g, 4.6 mmol) was added followed by addition of next alkylating agent (4.6 mmol). Stirring was continued at room temperature for another 24 h. Product was extracted in ethyl acetate (3 × 10.00 ml). Combined ethyl acetate extracts were dried over anhydrous sodium sulphate. Ethyl acetate was removed under reduced pressure and the crude product was purified on silica gel column chromatography using hexane-ethyl acetate solvent system.

RESULTS AND DISCUSSION

Initially, the methylation of 5,6-*O*-isopropylidene-L-ascrobic acid 1 in biphasic system (water:ethyl acetate, 1:1) using one equivalent of methyl iodide as alkylating agent and tetrabutyl ammonium iodide as phase transfer catalyst (PTC) was attempted. The reaction gave the 3-*O*-methyl-5,6-*O*-isopropylidene-L-ascorbic acid 2a in moderate yield as a sole product along with small amount of starting material. No trace of 2-*C*-alkylated product as well as 2,3-di-*O*-alkylated product was detected, either by thin layer or column chromatography. In order to obtain optimum reaction condition, further, we studied the effect of temperature and solvent system on product distribution and reaction time. At higher temperature (reflux) mixture of 2,3-di-*O*-alkylated and 3-*O*-alkylated products were detected. Replacing ethyl acetate by tetrahydrofuran as a solvent with water and changing ratio of water to ethyl acetate/tetrahydrofuran from 1:1 to 1:2, 1:3, 1:4 does not alter the product formation and/or reaction time. The results of alkylation with other alkyl halides are summarized in Table 1, which show that a variety of alkylating agents including activated as well as inactivated alkyl halides gave 3-*O*-alkyl-5,6-*O*-isopropylidene-L-ascrobic derivatives as sole product in moderate to good yields.

In order to test further scope of this methodology, 2,3-di-*O*-alkylation of 5,6-*O*-isopropylidene L-ascorbic acid was attempted under the PTC conditions using 2.1 equivalent alkylating agent. The 2,3-di-*O*-alkyl-5,6-*O*-isopropyldene L-ascorbic acids 3a-c were obtained in uniformly good yields (Table 2, Entry 1-3). Further, differentially protected 2,3-di-*O*-alkyl derivatives 3d-e were obtained in good yields (Table 2, Entry 4-6) in a single pot reaction by adding two different alkylating agents in a sequential manner. All the compounds synthesized were fully characterized by IR, NMR and Microanalysis (Tables 3 and 4).

Our protocol offers number of advantages over the existing methods (Wimalasena and Mahindaratne, 1994; Kulkarni and Thopate, 1996; Tahir and Hindsgaul, 2000; Beifuss et al., 2000; Kulkarni and Kate, 2004; Thierry et al., 2009). First, it gives exclusively 3-*O*-alkyl derivatives of 5,6-*O*-isopropylidene L-ascorbic acid without formation of 2,3-di-*O*-alkyl and/or 2-*C*-alkyl derivatives which makes purification process simple. The procedure does not need anhydrous reaction condition as water is used

Table 1. Chemoselective 3-*O*-alkylation of L-Ascorbic acid under phase transfer catalysis.

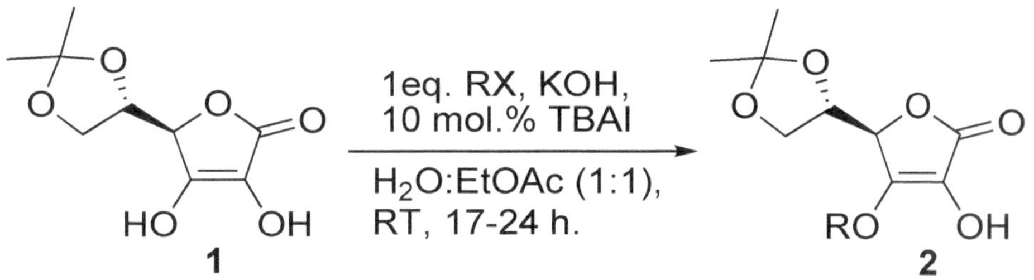

S/N	Product	R	Yield[a] (%)	M.P. (°C)
1	2a	Methyl	75	115-116
2	2b	Ethyl	65	Viscous oil
3	2c	Benzyl	75	108-110
4	2d	Allyl	70	Viscous oil
5	2e	*p*-Nitrobenzyl	72	Viscous oil
6	2f	2(Bromo-methyl) Naphthalene	77	128-130

[a] Isolated yield.

Table 2. Synthesis of 2,3-di-*O*-alkyl derivatives of 5,6-*O*-isopropylidene L-ascorbic.

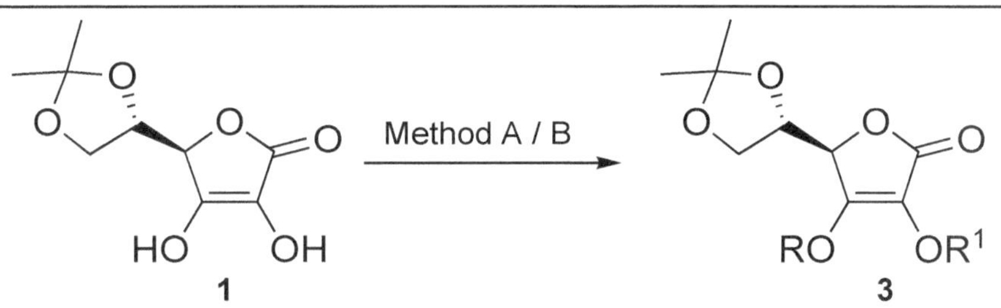

S/N	Product	R	R[1]	Method	Yield[a] (%)	M.P. (°C)
1	3a	Methyl	Methyl	A	94	100-102
2	3b	Benzyl	Benzyl	A	80	127-128
3	3c	Allyl	Allyl	A	85	Viscous oil
4	3d	Methyl	Benzyl	B	65	88-89
5	3e	Methyl	Allyl	B	65	Viscous oil
6	3f	Benzyl	Methyl	B	61	97-99

a. Isolated yield. Method A: 2.1eq. RX, KOH, 10 mol. % TBAI, H_2O:EtOAc (1:1), RT, 17-24 h. Method B: a) 1.1eq. RX, KOH, 10 mol. % TBAI, H_2O:EtOAc (1:1), RT, 17-24 h. b) 1.1eq. R^1X, KOH, 10 mol. % TBAI, H_2O:EtOAc (1:1), RT, 17-24 h.

as solvent. Finally, reaction proceeds under mild reaction conditions at room temperature.

Conclusion

In conclusion, we have developed a user friendly methodology for preparation of 3-*O*-alkyl derivatives of 5,6-*O*-isopropylidene L-ascorbic acid using phase transfer catalyst in biphasic system (water:ethyl acetate 1:1) at room temperature. This procedure permits access to these biologically and synthetically important derivatives in good yield and with good regio- and chemo- selectivity. Furthermore, present protocol can be used for preparation of 2,3-di-*O*-alkyl and differentially protected 2,3-di-*O*-alkyl derivatives at room temperature in a one-pot reaction.

ACKNOWLEDGEMENTS

Authors gratefully acknowledged the Department of Science and Technology (DST), New Delhi (No: SR/S1/OC-19/2007) for their financial support and also

Table 3. Optical rotation, IR spectral data, molecular formula and elemental analysis of synthesized compounds.

S/N	Product	$[\alpha]^{25}_D$	IR (cm^{-1})	Molecular formula	Elemental analysis	
					Required	Found
1	2a	+ 11.3 (c 1 $CHCl_3$)	1765, 1700	$C_{10}H_{14}O_6$	C, 52.17; H, 6.13	C, 52.19; H, 6.42
2	2b	+ 36.70 (c 0.2 $CHCl_3$)	1765, 1700	$C_{11}H_{16}O_6$	C, 54.09; H, 6.60	C, 54.15; H, 6.72
3	2c	+ 36.3 (c 1 methanol)	1759, 1702	$C_{16}H_{18}O_6$	C, 62.74; H, 5.92	C, 62.45; H, 5.91
4	2d	+ 33.62 (c 2 methanol)	1763, 1686	$C_{12}H_{16}O_6$	C, 56.24; H, 6.29	C, 56.12; H, 6.41
5	2e	+ 36.70 (c 0.6 $CHCl_3$)	1760, 1695	$C_{16}H_{17}NO_8$	C, 54.70; H, 4.88; N, 3.99	C, 54.65; H, 4.68; N, 3.98
6	2f	+ 11.66 (c 0.3 $CHCl_3$)	1755, 1699	$C_{20}H_{20}O_6$	C, 67.41; H, 5.66	C, 67.35 H, 5.72
7	3a	+ 11.34 (c 0.7 $CHCl_3$)	1760, 1679	$C_{11}H_{16}O_6$	C, 54.09; H, 6.60	C, 54.30 H, 6.68
8	3b	+ 23.4 (c 2 $CHCl_3$)	1750, 1676	$C_{23}H_{24}O_6$	C, 69.70; H, 6.10	C, 69.54 H, 6.05
9	3c	+ 47.52 (c 2 methanol)	1763, 1677	$C_{15}H_{20}O_6$	C, 60.80; H, 6.80	C, 60.72 H, 6.91
10	3d	+ 39.33 (c 2 methanol)	1745, 1675	$C_{17}H_{20}O_8$	C, 63.74; H, 6.29	C, 63.87 H, 6.35
11	3e	+ 16.70 (c 2 methanol)	1770, 1670	$C_{13}H_{18}O_6$	C, 57.77; H, 6.71	C, 57.54; H, 6.95
12	3f	+ 37.23 (c 2 methanol)	1754, 1681	$C_{17}H_{20}O_8$	C, 63.74; H, 6.29	C, 63.91; H, 6.39

Table 4. ^{1}H NMR and ^{13}C NMR chemical shifts of synthesized compounds.

S/No.	Product	^{1}H NMR (300 MHz, $CDCl_3$)	^{13}C NMR (75 MHz, $CDCl_3$)
1	2a	1.35 (s, 3 H); 1.38 (s, 3 H); 3.98-4.31 (m, 3 H); 4.45 (s, 3 H); 4.60 (d, 1 H, *J* = 3 Hz); 5.90 (bs, 1H)	25.45, 25.70, 59.78, 65.30, 74.15, 75.60, 110.30, 119.71, 150.35, 171.49
2	2b	1.37 (m, 6 H); 1.40 (s, 3 H); 4.03 (m, 1 H); 4.14 (m, 1 H); 4.25 (m, 1 H); 4.54 (m, 3 H); 6.1 (s, 1 H)	15.30, 25.51, 25.86, 65.27, 68.04, 74.28, 75.63, 110.24, 118.88, 148.79, 171.29
3	2c	1.36 (s, 3 H); 1.41 (s, 3 H); 4.11-4.42 (m, 3 H); 4.71 (d, 1 H, *J* = 3.8 Hz); 5.85 (s, 2 H); 7.65 (s, 5 H)	25.52, 25.89, 65.30, 73.45, 74.27, 75.76, 110.23, 121.91, 128.13, 128.61, 135.86, 149.30, 171.51
4	2d	1.37(s, 3 H); 1.43 (s, 3 H); 4.14-4.51 (m, 3 H); 4.74 (d, 1 H, *J* = 3.8 Hz); 5.17 (d, 2H, *J* = 5.1 Hz); 5.57 (m, 2 H), 6.2 (m, 1 H)	25.60, 25.90, 65.31, 72.30, 74.29, 75.68, 110.32, 119.15, 119.25, 132.23, 148.40, 171.35
5	2e	1.35 (s, 3 H); 1.36 (s, 3 H); 4.02-4.19 (m, 2 H); 4.32-4.39 (m 1 H); 6.64 (d, 1 H, *J* = 2.7 Hz); 5.62 (AB q, 2 H, *J* = 13.1 Hz); 7.58 (d, 2 H, *J* = 8.7 Hz), 8.24 (d, 2 H, *J* = 8.7 Hz)	25.43, 25.72, 65.18, 71.76, 73.60, 75.40, 110.39, 119.88, 123.77, 128.09, 142.89, 147.84, 147.95, 170.88
6	2f	1.32 (s, 3 H); 1.37 (s, 3H); 3.95-4.15 (m, 2 H); 4.26 (m, 1 H); 4.56 (d, 1 H, *J* = 3.6 Hz); 5.65 (s, 2H); 7.44-7.51 (m, 3 H), 7.77-7.86 (m, 4 H)	25.46, 25.81, 64.60, 65.25, 73.57, 73.72, 74.10, 110.25, 111.02, 125.46, 126.36, 126.46, 127.33, 127.49, 127.61, 127.67, 127.83, 133.01, 133.25, 171.06
7	3a	1.36 (s, 3 H); 1.39 (s, 3 H); 3.85 (s, 3 H); 4.01-4.13 (m, 2 H); 4.15 (s, 3 H); 4.25-4.31 (m, 1 H); 4.51 (d, 1 H, *J* = 3 Hz)	25.46, 25.73, 59.39, 65.14, 73.77, 74.39, 110.28, 123.13, 156.66, 168.86
8	3b	1.36 (s, 3 H); 1.40 (s, 3 H); 4.07 (m, 2 H); 4.23 (m, 1 H); 4.52 (d, 1 H, *J* = 3.0 Hz); 5.04-5.21 (m, 4 H); 7.18-7.40 (m, 10 H)	25.59, 25.85, 65.20, 73.48, 73.76, 73.87, 74.64, 110.24, 121.11, 127.73, 128.28, 128.55, 128.60, 128.63, 129.09, 130.08, 135.29, 135.87, 156.59, 169.14
9	3c	1.37(s, 3 H); 1.42 (s, 3 H); 4.14-4.48 (m, 3 H); 4.71 (d, 1 H, *J* = 2.5 Hz); 4.88 (d, 2 H, *J* = 6.4 Hz); 5.17 (d, 2 H, *J* = 5.1 Hz), 5.57(m, 4 H); 6.17 (m, 2 H)	25.62, 25.95, 65.36, 72.44, 72.65, 74.08, 74.68, 110.43, 118.97, 119.25, 121.54, 132.89, 155.90, 168.96
10	3d	1.41(s, 3H); 1.45 (s, 3 H); 3.90 (s, 3 H); 4.05 (m, 2 H); 4.25 (m, 1 H); 4.50 (d, 1 H, *J* = 2.5 Hz); 5.15 (q, 2 H, *J* = 11 Hz); 7.40 (s, 5 H)	25.45, 25.78, 59.76, 65.21, 73.26, 73.85, 74.55, 110.09, 123.06, 127.70, 128.80, 135.28, 155.46, 168.76
11	3e	1.30(s, 3 H); 1.38 (s, 3 H); 4.01-4.28 (m, 3 H); 4.35 (s, 3 H); 4.52 (d, 1 H, *J* = 3.2 Hz); 4.64 (d, 2 H, *J* = 6.4 Hz); 5.48 (m, 2 H); 6.14 (m, 1 H)	25.00, 25.36, 59.76, 65.21, 72.26, 73.76, 74.15, 109.79, 118.76, 120.98, 132.47, 157.12, 168.64
12	3f	1.36 (s, 3 H); 1.40 (s, 3 H); 3.80 (s, 3 H); 4.08-4.40 (m, 3 H); 4.56 (d, 1 H, *J* = 3.2 Hz); 5.48 (s, 2 H); 7.44 (s, 5 H)	25.45, 25.94, 59.75, 65.28, 73.46, 74.35, 75.68, 110.28, 119.57, 128.50, 128.65, 128.76, 135.7, 148.68, 171.27;

Professor Mukund G. Kulkarni (Department of Chemistry, University of Pune) and Dr. R. J. Barnabas (Ahmednagar College, Ahmednagar) for his helpful discussion and suggestions.

REFERENCES

Arnold R, Fodor G, George C, Karle I (1987). The Michael adduct of L-ascorbic acid to methyl vinyl ketone: Its remarkable self-condensation and other reactions. Can. J. Chem., 65: 131-136.

Beifuss U, Kunz O, Voss G (2000). Regioselective Synthesis of 3-*O*-Alkyl Ethers of Ascorbic Acid without Protecting Groups in a Single Step. Tetrahedron, 56(3): 357-361.

Bender J, Jepkens D, Hüsken H (2010). Ionic Liquids as Phase-Transfer Catalyst: Etherification Reaction of 1-Octanol with 1-Chlorobutane. Org. Process Res. Dev., 14(3): 716-721.

Fodor G, Arnold R, Mohacsi T, Karle I, Anderson JF (1983). A new role for l-ascorbic acid: Michael donor to α,β-unsaturated carbonyl compounds. Tetrahedron, 39(13): 2137-2145.

Fodor G, Sussangkarn K, Mathelier H, Arnold R, Karle I, George C (1984). New reaction of L-ascorbic acid: Unusual molecular complexes of the product. J. Org. Chem., 49(26): 5064-5069.

Jackson KGA, Jones JKN (1965).The *C*- and *O*-benzylation of L-ascorbic acid. Can. J. Chem., 43: 450-457.

Kato K, Terao S, Shimamoto N, Hirata M (1988). Studies on scavengers of active oxygen species. 1. Synthesis and biological activity of 2-O-alkylascorbic acids. J. Med. Chem., 31(4): 793-798.

Kim TS, Lee YJ, Jeong BS, Park HG, Jew SS (2006). Enantioselective Synthesis of (*R*)- and (*S*)-α-Alkylcysteines via Phase-Transfer Catalytic Alkylation. J. Org. Chem., 71(21): 8276-8278.

Kulkarni MG, Kate SD (2004). Chemistry of L-Ascorbic Acid. IV. Regioselective O-Alkylation of 5,6-*O*-Isopropylidene-L-ascorbic Acid Using CsF. Synth. Commun., 34(13): 2365-2370.

Kulkarni MG, Thopate SR (1996). Chemoselective alkylation of L-ascorbic acid. Tetrahedron, 52(4): 1293-1302.

Nihro Y, Miyataka H, Sudo T, Matsumoto H, Satoh T (1991). 3-O-alkylascorbic acids as free-radical quenchers: synthesis and inhibitory effect on lipid peroxidation. J. Med. Chem., 34(7): 2152-2157.

Olabisi AO, Wimalasena K (2004). Rational Approach to Selective and Direct 2-*O*-Alkylation of 5,6-*O*-Isopropylidine-L-ascorbic Acid J. Org. Chem., 69(21): 7026-7032.

Poss AJ, Belter RK (1988). The C-Alkylation of Ascorbic Acid. Synth. Commun., 18(4): 417-423.

Poss AJ, Smyth MS (1987). The total synthesis of piptosidin. Tetrahedron Lett., 28(45): 5469-5472.

Sussangkarn K, Fodor G, Karle I, George C (1988). Ascorbic Acid as a Michael Donor: Part II. Reaction with Alicyclic Enones. Tetrahedron, 44(23): 7047-7054.

Tahir H, Hindsgaul O (2000). Regio- and Chemoselective Alkylation of L-Ascorbic Acid under Mitsunobu Conditions. J. Org. Chem., 65(3): 911-913.

Thierry M, Paul H. Bang L (2009). Regio- and Chemoselective One-Step 3-O-Alkylenation of Unprotected Ascorbic Acid Using ω-Iodoalkanols. Synlett., 2: 217-220.

Tripathi RP, Singh B, Bisht SS, Pandey J (2009). L-Ascorbic acid in Organic Synthesis: An Overview. Curr. Org. Chem., 13: 99-122.

Wimalasena K, Mahindaratne MPD (1994). Chemistry of L-Ascorbic Acid: Regioselective and Stereocontrolled 2-C- and 3-C-Allylation via Thermal Claisen Rearrangement J. Org. Chem., 59(12): 3427-3432.

Yadav GD, Tekale SP (2010). Selective *O*-Alkylation of 2-Naphthol using Phosphonium-Based Ionic Liquid as the Phase Transfer Catalyst. Org. Process Res. Dev., 14(3): 722-727.

Complexes of Nile Red and β- or γ-cyclodextrin: A semiempirical-density functional theory-molecular dynamics route

Ruth Hojvat[1,2], Dora A. Barbiric[3] and Eduardo A. Castro[4]*

[1]Facultad de Ciencias Exactas y Naturales, Universidad de Belgrano, Zabala 1837 CABA (CP 1426DQG) Argentina.
[2]Escuela de Tecnología y Escuela de Ciencias Agrarias, Universidad Nacional del Noroeste de la Provincia de Buenos Aires, Sede Junin, R. Sáenz Peña 456, Junín B6000, Pcia de Buenos Aires, Argentina.
[3]Facultad de Ingeniería, Universidad de Buenos Aires, Av. Paseo Colón 850, C1063ACV, Buenos Aires, Argentina.
[4]Instituto de Investigaciones Fisicoquímicas Teóricas y Aplicadas, Universidad Nacional de La Plata, Suc. 4, C.C. 16, 1900La Plata, Buenos Aires, Argentina.

The association of Nile Red (NR) with β-, and γ-cyclodextrins (CDs) was analyzed by semiempirical parametric method 3 (PM3), density functional theory-Becke, three-parameter, Lee-Yang-Parr hybrid functional (DFT-B3LYP/6-31G), and molecular dynamics (MD) and Langevin dynamics (LD) calculations, including annealing and molecular mechanics (MM) geometry optimization. Our results suggest that inclusion complexes of NR and both CD are possible but occlusion complexes are more favourable. Differences in the intermolecular H-bonding yielded by the DFT-level of treatment could explain the distinct behaviour regarding deactivation to the ground state of NR in presence of β-, and γ-CD in fluorescence experiments. DFT optimization of NR in gaseous state rendered lower twisting angles and larger pyramidalization of the amino group than reported elsewhere.

Key words: Molecular modeling, cyclodextrins, Nile Red complexes.

INTRODUCTION

Nile Red (NR), that is, 9-diethylamino-5*H*-benzo[α]-phenoxazine-5-one, is a laser dye, highly sensitive to the polarity of the environment. In 1985, it was shown that NR will fluoresce in a nonpolar environment and could serve as a probe to detect nonpolar lipids in cells (Greenspan and Fowler, 1985). NR permeates all structures within a cell, but the characteristic yellow fluorescence (approximately 575 nm) only occurs when the dye is in a nonpolar environment, primarily neutral storage lipid droplets. Thanks to the photochemical stability and strong fluorescence nature of NR, it has been used as probe in many different applications in materials science and in biology, when the polarity of the medium needs to be explored (Deye et al., 1990). NR has thus been applied to determine the polarity of conventional solvents and binary mixtures (Dutta et al., 1996; Uusi-Penttilä et al., 1997; Kipkemboi and Easteal, 1994) and of supercritical fluids (Berger and Deye, 1990), and used to probe the microenvironment of polymers (Dutta et al., 1996), xerogels (Matsui and Nozawa, 1997), liquid crystals (Choi et al., 1997), zeolites (Meinershagen and Bein, 1999; Uppili et al., 2000), micelles (Datta et al., 1997; Maiti et al., 1997) and lipid bilayers (Freeman et al., 2001). It has also been incorporated, as the sensing element, in fiber-optic devices capable of identifying and quantifying a range of organic vapors (White et al., 1996; Johnson et al., 1997).

*Corresponding author. E-mail: eacast@gmail.com.

Abbreviations: B3LYP, Becke, three-parameter, Lee-Yang-Parr hybrid functional; **CAM-B3LYP,** Coulomb-attenuating method-B3LYP; **CD,** cyclodextrin; **DFT,** density functional theory; **H-bond,** hydrogen bond; **HOMO,** highest occupied molecular orbital; **LD,** Langevin dynamics; **LUMO,** lowest unoccupied molecular orbital; **MD,** molecular dynamics; **MM,** molecular mechanics; **NR,** Nile Red; **PICT,** planar intramolecular charge transfer; **PM3,** semiempirical parametric method 3; **TICT,** twisted intramolecular charge transfer.

On the other hand, cyclodextrins (CDs) are cyclic oligosaccharide produced by degradation of the amylose fraction of starch by glucosyltransferases, as one or several turns of the amylose helix are hydrolyzed off and their ends are joined together. A family of macrocycles with different numbers of glucose units are thus produced (Szejtli, 1988; Duchêne, 1987), but the most abundant are α-, β-, and γ-CD with six, seven and eight glucopyranose units, respectively. These doughnut shaped molecules have been investigated with the use of spectroscopic, kinetic and crystallographic methods (Szejtli, 1988; Duchêne, 1987; Saenger, 1980, 1984; Harata, 1991; Saenger et al., 1998). They are frequently characterized as truncated cones, with the secondary hydroxyl groups of the glucopyranose units situated on the wide edge of the cone, and the primary hydroxyls placed on the narrow edge or rim. The cavity is lined by the hydrogen atoms and the glycosidic oxygen bridges (Szejtli, 1998). In an aqueous solution, the slightly apolar CD cavity is occupied by high enthalpy water molecules. These water molecules can be readily substituted by less polar guest molecules. Most frequently the host:guest ratio is 1:1, but other stoichiometries are also possible.

NR is insoluble in water but freely soluble in organic solvents and it forms association complexes with β- and γ-CD which are soluble in water. According to experimental results (Srivatsavoy, 1999), the presence of multiple inclusion complexes of NR with β- and γ-CD was inferred, but with clear differences between the two CDs. For example, a much longer fluorescence lifetime in β-CD complexes (1.2 ns for the major component) as compared to γ-CD complexes (40 ps for the major component) was observed. This difference was ascribed to the effect of inclusion on the twisted intramolecular charge transfer (TICT) decay process in NR: its effect is negligible as regards β-CD, but substantial when involving γ-CD (Datta et al., 1997). The non-radiative deactivation of NR from its singlet excited state involves the electron donor diethylamino group and the electron acceptor carbonyl group attached to the same aromatic ring; in polar solvents, the flexible dialkylamino group could undergo twisting, which makes the donor orbital perpendicular to the acceptor orbital. The NR fluorescence lifetime reduction could be explained, if inclusion into γ-CD would somehow result in a considerable enhancement of the TICT decay process in NR.

However, there is literature laying stress on the possibility that in the case of NR, the TICT process does not occur. Cser et al. (2002), based on the fluorescent properties of NR in different aprotic and protic solvents, concluded that the fluorescence lifetime of NR is not sensitive to dielectric solvent–solute interactions but markedly decreases with the increase of the hydrogen bond (H-bond) donating ability in alcohols because vibrations associated with H-bonding are involved in the radiationless deactivation to the ground state process. Moreover, the negligible viscosity effect that they observed would indicate that twisting of the diethylamino moiety of NR does not play a significant role in the dissipation of the excitation energy. This appears coherent with the results reported by Dias et al. (1999, 2006), whose theoretical studies of NR by *ab initio* and semiempirical methods showed that NR is planar in the ground state with a high barrier to rotation of the diethylamino group by 90° and the charge transfer decreases after the twisting in contrast to the TICT predictions. Kawski et al. (2009) reached the same conclusion, after having analyzed the solvatochromic (Kawski et al., 2008) and thermochromic (Kawski et al., 2009) effects on absorption and fluorescence spectra of NR in several solvents; the difference between the dipole moments corresponding to the ground and excited states of NR is too small for a TICT state that should show nearly full charge transfer. Recent results (Guido et al., 2010) obtained by applying a time-dependent density functional theory (DFT) approach and taking into account bulk solvent effects appear to support this interpretation. The authors investigated the absorption and fluorescence spectra of NR by assessing both the planar intramolecular charge transfer (PICT) and TICT mechanisms. They used a panel of exchange correlation function including both global and range-separated hybrids, refined solvent models and the simulation of vibronic couplings. Accurate Coulomb-attenuating method-Becke, three-parameter, Lee-Yang-Parr hybrid functional (CAM-B3LYP) (Yanai et al., 2004) results indicated that the two peaks observed in fluorescence spectra of NR in apolar solvents, often interpreted as dual fluorescence, are in fact the result of a strongly active vibronic coupling. This supports the PICT, rather than the TICT process.

The geometry and mode of inclusion of NR guest into the β- and γ-CD hosts was studied by Wagner et al. (2003), who used ultraviolet-visible (UV-Vis) absorption, steady state fluorescence and electrospray mass ionization spectroscopies, as well as molecular modeling. They suggested an association complex involving capping of the larger rim of the γ-CD cavity by NR, as opposed to an inclusion complex with the NR guest within the β-CD cavity. The same was proposed by Hazra et al. (2004), who reported about the effect of both β- and γ-CDs on the TICT in NR, after having performed steady state and time resolved fluorescence spectroscopy studies. They observed a much higher retardation of non-radiative rate for TICT of NR in the presence of β- than γ-CD. The little retardation in presence of γ-CD was attributed to the formation of H-bonds between the -OH groups of γ-CD with NR molecules, also verified by the large increase in rotational relaxation time and quantum yield of NR in γ-CD system on addition of urea. The formation of 1:1 or 1:2 capped complexes between NR and γ-CD was finally proposed.

Given the different results and especially interpretations

as to the possible structures of NR:CD complexes, the purpose of this work was to inquire, from the theoretical standpoint, about the association of NR with β- and γ-CD, as well as the stabilization and features of the resulting structures. Our analysis comprehends from molecular mechanics (MM) to molecular dynamics (MD), passing through semiempirical calculations and DFT. We intend to offer some rationalization for the spectroscopic behaviour that different authors reported and the structures of complexes that we have obtained.

METHOD

We have modeled *in vacuo* inclusion and occlusion complexes (1:1 stoichiometry only) by firstly submitting the system to MM+ method and then to a semiempirical parametric method 3 (PM3) optimization, both contained in the Hyperchem Pro 6.0 package for Windows. PM3 program was chosen because it was asserted to model H-bonding at best (Li et al., 2000; Zheng and Merz, 1992; Morley et al., 1997; Morpurgo et al., 1998) and was later even improved (McNamara and Hillier, 2007; Korth, 2010). To perform our calculations, we used structures of β-CD, γ-CD and NR earlier optimized by PM3 method. The guest NR was located either facing the primary or the secondary rim of the CD. From the set of association structures that were generated, selected complexes (the three NR:β-CD complexes with lowest, similar energies; same criterion for NR:γ-CD complexes) were later submitted to DFT calculations. Some NR:γ-CD inclusion complexes were submitted to dynamics simulations as well, in order to assess their viability. A more detailed description of the procedures are discussed.

Inclusion complexes

Simulations of the association processes were initiated by locating the guest, NR, with its longitudinal molecular axis perpendicular to the cyclic rim of the CD, at ~2.5 Å from the center of the rim. Additionally, NR systematically presented either the amino group or the opposite extreme ring oriented towards the cyclic rim of the CD. Due to the C1 symmetry of the PM3-optimized structures of both CDs, several starting orientations of NR were tried by rotating the latter about its longitudinal axis. Each time the system was relaxed and it evolved via MM treatment, towards a complex structure, each resulting complex underwent a PM3 optimization.

Previous reports referred to the possibility that multiple types of complexes may form with both types of CDs (Srivatsavoy, 1999). We therefore chose a set of the most stable complexes among those yielded by the procedure just described. These complexes were structurally distinct but with very similar PM3 energies ($\Delta E<0.04\%$, in kcal/mol). They were three NR:β-CD complexes and three NR:γ-CD complexes. We finally applied the DFT-B3LYP/6-31G calculation to them as implemented in the Gaussian 03 program (2003).

Occlusion complexes

We followed a similar procedure involving first MM+ and then PM3 calculations, but this time we started by locating the guest with its molecular plane parallel either to the primary or the secondary circular rim of the β-CD and γ-CD, and over the cavity opening. According to bibliography (Wagner et al., 2003; Hazra et al., 2004), after spectroscopic studies, it was suggested that occlusion complexes (that is, NR capping the CD cavity opening) are more likely in the case of γ-CD. Therefore, we submitted the four more stable occlusion structures involving β-CD and same number of complexes involving γ-CD that we obtained in this stage, to subsequent DFT-B3LYP/6-31G calculations. In the case of β-CD, two out of four complexes had the guest occluding the secondary rim. In the case of γ-CD, one out of four showed NR occluding the secondary rim.

Dynamics

It was suggested (Wagner et al., 2003) that γ-CD and NR would probably form unstable inclusion complexes. Given that we did obtain NR:γ-CD inclusion structures, we chose to test further the stability of the three NR:γ-CD complexes previously mentioned. We submitted them to dynamics simulations combined with MM+, as implemented in the HyperChem package. Two situations were considered: MD *in vacuo* and Langevin dynamics LD in water (friction coefficient 5.691 ps^{-1}), both with annealing or cooling time. The simulation temperature was 500K and the heating rate 0.1 to 500 K in 10 ps. Annealing rate was set 500 to 300K in 200 ps. Additional parameter settings are shown in Table 1; application of the lower values will be specified in due course.

Nile Red

PM3 calculations rendered conformations of NR such that the amino-group presented each ethyl pointing to opposite sides of the molecular plane in a twisted conformation. Instead, DFT optimizations often yielded both extreme methyls on the same side of the molecular plane and almost perpendicular to it, approaching a planar conformation. So, we kept account of both the twisting (θ) and the pyramidalization (δ) angles (Guido et al., 2010) of NR in the complexes (Figure 1). The θ angle is the dihedral C_A-C_A-N-C, and the δ angle is defined as the deviation from planarity of the three atoms bound to the amino-N atom, that is, (180° - |improper(C_A-N-C-C´|). Correspondingly, the dihedral θ´: $C´_A$-C_A-N-C´ for the other ethyl is defined.

RESULTS

Inclusion calculations

Semiempirical

The structures described are shown in Figures 2 to 7, left side: Figures 2 to 4 refer to NR:β-CD structures, Figures 5 to 7 shows NR:γ-CD structures.

The three complexes NR:β-CD # selected after the PM3 treatment presented the following features (later modified by DFT calculation): the amino group of NR showed a shallow insertion in the CD cavity. In structures 1 and 3 (for labels, Table 2, column 1), the guest appeared emerging through the secondary rim of the cycle; in structure 2, through the primary rim. In the latter structure the amino-N showed hydrogen (H) interaction with a primary OH group of β-CD.

The three complexes NR:γ-CD selected after the PM3 treatment presented the following features (later modified by DFT calculation): in structures 4 and 6, the guest entered deeply the γ-CD cavity -by the side of its amino

Table 1. Parameter settings for MD and LD.

Parameter (ps)	MD *in vacuo*	LD (water)
Run time	200 or 100	200
Time step	10^{-3}	10^{-3} or 10^{-4}

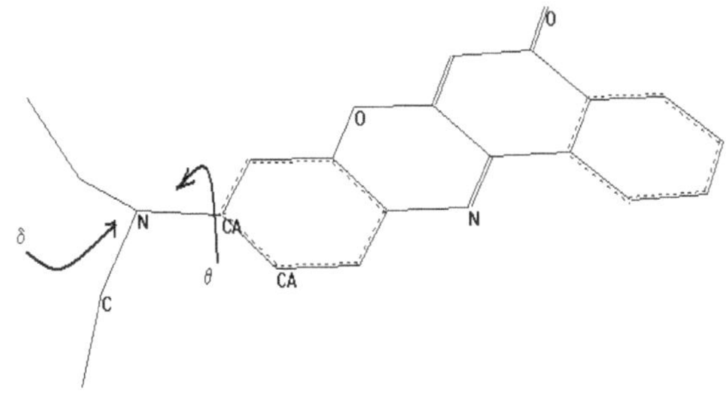

Figure 1. Nile Red and the twisting (θ) and pyramidalization (δ) angles.

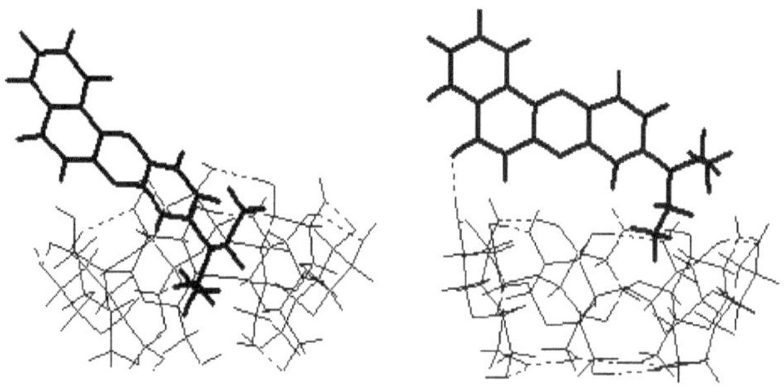

Figure 2. Complex NR:β-CD•1 according to PM3 (left) and DFT optimization (right); CD secondary rim up.

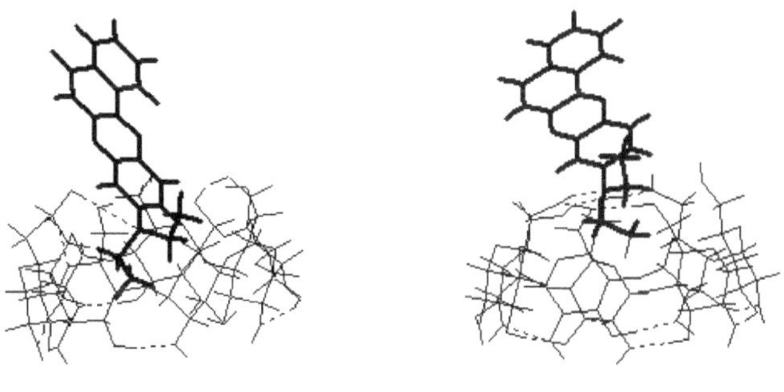

Figure 3. Complex NR:β-CD•2 according to PM3 (left) and DFT optimization (right); CD primary rim up.

group through the secondary rim (in 4) and the primary rim (in 6) of the cycle. Structure 6 showed H interaction between a primary OH group of γ-CD and the O-atom of the oxazinic heterocycle of NR. In complex 5, NR entered the cavity by the side of its condensed rings and showed deep insertion, too.

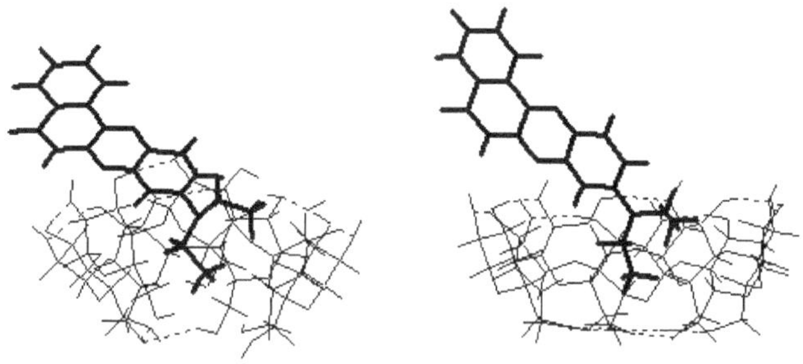

Figure 4. complex NR:β-CD•3 according to PM3 (left) and DFT optimization (right); CD secondary rim up.

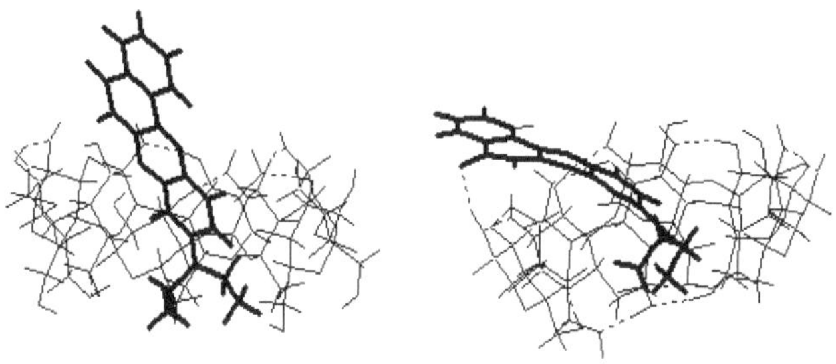

Figure 5. Complex NR:γ-CD•4 according to PM3 (left) and DFT optimization (right); CD secondary rim up.

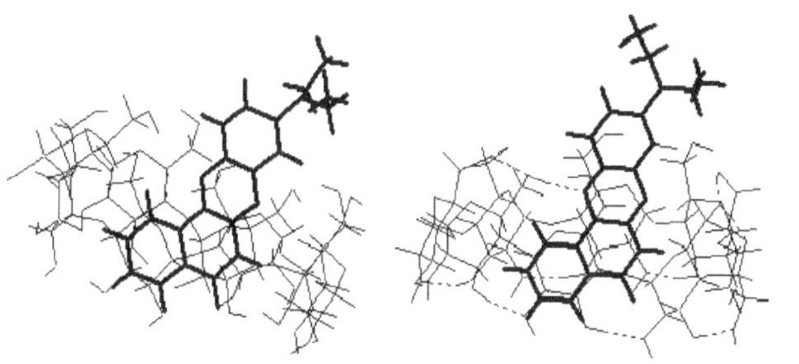

Figure 6. Complex NR:γ-CD 5 according to PM3 (left) and DFT optimization (right); CD primary rim up.

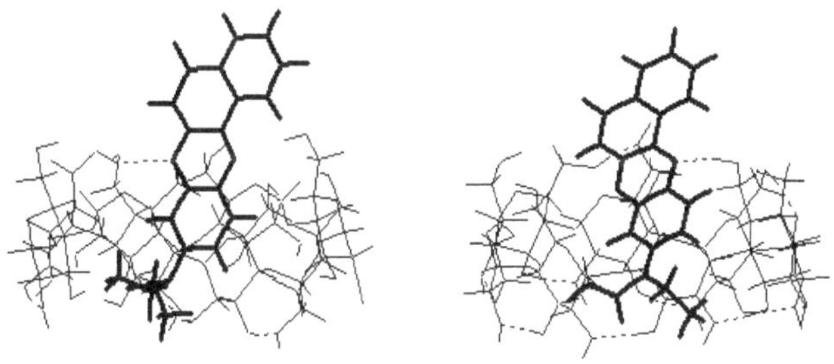

Figure 7. NR:γ-CD 6 complex according to PM3 (left) and DFT optimization (right); CD primary rim up.

The PM3 converged binding energies are shown in Table 2. The twisting θ and θ´ angles of NR, as well as the pyramidalization δ angle are shown in Table 3.

Table 2. Semiempirical PM3 binding energies and DFT energies (minima in *italics*) of inclusion complexes. Intermolecular H-bonds and interatomic distances.

NR in complex #	E(PM3) (kcal/mol)	Intermolecular H-bonding and X....H distance (Å)	E(DFT) (a.u.)	Intermolecular H-bonding and X....H distance (Å)
β-CD 1	-19072	-	-5306.60366283	C=O....HO 1.737
β-CD 2	-19069	Amino-N....HO 3.18	-5306.58017182	Amino-N....HO 1.849
β-CD 3	-19066	-	-5306.59130417	---
γ-CD 4	-21124	-	-5917.13957384	C=O....HO 1.983
γ-CD 5	-21122	-	-5917.14749763	Oxazine-N....HO 1.804
γ-CD 6	-21121	Oxazine-O....HO 2.64	-5917.13913407	Oxazine-N....HO 1.778

Table 3. Twisting (θ, θ') and pyramidalization (δ) angles (degrees) of the NR amino group in the inclusion complexes after PM3 (left column)/DFT (right column) calculations.

NR:CD complex #	θ:CA-CA-N-C		θ′:C′A-CA-N-C′		Improper angle CA-N-C-C′		δ	
β-CD 1	116.17	4.10	64.74	0.14	131	176	49	4
β-CD 2	36.94	27.72	85.45	71.20	-133	-136	47	44
β-CD 3	68.09	13.78	23.49	5.35	137	-172	43	8
γ-CD 4	116.21	3.17	149.85	12.09	-144	170	36	10
γ-CD 5	155.56	0.92	108.69	-4.17	133	-175	47	5
γ-CD 6	133.85	-10.02	89.05	-7.27	136	177	44	3

DFT

DFT treatment of the three complexes involving β-CD in Table 2, rendered NR:β-CD 1 the most stable, as shown in Table 2. However, the guest changed from a shallow inclusion to a rather occlusive location (Figure 2, right side). An intermolecular host-guest H-bond formed between carbonyl-O of NR and a secondary OH group of the cycle. The remote methyls of the amino group appeared both on the same side of the molecular plane of NR and the δ pyramidalization angle decreased notably (Table 3), towards a planar conformation. As for NR:β-CD 2, the guest kept the ethyl-groups oriented one above and the other below the molecular plane, the δ angle remained almost unchanged, and the amino-N tightened the H-bond with a primary OH group of the CD. This H-bond may explain the twisting and pyramidalization that the amino group preserved in this structure. After DFT treatment, complex NR:β-CD 3 changed mainly in the ethyls′ orientation; both last methyls appeared on the same side of the NR molecular plane, and the δ angle became much lower. The amino group could fit slightly twisted into the β-CD cavity.

Regarding NR:γ-CD complexes in Table 2, DFT optimization displaced NR:γ-CD 4 as the most stable conformer in this subset. The guest appeared less included, rather bent (Figure 5) and forming H-bond between the NR carbonyl-O and a secondary OH group of the cycle. Both amino-final-methyls of NR appeared on the same side of the molecular plane and the δ angle reveals a more sp^3-character of the amino-N. Structure 5 turned into the most stable one in this subset and an intermolecular H-bond between oxazine-N and a primary OH group was generated. The latter group appeared also involved in an intramolecular (intraCD) H-bond. This complex was the only one with the ring system of NR included in the cavity. The amino group approached a planar conformation. Finally, DFT optimization of

Table 4. Semiempirical PM3 binding energies and DFT energies (minima in *italics*) of NR:CD occlusion complexes. DFT intermolecular H-bonds and distances.

NR:CDocc complex #[a]	E(PM3) (kcal/mol)	E(DFT) (a.u.)	Intermolecular H-bonding[b]	O····H distance (Å)
β-CD o1/r2	-19065	-5306.60204537	C=O::::$(HO)_2$	1.663; 1.709
β-CD o2/r2	-19061	*-5306.62727285*	C=O::::$(HO)_2$	1.774; 1.791
β-CD o3/r1	-19058	-5306.57173199	-	-
β-CD o4/r1	-19055	-5306.59562289	C=O····HO	1.676
γ-CD o5/r2	-21119	*-5917.15033832*	C=O····HO	1.751
γ-CD o6/r1	-21116	-5917.13228360	-	-
γ-CD o7/r1	-21116	-5917.12238176	C=O····HO	1.708
γ-CD o8/r1	-21114	-5917.13089673	C=O::::$(HO)_2$	1.776; 1.807

[a] r1, r2 denote primary or secondary rim of CD. [b] C=O::::$(HO)_2$ denotes NR-carbonyl bound to two hydroxyl groups of CD.

Table 5. Twisting (θ, θ') and pyramidalization (δ) angles (degrees) of the amino group of NR in the occlusion complexes after PM3 (left column)/DFT (right column) calculations.

CD-NR occ complex #[a]	θ C_A-C_A-N-C		θ´ C'_A-C_A-N-C´		|improper angle|C_A-N-C-C´		δ	
β-CD o1/r2	-9.05	-2.14	-45.50	3.63	150	175	30	5
β-CD o2/r2	-27.44	4.95	-68.56	2.55	143	178	37	2
β-CD o3/r1	-101.0	1.24	-53.65	3.04	134	178	46	2
β-CD o4/r1	-46.91	7.25	-95.40	6.62	133	180	47	0
γ-CD o5/r2	-7.45	13.11	-44.01	-3.27	148	165	32	15
γ-CD o6/r1	27.06	-1.95	73.79	-4.05	136	179	44	1
γ-CD o7/r1	-6.01	-3.29	-44.47	-0.59	146	177	34	3
γ-CD o8/r1	42.87	5.66	5.15	1.45	147	176	33	4

[a] r1, r2 denote primary or secondary rim of CD.

Structure 6 changed the intermolecular H-bond between oxazine-O and a primary OH group of γ-CD, to oxazine-N and another primary OH group. As for the amino group, the δ angle was much decreased, and the terminal $-CH_3$ groups pointed at opposite directions one to the other, probably due to steric hindrance of the secondary rim to their rotation. The twist angles θ and θ', as well as δ appeared much lower.

Occlusion calculations

Table 4 (column 2) shows the energies yielded by our semiempirical calculations of the four more stable PM3 occlusion structures (labeled as occlusion #/rim #) involving NR with β-CD, and same number of structures involving NR with γ-CD. Each of the eight chosen complexes was later submitted to DFT optimization; the converged DFT energies and the H-bonds that were generated are also shown in Table 4. Table 5 shows the twisting θ and θ´, and the pyramidalization δ angles, according to PM3 and DFT optimizations of the eight complexes.

Figure 8 shows two DFT occlusion complexes of Table 4: β-CD o2/r2 and β-CD o1/r2, which are the most stable one (left structure) and the first less stable one (right structure). Similarly, Figure 9 shows NR:γ-CD occlusion complexes γ-CD o5/r2 and γ-CD o6/r1; r1 and r2 refer to interaction occurring either with primary or secondary rim. It thus becomes apparent that, regarding occlusion complexes and according to DFT results, interactions of NR with the secondary rim of both CDs tend to prevail. Additionally, interactions between NR and β-CD seem intensive, since double H-bonding between the C=O group and secondary hydroxyls occurred repeatedly. The most stable occlusion structure involving γ-CD showed one H-bond between NR and the host secondary rim. In this complex, the amino group least approached the planar conformation, probably due to interactions between the amino-N and one secondary hydroxyl (N····H distance: 2.172 Å), thus increasing the pyramidalization and twisting of the group.

Dynamics simulations

MD (*in vacuo*) and LD simulations of the three NR:γ-CD inclusion complexes shown in Table 2 led to the following results: complexes 4 and 5 converged, after the annealing, to the conformations shown in Figure 10 for

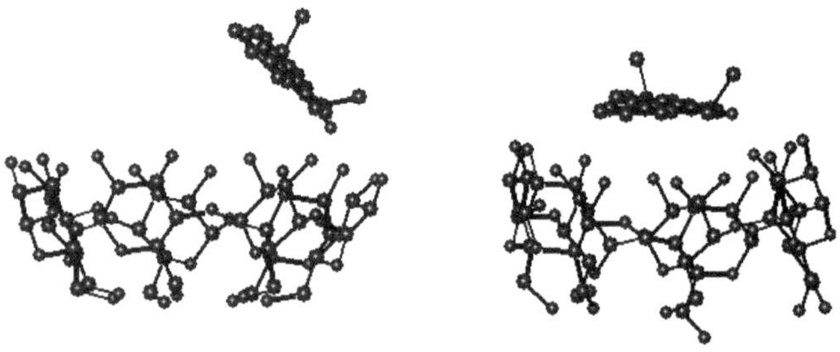

Figure 8. NR:β-CD o2/r2 (left) and NR:β-CD o1/r2 (right) occlusion complexes according to DFT optimization; H atoms are not shown (Table 4 for labeling).

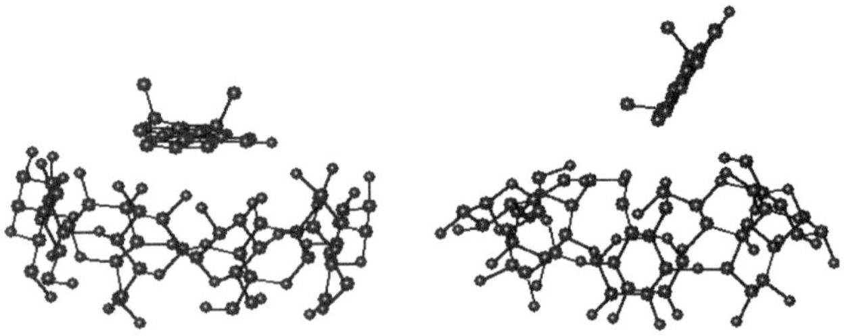

Figure 9. NR:γ-CD o5/r2 (left) and NR:γ-CD o6/r1 (right) occlusion complexes according to DFT optimization; H atoms are not shown.

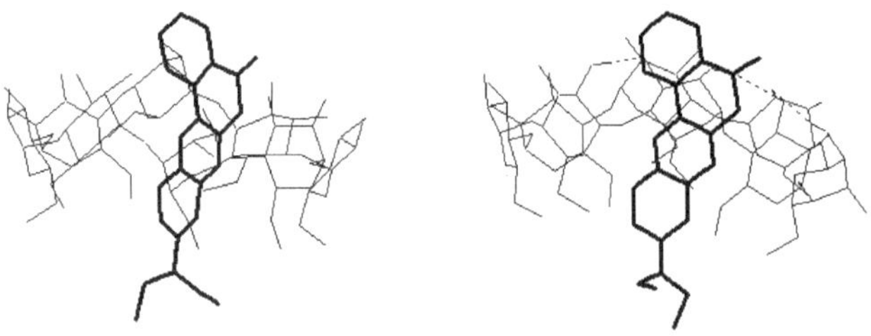

Figure 10. NR:g-CD 4 after MD (left) and LD (right); CD secondary rim up; H atoms not shown (Table 2 for labeling).

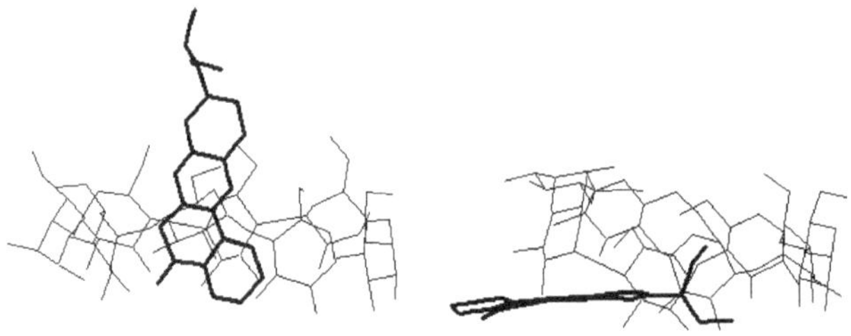

Figure 11. NR:γ-CD 5 after MD (left) and LD (right); primary rim up; H atoms not shown.

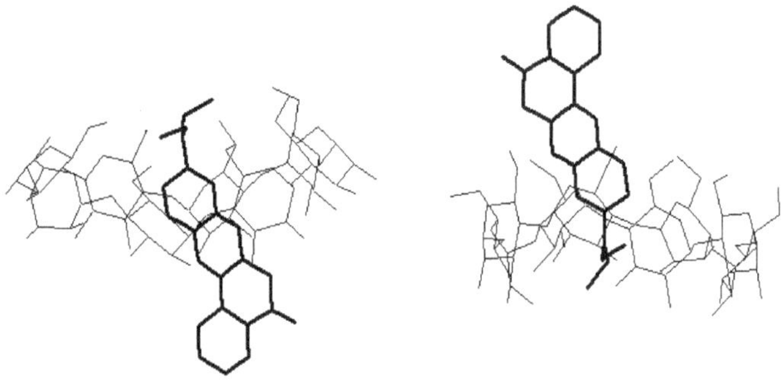

Figure 12. NR:γ-CD 6 after MD (left) and LD (right); primary rim up; H atoms not shown.

NR:γ-CD 4, and Figure 11 for NR:γ-CD 5. Figure 10 shows that both MD and LD yielded comparable inclusion structures. However, Figure 11 shows different results for MD and LD. The former preserved the inclusion complex and the intermolecular oxazine-N····HO bond (N···H distance, 2.20Å); but LD (with time step 10^{-4} ps) rendered an occlusion complex with NR capping the secondary rim of CD.

As for dynamics treatment of NR:γ-CD 6, we firstly report about LD results (with time step 10^{-4} ps): the complex remained stable and the oxazine-O····HO bond was preserved (O···H distance, 2.22 Å). Referring to the MD simulation, at 160 ps run time, the complex dissociated definitely. So we performed another MD simulation with 100 ps run time. The inclusion complex remained stable, as shown in Figure 12. It is worth mentioning that during the latter simulation, the guest emerged out of the CD through the secondary rim –that is, the lower one in Figure 12 moved several times back and fro, rotated and finally re-entered the CD cavity by the side of the amino group remaining inside until the end of the run.

Table 6 shows the energies of these structures, after dynamics simulations.

DFT-B3LYP/6-31G calculation of NR

At the early stage of our analysis, we had optimized NR semiempirically (PM3) and used this structure for modeling the PM3 complexes. This NR conformer exhibited the amino group in twisted conformation; instead, DFT optimization of NR yielded a planar conformation of this group. Table 7 presents, for comparison, structural features of both conformers; values of PM3 (twisted) were obtained by DFT-B3LYP/6-31G energy calculation of the conformer as yielded by the earlier PM3 optimization.

DISCUSSION

Results suggest that inclusion complexes of NR and both CDs are possible, but occlusion complexes are more favourable, that is, energetically more stable for

Table 6. PM3 and DFT energies of NR:γ-CD inclusion complexes after MD and LD.

CD-NR complex #	EMD (kcal/mol)	EMD (a.u.)	ELD (kcal/mol)	ELD (a.u.)
γ-CD 4	-20733	-5916.20561612	-20689	-5916.11890542
γ-CD 5	-20740	-5916.31356455	-20718	-5916.26226222
γ-CD 6	-20727	-5916.19696961	-20690	-5916.16468285

Table 7. Structural parameters of twisted (*t*) and planar (*p*) NR conformers.

Conformer →	PM3(*t*)	DFT (*p*)	Literature values
Energy (au)	-1032.6234076	-1032.6458161	-
HOMO (au)	-0.20654	-0.19625	-
LUMO (au)	-0.09709	-0.09315	-
ΔE(LU-HO) (au)	0.10945	0.10310	0.1563(*t*), 0.1481(*p*)a/0.1007b
Dipole moment (D)	5.9800	9.2235	6.13(*t*), 8.23(*p*)a/8.7b
θ / θ' twist (deg)	-67.80 / -18.13	-1.50 / 1.71	5.8(average)b
δ pyramid.(deg)	47	2.4	0.3b

[a] Dias et al. (1999); [b]planar, as yielded by DFT-B3LYP hybrid functional calculation (Guido et al., 2010).

both CDs. H-bonding interactions of the carbonyl-oxygen of NR with the secondary hydroxyls of the CDs are apparent as well. In the case of β-CD occluded by NR, two concurrent H-bonds with two adjacent rim hydroxyls were obtained; instead, only one intermolecular H-bond was formed in the case of γ-CD. These features offer a chance to rationalize the different behaviour reported about the deactivation process of NR associated to β- or γ-CD after photon excitation during fluorescence experiments, as will be next discussed.

Formation of intermolecular H-bonding has been found to have a significant influence on the photophysical behaviour of various carbonyl compounds and has been reported to lead to efficient energy dissipation. Morimoito et al. (2001) analyzed molecules such as amino- or piperidino-anthraquinones, aminofluorenones and some other aromatic compounds with at least one carbonyl group in their structure, in presence of alcohols used as solvents. They found that the intermolecular H-bonding in the excited state is anisotropic with respect to the molecular plane. Detailed kinetic analyses of the fluorescence dynamic decay revealed that the intermolecular H-bonding involved two species with different lifetimes. They proposed two modes of interaction of Hydrogen with the carbonyl group: (1) an in-plane mode with the p_y orbital of the carbonyl-oxygen, and (2) an out-of-plane mode with the $\pi^*(p_z)$ molecular orbital. The former mode would give rise to an emissive species in the excited state (deactivation lifetime in the order of *ns*). The latter mode species (deactivation lifetime in the order of *ps*), would undergo an efficient deactivation process through the out-of-plane bending motion of the carbonyl group. Such H-bond would thus, perform as effective in coupling the excited state and the ground state to its high frequency and would enable the radiationless transition, that is, internal conversion.

Within this framework, we obtained that NR:β-CD most stable occlusion complexes have saturated the H-bonding capacity of the C=O group in the ground state, since NR appears anchored by two adjacent hydroxyls of the CD rim which have already established contact with the oxygen lone pairs. These complexes, after photoexcitation, could only deactivate via photoemission occurring on the *ns* timescale. Instead, NR:γ-CD most stable occlusion complexes show one carbonyl-oxygen lone pair implicated in H-bond. There is thus room for the out-of-plane interaction with some other close hydroxyl group of the CD rim, in the excited state. This would explain the difference between the radiative and non-radiative rate constants reported by Hazra et al. (2004), whose values suggest that radiative deactivation is relatively more favourable for NR:β-CD complexes, whereas non-radiative deactivation is more important, even enhanced (Srivatsavoy, 1999), in NR:γ-CD complexes.

Regarding the most stable inclusion complexes, probably less abundant, NR shows one intermolecular H-bond involving the carboxyl in the case of β-CD, and none involving the carboxyl in the case of γ-CD, before photoexcitation. In both such complexes, NR could undergo any of the two modes of interaction as described by Morimoito et al. (2001) with some close CD hydroxyl in the excited state, giving way to either radiative or non-radiative deactivation to the ground state.

Our DFT optimization of free NR yielded lower twisting angles and larger pyramidalization than reported by Guido et al. (2010). Intermolecular interactions in the complexes demand these angles to adapt accordingly;

this was especially evident in the occlusion NR:γ-CD most stable structure. If not engaged, the nitrogen atom in an intermolecular H-bond, the tendency of the amino group towards planar configuration is prevalent, no matter how many or where else other H-bonds involving NR may be present.

Conclusions

Regarding inclusion complexes, NR in β-CD is only partially inserted and by the side of the amino group. Intermolecular H-bonding between NR and the rim hydroxyls of the CDs occurs either via the carbonyl-oxygen or the amino-nitrogen of the guest. Inclusion complexes of NR with γ-CD can occur either by insertion of the guest´s amino group or, alternatively, of its condensed rings. H-bonding can form either with the carbonyl-oxygen or the oxazine-nitrogen. In these inclusion complexes, H-bonds appear simple.

Occlusion complexes are, however, more stable. Additionally, in these complexes the guest´s carbonyl-oxygen tends to form double H-bonds with β-CD, but simple H-bonds with γ-CD. Double H-bonds might be directly connected to NR radiative deactivation to the ground state after photoexcitation, whereas simple (or absent) H-bonds might include also the non-radiative deactivation process of NR. Which way NR deactivates, would depend on the feasible modes of interaction, in the excited state, between the carbonyl-oxygen of NR and the CD´s hydroxyls: only in-plane mode if the H-bond is double; either in- or out-of-plane, if the H-bond is simple. Finally, the tendency of the amino group of NR towards planar configuration is prevalent, only modified when the nitrogen atom itself is involved in intermolecular H-bonding.

REFERENCES

Berger TA, Deye JF (1990). Composition and Density Effects using Methanol/Carbon Dioxide in Packed Column Supercritical Fluid Chromatography. Anal. Chem. 62:1181-1185.

Choi M, Jin D, Kim H, Kang TJ, Jeoung SC, Kim D (1997). Fluorescence Anisotropy of Nile Red and Oxazine 725 in an Isotropic Liquid Crystal. J. Phys. Chem. B 101:8092-8097.

Cser A, Nagy K, Biczók L (2002). Fluorescence lifetime of Nile Red as a probe for the hydrogen bonding strength with its microenvironment. Chem. Phys. Lett. 360:473-478.

Datta A, Mandal D, Pal SK, Bhattacharyya K (1997). Intramolecular Charge Transfer Processes in Confined Systems. Nile Red in Reverse Micelles. J. Phys. Chem. B 101:10221-10225.

Deye JF, Berger TA, Anderson AG (1990). Nile Red as solvatochromic dye for measuring solvent strength in normal liquids and mixtures of normal liquids with supercritical fluids and near critical fluids. Anal. Chem. 62:615-622.

Dias Jr. LC, Custodio R, Pessine FBT (1999). Theoretical studies of Nile Red by *ab initio* and semiempirical methods. Chem. Phys. Lett. 302:505-510.

Dias Jr. LC, Custodio R, Pessine FBT (2006). Investigation of the Nile Red Spectra by Semi-empirical Calculations and Spectrophotometric Measurements. Int. J. Quant. Chem. 106:2624-2632.

Duchêne D Ed. (1987). Cyclodextrin and their Industrial Uses; Editions de Santé:Paris. pp. 297-350.

Dutta AK, Kamada K, Ohta K (1996). Spectroscopic studies of nile red in organic solvents and polymers. J. Photochem. Photobiol. A: Chem. 93:57-64.

Freeman DM, Kroe RR, Ruffles R, Robson J, Coleman JR, Grygon CA (2001). A fluorescence method for probing the effect of small molecules on DPPC lipid bilayers using the hydrophobic dye, Nile Red. Biophys. J. 80:2375-Pos, p. 527a; last accessed January 2012 at:http://www.ncbi.nlm.nih.gov/pmc/articles/PMC1914015/pdf/biophysj00804-0558.pdf.

Gaussian 03, Revision A.1. Frisch MJ, Trucks GW, Schlegel HB, Scuseria GE, Robb MA, Cheeseman JR, Montgomery Jr. JA, Vreven T, Kudin KN, Burant JC, Millam JM, Iyengar SS, Tomasi J, Barone V, Mennucci B, Cossi M, Scalmani G, Rega N, Petersson GA, Nakatsuji H, Hada M, Ehara M, Toyota K, Fukuda R, Hasegawa J, Ishida M, Nakajima T, Honda Y, Kitao O, Nakai H, Klene M, Li X, Knox JE, Hratchian HP, Cross JB, Adamo C, Jaramillo J, Gomperts R, Stratmann RE, Yazyev O, Austin AJ, Cammi R, Pomelli C, Ochterski JW, Ayala PY, Morokuma K, Voth GA, Salvador P, Dannenberg JJ, Zakrzewski VG, Dapprich S, Daniels AD, Strain MC, Farkas O, Malick DK, Rabuck AD, Raghavachari K, Foresman JB, Ortiz JV, Cui Q, Baboul AG, Clifford S, Cioslowski J, Stefanov BB, Liu G, Liashenko A, Piskorz P, Komaromi I, Martin RL, Fox DJ, Keith T, Al-Laham MA, Peng CY, Nanayakkara A, Challacombe M, Gill PMW, Johnson B, Chen W, Wong MW, Gonzalez C, Pople JÁ (2003). Gaussian Inc, Pittsburgh PA.

Greenspan P, Fowler SD (1985). Spectrofluorometric studies of the lipid probe, nile red. J. Lipid Res. 26:781-789.

Guido CA, Mennucci B, Jacquemin D, Adamo C (2010). Planar *vs.* twisted intramolecular charge transfer mechanism in Nile Red:new hints from theory. Phys. Chem. Chem. Phys. 12:8016-8023.

Harata K (1991). Recent Advances in the X-ray analysis of cyclodextrin complexes. In Inclusion Compounds; Atwood JL, Davies JED, MacNicol DD, Eds.; Academic Press: London. 5:311-344.

Hazra P, Chakrabarty D, Chakraborty A, Sarkar N (2004). Intramolecular charge transfer and solvation dynamics of Nile Red in the nanocavity of cyclodextrins. Chem. Phys. Lett. 388:150-157.

HyperChem 6 Pro, Hypercube, Inc., Serial # 12-600-1500700081.

Johnson SR, Sutter JM, Engelhardt HL, Jurs PC, White J, Kauer JS, Dickinson TA, Walt DR (1997). Identification of Multiple Analytes Using an Optical Sensor Array and Pattern Recognition Neural Networks. Anal. Chem. 69:4641-4648.

Kawski A, Bojarski P, Kukliński B (2008). Estimation of ground- and excited-state dipole moments of Nile Red dye from solvatochromic effect on absorption and fluorescence spectra. Chem. Phys. Lett. 463:410-412.

Kawski A, Kukliński B, Bojarski P (2009). Photophysical properties and thermochromic shifts of electronic spectra of Nile Red in selected solvents. Excited states dipole moments. Chem. Phys. 359:58-64.

Kipkemboi PK, Easteal AJ (1994). Solvent Polarity Studies of the Water+t-Butyl Alcohol and Water+t-Butylamine Binary Systems with the Solvatochromic Dyes Nile Red and Pyridinium-N-phenoxide Betaine Refractometry and Permittivity Measurements. Aust. J. Chem. 47:1771-1781.

Korth M (2010). Third-Generation Hydrogen-Bonding Corrections for Semiempirical QM Methods and Force Fields. J. Chem. Theory Comput. 6:3808-3816.

Li XS, Mu TW, Guo QX (2000). A systematic quantum chemistry study on cyclodextrins. Monats. Chem. 131:849-855.

Maiti NC, Krishna MMG, Britto PJ, Periasamy N (1997). Fluorescence Dynamics of Dye Probes in Micelles. J. Phys. Chem. B 101:11051-11060.

Matsui K, Nozawa K (1997). Molecular Probing for the Microenvironment of Photonics Materials Prepared by the Sol-Gel Process. Bull. Chem. Soc. Jpn. 70:2331-2335.

McNamara JP, Hillier IH (2007). Semi-empirical molecular orbital methods including dispersion corrections for the accurate prediction of the full range of intermolecular interactions in biomolecules. Phys. Chem. Chem. Phys. 9:2362-2370.

Meinershagen JL, Bein T (1999). Optical Sensing in Nanopores. Encapsulation of the Solvatochromic Dye Nile Red in Zeolites. J. Am.

Chem. Soc. 121:448-449.

Morimoito A, Yatsuhashi T, Shimada T, Biczók L, Tryk D, Inoue H (2001). Radiationless Deactivation of an Intramolecular Charge Transfer Excited State through Hydrogen Bonding: Effect of Molecular Structure and Hard-Soft Anionic Character in the Excited State. J. Phys. Chem. A. 105:10488-10496 (and references therein).

Morley JO, Morley RM, Docherty R, Charlton MH (1997). Fundamental Studies on Brooker's Merocyanine. J. Am. Chem. Soc. 119:10192-10202.

Morpurgo S, Bossa M, Morpurgo GO (1998). Critical test of PM3-calculated proton transfer activation energies: A comparison with ab initio and AM1 calculations. J. Molec. Struct. (THEOCHEM) 429:71-80.

Saenger W (1980). Cyclodextrins in research and industry. Angew. Chem. Int. Ed. Engl. 19:344-362.

Saenger W (1984). Structural aspects of cyclodextrins and their inclusion complexes. Inclusion Compounds; Atwood JL, Davies JED, MacNicol DD, Eds.; Academic Press: London. 2:231-260.

Saenger W, Jacob J Gessler K, Steiner T, Hoffmann D, Sanbe H, Koizumi K, Smith SM Takaha T (1998). Structures of the Common Cyclodextrins and Their Larger Analogues - Beyond the Doughnut. Chem. Rev. 98:1787-1802.

Srivatsavoy VJP (1999). Enhancement of excited state nonradiative deactivation of Nile Red in γ-cyclodextrin:evidence for multiple inclusion complexes. J. Lumin. 82:17-23.

Szejtli J (1988). Cyclodextrin Technology; Kluwer Academic Publishers: Dordrecht.

Szejtli J (1998). Introduction and General Overview of Cyclodextrin Chemistry. Chem. Rev. 98:1743-1753.

Uppili S, Thomas KJ, Crompton EM, Ramamurthy V (2000). Probing Zeolites with Organic Molecules:Supercages of X and Y Zeolites Are Superpolar. Langmuir 16:265-274.

Uusi-Penttilä MS, Richards RJ, Torgerson BK, Berglund KA (1997). Spectroscopically Determined Dielectric Constants for Various Esters. Ind. Eng. Chem. Res. 36:510-512.

Wagner BD, Stojanovic N, LeClair G, Jankowski CK (2003). A Spectroscopic and Molecular Modelling Study of the Nature of the Association Complexes of Nile Red with Cyclodextrins. J. Incl. Phenom. Macrocycl. Chem. 45:275-283.

White J, Kauer JS, Dickinson TA, Walt DR (1996). Rapid Analyte Recognition in a Device Based on Optical Sensors and the Olfactory System. Anal. Chem. 68:2191-2202.

Yanai T, Tew DP, Handy NC (2004). A new hybrid exchange-correlation functional using the Coulomb-attenuating method (CAM-B3LYP). Chem. Phys. Lett. 393:51-57.

Zheng YJ, Merz Jr. KM (1992). Study of hydrogen bonding interactions relevant to biomolecular structure and function. J. Comp. Chem. 13:1151-1169.

Theoretical study of methyl chloride conversion over H-ZSM5 zeolite

C. Boukratem[1]*, Y. Bouhadda[2] and T. Fergoug[2]

[1]Institute of Chemistry. University of Oran, 31000, Algeria.
[2]Department of Chemistry, University of Mascara, 29000, Algeria.

We note in the reaction that the main molecules included in the process and explained that the transformations (Boukratem, 1997, 2003) are CH_3Cl, C_2H_4 and HCl. Other studies were led (Jaumain and Su, 2002; Wei et al., 2006) on the same subject. With these three elements, we constitute a mathematical development relating the thermodynamic study. The final equation what was found through the fundamentals is a good theoretical representation.

Key words: Thermodynamic, catalytic activity, zeolite, theoretical study, mechanism.

INTRODUCTION

It is very interesting to have finally, tools to approach this reaction in its thermodynamic aspect. Using terms of catalytic activity and fundamental laws, we obtain finally a useful equation.

CATALYTIC ACTIVITY

The catalytic material is synthesized by the method described by a patent of Mobil (Argauer et al., 1972). It was identified by the diffraction pattern. The reaction is carried out at atmospheric pressure with a fixed bed continuous microreactor (0.8 × 25 cm quartz tube) with 200 mg catalyst. Results are obtained with different values of Si/Al which traduce different values of acidic form H-ZSM5. The solid contained in the reactor tube was heated in a programmable electric furnace under oxygen till 773K.

RESULTS AND DISCUSSION

Physical elements like polarized (Peter and Vollhardt, 1987) atoms or physical constraint factors of molecular groups explain the reaction. In methyl chloride molecule, the dimension of the orbital p of chlorine is important and the electronic cloud around the halogen becomes more diffused and more polarized. The recovery in transition state is more efficient, because the orbital p of Cl is in recovering with the hybrid orbital sp^3 of carbon. However, we note a decay of the force of the C-Cl bond inside the cavity of the solid, because of the new electronic environment. The dipolar moment of the bond controls the chemical behavior of chlorine. In the acidic site Si-OH, the oxygen which is rich in electrons will attack the carbon which is positively polarized. In line with this phenomenon, the aluminum atom which is deficient in electrons will constitute a new bond with chlorine. Finally, we have exactly the manner of the brooking of C-Cl bond in methyl chloride in the solid, where Si-OH and Al atom undertake together; we also have a dorsal substitution. This one is in order 2, where there is no physical constriction (Scheme 1). In Scheme2, we have the steps (Boukratem et al., 2003) of the reaction-giving hydrocarbons. The last form is equivalent to Scheme 3.

We are interested by the reaction of conversion of methyl chloride to ethylene and propylene. About the mechanism, infrared (Boukratem, 2003; Salvador and Kladning, 1977) studies have been realized showing the nature of the brooking of the C-Cl bond. We note through

*Corresponding author. E-mail: chikh57@hotmail.fr.

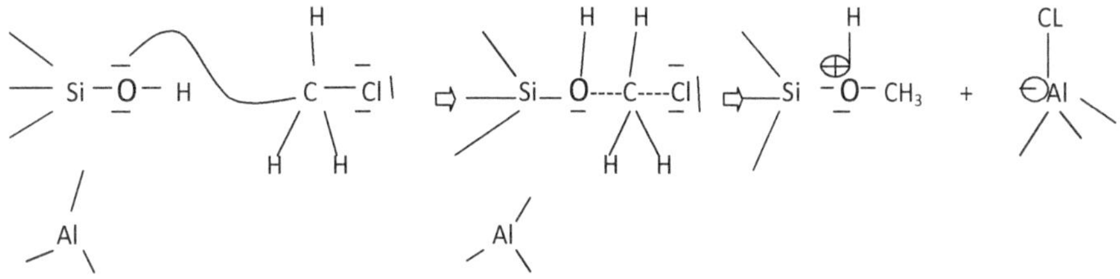

Scheme 1. First step of conversion showing where there is no physical constriction.

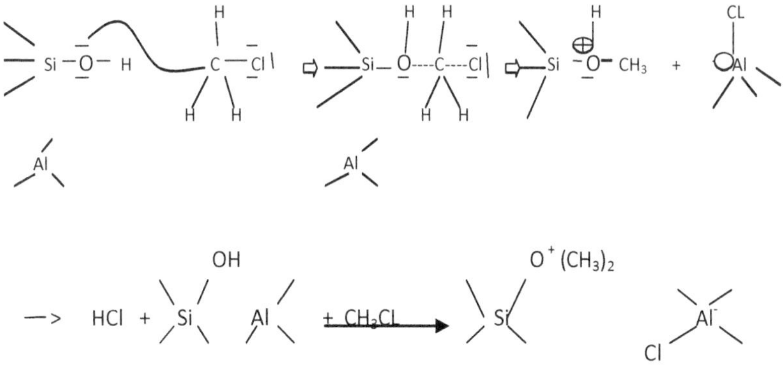

Scheme 2. Essential steps of the reaction-giving hydrocarbons.

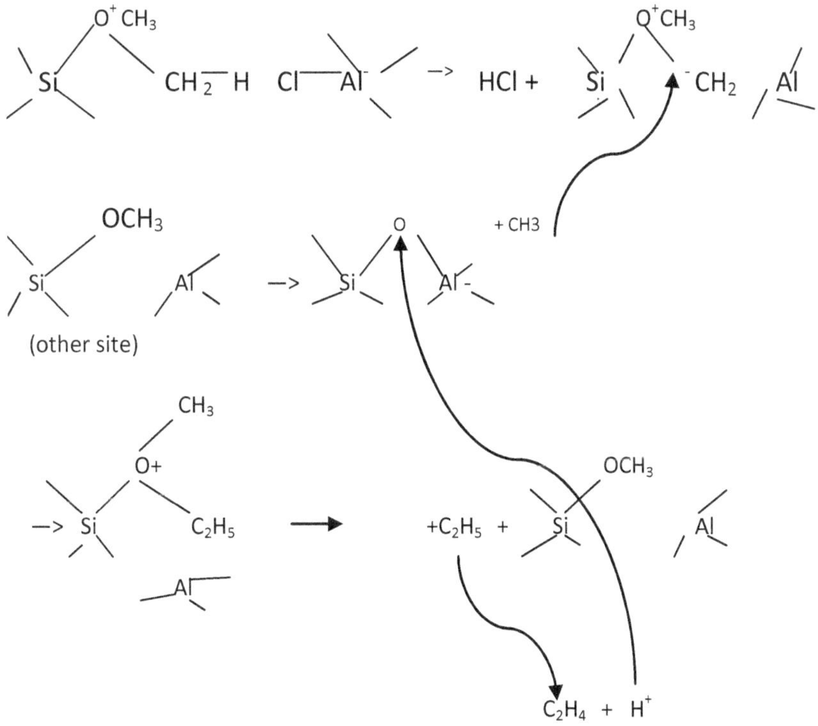

C_2H_4 + CH_3^+ —> C_3H_6 + H^+ —>......... linear and aromatic hydrocarbons

Scheme 3. Final form of reaction in conversion of methyl chloride to ethylene and propylene.

the thermodynamic study that we can obtain one usual relationship for free enthalpy ΔG. Any reaction is managed by this function including other thermodynamic elements. The ΔG is a function of state. Therefore, we can take into account only the beginning and final states for the reaction whichever way followed by the chemical transformation. Likewise, we do not need details of intermediate steps of all the reaction. We resume the reaction by:

$2CH_3Cl_{(v)} \rightarrow C_2H_{4(v)} + 2HCl_{(v)}$

The equation of Gibbs-Helmotz $G = H + T\,(\partial G/\partial T)_P$ permits the following development:

$$[\partial(G/T)/\partial T]_P = 1/T\,(\partial G/\partial T)_P - G/T^2 = 1/T^2[T\,(\partial G/\partial T)_P - G] = -H/T^2$$

$$[\partial\,(G/T)/\partial T]_P = -H/T^2 \Rightarrow \partial\,(\Delta G_T/T)\,/\,\partial T]_P = -\Delta H_T\,/T^2 \quad (1)$$

In this catalytic testing, we have a dynamic regime under one atmosphere which means a constant pressure.

$$\Delta H_T = \Delta H° + \int \Delta C_P\, d\,T = \Delta H° + \int (A + BT)\, d\,T = \Delta H° + AT + (B/2)\,T^2$$

with

$A = \sum Ap - Ar$ (relatively to products and reactive of the reaction)
$B = \sum Bp - Br$

Therefore, the expression (Equation 1) becomes:

$$\Delta G_T/T - \Delta G°/T° = \int (\Delta H° + AT + (B/2)\,T^2)\,d\,T\,/\,T^2$$
$$= \Delta H°\,(1\,/\,T° - 1/T) + A\,\mathrm{Log}\,(T/T°) + (B/2)\,(T - T°) \quad (2)$$

Before giving values of ΔH° and Cp, relatively of terms of A and B, we must know that, in the solid, the reactive is in a gaseous state. The dimension of the main cavity of the Mobil five is 9 Å. In addition, the middle distance between molecules is ten times the diameter of a molecule under one atmosphere. In literature, we have standard values needed in our reaction at 298K.

$2CH_3Cl \longrightarrow C_2H_4 + 2HCl$

$\Delta H°_R = \Delta H°_{f\,(C2H4)} + 2\Delta H°_{f\,(HCL)} - 2\Delta H°_{f\,(CH3Cl)} = 52.28 + 2(-92.31) - 2(-80.83) = 29.32$ kcal/mol

$\Delta S°_R = S°_{f\,(C2H4)} + 2S°_{f\,(HCL)} - 2S°_{f\,(CH3CL)} = 219.6 + 2\times186.8 - 2\times234.8 = 123.6$ cal/mol

$\Delta G°_R = \Delta H°_R - T°\Delta S°_R = 29320 - 298 \times 123.6 = -7510$ cal/mol

Before developing the expression of ΔG, the term B, in expression (Equation 2), is estimated to be equal to 10^{-4}. Therefore, we can neglect the term (B/2) (T° - T) in the Equation 2 of ΔG.

Values (McQuarrie, 1992) of C_P:

$C_{P\,(CH3Cl)} = 35.7$ J/K/mol

$C_{P\,(C2H4)} = 42$ J/K/mol

$C_{P\,(HCl)} = 9$ cal/K/mol (diatomic gas)

$$\Delta G/T = \Delta G°/T° + \Delta H°\,(1/T° - 1/T) + A\,\mathrm{Log}\,(T/T°) + (B/2)\,(T-T°)$$

$$\Delta G/T = \Delta G°/T° + \Delta H°\,(1/T° - 1/T) + A\,\mathrm{Log}\,(T/T°)$$

$$\Delta G/T = -7.51 \times 10^3/298 + 29.32 \times 10^3(1/298 - 1/T) + (1/4.18)(2 \times 9 \times 4.18 + 42 - 2 \times 35.7)\,\mathrm{Log}(T/298)$$

In the catalytic testing, we note that methyl chloride begins to be converted at 573K. This value changes slightly according to operating parameters. The resolution of the equation ΔG=f (T) gives the theoretical value of the temperature reaction.

We do not forget all the approximations we did to reach this result. The mathematic manner to resolve the equation of ΔG is to apply the theorem of the intermediate values.

Conclusion

Using a theoretical way with a thermodynamic approach, we realize the circumstances of the reaction which occur in the solid. This model could serve for any similar reaction.

REFERENCES

Argauer RT, Landolt RG, Patent US (1972). Zeolites and ordered mesoporous materials: Progress and prospective, 3: 702, 886.
Boukratem C (1997). Influence of some parameters on methyl chloride conversion over various zeolites: H-ZSM5, H-ZSM12 and Mordenite. ROUMANIAN Rev. Chem., 42: 12.
Boukratem C (2003). Methyl chloride conversion over H-ZSM5: Study of the variation of the ratio $S(C_3^=)/S(C_2^=)$ at low activity. Remarks on the mechanism. J. Chin. Chem. Soc., (Taipei), 50: 2.
Jaumain D, Su BL (2002). Conversion directe du chlorométhane en hydrocarbures sur différents matériaux acido-basiques de types micro et mésoporeux. CATAL. TODAY.
McQuarrie R (1992). General Chemistry. Published in USA by FREEMAN WH ,Co. New York, Oxford, p. 267.
Peter K, Vollhardt C (1987). Organic Chemistry University of California, Berkeley. Published in the USA by FREEMAN WN, Co. New York Oxford, p. 216.
Salvador P, Kladning W (1977). The design of catalyst using molecular modeling. Faraday Trans., 73: 1153.
Wei Y, Zhang D, Liu Z, Su BL (2006). The methyl halide to hydrocarbon reaction over H-SAPO-34. J. Catal., 241: 243-254.

Influence of pH on the uptake of toluene from water by the composite poly (4-vinylpyridinium)-maghnite

M. A. Zenasni[1]*, S. Benfarhi[2], A. Mansri[3], H. Benmehdi[1], B. Meroufel[1], J. Desbrieres[4] and R. Dedriveres[4]

[1]Department of Sciences, Faculty of Sciences and Technology, University of Bechar, 08000 Bechar, Algeria.
[2]Department of Chemistry, Faculty of Sciences, University of Batna, 05000 Batna, Algeria.
[3]Department of Chemistry, Faculty of Sciences, University of Tlemcen, 13000 Tlemcen, Algeria.
[4]University of Pau, Helioparc Pau Pyrenees, 64053 Pau Cedex 09, France.

The objective of this study is to investigate the effects of the degree of α ionization on the rate uptake of organic pollutant (toluene) by poly (4-vinylpyridinium)-maghnite. Two-step experiment was carried out where the organophilic maghnite adsorbent was prepared in the first step and toluene was adsorbed on maghnite in the second step. This material organophilic maghnite was prepared by exchanging the organic polycation as poly (4-vinylpyridinium) chloride (PVPS) for sodium (Na^+) on the layer surface of clay. This process of wastewater treatment is of interest for the wastes stabilisation in which toluene contaminations occurs. Batch adsorption studies were conducted to evaluate the effects of initial toluene concentration. The results show that the two-step process is an efficient, simple and low cost technological way for the removal of organic pollutants from water.

Key words: Adsorption, organic pollutant, P4VP, maghnite, organoclay.

INTRODUCTION

The development of technological ways for the decontamination of soils and waters polluted by hydrophobic organic compounds encouraged re-searchers to use polycationic surfactant as potential agents for the enhanced solubilisation and the removal of contaminants from soils and sediments (Fournanis et al., 2001; Zadaka et al., 2009; Liu, 2007). Pollution caused by organic compounds is a common problem faced by many countries and is likely the cause of health hazards, harm to ecology, damage to structures or amenities, and interference with the legal use of water. Organic contaminants include many categories of compounds, of which benzylics are recognised as hazardous chemical pollutants. Many methods are possible to taking away benzylics from effluents that are chemical oxidation, ozonolysis, flocculation, reverse osmosis, adsorption, ion exchange, and biological techniques. Adsorption techniques are rapidly gaining prominence as a treatment process for providing high-quality effluents that are low in concentration of dissolved organics (Walker et al., 1998). A number of adsorbents, such as activated carbons, peat moss, brown coal, industrial waste, agricultural products, polymers, chitosan, and fly ash, are used for organics removal (Chang et al., 2004). Cost effectiveness, availability, and adsorption properties are the main criteria for choosing an adsorbent to take out pollutants. Some work (Li et al., 2007; Vidal et al., 2009; Breen, 1998, 1999) have shown that changes in smectites polycation could make them sufficientlyhydrophobic and easily allow the retention of aromatic hydrocarbons in water.

The aromatic hydrocarbons benzene, toluene, ethylbenzene and xylene (that is, "BTEX compounds") are of concern as water contaminants resulting from petrochemical spills and leakages. A number of workers (Jiang et al., 2002; Crini, 2005; Durand-Piana et al., 1987; Billingham et al., 1997) also showed that smectite/ polycation complexes can develop a net positive charge (Becker et al., 2000) found that anionic dyes were

*Corresponding author. E-mail: am.zenasni@gmail.com.

removed from wool dyeing effluent water by bentonite and cationic polyelectrolytes in combination, but not by either of these components alone. A likely interpretation of this result is that attachment of the polycations to the aluminosilicate in appropriate proportions provided a positively charged solid phase for adsorption of anionic dye molecules. These various studies indicate a need for further work to devise principles for the use of smectite–polycation complexes for water pollutant removal. In the present study poly (4-vinylpyridinium)-maghnite with degree of α ionization different were employed for removal of toluene and used as an effective adsorbent in the wastewater treatment.

MATERIALS AND METHODS

Bentonite

The bentonite (maghnite) used in this work comes from a quarry located in Maghnia (West of Algeria) and was supplied by the company "ENOF" (an Algerian manufacture specialized in the production of nonferric products and useful substances). The different chemical elements of the native bentonite were transformed into oxides and analysed by X-ray fluorescence ENOF. Its chemical composition was found to be as follows: 54.90% SiO_2, 1.85% MgO, 27.71% Al_2O_3, 0.08% K_2O, 0.08% CaO, 2.82% F_2O_3, 3.14% Na_2O, 9.4% loss on ignition. These results confirm that the bentonite used consists essentially of montmorillonite, since the ratio SiO_2/Al_2O_3 is equal to 3.77 and thus belongs to the family of the phyllosilicates (Khenifi et al., 2009; Boufatit et al., 2008). These bentonite form stable suspensions in water and have flat platelets or needle-like structures. The Na-exchanged form of bentonite was prepared by stirring samples for 24 h with 1 M NaCl. This was followed by several washes with distilled water and filtration to remove excess NaCl and other exchangable cations from the clay. The clay was then resuspended and filtered until a negative chloride test was obtained with 0.1 M $AgNO_3$ (Gupta et al., 2006).

Preparation of poly (4-vinylpyridinium)-maghnite

The solutions of Poly(4-vinylpyridinium) chloride salts (P4VPS) with degree of α ionization different equal to 0.10, 0.25, 1.00 and 1.50 were prepared by adding 0.25 g P4VP for concentrations of different solution for hydrochloric acid (HCl) equal to 0.2, 0.5, 2 and 3 M. An amount of 2.5 g of Na-maghnite per 30 mL aqueous solutions of P4VP chloride salt was contacted for 24 h. The solids were washed with distilled water to remove excess salt, separated, and dried at 80°C. The P4VP–Maghnites were named P4VPS-Mag 1, 2, 3 and 4. The values of degree of α ionization mentioned in the Table 1 are calculated from the following equation: α is degree of ionization defined as the report of number of mole of the HCl in it of the monomer 4-vinylpyridine (4VP).

$$\alpha = \frac{n_a(HCl)}{n_b(4VP)}$$

Adsorbate

The toluene used in this study was an analytical grade reagent of 99% purity from the Aldrich.

Adsorption experiments

Preliminary adsorption experiments were performed at 25°C to determine the influence of pH on the uptake of toluene from water by the composite poly (4-vinylpyridinium)-maghnite. Batch adsorption experiments were conducted in 10 ml centrifuge tubes with 0.1 g of organo-maghnite combined with a 5 ml solution containing an appropriate concentration of a solute (250 and 500 ppm). After centrifugation, the concentration of the supernatant was determined. Toluene uptake by organo-maghnite was calculated from the difference between the quantity of toluene added and the quantity remaining in the equilibrium solution after 24 h of contact. The amount of toluene was determined by measuring of the absorbance at 271 nm in a Unicam Ultraviolet UV1 spectrophotometer. The amount of species adsorbed by surfactant-modified bentonite, q_e (mg/g), could be calculated as

$$q_e = \frac{V(C_0 - C_e)}{W}$$ (Rawajfih et al., 2006).

Where C_0 and C_e (mg/*l*) are the initial and final concentrations of toluene, respectively, V the volume of toluene solution and W is the weight (g) of organo-maghnite adsorbent.

Instrumentation

The distance between the layers of the organo-maghnite, the basal spacing d_{001}, was determined by X-ray diffraction analysis. X-ray diffraction (XRD) measurements were performed using an Inel automatic diffractometer (Detector CPS 120 and Source XRG 3000) with a Cu tube source (λ = 1.54 Å); generator tension was 50 kV at a current of 20 mA. Scans were taken from 2θ = 1.0 to 10, step size = 0.1 and scan time per step of 10 s using the high-resolution mode (Witthuhn et al., 2005). The d_{001} spacing of clay particles was calculated according to Bragg's law

$$d = \frac{\lambda}{2\sin\theta}$$

Thermogravimetric analysis, (TGA), was performed on a TA instruments 2950 unit under a flowing nitrogen atmosphere at a scan rate of 10°C per minute from 20 to 800°C (Chigwada et al., 2006).

RESULTS AND DISCUSSION

Characterization of organophilic maghnite

It appears, in Figure 2, that the preparation of the P4VP with HCl gives an organic phase slightly fixed on maghnites. The reaction with the HCl creates a positive load on nitrogen of polymer (P4VPS), which induces an increase in average dimensions polymers chains; this is due to the internal electrostatic repulsions. This effect makes it difficult to penetrate the polymer between the layers. The examination of these results shows that the amount of intercalated polymer P4VP are found between 6 to 10% by weight of maghnite, for an initial report P4VP/BC 10%. However, if we compare these values for α different degree, we find that the ratio between the

Table 1. A physicochemical characteristic of the different supports P4VPS – Maghnite and their sorption capacities toluene at 250 ppm and 500 ppm in batch tests.

Variable					Rate of retention of toluene			
					C_0 = 500 ppm		C_0 = 250 ppm	
Sample	α	pH solution	P4VPinter (%)	d_{001} Å	q_e (mg g^{-1})	% of uptake toluene/ at C_0	q_e (mg g^{-1})	% of uptake toluene/ at C_0
Maghnite	0.00	-	0.0	13.8	-	-	-	-
P4VPS-Mag 1	0.10	3.8	8.6	14.7	0.1	0.4	2.45	19.6
P4VPS-Mag 2	0.25	2.5	9.5	14.8	1.05	4.2	3	24
P4VPS-Mag 3	1.00	1.2	9.6	15.0	2.9	11.6	4.7	37.6
P4VPS-Mag 4	1.50	0.5	7.1	15.0	4.15	16.6	5.85	46.8

minimum amount of P4VPS inserted for α ≈ 0.7-1 and the minimum amount of P4VPS inserted for α ≥ 1.5 exceeds 25%. This finding shows the importance of the parameter ionization on the exfoliation of maghnite by the P4VPS. The spacing interlayer of maghnite is $d_{(001)}$ = 1,38 Å. Maghnite modified by cationic polymer such as poly(4-vinylpyridinium) salt (P4VPS) gives organophilic clay. As it is shown in Table 1 the canionics polymers were intercalated in the interlayer space of the maghnite and the spacing increased gradually with to increase degree of α ionization (to increase rate of quarterisation P4VP) (Figure 3). The addition of P4VPS with low degree of α ionization equal 0,1, the spacing was around 14.7Å, Table 1, indicated that low rate quarterisation N^+ from P4VP is accommodated in the interlayer. Subsequent addition of P4VPS with increase degree of α ionization (α = 1, 5) into the layers led to an increase in $d_{(001)}$ spacing up to around 15.0 Å, (Table 1). Polycations are adsorbed via coulombic interactions between the cationic groups on the polymer and the negatively charged clay surface.

It is generally accepted that there is instantaneous collapse of the polycation onto the clay surface and that few loops are formed. The amount of species adsorbed by surfactant-modified bentonite, q_e (mg/g), could be calculated as

$$q_e = \frac{V(C_0 - C_e)}{W}$$

where C_0 and C_e (mg/l) are the initial and final concentrations of toluene, respectively; V the volume of dye solution and W is the weight (g) of organobentonite adsorbent. Degree of α ionization (quaternisation for P4VP by various concentrations of HCl).

$$\alpha = \frac{(C_{HCl} \times V_{HCl}) \times M_{4VP}}{m_{P4VP}}$$

Where C_{HCl} (mol/l) Concentration of HCl, V_{HCl} (L) Volume of HCl, M_{4VP} (g/mol) molecular weight of monomer 4-vinylpyridine, m_{P4VP} (g) weight of polymer poly(4-vinylpyridine).

The effect of the ionisation's degree of P4VP on toluene's adsorption by organophilic maghnite

Figure 4 shows the influence of degree of α ionization of P4VPS on the percentage removal of toluene on organophilic maghnite (P4VPS-Maghnite). It was reported that natural maghnite do not show adsorption capacity for toluene, because the contribution of physical adsorption of toluene by surface silicate layers is negligible (Jock-Churchman, 2002).

Hence, the adsorption increment of toluene showed by the sodium-exchanged maghnite, can be attributed to the presence of the hydrophobic sites on the surface developed by siloxane groups of the intercalated poly(4-vinylpyridinium) cations. These results encourage us to continue the measurements lower concentration for toluene. Indeed, the use of these supports for the decontamination of drinking waters will require a total retention of the pollutants starting from the much diluted solutions. In our case, we use these tests to measure the capacities retention of our organophilic supports. For P4VPS-Mag 4 containing only 7,1% of P4VP and obtained with a value of α = 1,5, the rate of retention is maximum whatever the concentration initial C0 toluene (Table 1).

This observation shows that the pH, system with a proton H^+ is directly related to the polar nature of space and organophilic interlayer where is the adsorption of toluene. In Figure 5, the retention of toluene by the studied supports increases on the one hand with the reduction in the initial toluene concentration, and on the other hand with the increase in the degree of α ionization. The observation of this result is that the load factor of adsorbed polymer salt P4VPS modifies the electrostatic state of space interlayers, place of the adsorption of the toluene molecule. Indeed, the screening charges of maghnite by quaternised P4VP make space interlayers less polar and thus favorable to the adsorption of the absorbent molecule. The P4VP cannot, in any case, be solely responsible for the retention of as much more support high in P4VP retain amounts considerably lower in toluene.

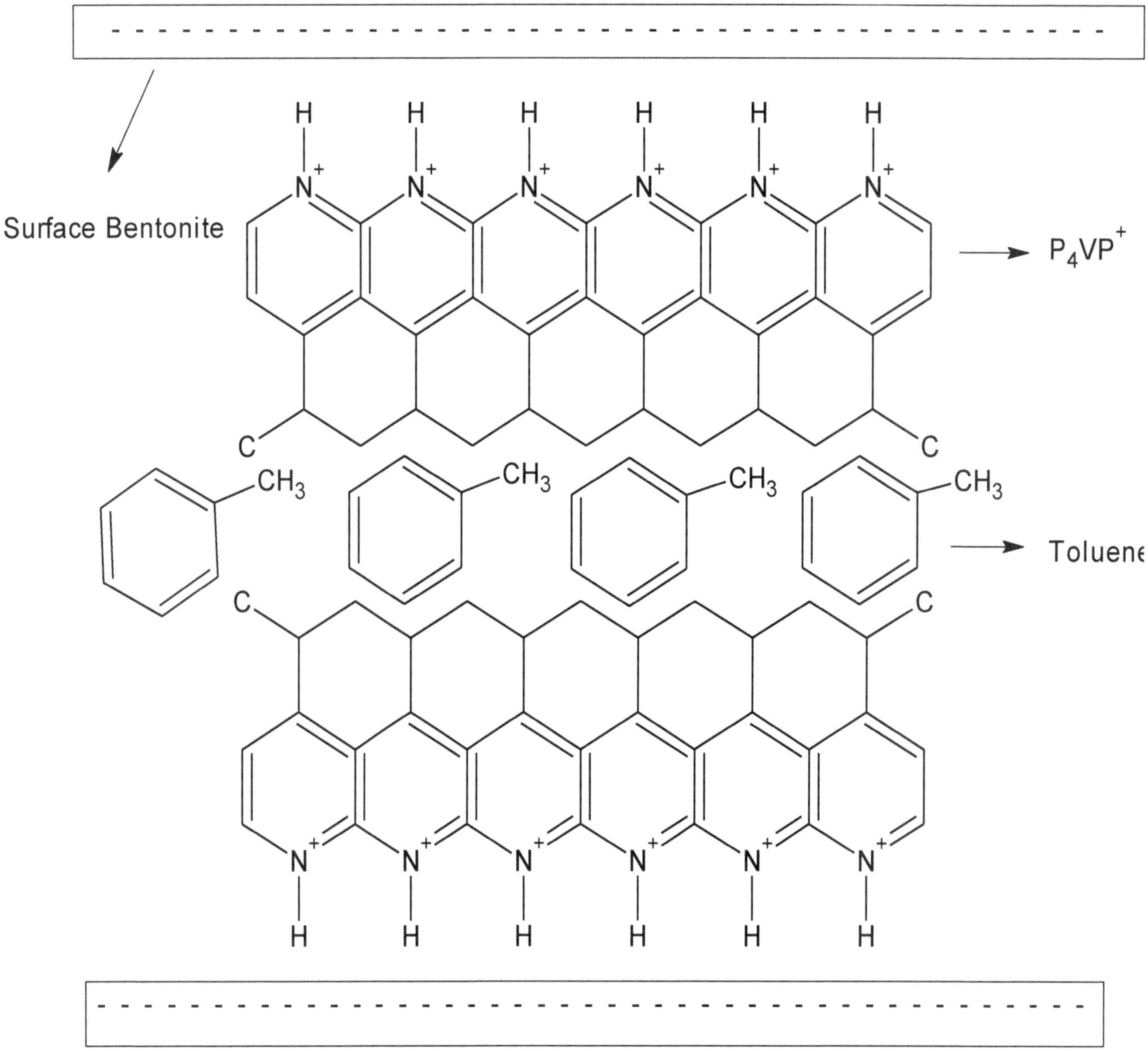

Figure 1. Proposed microstructures in the interlayer of P4VPS-Mag.

Effect of pH

The pH is the most important factor affecting the adsorption process. To study the influence of pH on the adsorption capacity of P4VPS-Maghnite, experiments were performed using various initial pHs varying from 0, 5 to 3, 8. The uptake of toluene decreased with increasing pH (Figure 6). It was observed that the adsorption is highly dependent on the pH of the solution, which affects the surface charge of the adsorbent and the ionisation's degree and speciation of the adsorbate. The quaternised polymer is adsorbed on the surface of maghnite; it associated with a surface of the clay negative charges in order to compensate for its positive loads N^+. The observation of this result is that if quaternised the polymer load factor increases, the polarity of interlayers spaces decreases more and more, and by consequence the toluene retention becomes important. For poly(4-vinylpyridinium)-maghnite (P4VPS-Mag), the attractive forces to organic solutes were composed of the affinity of exposed siloxane surfaces and adsorbed N^+ from polymer, and adsorption toluene as described in Figure 1.

Conclusion

This paper deals with the removal of toluene from aqueous solution by two-step; first step organo-maghnite

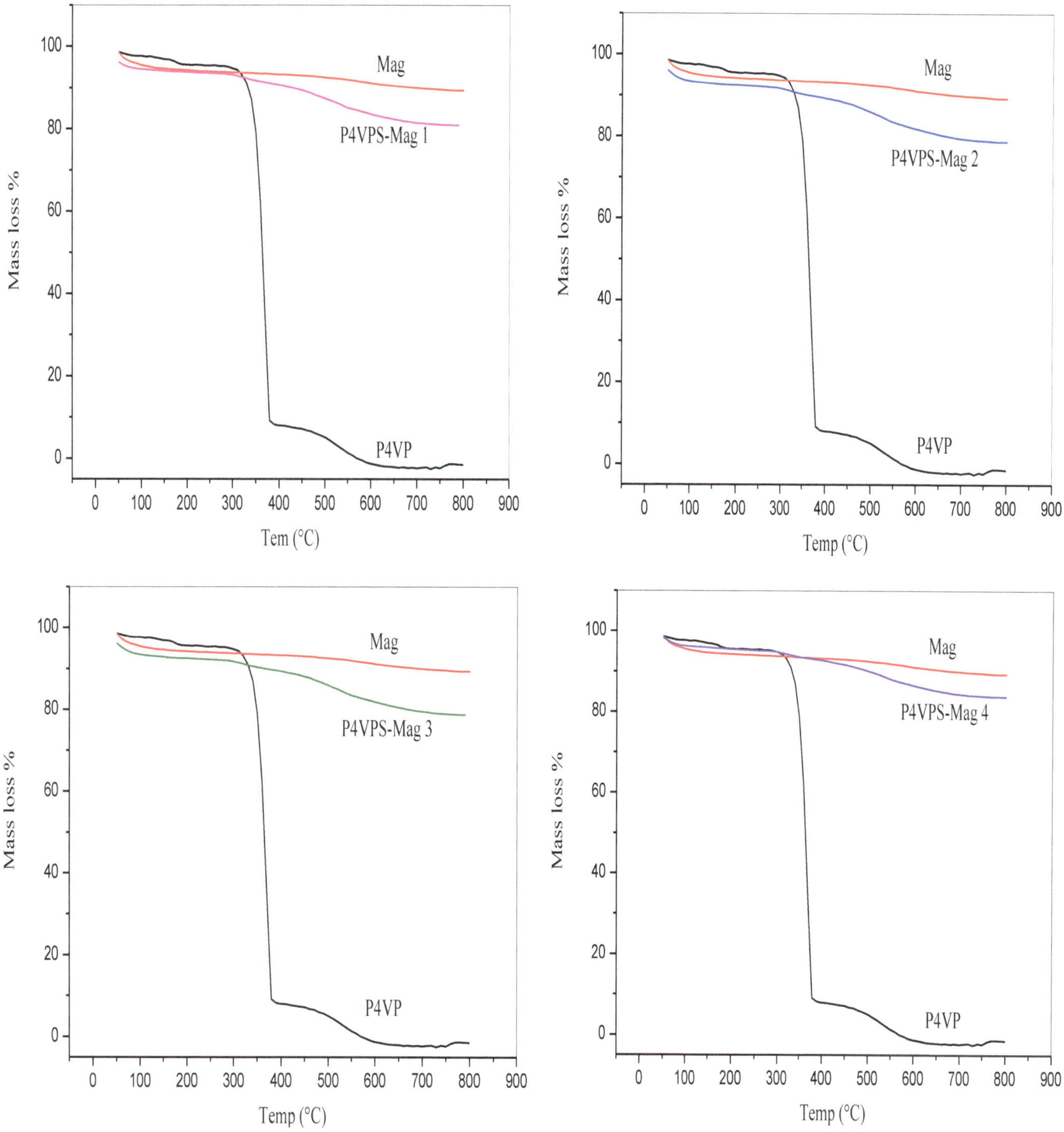

Figure 2. TGA curves for P4VP, Mag and P4VPS– Maghnite.

(poly(4-vinylpyridinium)-maghnite) which is a simple bentonite exchanged with polycationic poly(4-vinylpyridinium) salt (P4VPS) and the second step is the adsorption of the pollutant toluene by organo-maghnite. The results of this study indicate that the method improves greatly the adsorption behaviour of toluene and could be quickly removed from water by P4VPS-Maghnite. The maximum uptake of toluene occurred at higher degree of α ionization because of the high rate quaternisation polymer P4VP. These results indicate the interest of the degree of ionization α consideration in the preparation of organo-maghnite based P4VPS. This

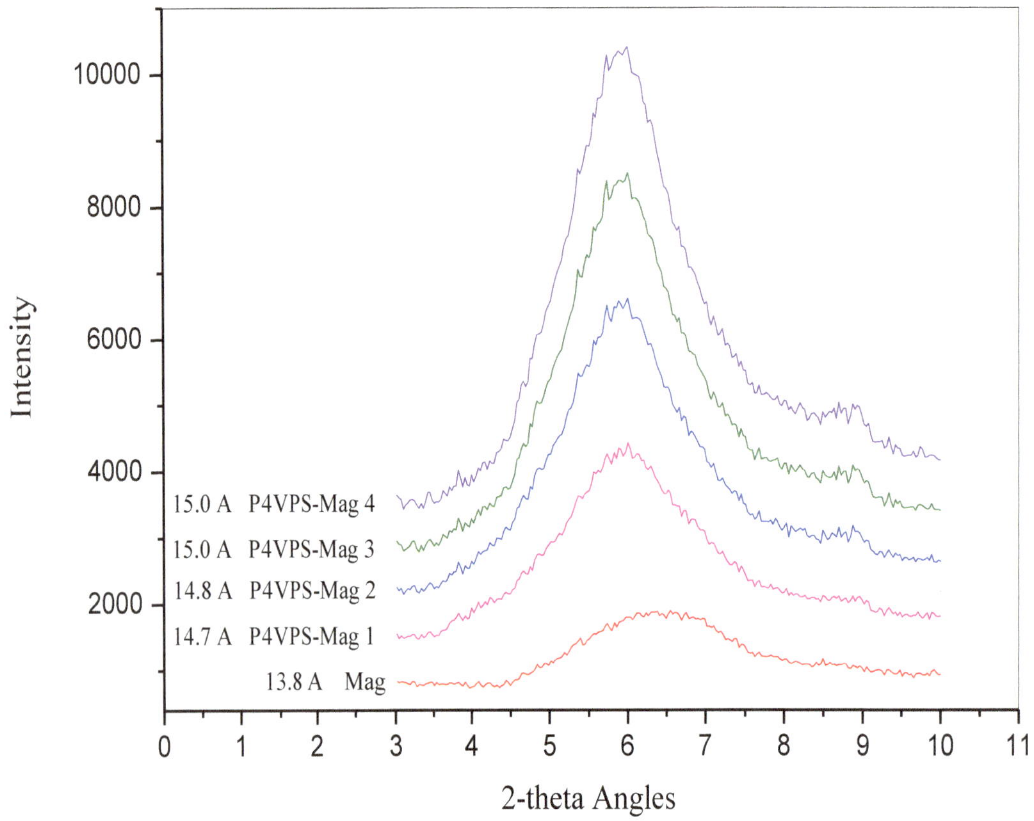

Figure 3. X-ray diffraction patterns of curves Maghnite and P4VPS – Maghnite.

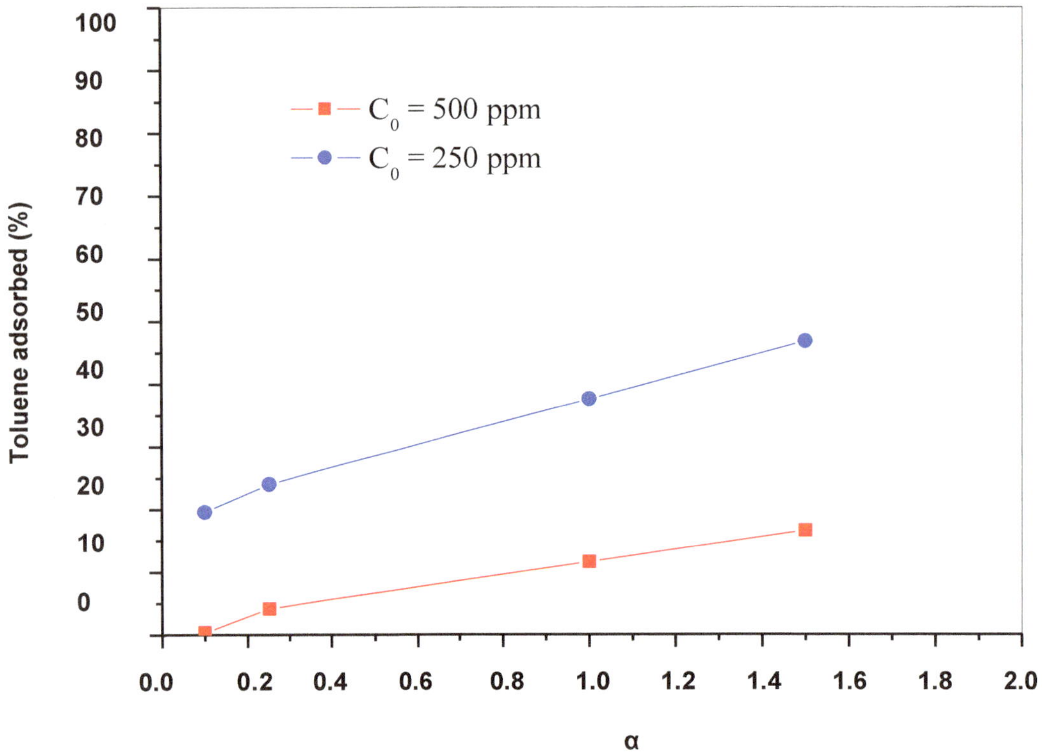

Figure 4. The effect of ionisation degree on the percentage retention of toluene.

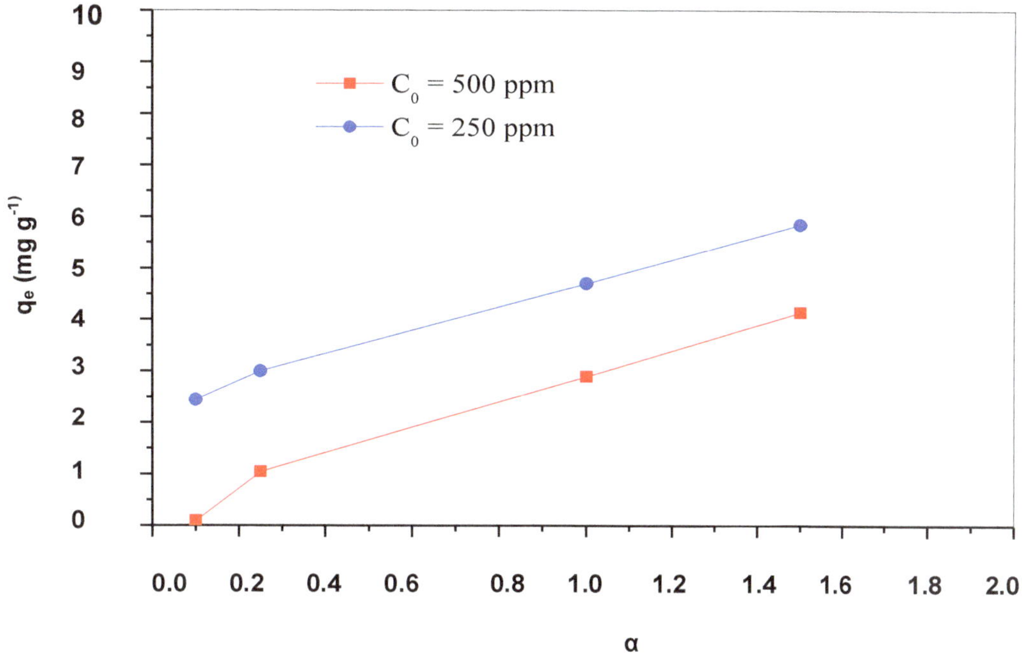

Figure 5. The effect of ionisation degree on the retention of toluene.

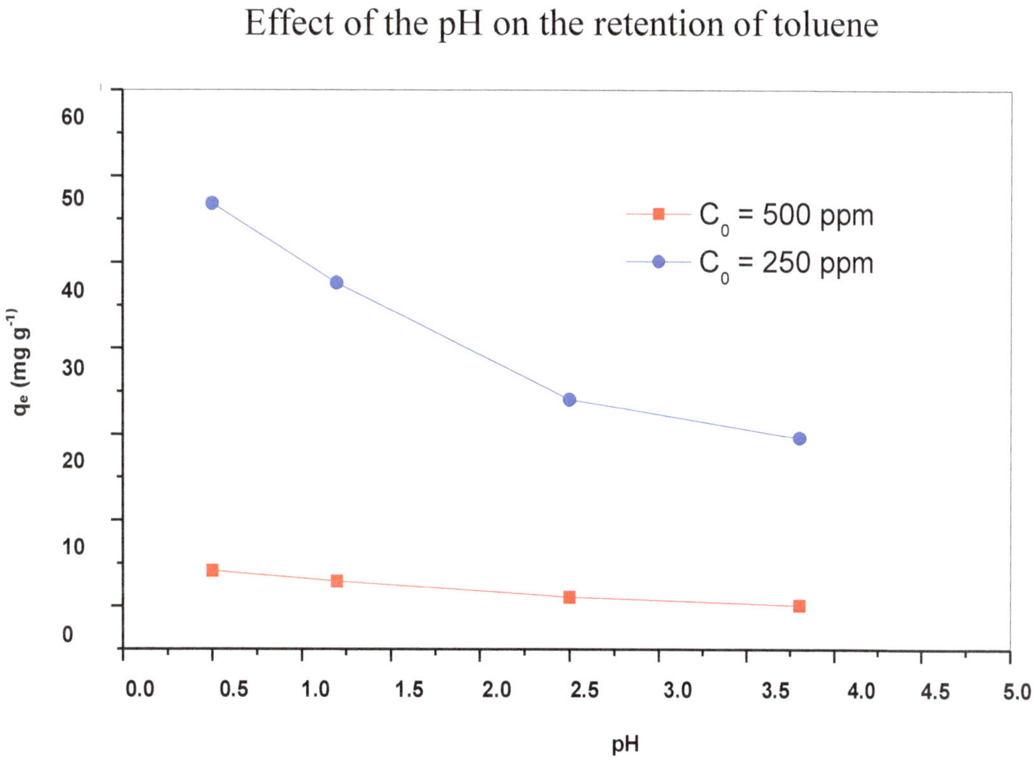

Figure 6. The effect of the pH on the retention of toluene.

knowledge will be useful in selecting the concentration of P4VPS that will be the most appropriate for application in remediation technologies for organo-bentonite or water polluted by hydrophobic organic compounds, thus avoiding excessive use of polymer. The findings of this study were useful in determining optimum conditions (degree of α ionisation for toluene removal both from water on organophilic maghnite (presence of cationic polymer).

REFERENCES

Becker N, Chin H, Cumming J, Dickinson G, Kolarik L, Stanko J, West J, Yap T (2000). Cleaner production in wool dyebath recycling. Environ. Technol. (submitted for publication).

Billingham J, Breen C, Rawson JO, Yarwood J, Mann BE (1997). The adsorption of polycations on clays. A comparative insitu study using 133Cs and 23Na solution phase NMR. J. Colloid Interface Sci., 193: 183-189.

Boufatit B, Ait-Amar H, Mc Whinnie WR (2008). Development of an algerian material montmorillonite clay - Intercalation with selective long chain alkylammonium cations (Octadecytrimethylammonium, Cetylpyridium and Tetrabutylammonium) and with tellerium complexes. Desalination, 223: 366-374.

Breen C, Watson R (1998). Polycation-exchanged clays as sorbents for organic pollutants. Influence of layer charge on sorption capacity. J. Colloid Interface Sci., 208: 422-429.

Breen C (1999). The characterisation and use of polycation-exchanged bentonites. Appl. Clay Sci. 15: 187-219.

Chang MY, Juang RS (2004). Adsoprtion of tannic acid, humic acid, and dyes from water using the composite of chitosan and activated clay . J. Colloid Interface Sci., 278: 18.

Chigwada G, Wang D, Wilkie CA (2006). Polystyrene nanocomposites based on quinolinium and pyridinium surfactants, Polymer Degradation and Stability, 91: 848-855.

Crini G (2005). Recent developments in polysaccharide-based materials used as adsorbents in wastewater treatment, Prog. Polym. Sci., 30: 38-70.

Durand-Piana G, LaFuma F, Audebert R (1987). Flocculation and adsorption properties of cationic polyelectrolytes toward Na - montmorillonite dilute suspensions. J. Colloid Interface Sci., 119: 474-480.

Fournaris KG, Boukos N, Petridis D (2001). Aqueous polymerization of protonated 4-vinylpyridine in montmorillonite, Appl. Clay Sci., 19: 77-88.

Gupta SS, Bhattacharyya KG (2006). Adsorption of Ni (II) on clays. J. Colloid Interface Sci., 295: 21-32.

Jiang JQ, Cooper C, Ouki S (2002). Comparison of modified montmorillonite adsorbents Part I: preparation, characterization and phenol adsorption, Chemosphere, 47: 711-716.

Jock-Churchman G (2002). Formation of complexes between bentonite and different cationic polyelectrolytes and their use as sorbents for non-ionic and anionic pollutants, Appl. Clay Sci., 21: 177- 189.

Khenifi A, Bouberka Z, Bentaleb K, Hamani H, Derriche Z (2009). Removal of 2,4-DCP from wastewater by CTAB/bentonite using one-step and two-step methods: A comparative study. Chem. Engr. J., 146: 345-354.

Li Q, Yue QY, Su Y, Gao BY, Fu L (2007). Cationic polyelectrolyte/bentonite prepared by ultrasonic technique and its use as adsorbent for Reactive Blue K-GL dye. J. Hazardous Mater., 147: 370-380.

Liu P (2007). Polymer modified clay minerals: A review. Appl. Clay Sci., 38: 64-76.

Rawajfih Z, Nsour N (2006). Characteristic of phenol and chlorinated phenol sorption onto surfactant-modified bentonite, J. Colloid Interface Sci., 298: 39-49.

Vidal NC, Volzone C (2009). Analysis of tetramethylammonium-montmorillonite and retention of toluene from aqueous solution. Appl. Clay Sci., 45: 227-231.

Walker GM, Weatherley LR (1998). Fixed bed adsorption of acid dyes onto activated carbon. Environ. Pollut., Environ. Pollut., 99: 133.

Witthuhn B, Pernyeszi T, Klauth P, Vereecken H, Klumpp E (2005). Sorption study of 2,4-dichlorophenol on organoclays constructed for soil bioremediation, Colloids and Surfaces A: Physicochem. Engr. Aspects, 265: 81-87.

Zadaka D, Nir S, Radian A, Mishael YG (2009). Atrazine removal from water by polycation-clay composites: Effect of dissolved organic matter and comparison to activated carbon. Water Res., 43: 677-683.

Spectrophotometric determination of Chromium(III) in Egyptian ilmenite from phosphate solution using Egyptian white sand (EWS) as a selective adsorbent

Abdou A. Abdou

Nuclear Materials Authority (NMA) Egypt. E-mail: abdouchem76@gmail.com.

Egyptian white sand (EWS) was used as an adsorbent for the selective separation of Cr(III) from phosphate solutions which is to be determined colorimetrically using 1,5-diphenylcarbazide. The experimental factors affecting the maximum selective adsorption/desorption of Cr(III) on/from EWS where the amount of adsorbent (g), pH of initial concentration of Cr(III) in solution, contact time, temperature and initial concentration of Cr(III) in solution were investigated. No effect of $[PO_4]^{3-}$, Ti(IV) and Fe(III) on Cr(III) sorption were verified. Accuracy and relative standard deviations (RSD) were acceptable for all the analyses. The maximum sorption capacity, Q_e, of adsorbed Cr(III) per gram of sand was 93.5 mg/g, at EWS concentration 20 g/L, pH 0.8, contact time 120 min and temperature 30°C for adsorption of Cr(III) on EWS, and pH 0.4, contact time 6 min and temperature 25°C for desorption of Cr(III) from EWS.

Key words: Ilmenite, phosphoric acid, Chromium(III), Egyptian white sand (EWS), adsorption, desorption.

INTRODUCTION

It was reported that about 15% of the total world supply of titanium are supplied from the mineral rutile. Their reserves are limited with titanium demand increasing. Ilmenite, on the other hand, (assaying 50 to 60% TiO_2) is nowadays on adequate supply along worldwide. Most of the ilmenite mining is nowadays consumed to make pigments, TiO_2, which accounts about 95% of the total produced titanium (Mackey, 1974).

Several commercial processes have been set up, including mainly ilmenite smelting and roasting processes. Smelting produces a high titanium slag and a low manganese iron by product as in South Africa, Canada and Norway (Mackey, 1974). Ilmenite roasting to synthetic rutile (91 to 95% TiO_2) involves two main variants, namely, the Becher process (Australia) where they use ilmenite grade of 57 to 63% TiO_2 and the Benelite process in which an acid leaching step follows the roasting process USA, India and Malaysia can treat different grades of ilmenite 50 to 63% TiO_2 (Becher, 1963; Benelite, 1974). The manufacture of pigment is via two main routes; namely, the sulfate route and the chloride route. Each of these two routes requires different feedstock. In the sulphate route, the finely ground feedstock of ilmenite concentrate with minimum assay of 44% TiO_2 or sulfatable titanium slage of 78% TiO_2 are digested in concentrated sulphuric acid yielding a titanium sulfate solution which is later hydrolyzed and precipitated to form TiO_2 pigment and a solid waste consisting mostly of ferrous sulfate heptahydrate crystals.

In the production of titanium dioxide pigments by the "sulfate" process, ilmenite is dissolved in sulfuric acid and chromium is present in this solution as Cr(III) (James, 1970).

In the chloride process, the rutile reacts with chlorine gas to form titanium tetrachloride which is then converted directly to the TiO_2 pigment with the liberation of chlorine gas by heating the vapour in admixture with oxygen or air at high temperatures.

Among trace impurities normally found in white titanium dioxide pigments, chromium has been indicated to have the most adverse effect on whiteness. Stonehill (1958) indicated that chromium in the 10 mg/L range decreased the whiteness and McTaggert and Bear (1955) indicated that traces of chromium enhanced the undesirable photochemical properties of titanium dioxide pigments. James (1970) claimed that the decrease in pigment whiteness could be visually detected when the chromium content was increased to 1 mg/L. He also showed that the reflectance decreased throughout the visible spectra when the chromium content was increased to 2.2 mg/L. Other impurities affected whiteness only at higher concentrations. Iron, for example, had no visual effect up to 22 mg/L. The chromium impurity is reduced somewhat during the processing of titanium ore by the "sulfate" process, but titanium dioxide pigments may still contain 1 to 10 mg/L of chromium. Many other impurities, including iron, are reduced during normal processing to low concentrations where they do not affect whiteness (Sandell, 1959).

In the evaluation of ilmenite concentrates as a source of titanium, the chromium content is of particular significance owing to its effect on the colour of pigment-grade titanium dioxide (Mctaggart and Bear, 1955). Several spectrophotometric procedures for the determination of chromium arc available, the most sensitive and most frequently employed being the diphenylcarbazide method (Blundy, 1958) but few of these procedures have been concerned with materials containing relatively large amounts of iron and titanium as well as vanadium at a similar level to chromium. Iron causes serious interference in the diphenylcarbazide - Cr(VI) reaction (Dean et al., 1948). One common procedure to avoid iron interference is to oxidize Cr(III) to Cr(VI) in a nitrate or peroxide alkaline fusion and filter the iron oxide residue from the aqueous leach to collect the chromium in an iron-free filtrate. This approach has a number of disadvantages, where removal of chromium from the oxide residue may not be absolutely complete (Blundy, 1958); a portion of the iron can remain dispersed in the alkaline filtrate (Dean et al., 1948), peroxide or nitrite from the fluxing materials can cause reduction of Cr(VI) when the solution is acidified, and vanadium, which also interferes in the diphenylcarbazide procedure, is not separated from chromium (Pilkington and Smith, 1967).

Methods in the literature for separating and concentrating chromium by ion exchange before its determination are ultraviolet visible spectrometry, isotope dilution mass spectrometry (IDMS) and graphite furnace atomic absorption spectrometry (GFAAS). However, none of these measurement techniques has the requirement for virtually complete separation of the analyte from its matrix.

Ion exchange chromatography has been applied for the separation of chromium from a variety of matrices (Zmijewska et al., 1984) used batch anion exchange to separate chromium from its matrix prior to its determination by gamma-ray spectrometry. Some workers have succeeded in separating chromium from its matrix using chelating resin (Lee et al., 1977), but many others (Sturgeon et al., 1980; Isozaki et al., 1983) have reported poor recoveries. Since 1988, researchers with the specific objective of preparing chromium for mass spectrometric determination have used, among other techniques, electrolytic separation (Gotz and Heumann, 1988) cation-exchange separation of Cr(III) (Birck and Lugmair, 1988) and anion-exchange separation of Cr(VI) (Volkening and Heumann, 1989).

Several recent publications utilized different inexpensive and locally abundantly available adsorbents for chromium separation such as activated carbon (Leyva-Ramos et al., 1995), agricultural by-products (Samantaroy et al., 1997), waste materials (Namasivayam and Yamuna, 1995), and charge minerals (Singh et al., 1992) which are efficient adsorbent for chromium separation from its matrix but suffers a disadvantage that it is expensive.

Several reports deal with the use of white sand as an inexpensive and efficient adsorbent of heavy metals from industrial effluents such as chromium (Baig et al., 2003; Aslam et al., 2004).

White sand can be formed in nature by natural weathering of sandstone and quartzite or mechanically by crushing a sandstone and quartzite or by flotation process, whereby the various constituents in a pegmatite or kaolin mixture are separated. The occurrence of silica sand is widespread and extensive. Silica is an adsorbent widely used in the purification and separation processes due to its adsorption properties, high surface area of active sites and porosity (Unob et al., 2007).

In this paper, the Egyptian white sand (EWS), is a low-cost locally available adsorbent, where several factors were studied to determine the efficiency for separation of Cr(III) from Egyptian ilmenite mineral solution after complete dissolution using phosphoric acid. The amount of adsorbent (g), effects of pH, contact time and temperature, beside factors affecting on desorption of Cr(III) from EWS were investigated by varying any one of the process parameters and holding the other parameters constant. Throughout the study, the amount of adsorbent from 4 to 30 g/L, the contact time was varied from 30 to 300 min, the pH from 0.2 to 1.5, temperature from 25 to 60°C and the initial Cr(III) concentration from 20 to 100 mg/L.

EXPERIMENTAL

Materials

All primary chemicals used were of analytical reagent grade. $CrCl_3 .6H_2O$, $FeCl_3.6H_2O$, TiO_2, NH_4F, H_3PO_4, H_2SO_4 and 1,5 diphenylcarbazide were purchased from Merck, Germany.

Preparation of the adsorbates and working solutions

A stock solutions (5000 mg/L) of Cr(III), Ti(IV) and Fe(III) were prepared by dissolving 2.56 g of AR grade $CrCl_3 .6H_2O$, 0.835 g of AR grade TiO_2 and 2.42 g of AR grade $FeCl_3.6H_2O$, respectively in 100 ml of distilled water. Standard solutions of the required Cr(III),

Ti(IV) and Fe(III) were prepared by appropriate dilution.

Preparation of the adsorbents (EWS) and their washing solutions

The EWS was collected from Southwestern Sinai Desert, Egypt which was sieved to obtain grain size of (50 to 200 mesh), the sand was washed repeatedly with hot water, 6 M hydrochloric acid, washing solution, filtered off and washed with distilled water. It was then dried at 110°C for 2 h. The washing solution was prepared by mixing 5 ml of 0.5 mol L^{-1} sodium chloride, 3 ml of 0.3 mol L^{-1} ammonium chloride, 2 ml of 0.015 mol L^{-1} ethylene diamine tetraacetic acid (EDTA) and 2 ml of 0.015 mol L^{-1} tartaric acid. This solution was used for washing EWS for regenerating it before use (Zdenek and Vaclav, 1971).

Ilmenite concentrate

A technological sample of Rosetta beach ilmenite concentrate was kindly provided by the Black Sands Project of Nuclear Materials Authority (NMA) (Abdou, 2011), A reference ilmenite sample (Rosetta-Refer.) (Abdou, 2011), beside international standard sample (SARM 60) (Hansen, 1996) was also included for controlling analysis confidence in the present work.

Instrumentations

1) UV-Double Beam Spectrophotometer (UNICAM) was used with 1 cm cells. The optical system was checked automatically and the instrument was periodically calibrated, using arsenazo III (Marczenko, 1990).
2) pH-meter (SCHOTT GERATE, Germany), was applied for all experiments in the present work. Calibration of the pH-meter was carried out before each experiment by using two successive buffer solutions (pH 4 and 7, pH 7 and 10).

Experimental methods

Loading of Cr(III) on the EWS

Different concentrations [20 to 100 mg/L Cr(III)] were mixed separately with 500 mg/L Ti(IV), 600 mg/L Fe(III) and 1 ml of 1 mol L^{-1} NH_4F which acts as masking agent for ferric ions (Marczenko, 1990), then completed to 100 ml volumetric flask using diluted H_3PO_4 (<0.2 mol L^{-1}). Each solution mixture was then loaded, by using batch adsorption process in duplicate to establish the accuracy of the procedure, on a known amount of EWS adsorbent using 250 ml conical flasks containing 100 ml of the test solutions at the desired initial chromium concentration using the previously optimum conditions for adsorption. The contents of the flask were filtered, Ti(IV) and Fe(III) in each effluent solution (unabsorbed) were then determined.

All the above experiments were carried out using 60 mg/L Cr(III) only without adding constant concentrations from Ti(IV) and Fe(III) to study factors affecting on loading of Cr(III) on the EWS. The concentrations of Cr(III) ions adsorbed were obtained by difference. The concentration of Cr(III) before and after adsorption was obtained using the calibration curve.

Elution of Cr(III) from EWS

Elution was performed with 10 ml of 1 mol L^{-1} H_2SO_4 solution and collected in beakers for chromium determination using 1,5-diphenylcarbazied procedure.

Determination of chromium

In the determination of chromium concentration by UV-VIS Spectrometric method, Cr(III) is converted to Cr(VI) using permanganate and 30% H_2O_2. The Cr(III) is oxidized to $Cr_2O_7^{2-}$ (dichromate) with $KMnO_4$ and determined colorimetrically using 0.25% 1,5-diphenylcarbazide to form a purple-violet colored complex. The absorbance of the coloured complex was measured using a UV-VIS double beam spectrophotometer model UNICAM, at the maximum absorption wavelength of 540 nm, and chromium concentration of tested samples was determined by a calibration curve prepared using standard solutions of concentrations of Cr(III) between 0.15 and 0.7 mg/L (Greenberg et al., 1992).

Data analysis

The percentage adsorption of chromium was calculated as follows:

$$\%\ \text{Adsorption of Cr(III)} = \frac{(C_{int} - C_{fin})}{C_{int}} \times 100 \quad (1)$$

Where C_{int} and C_{fin} are the initial and final chromium concentrations (mg/L), respectively.

Efficiency of EWS (sorption capacity) of adsorbed chromium per gram of EWS, Q_e, was calculated from Equation 2:

$$Q_e\ (\text{mg/g}) = \frac{(C_{int} - C_{fin}) \times V}{m} \quad (2)$$

Where V and m are the volume of the solution (ml) and the mass of adsorbent (g) (Demirbas et al., 2004).

RESULTS AND DISCUSSION

Factors affecting the loading of chromium on EWS

Effect of the amount of adsorbent (g) on the adsorption of Cr(III) ions

EWS was used at concentration ranging from 4 to 30 g/L in a batch adsorption technique, while the other parameters were kept constant. At 4 g/L of adsorbent, the adsorption of Cr(III) was found to be 34.2%. Increasing the adsorbent concentration resulted in an increase in the separation percentage of Cr(III) (Figure 1). At 20 g/L of EWS, the adsorption of Cr(III) from solution was found to be 91.5%. The variation in the sorption capacity of EWS adsorbents could be related to the type and concentration of surface groups responsible for interaction with the metal ions Cr(III). The percent adsorption of Cr(III) by EWS adsorbent increased with increasing the amount of adsorbent. This may be attributed to increased surface area and the availability of more adsorption active sites. Any further addition of the adsorbent more than 20 g/L beyond this did not cause any significant change in the adsorption may be due to the formation of clusters of adsorbent particles resulting in decreased surface area

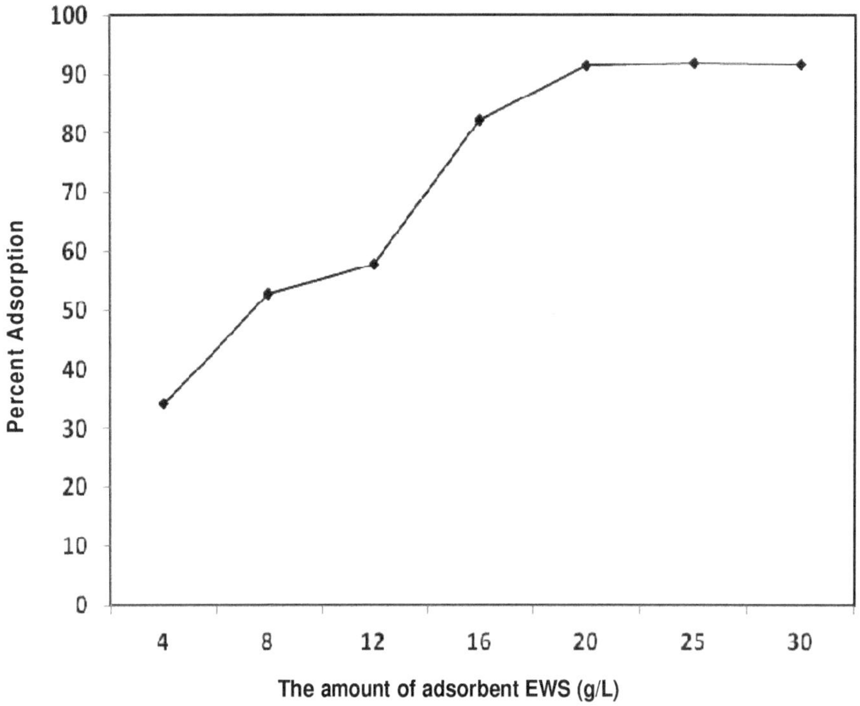

Figure 1. Effect of adsorbent concentration on Cr(III) adsorption using EWS: At a Cr(III) concentration of 60 mg/L.

Table 1. Chemical analysis of natural Egyptian white silica sand (EWS).

Element oxide	Weight percent
SiO_2	98.5
TiO_2	0.12
Al_2O_3	0.31
Na_2O	0.11
Fe_2O_3	0.22
H_2O	0.11
Total	99.37

(Al-Anber, 2010).

The composition of sand is highly variable, depending on the local rock sources and conditions. Egyptian natural white silica sand was subjected to chemical analysis before starting the experiments after physical separation of impurities as shown in Table 1. The sand is mainly composed of SiO_2 (>98% quartz).

The sorption of metal ions can take place by the cation exchange reaction through the substitution of protons from silanol groups on the surface by the adsorbed metal ions from the solution, as follows:

$$\mathrm{m(SiOH)} \leftrightarrow \mathrm{m(SiO^-)} + \mathrm{mH^+} \quad (3)$$

$$\mathrm{M^{n+}} + \mathrm{m(SiO^-)} \leftrightarrow \mathrm{M(OSi)_m^{(n-m)+}} \quad (4)$$

The overall reaction can thus be represented as:

$$\mathrm{M^{n+}} + \mathrm{m(SiOH)} \leftrightarrow \mathrm{M(OSi)_m^{(n-m)+}} + \mathrm{mH^+} \quad (5)$$

Where M^{n+} is the metal ion with n^+ charge such as Cr^{3+}, SiOH = silanol group on SiO_2 surface, mH^+ = number of protons released. Although the ion exchange reaction is able to explain the sorption process, metal sorption on the white silica sand takes place through complex formation on the cell surface after interaction between the metal and active groups present on white silica sand surface (Lee and Allen, 2001).

In order to understand the adsorption of chromium on sand, there should be consideration of the types of associations among adsorbing metal ions and silica. Silica (SiO_2) has a structure composed of infinite three-dimensional framework of tetrahedron (Murray, 1994). Each silicon atom forms four single bonds with four oxygen atoms located at the four corners of a tetrahedron.

A surface functional group in silicates plays a significant role in the adsorption process. It is a plane of oxygen atoms bound to the silica tetrahedral layer and hydroxyl groups that are associated with the edges of the silicate structural units (Donald, 1998). These functional groups provide surface sites for the chemisorptions of transition and heavy metals such as M^{n+} (Murray, 1994). The surface functional groups can be represented as

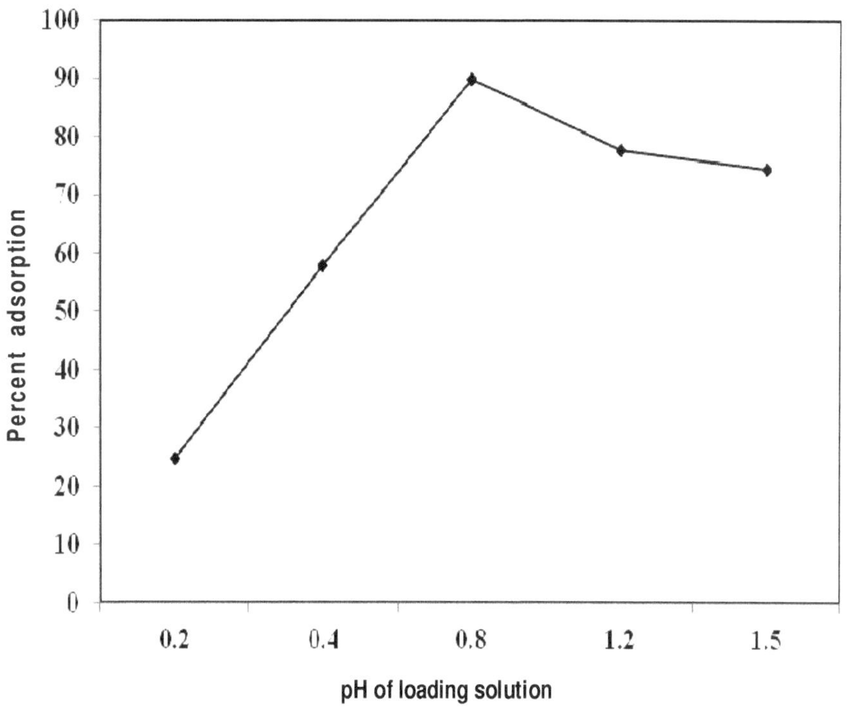

Figure 2. Effect of pH on Cr(III) adsorption using EWS at a Cr(III) concentration of 60 mg/L and adsorbent concentration (EWS) 20 g/L.

shown in Equation 6:

$$\equiv Si-OH + M^{n+} \rightleftharpoons \equiv Si-OM^{(n-1)+} + H^{+} \quad (6)$$

Effect of pH on the adsorption of Cr(III) ions

There was an increase in the pH of the solution from the initial value, pH_{int}, after stirring of the batch mixture for 60 min with a sample solution containing 60 mg/L Cr(III) while the other parameters were kept constant. The percentage adsorption of Cr(III) from solution was affected dramatically by the change in pH_{int}. The adsorption of Cr(III), increases from lower pH values till it reaches maximum at pH 0.8 after which it decreases again (Figure 2). At the later pH, the number of negatively charged groups on the adsorbent matrix increases and enhances the adsorption of the Cr(III) species by coulombic attraction, as shown in Equation 7.

$$2(\equiv O^{-}) + Cr(OH)_2^{+} \leftrightarrow (\equiv O^{-})_2\,Cr(OH)_2^{+} \quad (7)$$

The results for the adsorption of Cr(III) is shown in (Figure 2). Generally, Cr(III) is better adsorbed at low pH values. This could be due to the fact that at low pH, Cr exist as Cr(III) and can therefore be better adsorbed by the C=O functional group on the adsorbent since oxygen is electronegative and can attract ions that are positively charged (Sharma and Forster, 1994).

Effect of contact time on the adsorption of Cr(III) ions

The effect of contact time on adsorption of Cr(III) was studied and results are shown in Figure 3, while the other parameters were kept constant. 60 mg/L of Cr(III) solution and 20 g/L of the EWS adsorbent were used for this experiment. A pH of 0.8 was used for Cr(III) solution. It was observed that the EWS adsorbent consistently have an acceptable percentage of adsorption for chromium through all the contact times used which means equilibrium was already reached as early as 120 min. The percentage of Cr(III) adsorbed increased till it reached maximum at 120 min and it significantly decreased after that, which could be as a result of desorption of the metal ions.

Effect of temperature on the adsorption of Cr(III) ions

The effect of temperature on adsorption of Cr(III), using EWS was studied from 25 to 60°C, while the other parameters were kept constant. As shown in Figure 4, the decrease in adsorption of Cr(III) as temperature increases may be due to the relative increase in the escaping tendency of the metal ions from the solid phase to the bulk phase; or due to the weakness of adsorptive forces between the active sites of the adsorbents and the

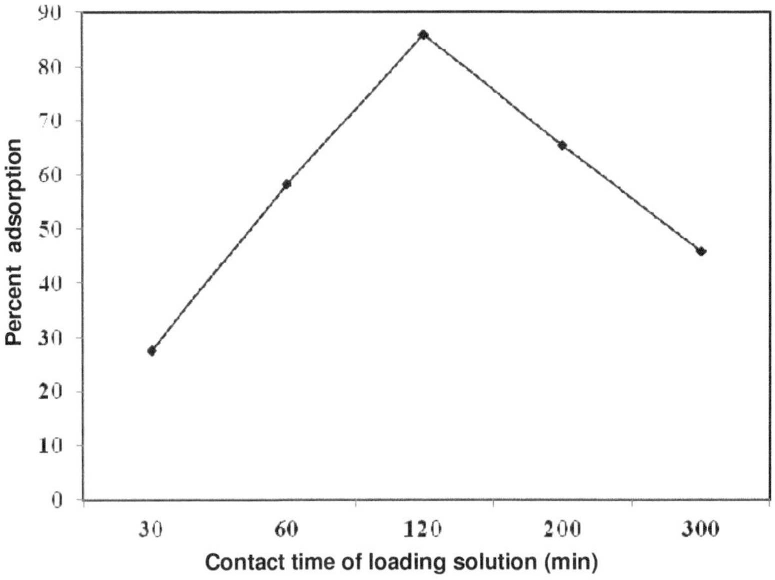

Figure 3. Effect of contact time on Cr(III) adsorption using EWS at a Cr(III) concentration of 60 mg/L; adsorbent concentration (EWS) 20 g/L and pH 0.8.

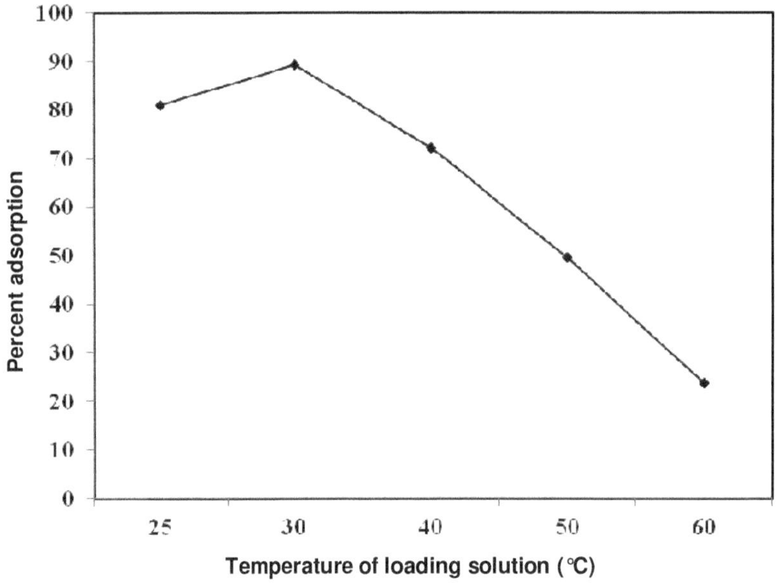

Figure 4. Effect of temperature on Cr(III) adsorption using EWS at a Cr(III) concentration of 60 mg/L; adsorbent concentration (EWS) 20 g/L; pH 0.8 and contact time 120 min.

adsorbate species and between the adjacent molecules of adsorbed phase (Meena et al., 2005).

Factors affecting the desorption of Cr(III) from the EWS

Effect of different eluent reagents

Optimizing the concentration of eluting reagent such as H_2SO_4, HNO_3, HCl and NaOH to reach the maximum desorption of Cr(III) from EWS was studied, keeping the other parameters constant. Figure 5 shows the percent of Cr(III) desorption increased with increase in sulfuric acid concentrations from 0.2 to 1.0 mol L^{-1} which is greater than the other eluent.

For the desorption of Cr(III) sorbed on EWS, an acidic sulfate solution of 1 mol L^{-1} H_2SO_4 was found to be the best eluent reagent in comparison with the other eluents

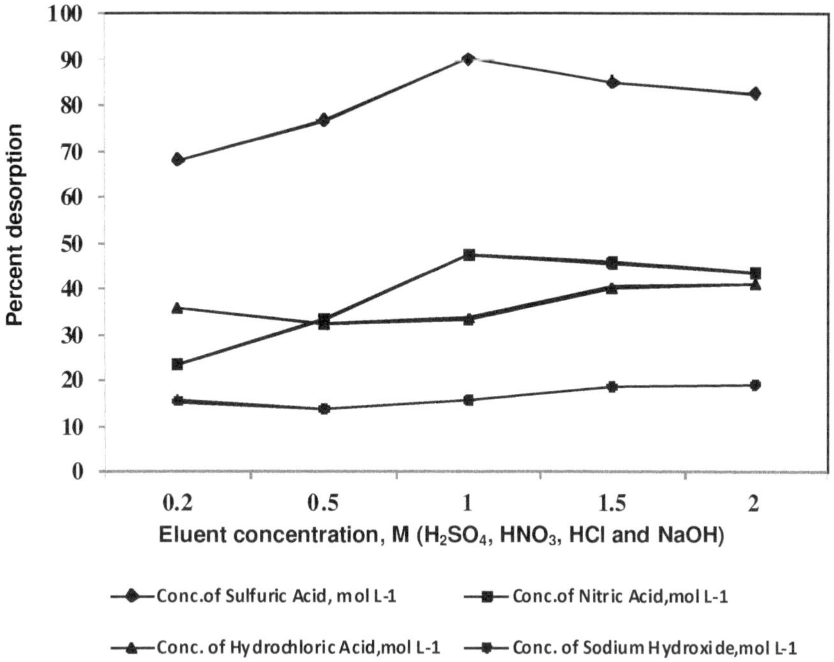

Figure 5. Optimization of the effective eluent for Cr(III) desorption from EWS: [Ti(IV) 500 mg/L; Fe(III) 600 mg/L; 1 ml of 1 mol L^{-1} NH_4F; Cr(III) 60 mg/L; adsorbent concentration (EWS) 20 g/L and 10 ml of 0.2 to 2.0 mol L^{-1} (H_2SO_4 or HNO_3 or HCl or NaOH)].

such as HNO_3, HCl and NaOH. Desorption or elution takes place because of the formation of ion exchange between Cr(III) sorbed on EWS and acidic sulfate solution of 1 mol L^{-1} H_2SO_4. The desorption effect of the acidic solution of H_2SO_4 depends on a number of factors, such as pH of desorbed solution, contact time and temperature were studied to give the best optimum conditions for Cr(III) desorption.

The effect of pH of desorbed solution on the desorption of Cr(III) from the EWS using 10 ml of 1M H_2SO_4

The effect of pH on desorption of Cr(III) from EWS was studied and results are shown in Figure 6, while the other parameters were kept constant. 60 mg/L of Cr(III) solution and 20 g/L of the EWS adsorbent were used forthis experiment. It was observed that the percentage of Cr(III) desorbed increased till it reached maximum at 0.4 pH and it significantly decreased after that; this is because the pH of the initial eluting solution is significantly influent to the desorption process as it controls the electrostatic interactions between the adsorbent and the adsorbate. When pH of Cr(III) increased above 0.4, the precipitates were formed, because high hindrance between Cr(III) and hydroxyl group of surface bond of adsorbent and other inferences in the solution could hinder the diffusion of chromium ion into the surface and pore of EWS. Thus, the optimum pH for Cr(III) desorption on EWS was found to be 0.4.

The effect of contact time on the desorption of Cr(III) from the EWS using 10 ml of 1 mol L^{-1} H_2SO_4

The effect of contact time on desorption of Cr(III) from EWS was studied and results are shown in Figure 7. It was observed that the percent of Cr(III) desorption increased with increase in the contact time from 2 to 6 min and reached a maximum value after 6 min.

The effect of temperature on the desorption of Cr(III) from the EWS

The effect of temperature on the desorption of Cr(III) from EWS was studied and results are shown in Figure 8. It was observed that the percent of Cr(III) desorption has maximum value at 25°C and decreased with increase in the temperature from 30 to 60°C.

The effect of Cr(III) concentration on desorption from phosphate solution (Construction calibration curve)

The effect of Cr(III) ion concentration adsorbed is shown in Figure 9. The desorption of Cr(III) from adsorbent EWS

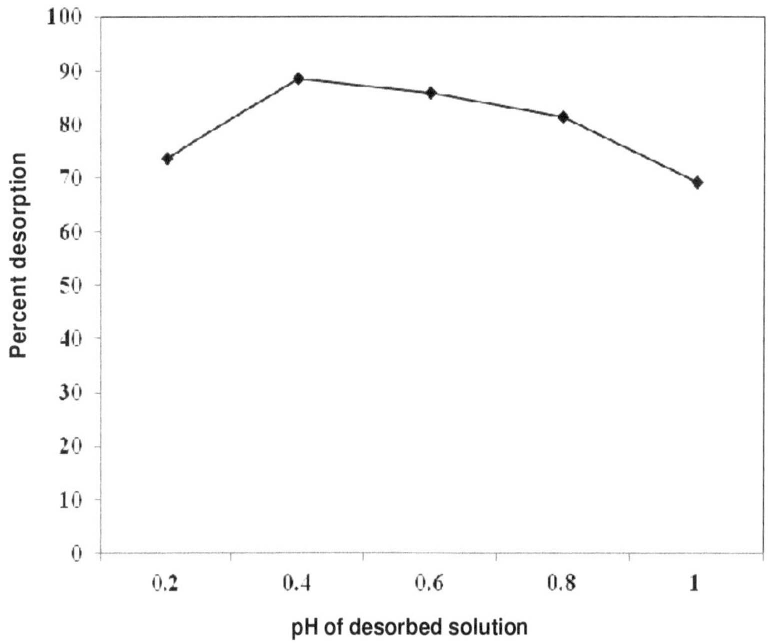

Figure 6. Effect of pH for Cr(III) desorption from adsorbent EWS: [Ti(IV) 500 mg/L; Fe(III) 600 mg/L; 1 ml of 1 mol L^{-1} NH_4F; Cr(III) 60 mg/L; adsorbent concentration (EWS) 20 g/L and 10 ml of 1 mol L^{-1} H_2SO_4].

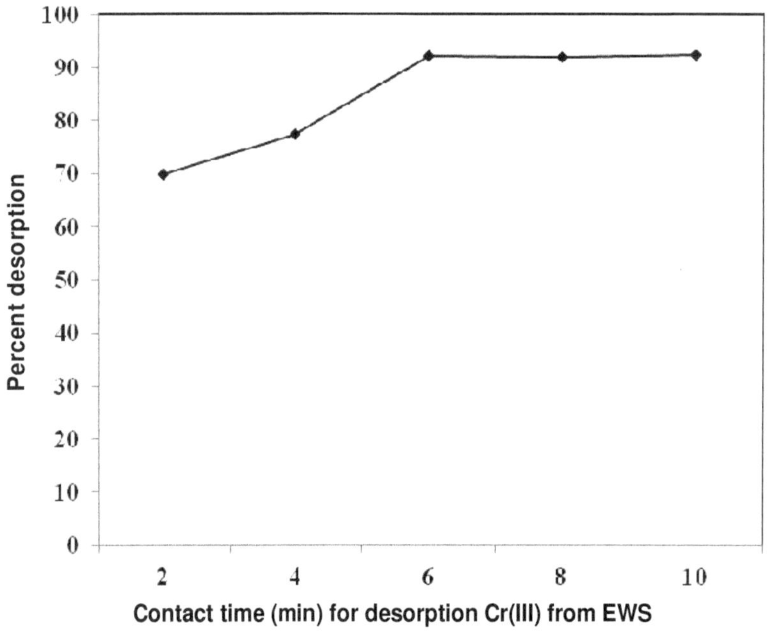

Figure 7. Effect of contact time on Cr(III) desorption from EWS: [Ti(IV) 500 mg/L; Fe(III) 600 mg/L; 1 ml of 1 mol L^{-1} NH_4F; Cr(III) 60 mg/L; adsorbent concentration (EWS) 20 g/L;10 ml of 1 mol L^{-1} H_2SO_4 and 0.4 pH].

increased as the initial concentration of the Cr(III) ion was increased. The EWS was able to desorbs a concentration of 87.6 mg/L, when an initial concentration of 100 mg/L solution of Cr(III) ion was used.

Optimum conditions for Cr(III) adsorption from phosphate solution are as follows:

1) Concentration of adsorbent (EWS) 20 g/L - pH of

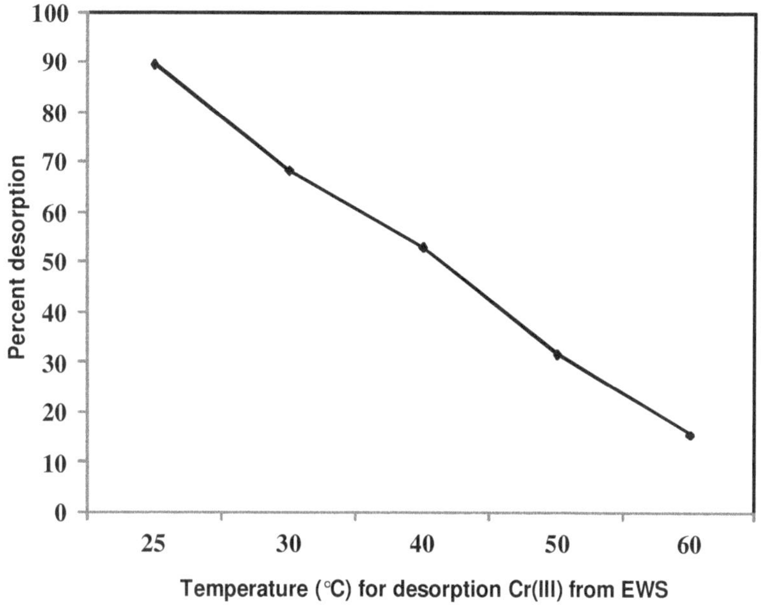

Figure 8. Effect of temperature (°C) on Cr(III) desorption from EWS: [Ti(IV) 500 mg/L; Fe(III) 600 mg/L; 1mL of 1 mol L^{-1} NH_4F; Cr(III) 60mg/L; adsorbent concentration (EWS) 20 g/L;10 ml of 1 mol L^{-1} H_2SO_4; 0.4 pH and contact time 6 min].

loading solution 0.8,
2) Contact time 120 min - Temperature of loading solution 30°C Optimum conditions for Cr(III) desorption from phosphate solution are as follows:

i) Eluting reagent 10 ml of 1 mol L^{-1} H_2SO_4 - pH of desorbed solution 0.4
ii) Contact time 6 min - Temperature of desorbed solution 25°C.

Studying the effect of adding constant concentration of Ti(IV) and Fe(III) on the desorption of Cr(III) on EWS

Effect of adding constant concentration of Ti(IV) and Fe(III) on the desorption of Cr(III) from EWS is shown in Table 2 which illustrated that there is no effect of Ti(IV) and Fe(III) concentration on the desorption of Cr(III) after adding 1 ml of 1 mol L^{-1} NH_4F for masking Fe(III) to avoid adsorption of Fe(III) on EWS and this adsorbent is more selective for Cr(III).

Application

The proposed dissolution method of the ilmenite samples for Cr analysis

Dissolution of the samples (Rosetta-Ilmenite, Rosetta-Refer and international standard sample) SARM 60 was proposed by H_3PO_4 as follows: 0.5 g of each dried sample was wetted with few drops of water in a conical flask; 80 ml of 85% H_3PO_4 was consequently added. The system was heated, with continuously stirring, at 200°C for 15 min to promote the sample dissolution in the evacuated system for keeping chromium in its reduced form Cr(III), leave to cool and complete up to 100 ml volumetric flask with distilled water. A suitable sample volume from the sample solution was mixed with 2 g of EWS, completed up to 100 ml conical flask, adjust the pH to 0.8. This solution was capped and shaken in the flask shaker at 200 rpm for 2 h at 30°C. The sample was filtered and the effluent concentration of Ti and Fe were determined. Desorption of Cr(III) from EWS was performed by using 10 ml of 1 mol L^{-1} H_2SO_4, 0.4 pH, 6 min and 25°C as shown in Table 3.

RECOMMENDATIONS STEPS FOR THE SELECTIVE SEPARATION OF Cr(III)

1) The EWS was first washed with hot water, 6 mol L^{-1} hydrochloric acid, washing solution, filtered off and washed with distilled water, dried at 110°C for 2 h for loading before use.
2) A weight 2.0 g of EWS adsorbent was taken in 100 ml of Cr(III) in phosphate solution after complete dissolution of ilmenite samples with H_3PO_4.
3) The conical flasks were capped and shaken in the flask shaker at 200 rpm for 2 h.

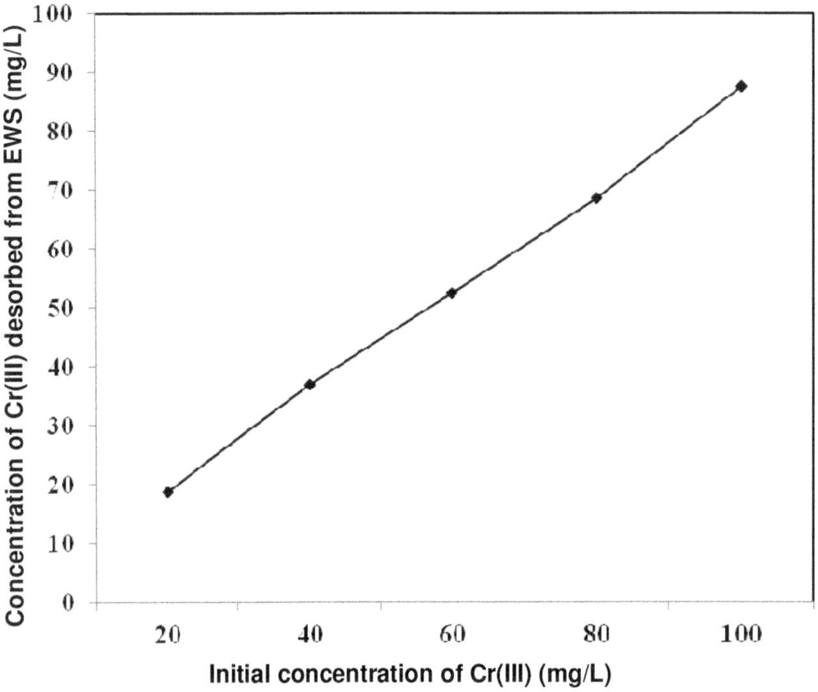

Figure 9. Effect of concentration of Cr(III) on desorption from EWS: [Ti(IV) 500 mg/L; Fe(III) 600 mg/L; 1 ml of 1 mol L^{-1} NH_4F; Cr(III) (20 to 100) mg/L; adsorbent concentration (EWS) 20 g/L; pH 0.8; 120 min; 30℃ and 10 ml of 1 mol L^{-1} H_2SO_4; 0.4 pH; 6 min and 25℃].

Table 2. Effect of constant concentration of Ti(IV) and Fe(III) on the desorption of Cr(III) from EWS: [Ti(IV) 500 mg/L; Fe(III) 600 mg/L; 1 ml 1 mol L^{-1} NH_4F; Cr(III) 20 to 100 mg/L; adsorbent concentration (EWS) 20 g/L; pH 0.8; 10 ml of 1 mol L^{-1} H_2SO_4; 0.4 pH; contact time 6 min and temperature 25℃].

Different concentration of Cr(III) (mg/L)	Concentration of Ti(IV) (mg/L) unabsorbed solution (Effluent)	Concentration of Fe(III) (mg/L) unabsorbed solution (Effluent)	Percent desorption of Cr(III)
20	491.1	582.8	93.5
40	492.5	576.4	92.0
60	489.7	578.6	87.3
80	493.5	575.6	85.7
100	486.8	573.0	87.6

Table 3. Spectrophotometric determination of Cr(III), Ti(IV) and Fe(III) in the ilmenite samples (Rosetta-Ilmenite, Rosetta-Refer and international standard SARM 60).

Sample name	Calculated (C_{int}-C_{fin}) concentration of Cr(III) (mg/L) (Effluent)	Concentration of Cr(III) (mg/L) desorbed solution	Concentration of Ti(IV) (mg/L) unabsorbed solution (Effluent)	Concentration of Fe(III) (mg/L) unabsorbed solution (Effluent)	Percent desorption of Cr(III)
Rosetta-Ilmenite	50	(1050)[a]	(22.61×10^4)[a]	(41.1×10^4)[a]	95.45
Rosetta-Reference	20	(980)[b]	(27.9×10^4)[b]	(30.2×10^4)[b]	98.0
SARM (60)	30	(450)[c]	(28.6×10^4)[c]	(35.5×10^4)[c]	93.75

[a], Obtained earlier out of Ph.D. (Abdou, 2011); Cr(III) = 1100 mg/L, Ti(IV) = 22.59×10^4 mg/L and Total Fe = 41.39×10^4 mg/L. [b], obtained earlier out of many laboratories; Cr(III) = 1000mg/l, Ti(IV) = 27.8×10^4mg/L and Total Fe – 30.8×10^4 mg/l . [c], certified value (SARM 60); Cr(III) = 480 mg/L, Ti(IV) = 28.7×10^4 mg/L and Total Fe = 36.1×10^4 mg/L. Statistical calculations were carried out for SARM 60 and results obtained are: Sd = 0.365 and RSD = 0.1 (for Cr); Sd = 0.291 and RSD = 1.02 (for Ti); Sd = 0.311 and RSD = 0.88 (for Total Fe).

4) The pH for loading solution was adjusted at 0.8 and the temperature was fixed at 30°C.
5) The sample was filtered through filter paper and the effluent concentration of Ti and Fe were determined.
6) Desorption of Cr(III) from EWS was performed by using 10 ml of 1 mol L^{-1} H_2SO_4, 0.4 pH, contact time 6 min and temperature 25°C.
7) Previous steps were repeated for statistical evaluation of the results produced.

Conclusion

Egyptian white sand (EWS) from South Western Sinai is used as a selective adsorbent for the separation of Cr(III) from the phosphate solution of Rosetta ilmenite leaving behind both Ti and Fe. Interference from the phosphate medium has not also been verified. After its desorption, Cr(III) was spectrophotometrically determined using 1,5-diphenylcarbazide. The relevant factors of both adsorption and desorption have been studied using synthetic solutions. Accordingly, it was found that the optimum adsorption factors involved a pH of 0.8, a contact time of 120 min at 30°C using a synthetic solution assaying 20 mg Cr(III)/l and in which the EWS amounted to 20 g for one liter. Under these conditions, it was revealed that the maximum sorption capacity, Qe, attained 93.5 mg/g EWS. For desorption, 1M H2SO4 (pH 0.4) found efficient during 6 min at room temperature.

REFERENCES

Abdou AA (2011). Improved Analytical Methods for Assaying Some Economic Egyptian Black Sand Minerals Concentrates, Ph.D. Thesis, Fac. Sci. Tanta Univ. Egypt.

Al-Anber AM (2010). Removal of high-level Fe3+ from aqueous solution using natural inorganic materials: Bentonite (NB) and quartz (NQ). Desalination 250:885.

Aslam MM, Hassan I, Malik M (2004). Sand as adsorbent of zinc from industrial effluents. Electon. J. Agric. Food Chem. 3:792.

Baig MA, Mehmoud B, Martin A (2003). Removal of chromium from industrial effluents by sand filtration. Electon. J. Agric. Food Chem. 2:374.

Becher RG (1963). Australian US Patent NO. 247:110.

Benelite Corp of America (1974). Beneficiation of titaniferous ores. U.S. Patent No. 3:825-419.

Birck JL, Lugmair GW (1988). Nickel and chromium isotopes in Allende inclusions. Earth Planet. Sci. Lett. 90:131.

Blundy PD (1958). The determination of chromium by a solvent-extraction method. Analyst 83:555.

Dean JA, Bevlrly ML, Geurts IJ (1948). The colorimetric determination of nickel, chromium, and manganese in steel. Anal. Chem. Acta 21:377.

Demirbas E, Kobya M, Senturk E, Ozkan T (2004). Adsorption kinetics for the removal of chromium (VI) from aqueous solutions on the activated carbons. ISSN 0378-4738 = Water SA, 30:533.

Donald LS (1998). Kinetics of sorption/release reactions on natural particles. Environmental Soil Chemistry [M]. San Diego: Academic Press.

Gotz A, Heumann KG (1988). Chromium trace determination in inorganic. Fresenius Z. Anal. Chem. 331:123.

Greenberg AE, Clesceri LS, Eaton AD (1992). Standard Methods for the Examination of Water and Wastewater, 18th ed., American public Health Association, Washington, DC, P. 358.

Hansen RG (1996). The preparation and Certification of Six Reference Materials for The heavy Minerals Industry, Randurg, MINTEK (Council for Mineral Technology), Mintek Report M412.

Isozaki A, Kumagai K, Utsumi S (1983). An atomic absorption spectrometric method for the individual determination of chromium(III) and chromium(VI) by atomization of chromium from a chelating resin in a graphite tube. Anal. Chim. Acta 153:15.

James PB (1970). Titanium dioxide pigment liquid-liquid extraction of chromium. Ind. Eng. Chern. Prod. Res. Dev. 9:3.

Lee C, Kim NB, Lee IC, Chung KS (1977). The use of a chelating resin column for preconcentration of trace elements from sea-water in their determination by neutron-activation analysis. Talanta 24:241.

Lee SM, Allen PD (2001). Removal of Cu(II) and Cd(II) from aqueous solution by sea food processing waste sludge. Water Res. 35:534.

Leyva-Ramos R, Fuentes-Rubio L, Guerrero-Coronado R, Mendoza-Barron J, (1995). Adsorption of trivalent chromium from aqueous solutions onto activated carbon. J. Chem. Technol. Biotechnol. 62:64.

Mackey TS (1974). Acid leaching of ilmenite into synthetic rutile. Ind. Eng. Chem. Prod. Res. Dev. 13:9.

Marczenko Z (1990). Spectrophotometric Determination of Elements. 4th Edition, John Wiley and Sons Inc., New York.

McTaggert FK, Bear J (1955). Phototropic Effects in Oxides. J. Appl. Chem. (London) 5:643.

Meena AK, Mishra GK, Rai PK, Rajagopal C, Nagar PN (2005). Removal of heavy metal ions from aqueous solutions using carbon aerogel as an adsorbent. J. Hazard. Mater. B122:161.

Murray BM (1994). Environmental chemistry of soils [M]. USA: Oxford University Press.

Namasivayam C, Yamuna RT (1995). Adsorption of chromium(VI) by a low-cost adsorbent: biogas residual slurry. Chemosphere 30: 561.

Pilkington ES, Smith PR (1967). The spectrophotometric determination of chromium in ilmenite. Anal. Chim. Acta 39:312.

Samantaroy S, Mohanty AK, Misra M (1997). Removal of hexavalent chromium by kendu fruit gum dust. J. Appl. Polym. Sci. 66:1485.

Sandell EB (1959). "Colorimetric Determination of Traces of Metals," 3rd ed., Interscience, New York. P. 392.

Sharma DC, Foster CF (1994). The treatment of chromium wastewaters using the sorptive potential of leaf mould. Biores. Technol. 49: 31.

Singh DB, Rupainwar DC, Prasad G (1992). Studies on the removal of Cr(VI) from waste water by feldspar. J. Chem. Technol. Biotechnol. 53:127.

Stonehill LG (1958). Reduction of hexavalent chromium in water samples acidified for preservation. Chem. Analyst 47:68.

Sturgeon RE, Berman SS, Desaulniers JAH, Russell DS (1980). Pre-concentration of trace metals from sea-water for determination by graphite-furnace atomic-absorption spectrometry. Talanta 27:85.

Unob F, Wongsiri B, Phaeon N, Puanngam M, Shiowatana J (2007). Reuse of waste silica as adsorbent for metal removal by iron oxide modification. J. Hazard. Mater. 142:455.

Volkening J, Heumann KG (1989). Determination of heavy metal traces in metallic materials with IDMS. Part 1. Analysis of Cr, Ni, Zn, Cd, and Pb in pure copper. Fresenius Z. Anal. Chem. 335:478.

Zdenek S, Vaclav S (1971). The use of silica gel for the separation of traces of uranium. Anal. Chim. Acca 53:335.

Zmijewska W, Polkowska M, Stokowska H (1984). Preconcentration of trace elements from water by coprecipitation and ion exchange. J. Radioanal. Nucl. Chem. 84:319.

Synthesis, spectroscopy and electrochemistry of 2-(mercapto/hydroxyl)pyridine-bis-{1-(alkyl)-2-(arylazo) imidazole}ruthenium(II)

Prithwiraj Byabartta[1,2]

[1]Departmento de Quimica Inorganica-Instituto de Ciencia deMateriales de Aragon, Universidad de Zaragoza-CSIC, Zaragoza-50009, Spain.

[2]Department of Chemistry, Jogesh Chandra Chaudhuri College, 30-Prince Anwar Shah Road, Kolkata-700033, India.

Ruthenium–pyridine complexes incorporating azoimidazole ligands, L, of the type [Ru^{II}(RaaiR′)$_2$-(X-pyridine)] 1–6 [RaaiR′ = *p*-R-C_6H_4-N=N-C_3H_2-NN-1-R′, (*1-6*), abbreviated as N,N′-chelator, where N(imidazole) and N(azo) represent N and N′, respectively; R=H (*a*), Me (*b*), Cl (*c*) and R′ = Me (*1,4*), CH_2CH_3 (*2,5*), Bz(3,6), X=S, 2-mercaptopyridine(1-3), X=O, 2-hydroxy-pyridine(4-6)] have been synthesized and their spectroelectrochemical aspects investigated. The complexes show intense charge-transfer bands in the UV-visible region which have been found to be reasonably blue shifted. The complexes systematically exhibit two oxidation processes and two successive one-electron reductions.

Key words: Ruthenium (II), arylazoimidazole, Infrared spectroscopy (IR), Nuclear magnetic resonance (NMR), Electrospray Ionization Mass Spectrometry (ESIMS), Cyclic Voltammetry (CV).

INTRODUCTION

Ruthenium is a component of mixed-metal oxide (MMO) anodes used for cathodic protection of underground and submerged structures and for electrolytic cells for chemical processes such as generating chlorine from salt water. The fluorescence of some ruthenium complexes is quenched by oxygen which has led to their use as optode sensors for oxygen. Ruthenium red, [$(NH_3)_5$Ru-O-Ru$(NH_3)_4$-O-Ru$(NH_3)_5$]$^{6+}$ is a biological stain used to stain polyanionic molecules such as pectin and nucleic acids for light microscopy and electron microscopy (Wang et al 2005). The beta-decaying isotope 106 of ruthenium is used in radiotherapy of eye tumors mainly malignant melanomas of the uvea. Ruthenium-centered complexes are being researched for possible anticancer properties. Compared with platinum complexes, those of ruthenium show greater resistance to hydrolysis and more selective action on tumors. NAMI-A and KP1019 are two drugs undergoing clinical evaluation against metastatic tumors and colon cancers. Because of its ability to harden platinum and palladium, ruthenium is used in platinum and palladium alloys to make wear-resistant electrical contacts. In this application, only thin plated films are used to achieve the necessary wear-resistance. Because of its lower cost and similar properties compared to rhodium, the use as plating material for electric contacts is one of the major applications. The thin coatings are either applied by electroplating or sputtering. Ruthenium dioxide and lead and bismuth ruthenates are used in thick-film chip resistors. These two electronic applications account for 50% of the ruthenium consumption. Only a few ruthenium alloys are used other than those with other platinum group metals. Ruthenium is often used in small quantities in those alloys to improve some of their properties. The beneficial effect on the corrosion resistance of titanium alloys led to the development of a special alloy containing 0.1% ruthenium. Ruthenium is also used in some advanced high-temperature single-crystal superalloys with applications including the turbine blades in jet engines. Several nickel based superalloy compositions are described in the literature. Among them are EPM-102 (with 3% Ru) and TMS-162 (with 6% Ru) as well as TMS-138 and TMS-174, both containing 6% rhenium. Fountain pen nibs are frequently tipped with alloys containing ruthenium. From 1944 onward, the famous Parker 51 fountain pen was fitted with the "RU" nib, a 14K gold nib tipped with 96.2% ruthenium and 3.8% iridium.

A wide number of ruthenium complexes containing heterocyclic nitrogeneous molecules and related ligands have been reported to date (Maiti et. al, 2001; Pal et al 1996; Chakraborty et al 2005). They are of considerable interest primarily due to variable oxidation states, building blocks for supramolecular assemblies, photo-physical properties, directional electron and energy transfer, potential anticancer property. Modification of heterocyclic ligands may be carried out by incorporating new donor centers, spectator groups, change of ring size and number of Heteroatoms. They can significantly influence the physical and chemical properties of the complex molecules. Since the discovery of important redox, photophysical and photochemical properties of ruthenium complexes having 2,2-bipyridine (bipy) as ligand, there has been continuous research activity in the direction of developing newer ruthenium–bipyridine systems with the perspective of interesting physicochemical properties (Ghosh et al 2013; Shivakumar et al 2000; Pramanik et al 2000; Das et al 2002; Shivakumar et. al 1998). In this context, different kinds of mixed ligand ruthenium–bipyridine complexes have been synthesized and studied over the last fifteen years (Rasmussen et al 1996; Krause and Kirsten 1984). The basic strategies behind all these activities are either to incorporate different groups within the bipyridine moiety itself or use other types of donor sites like azoimine function along with the $Ru(bipy)_2$ core to form mixed-ligand tris-chelates to modulate the photo-redox activities of this class of complexes (Santra et al., 2002). The present work originates in preparing new mixed-ligand ruthenium–arylazoimidazole complexes of type $[Ru(RaaiR)_2L]$, where L is a ligand which can form a four membered chelate ring on co-ordination and in studying the redox and spectroscopic properties of the $Ru(RaaiR)_2$ core (Bag et al., 1988). The author has chosen pyridine-2-thiol and pyridin-2-ol as ligand L. This work demonstrates example of $[Ru(RaaiR)_2(X\text{-Pyridine})]$ [$RaaiR'$=*p*-R-C_6H_4-N=N-C_3H_2-NN-1-R', (*1-6*), abbreviated as N,N$'$-chelator, where N(imidazole) and N(azo) represent N and N$'$, respectively; R=H (*a*), Me (*b*), Cl (*c*) and R'=Me (*1,4*), CH_2CH_3 (*2,5*), Bz(3,6), X=S, 2-mercaptopyridine(1-3), X=O, 2-hydroxy-pyridine(4-6)] systems. Herein, the author reports the synthesis of two complexes having RuN_5S and RuN_5O chromophores, their spectroscopic characterisation, electron-transfer properties, preliminary photophysical aspects.

EXPERIMENTAL

Materials and physical measurements

Published methods were used to prepare ctc-$RuCl_2(RaaiR)_2$, ctc-$[Ru(OH_2)_2(RaaiR)_2](ClO)_2$, (Bag et al., 1988). Microanalytical data (C, H, N) were collected using a Perkin Elmer 2400 CHN instrument. Solution electronic spectra were recorded on a JASCO UV-VIS-NIR V-570 spectrophotometer. I.r. spectra were obtained using a JASCO 420 spectrophotometer (using KBr disks, 4000-200 cm). The H nmr spectra in $CDCl_3$ were obtained on a Bruker 500 MHz FT n.m.r spectrometer using $SiMe_4$ as internal reference. Solution electrical conductivities were measured using a Systronics 304 conductivity meter with solute concentration 10 M in acetonitrile. Electrochemical work was carried out using an EG and G PARC Versastat computer controlled 250 electrochemical system. All experiments were performed under a N_2 atmosphere at 298K using a Pt-disk milli working electrode at a scan rate of 50 mVs^{-1}. All results were referenced to a saturated calomel electrode (SCE).

Preparation of the complexes

Preparation of 2-mercaptopyridine-bis-{-2-(p-tolylazopyrimidine}ruthenium(II), $[Ru(RaaiR')_2(SHPy)]$

Initially a solution of $[Ru(RaaiR')_2(EtOH)_2]^{2+}$ was prepared by stirring under reflux condition, a mixture of $Ru(RaaiR')_2Cl_2$ (100 mg, 0.186 mmol) and $AgNO_3$ (79 mg, 0.465 mmol) in dry ethanol (20 cm^3) for 1 h and removing the AgCl precipitate. To the filtrate were added 2 mercaptopyridine (65 mg, 0.471 mmol) and NaOMe (38 mg, 0.95 mmol) and the solution was heated to reflux for 12 h under a dinitrogen atmosphere, during which time changed from purple to blue-violet. The solvent was then removed under reduced pressure and the solid mass thus obtained purified by using a alumina (neutral) column. A red-violet band corresponding was eluted with acetonitrile. On evaporation, the solid complex was obtained in 72% yield (81 mg). Finally, the product was recrystallised from 1:4 v/v dichloromethane–light petroleum. Analysis for $[Ru(S\text{-}Pyridine)(HaaiMe)_2](ClO_4)$, ***(1a)***, Found: C, 43.98, H, 3.6, N, 18.4, Calcd for $[C_{25}H_{24}N_9SRu](ClO_4)$, C, 43.93, H, 3.5, N, 18.8, IR ν(N=N) 1379, ν(C=N) 1598, ν(S-Py) 1600,1450, ESIMS, 682.5[M^+], 583[M-ClO_4]; $^{13}C\{^1H\}$NMR, ppm, 134(C2), 124(C4), 125(C5), 125(C7,11), 129(C8,10), 134(C6); Analysis for $[Ru(S\text{-}Pyridine)(MeaaiMe)_2](ClO_4)$, ***(1b)***, Found: C, 45.98, H, 3.9, N, 17.74, Calcd for $[C_{27}H_{28}N_9SRu](ClO_4)$, C, 45.93, H, 3.9, N, 17.8, IR ν(N=N) 1370, ν(C=N) 1590, ν(S-Py) 1600,1450, ESIMS, 710.5[M^+], 611[M-ClO_4]; Analysis for $[Ru(S\text{-}Pyridine)(ClaaiMe)_2](ClO_4)$, ***(1c)***, Found: C, 39.98, H, 2.96, N, 16.84, Calcd for $[C_{25}H_{22}N_9SRuCl_2](ClO_4)$, C,

39.93, H, 2.5, N, 16.8, IR ν(N=N) 1370, ν(C=N) 1590, ν(S-Py) 1600,1459, ESIMS, 751.5[M^+], 652[M-ClO_4]; Analysis for [Ru(S-Pyridine)(HaaiEt)$_2$](ClO_4), (**2a**), Found: C, 45.98, H, 3.9, N, 17.74, Calcd for [$C_{27}H_{28}N_9SRu$](ClO_4), C, 45.93, H, 3.9, N, 17.8, IR ν(N=N) 1370, ν(C=N) 1590, ν(S-Py) 1600,1450, ESIMS, 710.5[M^+], 611[M-ClO_4]; Analysis for [Ru(S-Pyridine)(MeaaiMe)$_2$](ClO_4), (**2b**), Found: C, 47.18, H, 4.39, N, 17.14, Calcd for [$C_{29}H_{32}N_9SRu$](ClO_4), C, 47.93, H, 4.3, N, 17.18, IR ν(N=N) 1379, ν(C=N) 1599, ν(S-Py) 1609, 1450, ESIMS, 738.5[M^+], 639[M-ClO_4]; Analysis for [Ru(S-Pyridine)(ClaaiEt)$_2$](ClO_4), (**2c**), Found: C, 41.68, H, 3.39, N, 16.24, Calcd for [$C_{27}H_{26}N_9SRuCl_2$](ClO_4), C, 41.63, H, 3.3, N, 16.28, IR ν(N=N) 1379, ν(C=N) 1597, ν(S-Py) 1607,1450, ESIMS, 779.5[M^+], 680[M-ClO_4]; Analysis for [Ru(S-Pyridine)(HaaiBz)$_2$](ClO_4), (**3a**), Found: C, 53.28, H, 3.9, N, 15.14, Calcd for [$C_{37}H_{32}N_9SRu$](ClO_4), C, 53.23, H, 3.9, N, 15.18, IR ν(N=N) 1371, ν(C=N) 1599, ν(S-Py) 1600,1450, ESIMS, 834.5[M^+], 735[M-ClO_4]; Analysis for [Ru(S-Pyridine)(MeaaiBz)$_2$](ClO_4), (**3b**), Found: C, 54.38, H, 4.29, N, 14.74, Calcd for [$C_{39}H_{36}N_9SRu$](ClO_4), C, 54.39, H, 4.19, N, 14.8, IR ν(N=N) 1379, ν(C=N) 1590, ν(S-Py) 1609,1450, ESIMS, 862.5[M^+], 763[M-ClO_4]; Analysis for [Ru(S-Pyridine)(ClaaiBz)$_2$](ClO_4), (**3c**), Found: C, 49.18, H, 3.3, N, 13.74, Calcd for [$C_{37}H_{30}N_9SRuCl_2$](ClO_4), C, 49.13, H, 3.3, N, 13.8, IR ν(N=N) 1373, ν(C=N) 1593, ν(S-Py) 1603,1450, ESIMS, 903.5[M^+], 804[M-ClO_4].

Preparation of 2-hydroxypyridine-bis-{-2-(p-tolylazopyrimidine}ruthenium(II), [Ru(RaaiR´)$_2$(OHPy)]

Initially, a solution of [Ru(RaaiR´)$_2$(EtOH)$_2$]$^{2+}$ was prepared by stirring under reflux condition, a mixture of Ru(RaaiR´)$_2$$Cl_2$ (100 mg, 0.186 mmol) and $AgNO_3$ (79 mg, 0.465 mmol) in dry ethanol (20 cm^3) for 1 h and removing the AgCl precipitate. To the filtrate were added 2 hydroxypyridine (65 mg, 0.471 mmol) and NaOMe (38 mg, 0.95 mmol) and the solution was heated to reflux for 12 h under a dinitrogen atmosphere, during which time changed from purple to blue-violet. The solvent was then removed under reduced pressure and the solid mass thus obtained purified by using a alumina (neutral) column. A red-violet band corresponding was eluted with acetonitrile. On evaporation, the solid complex was obtained in 72% yield (81 mg). Finally the product was recrystallised from 1: 4 v/v dichloromethane–light petroleum. Analysis for [Ru(O-Pyridine)(HaaiMe)$_2$](ClO_4), (**4a**), Found: C, 43.98, H, 3.6, N, 18.4, Calcd for [$C_{25}H_{24}N_9ORu$](ClO_4), C, 43.93, H, 3.5, N, 18.8, IR ν(N=N) 1379, ν(C=N) 1598, ν(O-Py) 1600,1450, $^{13}C\{^1H\}$NMR, ppm, 134(C2), 124(C4), 125(C5), 125(C7,11), 129(C8,10), 134(C6); Analysis for [Ru(O-Pyridine)(MeaaiMe)$_2$](ClO_4), (**4b**), Found: C, 45.98, H, 3.9, N, 17.74, Calcd for [$C_{27}H_{28}N_9ORu$](ClO_4), C, 45.93, H, 3.9, N, 17.8, IR ν(N=N) 1370, ν(C=N) 1590, ν(O-Py) 1600,1450, $^{13}C\{^1H\}$NMR, ppm, 134(C2), 124(C4), 125(C5), 125(C7,11), 129(C8,10), 134(C6); Analysis for [Ru(O-Pyridine)(ClaaiMe)$_2$](ClO_4), (**4c**), Found: C, 39.98, H, 2.96, N, 16.84, Calcd for [$C_{25}H_{22}N_9ORuCl_2$](ClO_4), C, 39.93, H, 2.5, N, 16.8, IR ν(N=N) 1370, ν(C=N) 1590, ν(O-Py) 1600,1459, $^{13}C\{^1H\}$NMR, ppm, 134(C2), 124(C4), 125(C7,11), 129(C8,10), 134(C6); Analysis for [Ru(O-Pyridine)(HaaiEt)$_2$](ClO_4), (**5a**), Found: C, 45.98, H, 3.9, N, 17.74, Calcd for [$C_{27}H_{28}N_9ORu$](ClO_4), C, 45.93, H, 3.9, N, 17.8, IR ν(N=N) 1370, ν(C=N) 1590, ν(O-Py) 1600,1450, Analysis for [Ru(O-Pyridine)(MeaaiMe)$_2$](ClO_4), (**5b**), Found: C, 47.18, H, 4.39, N, 17.14, Calcd for [$C_{29}H_{32}N_9ORu$](ClO_4), C, 47.93, H, 4.3, N, 17.18, IR ν(N=N) 1379, ν(C=N) 1599, ν(O-Py) 1609,1450, Analysis for [Ru(O-Pyridine)(ClaaiEt)$_2$](ClO_4), (**5c**), Found: C, 41.68, H, 3.39, N, 16.24, Calcd for [$C_{27}H_{26}N_9ORuCl_2$](ClO_4), C, 41.63, H, 3.3, N, 16.28, IR ν(N=N) 1379, ν(C=N) 1597, ν(O-Py) 1607,1450, Analysis for [Ru(O-Pyridine)(HaaiBz)$_2$](ClO_4), (**6a**), Found: C, 53.28, H, 3.9, N, 15.14, Calcd for [$C_{37}H_{32}N_9ORu$](ClO_4), C, 53.23, H, 3.9, N, 15.18, IR ν(N=N) 1371, ν(C=N) 1599, ν(O-Py) 1600,1450, $^{13}C\{^1H\}$NMR, ppm, 134(C2), 124(C4), 125(C5), 125(C7,11), 129(C8,10); Analysis for [Ru(O-Pyridine)(MeaaiBz)$_2$](ClO_4), (**6b**), Found: C, 54.38, H, 4.29, N, 14.74, Calcd for [$C_{39}H_{36}N_9ORu$](ClO_4), C, 54.39, H, 4.19, N, 14.8, IR ν(N=N) 1379, ν(C=N) 1590, ν(O-Py) 1609,1450, Analysis for [Ru(O-Pyridine)(ClaaiBz)$_2$](ClO_4), (**6c**), Found: C, 49.18, H, 3.3, N, 13.74, Calcd for [$C_{37}H_{30}N_9ORuCl_2$](ClO_4), C, 49.13, H, 3.3, N, 13.8, IR ν(N=N) 1373, ν(C=N) 1593, ν(O-Py) 1603,1450, $^{13}C\{^1H\}$NMR, ppm, 134(C2), 124(C4), 125(C5), 125(C7,11), 129(C8,10);

RESULTS AND DISCUSSION

Synthesis and formulation

The two ligands pyridine-2-thiol and pyridin-2-ol are good chelating and the anionic forms of the ligands (L) bind to the metal ion in a bidentate N,S and N,O manner respectively forming four-membered chelate rings. The complex [Ru(RaaiR´)$_2$(X-Pyridine)], **1-6**, [RaaiR$^/$=*p*-R-C_6H_4-N=N-C_3H_2-NN-1-R$^/$, (*1-6*), abbreviated as N,N$^/$-chelator, where N(imidazole) and N(azo) represent N and N$^/$, respectively; R = H (*a*), Me (*b*), Cl (*c*) and R$^/$ = Me (*1,4*), CH_2CH_3 (*2,5*), Bz(3,6), X=S, 2-mercaptopyridine(**1-3**), X=O, 2-hydroxy-pyridine(**4-6**)] has been synthesized from [Ru(RaaiR´)$_2$$Cl_2$] following the synthetic route shown in Scheme 1 (Bhattacharya et al., 1993). The complex cation was precipitated directly from the reaction mixture as its perchlorate salt. The yield of the complexes was approximately 80% in each case. The unsymmetric nature of X-Pyridine functions lead to the possibility of coexistence of four geometrical isomers (**A–D**). However, the reported complexes incorporating the Ru(RaaiR´)$_2$ moiety, it may be assumed that the reactions in Scheme 1 also follow the stereoretentive pathway (structure **A**). The complexes are soluble in both polar and non-polar solvents and slightly in water. Their microanalytical data are in good agreement with the calculated values and thus confirm the composition of the mixed-ligand tris-chelates. The complexes act as 1: 1 electrolytes in acetonitrile solution. Solid-state magnetic moment measurements at room temperature show that the monocations 1 and 2 are diamagnetic ($t_{2g}{}^6$, idealised, *S*=0) (Tokel-Takvoryan et al., 1973).

Spectral and redox study

The Fourier-transform IR spectra for the complexes were recorded as KBr discs in the range 4000–400 cm^{-1} and display several sharp bands of different intensities. A very strong and broad band near 1100 cm^{-1} and a strong and sharp vibration band near 630 cm^{-1} are observed for both complexes due to the presence of ionic perchlorate (Goswami et al., 1983). The other expected vibrations due to arylazoimidazole and X-Pyridine are systematically

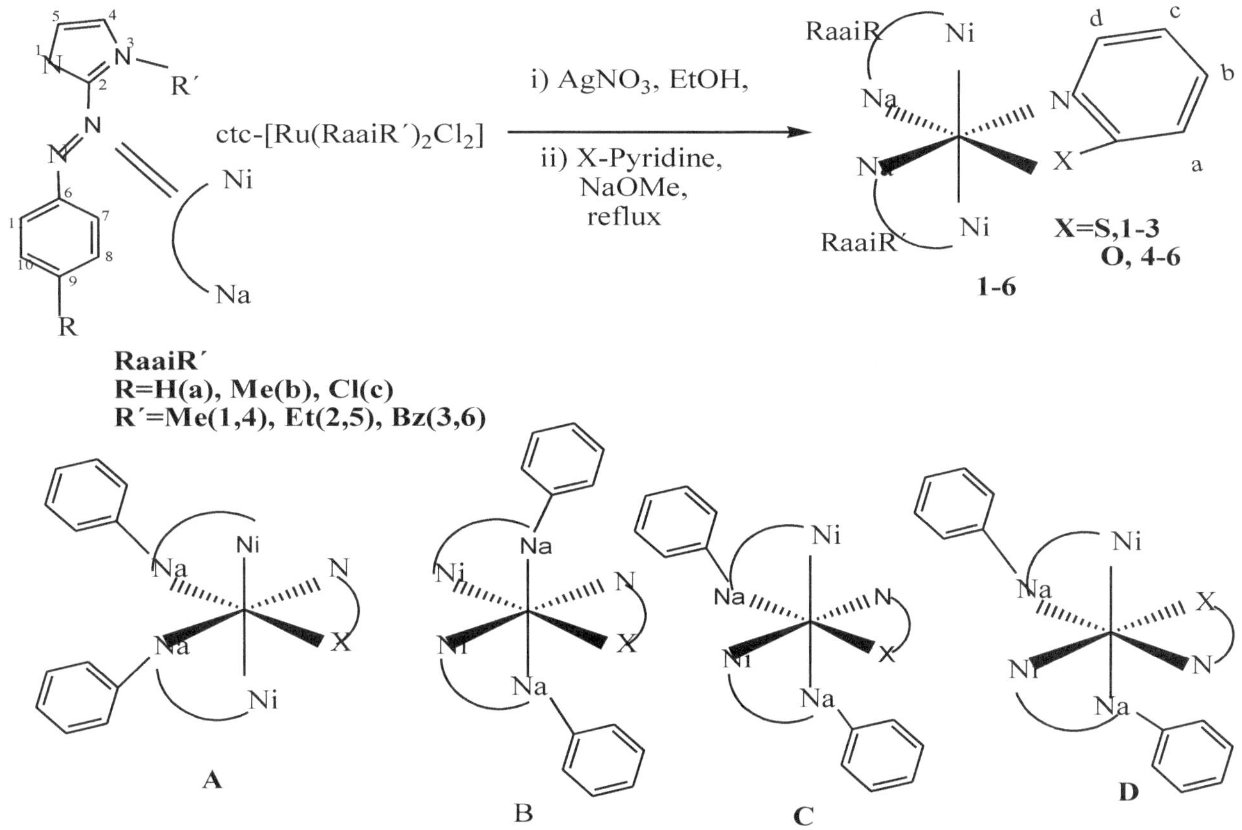

Scheme 1. Synthetic route.

present in the spectra and are therefore not specifically reported here (Ghosh et al., 2013). Solution electronic spectra of the complexes were studied in acetonitrile solvent in the UV/VIS region (200 to 700 nm). Multiple absorptions may arise due to the presence of different acceptor levels in the complexes. Complex 1-3 exhibits one band at 510 nm associated with a shoulder at 456 nm and another band near 350 nm (Tables 1 and 2) (Rao et al., 1996). In the visible region the spectrum of complexes, 4-6 is virtually identical to that of 1-3, except the lowest-energy band does not have an associated shoulder and the bands are slightly blue shifted. The bands in the UV region are of intra-ligand (p–p*) type or charge-transfer transitions involving levels which are higher in energy than those of the ligand lowest unoccupied molecular orbital (LUMO). The ^{1}H NMR spectra of the complexes 1–6 were recorded in $CDCl_3$ solvent (Tables 3 and 4). The presence of the unsymmetric X-Pyridine group in **1–6** makes all the five aromatic rings non-equivalent. The ^{13}C NMR spectra (measured in $CDCl_3$) of (***1-6***) complexes were done to assign molecular skeleton. The electron-transfer properties of both complexes have been studied by cyclic voltammetry in acetonitrile solvent. All the potentials are referenced to the saturated calomel electrode (SCE) (Byabartta et al., 2002; Byabartta et al., 2001; Byabartta et al., 2003; Santra et al., 2002; Jasimuddin et al., 2004). Both complexes are electroactive with respect to the metal as well as the ligand centres and display the same four redox processes in the potential range ±2 V at 298 K. The assignments of the responses to specific couples are based on the following considerations. Complexes exhibit one quasi-reversible oxidative response with E_{298} values of 0.54 and 0.64 V, respectively. The anodic and cathodic peak heights are approximately equal and vary with the square root of the scan rate. The peak potentials *E*pa and *E*pc are virtually independent of the scan rate. This quasi-reversible oxidative process is assigned to the ruthenium(III)–ruthenium(II) couple:

$[Ru^{II}(RaaiR')_2\text{-}(X\text{-pyridine})]^{+}$ ==== $[Ru^{III}(RaaiR')_2\text{-}(X\text{-pyridine})]^{2+}$ + e .... (1)

$[Ru^{II}(RaaiR')_2\text{-}(X\text{-pyridine})]^{+}$ + e ==== $[Ru^{III}(RaaiR')(RaaiR'^{-})\text{-}(X\text{-pyridine})]^{+}$ .... (2)

$[Ru^{III}(RaaiR')(RaaiR'^{-})(X\text{-pyridine})]^{+}$+e ====$[Ru^{III}(RaaiR'^{-})(RaaiR'^{-})(X\text{-pyridine})]^{+}$ .... (3)

Table 1. Cyclic Voltammetric[b] and UV-Vis[a] spectral data of , $[Ru^{II}(RaaiR')_2\text{-}(S\text{-pyridine})]$, **1.**

Compd	Ru^{II}-Ru^{III} Couple-I	Ru^{IV}-Ru^{III} Couple-II	RaaiR´ Couple-III	RaaiR´ Couple-IV	UV-Vis [λ_{max}(nm) (10^{-3} ε/dm^3 mol^{-1} cm^{-1})]
(1a)	0.50(90)	1.37(140)	-1.61(80)	-1.91(90)	510(10200), 456(3908)[d], 346(11566), 294(46709), 245(30222), 215(21333)
(1b)	0.51(90)	1.31(130)	-1.63(80)	-1.94(90)	512(10200), 452(3922)[d], 342(11226), 292(46229), 242(30222), 212(21333)
(1c)	0.50(90)	1.37(140)	-1.61(80)	-1.91(90)	510(10200), 456(3908)[d], 346(11566), 294(46709), 245(30222), 215(21333)

Table 2. Cyclic Voltammetric[b] and UV-Vis[a] spectral data of , $[Ru^{II}(RaaiR')_2\text{-}(O\text{-pyridine})]$, **4.**

Compd	Ru^{II}-Ru^{III} Couple-I	Ru^{IV}-Ru^{III} Couple-II	RaaiR´ Couple-III	RaaiR´ Couple-IV	UV-Vis [λ_{max}(nm) (10^{-3} ε/dm^3 mol^{-1} cm^{-1})]
(4a)	0.64(100)	1.03(150)	-1.57(90)	-1.92(90)	500(6100), 338(10500), 292(49500), 245(26100), 218(21250)
(4b)	0.61(100)	1.01(140)	-1.51(90)	-1.91(90)	501(6100), 331(10500), 291(49110), 241(26101), 211(21150)
(4c)	0.64(100)	1.03(150)	-1.57(90)	-1.92(90)	500(6100), 338(10500), 292(49500), 245(26100), 218(21250)

[a]Solvent dry MeCN; [d]shoulder; [b]Solvent dry MeCN, supporting electrolyte $[^nBu_4N][ClO_4]$ (0.1M), w.e. Pt-disk, a.e. Pt-wire, r.e. SCE, solute conc. ~10^{-3} M, scan rate 50 mVs^{-1}, $\Delta E_p = | E_{pa} - E_{pc} |$ where E_{pa} = anodic peak potential and E_{pc} = cathodic peak potential.

Table 3. ^{1}H-n.m.r. spectral data, δ (J/Hz), ppm in $CDCl_3$ of the complexes, **1-3.**

Compd	aH[c]	bH[d]	cH[d]	4-H[c]	5-H[c]	7,11-H[c]	8,10-H	$N\text{-}CH_3$	$N\text{-}CH_2$	dH[c]
(1a)[a]	8.52(5)	8.81(7)	8.63(4)	7.55(7.5)	7.46 (7.5)	8.03 (8.1)	7.75 (8.1)[d]	2.09[f]		8.4(9)
(1b)	8.62(7)	8.81(7)	8.73(8)	7.58 (7.5)	6.96 (7.5)	8.07 (8.1)	7.84 (8.1)[c]	2.17[f]		8.3(3)
(1c)	8.52(5)	8.81(7)	8.63(4)	7.3 (8.1)	7.42 (8.1)	8.05 (7.8)	7.85 (7.8)[c]	2.16[f]		8.4(5)
(2a)[a]	8.22(5)	8.91(8)	8.63(9)	7.4 (7.5)	7.00 (7.5)	8.01 (7.8)	7.75 (8.1)[d]	1.52(8.1)[d]	4.65(10.0)[e]	8.1(9)
(2b)	8.52(5)	8.81(7)	8.63(4)	7.53 (8.1)	7.24 (8.1)	8.01 (7.5)	7.72 (7.5)[c]	1.58(8.1)[d]	4.56(10.0)[e]	8.4(3)
(2c)	8.52(5)	8.41(3)	8.13(5)	7.34 (8.1)	7.36 (8.1)	8.04 (7.5)	7.76 (7.5)[c]	1.55(8.1)[d]	4.53(11.0)[e]	8.1(9)
(3a)[a]	8.52(5)	8.81(7)	8.63(4)	7.53 (8.1)	7.24 (8.1)	8.01 (7.5)	7.72 (7.5)[c]	1.58(8.1)[d]	5.26(10.0)[e]	8.4(3)
(3b)	8.52(5)	8.81(7)	8.63(4)	7.53 (8.1)	7.24 (8.1)	8.01 (7.5)	7.72 (7.5)[c]	1.58(8.1)[d]	5.56(8.0)[e]	8.4(3)
(3c)	8.52(5)	8.81(7)	8.63(4)	7.53 (8.1)	7.24 (8.1)	8.01 (7.5)	7.72 (7.5)[c]	1.58(8.1)[d]	5.56(10.0)[e]	8.4(3)

[a] δ(9-H) 7.45 ppm(m); [b] δ (9-Me); [c] doublet; [d] triplet; [e] AB type sextet, geminal coupling constant; [f] 1-Me, singlet.

Equation (1) is the one-electron nature of this process for both complexes was confirmed by constant-potential coulometry. The complexes have an identical RuN_5 core but the sixth co-ordination sites in nine cases are a thiolato group and in the other it is a phenolato group.

Table 4. ^{1}H-n.m.r. spectral data, δ (J/Hz), ppm in $CDCl_3$ of the complexes, 4-6.

Compd	aH^c	bH^d	cH^d	4-H^c	5-H^c	7,11-H^c	8,10-H	N-CH_3	N-CH_2	dH^c
(4a)^a	8.42(5.8)	8.84(7)	8.64(4)	7.54(7.5)	7.44(7.5)	8.04(8.1)	7.75(8.1)^d	2.09^f		8.4(9)
(4b)	8.62(7)	8.81(7)	8.73(8)	7.58(7.5)	6.96(7.5)	8.07(8.1)	7.84(8.1)^c	2.17^f		8.3(6.6)
(4c)	8.54(5.7)	8.84(7)	8.64(4)	7.3 4(8.1)	7.42(8.1)	8.05(7.8)	7.85(7.8)^c	2.14^f		8.4(5)
(5a)^a	8.22(5)	8.91(8)	8.62(9)	7.42(7.5)	7.02(7.5)	8.04(7.8)	7.75(8.1)^d	1.52(8.1)^d	4.65 (10.0)^e	8.1(9)
(5b)	8.52(5.5)	8.81(7.2)	8.63(4.9)	7.52(8.1)	7.24(8.1)	8.01(7.5)	7.72(7.5)^c	1.58(8.1)^d	4.56 (10.0)^e	8.4(7.9)
(5c)	8.52(5)	8.41(6.2)	8.13(5.9)	7.34(8.1)	7.36(8.1)	8.04(7.5)	7.76(7.5)^c	1.55(8.1)^d	4.53 (11.0)^e	8.1(9)
(6a)^a	8.52(5.2)	8.81(7.7)	8.63(4.7)	7.53(8.1)	7.24(8.1)	8.01(7.5)	7.72(7.5)^c	1.58(8.1)^d	5.26 (10.0)^e	8.4(3)
(6b)	8.52(5)	8.81(7)	8.63(4)	7.53(8.1)	7.24(8.1)	8.01(7.5)	7.72(7.5)^c	1.58(8.1)^d	5.26 (8.0)^e	8.4(6.6)
(6c)	8.52(5.5)	8.81(7,6)	8.63(4.8)	7.53(8.1)	7.24(8.1)	8.01(7.5)	7.72(7.5)^c	1.58(8.1)^d	5.26 (10.0)^e	8.4(7.7)

^{a} δ(9-H) 7.45 ppm(m); ^{b} δ (9-Me); ^{c} doublet; ^{d} triplet; ^{e} AB type sextet, geminal coupling constant; ^{f} 1-Me, singlet.

The monoanionic thiolate and phenolate ligands reduce the overall charge of the complex cation. The complexes display a second quasi-reversible oxidation above 1 V. For the thiolato complex 1, it appears near 1.41 V and for the phenolato complex near 1.03 V.

The one-electron nature of this oxidation process for both complexes is confirmed by direct comparison of the current height of these second processes with those of the previous one-electron Ru^{III}/Ru^{II} reduction process. Both complexes display two successive reversible one-electron reductions near -1.6 and -1.9 V (Tables 1 and 2). The ligand X-Pyridine does not exhibit any ligand reduction within the above-mentioned potential region.

Thus, the above reductions are assigned to the co-ordinated arylazoimidazole ligands. Complexes contain two ligands so that four successive reductions are expected from each.

In practice two, one-electron reductions have been observed for each complex, which are assigned to the azoimine groups of the co-ordinated RaaiR ligands as shown in Equations (2) and (3).

Conclusion

The present work demonstrates X-Pyridine function leading to the formation of a new class of coordinated complex. The complexes are well characterised by UV-Vis, i.r., ^{1}H n.m.r., C.V and ESI mass spectrometry. The complexes systematically exhibit two oxidation processes and two successive one-electron reductions. The stepwise electrochemical oxidations indicate that the initial one-electron oxidation process corresponds to stereoretentive oxidation of the ruthenium(II) centre to ruthenium(III).

Conflict of Interest

The authors have not declared any conflict of interest.

ACKNOWLEDGEMENT

The Department of Science and Technology (DST) is thanked for its financial support. (FAST TRACK Grand No. SERB/F/4888/2012-13 Dated 30.11.2012, Project Title: GOLD(I) and GOLD(III) ARYLAZOIMIDAZOLE (N, N DONAR) and OXO COMPLEXES : SYNTHESIS, STRUCTURE, SPECTRAL STUDY, ELECTROCHEMISTRY AND CHEMICAL REACTIVITY).

REFERENCES

Byabartta P, Dinda J, Santra PK, Sinha C, Panneerselvam K, Liao FL, Lu TH (2001). Heteroleptic tris-chelates of ruthenium(II)-bipyridine-arylazoimidazoles: Synthesis, spectral characterisation and electrochemical properties. Single crystal X-ray structure of [(2,2/-bipyridine)-bis-{1-methyl-2-(p-tolylazo)imidazole} ruthenium(II)] perchlorate J. Chem. Soc. Dalton Trans. pp. 2825-2832. http://dx.doi.org/10.1039/b103782j

Byabartta P, Jasimuddin Sk, Ghosh BK, Sinha C, Slawin AMZ, Woollins JD (2002). Nitro-ruthenium(II)-arylazoimidazoles: Synthesis, spectra, crystal structure and electrochemistry of dinitro-bis-{1-alkyl-2-(arylazo)imidazole} ruthenium(II). Nitro-nitroso derivatives and reactivity of electrophilic {Ru-NO}6 system. New J. Chem.26:1415-1424. http://dx.doi.org/10.1039/b204442k

Byabartta P, Jasimuddin Sk, Mostafa G, Lu TH, Sinha C (2003). The synthesis, spectral studies and electrochemistry of 1,10-(phenanthroline)-bis-{1-alkyl-2(arylazo)imidazole}ruthenium(II) perchlorate. Single crystal X-ray structure of [Ru(phen)(HaaiMe)2](ClO4)2 [phen = 1,10-phenanthroline, HaaiMe = 1-methyl-2-(phenylazo)imidazole] Polyhedron

22(6):849-859. http://dx.doi.org/10.1016/S0277-5387(03)00004-4

Chakraborty I, Panda BK, Gangopadhyay J, Chakravorty (2005). A Rhenium Chemistry of Azooximes: Oxygen Atom Transfer, Azoimine Chelation, and Imine−Oxime Contrast. Inorg. Chem. 44(4):1054-1060. http://dx.doi.org/10.1021/ic0486116

Das C, Saha A, Hung C, Lee G, Peng S, Goswami S (2002). Ruthenium Complexes of 2-[(4-(Arylamino)phenyl)azo]pyridine Formed via Regioselective Phenyl Ring Amination of Coordinated 2-(Phenylazo)pyridine: Isolation of Products, X-ray Structure, and Redox and Optical Properties. Inorg. Chem. 42(1):198-204. DOI: 10.1021/ic0203724. http://dx.doi.org/10.1021/ic0203724

Ghosh P, Samanta S, Roy SK, Joy S, Krämer T, McGrady JE, Goswami S (2013). Redox Noninnocence in Coordinated 2-(Arylazo)pyridines: Steric Control of Ligand-Based Redox Processes in Cobalt Complexes, Inorg. Chem. 52(24):14040-14049. http://dx.doi.org/10.1021/ic4018079

Jasimuddin Sk, Byabartta P, Mostafa G, Lu TH, Sinha C (2004). Synthesis, spectral studies, crystal structure and redox properties of homoleptic tris-chelated ruthenium(II)– arylazoimidazoles. Polyhedron 23(5):727-733. http://dx.doi.org/10.1016/j.poly.2003.10.018

Krause RA, Krause K (1984). Chemistry of bipyridyl-like ligands.3. Complexes of ruthenium(II) with2-((4-nitrophenyl)azo)pyridine. Inorg. Chem. 23(14):2195-2198. DOI: 10.1021/ic00182a042. http://dx.doi.org/10.1021/ic00182a042

Maiti N, Pal S, Chattopadhyay S (2001). Reaction of 2-(Phenylazo)aniline with Na_2PdCl_4: Formation of a 2-(Phenylazo)imino Complex of Bivalent Palladium. Inorg. Chem. 40(9):2204-2205. http://dx.doi.org/10.1021/ic000369p

Pal CK, Chattopadhyay S, Sinha C, Chakravorty A (1996). A Bis(azo−imine)palladium(II) System with 10 Ligand π Electrons. Synthesis, Structure, Serial Redox, and Relationship to Bis(azooximates) and Other Species. Inorg. Chem. 35(9):2442-2447. http://dx.doi.org/10.1021/ic950361t

Pramanik K, Shivakumar M, Ghosh P, Chakravorty A (2000). Azo Anion Radical Complexes Of Osmium And Related Nonradical Species. Inorg. Chem. 39(2):195-199. http://dx.doi.org/10.1021/ic9908322

Rasmussen SC, Thompson DW, Singh V, Petersen JD (1996). Controlled Synthesis of a New, Soluble, Conjugated Metallopolymer Containing Ruthenium Chromophoric Units. Inorg. Chem. 35(12):3449-3450. http://dx.doi.org/10.1021/ic9502786

Santra PK, Byabartta P, Falvello LR, Sinha C (2002). Chemistry of azopyrimidines: Couplings of aromatic amines with the ortho C-H-function of the pendant phenyl ring in dichloro{2-(phenylazo)pyrimidine}palladium(II) complex. Eur. J. Inorg. Chem. Pp. 1124-1131.

Shivakumar M, Pramanik K, Bhattacharyya I, Chakravorty A (2000). Chemistry of Metal-Bound Anion Radicals. A Family of Mono- and Bis(azopyridine) Chelates of Bivalent Ruthenium. Inorg. Chem. 39(19):4332-4338.

Shivakumar M, Pramanik K, Ghosh P, Chakravorty A (1998). Isolation and Structure of the First Azo Anion Radical Complexes of Ruthenium. Inorg. Chem. 37(23):5968-5969. PMID: 11670728

Wang Y, Yang Z (2005). Synthesis Characterization and DNA-binding Properties of Three 3d Transition Metal Complexes of the Schiff Base Derived from Diethenetriamine with PMBP. Transition Metal Chem. 30(7):902-906. http://dx.doi.org/10.1007/s11243-005-6298-y

Preconcentration of Cd(II) by chemically converted graphene sheets adsorbed on surfactant- coated C_{18} before determination by flame atomic absorption spectrometry (FAAS)

Ali Moghimi[1] and Majid Farahmandjou[2]

[1]Department of Chemistry, Varamin (Pishva) Branch Islamic Azad University, Varamin, Iran.
[2]Department of Physics, Varamin Pishva Branch, Islamic Azad University, Varamin, Iran.

A simple, highly sensitive, accurate and selective method for determination of trace amounts of Cd(II) in water samples is presented. A novel chemically converted graphene sheets (CCG) solid-phase extraction adsorbent was synthesized by CCG sheets. The stability of a CCG especially in concentrated hydrochloric acid was studied which used as a recycling and preconcentration reagent for further uses of CCG. The method is based on CCG of Cd(II) on surfactant-coated C_{18}, modified with a CCG. The retained ions were then eluted with 4 ml of 4 M nitric acid and determined by flame atomic absorption spectrometry (FAAS) at 283.3 nm for Cd. The influence of flow rates of sample and eluent solutions, pH, breakthrough volume, effect of foreign ions on chelation and recovery were investigated. 1.5 g of surfactant-coated C_{18} adsorbs 40 mg of the Schiff's base which in turn can retain 15.2 ± 0.8 mg of each of the two ions. The limit of detection (3σ) for Cd(II) was found to be 3.20 ng L^{-1}. The enrichment factors for both ions are 100. The mentioned method was successfully applied on determination of Cd in different water samples. The ions were also speciated by means of three-column system.

Key words: Determination of Cd, preconcentration, chemically converted graphene sheets (CCG), flame atomic absorption spectrometry (FAAS).

INTRODUCTION

Cd at trace concentrations acts as both a micronutrient and a toxicant in marine and fresh water systems (Izatt et al., 1985, 1991, 1995; Blake et al., 1996; Arca et al., 2001; Ghoulipour and Husain 2002; Hashemi et al., 2001; Shcherbinina et al., 1990). This element is needed by plants at only very low levels and is toxic at higher levels. At these levels, Cd can bind to the cell membrane and hinder the transport process through the cell wall. Cd at nearly 40 ng ml^{-1} is required for normal metabolism of many living organisms (Gomes-Gomes et al., 1995; Unger, 1979). On the other hand, Cd is an important element in many industries. Thus, the development of new methods for selective separation, concentration and determination of it in sub-micro levels in different industrial, medicinal and environmental samples is of continuing interest. The determination of Cd is usually carried out by flame and graphite furnace atomic absorption spectrometry (AAS), (Boudreau and Cooper, 1989) as well as spectrometric methods (Bruening et al., 1991; Mahmoud 1997a).

Solid phase extraction (SPE) methods are the best alternatives for traditional classic methods due to

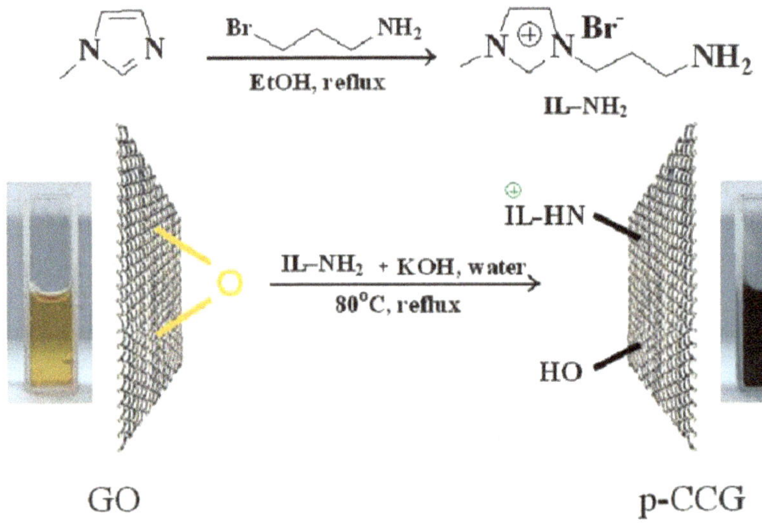

Scheme 1. Illustration of the preparation of CCG.

selective removal of trace amounts of metal ions from their matrices. SPE determinations can be carried out on different efficient ways. One of the most appropriative performance features of SPE is achieved by using octadecyl silica membrane disks. SPE reduce the use of toxic solvent, disposal costs, and extraction time (Mahmoud and Soliman, 1997; Mahmoud 1997b). The octadecyl silica membrane disks involves shorter sample processing time and decreased plugging due to the large cross-sectional area of the disk and small pressure drop which allows higher flow-rates; reduced channeling resulting from the use of sorbent with smaller particle size and a greater mechanical stability of the sorbent bed (Tong et al., 1990).

In our previous attempts, we modified SPE membrane disks with suitable compounds for selective determination of chromium (Dadler et al., 1987; Ahmad et al., 2011; Moghimi, 2007) and Lead (Mahmoud, 1999). Meanwhile, other investigators have successfully utilized these sorbents for quantitative extraction and monitoring trace amounts of Lead (Leyden et al., 1976; Moghimi et al., 2009; Liu et al., 1992), Copper (Liu et al., 1996; Mishenina et al., 1996; Wang et al., 1999), Silver (Wang et al., 1997; Zhang et al., 1982), Mercury (Zhou et al., 1983; Zargaran et al., 2008), Cadmium (Tabarzadi et al., 2010), Palladium (Shin et al., 2004), Ce (Nayebi and Moghimi et al., 2006) and UO_2 (Mahmoud, 1998).

The main goal of the present work is development of a fast, sensitive and efficient way for enrichment and extraction of trace amounts of Cd(II) from aqueous media by means of a surfactant- coated C_{18} modified with chemically converted graphene sheets (CCG) (Scheme 1) (Yang et al., 2009).

Such a determination has not been reported in the literature. The structure of CCG is shown in Scheme 1. The chelated ions were desorbed and determined by flame atomic absorption spectrometry (FAAS). The modified solid phase could be used at least 50 times with acceptable reproducibility without any change in the composition of the sorbent, CCG or Sodium dodecyl sulfate (SDS). On the other hand, in terms of economy it is much cheaper than those in the market, like C_{18} SPE mini-column.

EXPERIMENTAL

Reagents and apparatus

Graphite oxide was prepared from purified natural graphite (SP-1, Bay Carbon, Michigan, average particle size 30 nm) by the Hummers method and dried for a week over phosphorus pentoxide in a vacuum desiccator before use. 4-Isocyanatobenzenesulfonyl azide was prepared from 4-carboxybenzenesulfonyl azide via a published procedure (Tong et al., 1990). All solutions were prepared with doubly distilled deionized water from Merck (Darmstadt, Germany). The C_{18} powder for chromatography with diameter of about 50 µm was obtained from Katayama Chemicals from supelco. It was conditioned before use by suspending in 4 M nitric acid for 20 min, and then washed 2 times with water. SDS was obtained from Merck (Darmstadt, Germany) and used without any further purification.

Synthetic procedures

Preparation of CCG

The preparation of CCG functionalized by $IL\text{-}NH_2$ is illustrated in Scheme 1. Details of the synthesis can be found in Yang et al. (2009).

Column preparation

CCG (40 mg) were packed into an SPE mini-column (6.0 cm × 9 mm i.d., polypropylene). A polypropylene frit was placed at each

Table 1. The operational conditions of flame for determination of Cd.

Slit width	0.7 nm
Operation current of HI-HCL	10 mA
Resonance fine	283.3
Type of background correction	Deuterium lamp
Type of flame	Air/acetylene
Air flow	7.0 $ml.min^{-1}$
Acetylene flow	1.7 $ml.min^{-1}$

end of the column to prevent loss of the adsorbent. Before use, 0.5 mol L^{-1} HNO_3 and DDW were passed through the column to clean it.

Apparatus

The pH measurements were conducted by an ATC pH meter (EDT instruments, GP 353) calibrated against two standard buffer solutions of pH 4.0 and 9.2. Infrared spectra of CCG were carried out from KBr pellet by a Perkin-Elmer 1430 ratio recording spectrophotometer. Atomic absorption analysis of all the metal ions except Zn(II) were performed with a Perkin-Elmer 2380 FAAS. Zn(II) determinations were performed by a Varian Spect AA-10. Raman spectrophotometer analysis was performed with a Perkin-Elmer.

Preparation of admicell column

To 40 ml of water containing 1.5 g of C_{18}, 150 mg of the above Schiff base-chitosan- grafted multi walled carbon nanotubes was loaded after washing acetone, 1 mol L^{-1} HNO_3 solution and water, respectively, solution was added. The pH of the suspension was adjusted to 2.0 by addition of 4 M HNO_3 and stirred by mechanical stirrer for 20 min. Then the top liquid was decanted (and discarded) and the remained C_{18} was washed 3 times with water, then with 5 ml of 4 M HNO_3 and again 3 times with water. The prepared sorbent was transfered to a polypropylen tube (i.d 5 mm, length 10 mm). Determination of Cd^{2+} contents in working samples were carried out by a Varian spectra A 200 model atomic absorption spectrometer equipped with a high-intensity hallow cathode lamp (HI-HCl) according to the recommendations of the manufacturers. These characteristics are tabulated in Table 1. A metrohm 691 pH meter equipped with a combined glass calomel electrode was used for pH measurements.

Procedure

The pH of a solution containing 100 ng of each Cd(II) was adjusted to 2.0. This solution was passed through the admicell column with a flow rate of 5 ml min^{-1}. The column was washed with 10 ml of water and the retained ions were desorbed with 1 ml of 4 M HNO_3 with a flow rate of 2 ml min^{-1}. The desorption procedure was repeated 3 more times. All the acid solutions (4 ml all together) were collected in a 10 ml volumetric flask and diluted to the mark with water. The concentrations of Cd in the solution were determined by FAAS at 283.3.

Determination of Cd in water samples

Polyethylene bottles, soaked in 1 M HNO_3 overnight, and washed 2 times with water were used for sampling. The water sample was filtered through a 0.45 µm pores filter. The pH of a 1000 ml portion of each sample was adjusted to 2.0 (4 M HNO_3) and passed through the column under a flow rate of 5 ml min^{-1}. The column was washed with water and the ions were desorbed and determined as the above mentioned procedure.

Speciation of Cd in water samples

This procedure is reported in several articles. The method has been evaluated and optimized for speciation and its application on complex mixtures (Wang et al., 1997; Zhang et al., 1982). The chelating cation exchanger (Chelex-100) and anion exchanger, Dowex 1X-8 resins were washed with 1 M HCl, water, 1 M NaOH and water, respectively. 1.2 g of each resin was transfered to separate polyethylene columns. Each column was washed with 10 ml of 2 M HNO_3 and then 30 ml of water. The C_{18} bounded silica adsorber in a separate column was conditioned with 5 ml of methanol, then 5 ml of 2 M HNO_3 and at the end with 20 ml of water. 5 ml of methanol was added on top of the adsorber, and passed through it until the level of methanol reached just the surface of the adsorber. Then water was added on it and connected to the other two columns. A certain volume of water sample was filtered through a 0.45 µm filter and then passed through the three-column system, Dowex 1X-8, RP-C18 silica adsorber and Chelex-100, respectively. The columns were then separated. The anion and cation exchanger columns were washed with 10 ml of 2 M HNO_3 and the C_{18} column with 10 ml of 1 M HCl. The flow rate of eluents was 1 ml min^{-1}. The Cd content of each eluted solution was determined by FAAS.

RESULTS AND DISCUSSION

Figure 1 is a typical Atomic Force Microscope (AFM) image of a CCG dispersion in water (0.25 mg ml^{-1}) after deposition on a freshly cleaved mica surface through drop-casting. The AFM analysis reveals that the average interlayer spacing for exfoliated CCG sheets obtained in this work was ca. 1.49 nm. When compared with well-exfoliated GO sheets, with a spacing of ca. 0.96 nm (Figure S1), the distance between CCG sheets is greater, as would be expected. This is due to both the presence of the ionic liquid chains grafted onto both sheet sides, as well as the electrostatic repulsion between the p-CCG sheets. Moreover, it is unavoidable that some solvent molecules are still trapped between the CCG sheets after drying (at ambient conditions for 24 h) and these

Figure 1. AFM image (left) of a p-CCG dispersion in water (0.25 mg ml^{-1}) on freshly cleaved mica surface through drop-casting, and height profiles (right) along the lines shown in AFM image.

Figure 2. TEM image of p-CCG sheets.

molecules will have also contributed to the measured interlayer spacing. Similarly, an X-ray diffraction (XRD) analysis (Figure S2) was carried out in order to investigate and compare the exfoliation of GO and CCG. The (002) diffraction peak of graphite (Figure S2a) appears at *ca.* 26.6°, and the interlayer spacing is *ca.* 0.34 nm (Yang et al., 2009). The characteristic diffraction peak (001) of GO (Yang et al., 2009) was observed in Figure S2b, and corresponds to a GO interlayer spacing of ~0.75 nm, as expected for well exfoliated sheets. However, it is worthwhile to note that, no obvious peaks attributed to graphite or GO could be found for CCG (Figure S2c), indicating that exfoliation of the layered CCG was obtained successfully in this work (Yang et al., 2009). In conclusion, well exfoliated GO and CCG sheets were successfully obtained in this work.

It was found that the weight loss at around 200°C for the p-CCG composites was much lower than that of the GO composites, indicating a decreased amount of oxygenated functional groups. Finally, Transmission Electron Microscope (TEM) imaging of the CCG sheets was performed, and Figure 2 shows some multi- CCG flakes.

Stability studies

The stability of the newly synthesized CCG phases was

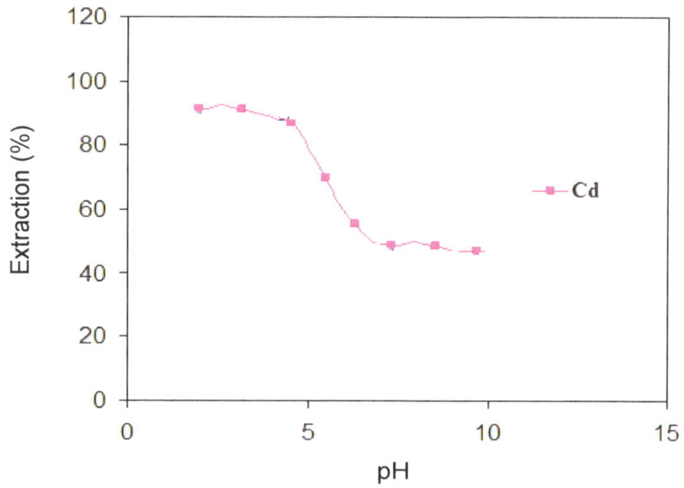

Figure 3. Extraction percentage of Cd(II) against pH.

performed in different buffer solutions (pH 1, 2, 3, 4, 5, 6 and 0.1 M sodium acetate) in order to assess the possible leaching or hydrolysis processes. Because the metal capacity values determined in the study revealed that the highest one corresponds to Cd(II)s, this ion was used to evaluate the stability measurements for the CCG phase (Mahmoud and Soliman 1997). The results of this study proved that the CCG is more resistant than the chemically adsorbed analog especially in 1.0, 5.0 and 10.0 M hydrochloric acid with hydrolysis percentage of 2.25, 6.10 and 10.50 for phase, respectively.

Thus, these stability studies indicated the suitability of phase for application in various acid solutions especially concentrated hydrochloric acid and extension of the experimental range to very strong acidic media which is not suitable for other normal and selective chelating ion exchangers based on a nano polymeric matrix (Gomes-Gomes et al., 1995). Finally, the CCG phases were also found to be stable over a range of 1 year during the course of this work.

Primary investigations revealed that surfactant-coated C_{18} could not retain Cd(II) cations, but when modified with the CCG retains these cations selectively. It was then decided to investigate the capability of the CCG as a ligand for simultaneous preconcentration and determination of Cd on admicell.

The C_{18} surface in acidic media(1 < pH < 6) attracts protons and becomes positively charged. The hydrophyl part of SDS ($-SO_3^-$), is attached strongly to these protons. On the other hand, the CCG are attached to hydrophobe part of SDS and retain small quantities of metallic cations (Smith, 2009; Moghimi et al., 2009).

Non occurrence of the effect of pH

The effect of pH of the aqueous solution on the extraction of 100 ng of each of the cations Cd(II) was studied in the pH rang of 1 to 10. The pH of the solution was adjusted by means of either 0.01 M HNO_3 or 0.01 M NaOH. The results indicate that complete chelation and recovery of Cd(II) occurs in pH range of 2 to 4 and that of 2 to 8 and are shown in Figure 3. It is probable that at higher pH values, the cations might be hydrolyzed and complete desorption occur. Hence, in order to prevent hydrolysis of the cations and also keeping SDS on the C_{18}, pH = 2.0 was chosen for further studies.

Effect of flow rates of solutions

Effect of flow rate of the solutions of the cations on chelation of them on the substrate was also studied. It was indicated that flow rates of 1 to 5 ml min^{-1} would not affect the retention efficiency of the substrate. Higher flow rates cause incomplete chelation of the cations on the sorbent. The similar range of flow rate for chelation of cations on modified C_{18} with SDS and a CCG has been reported in literature (Smith, 2009; Moghimi et al., 2009). Flow rate of 1 to 2 ml min^{-1} for desorption of the cations with 4 ml of 4 M HNO_3 has been found suitable. Higher flow rates need larger volume of acid. Hence, flow rates of 5 and 2 ml min^{-1} were used for sample solution and eluting solvent throughout respectively.

Effect of the CCG quantity

In order to study optimum quantity of the CCG on quantitative extraction of Cd, 50 ml portions of solutions containing 100 ng of each cation were passed through different columns of the sorbent of which were modified with various amounts, between 10 to 50 mg of the CCG. The best result was obtained on the sorbent which was

Table 2. Effect of foreign ions on the recovery of 100 ng of Cd.

Diverse ion	Amounts taken (mg) added to 50 ml	% Found	% Recovery of Cd^{2+} ion
Na^+	92.2	1.19(2.9)[a]	98.6s(1.9)
K^+	92.2	1.38(2.1)	98.7(2.2)
Mg^{2+}	13.5	0.8(1.8)	96.9(2.7)
Ca^{2+}	23.3	1.29(2.0)	95.4(1.9)
Sr^{2+}	3.32	2.81(2.2)	98.2(2.1)
Ba^{2+}	2.26	3.16(2.4)	98.3(2.0)
Mn^{2+}	2.44	1.75(2.3)	98.5(1.8)
Co^{2+}	2.37	1.4(2.3)	98.1(2.2)
Ni^{2+}	2.25	2.0(2.14)	98.4(2.4)
Zn^{2+}	2.44	1.97(2.1)	98.7(2.2)
Bi^{3+}	2.30	2.7(1.4)	98.4(2.7)
Cu^{2+}	2.56	2.81(2.3)	97.7(2.5)
Fe^{3+}	2.40	3.45(2.4)	97.6(2.8)
Cr^{3+}	1.30	2.92(2.2)	96.3(2.4)
UO^{2+}	2.89	1.3(2.2)	97.3(2.2)
NO_3^-	5.5	2.3 (2.3)	96.4(2.6)
$CH3COO^-$	5.3	2.2(2.6)	95.5(2.2)
SO_4^{2-}	5.0	2.9(3.0)	98.4(2.1)
CO_3^{2-}	5.4	1.8(2.5)	96.3(2.5)
PO_4^{3-}	2.6	2.1(2.0)	98.9(2.0)

[a], Values in parenthesis are CVs based on three individual replicate measurements.

modified with 40 mg of the CCG.

Figures of merit

The breakthrough volume is of prime importance for SPEs. Hence, the effect of sample volume on the recovery of the cations was studied. 100 ng of each cation was dissolved in 50, 100, 500 and 1000 ml of water. It was indicated that in all the cases, chelation and desorption of the cations were quantitative. It was then concluded that the breakthrough volume could be even more than 1000 ml. Because the sample volume was 1000 ml and the cations were eluted into 10 ml solution, the enrichment factor for both cations is 100, which is easily achievable.

The maximum capacity of 1.5 g of the substrate was determined as follows: 500 ml of a solution containing 50 mg of each cation was passed through the column. The chelated ions were eluted and determined by FAAS. The maximum capacity of the sorbent for three individual replicates was found to be 15.2 ± 0.8 μg of each cation. The limit of detection (3σ) for the cations (Zargaran et al., 2008) was found to be 3.20 ngl^{-1} for Cd ions. Reproducibility of the method for extraction and determination of 100 ng of each cation in a 50 ml solution was examined. As the results of seven individual replicate measurements indicated, they were 2.85 and 2.98% for Cd(II).

Effect of foreign ions

Effect of foreign ions was also investigated on the measurements of Cd. Here, a certain amount of foreign ion was added to 50 ml of sample solution containing 100 ng of each Cd(II) with a pH of 2.5. The amounts of the foreign ions and the percentages of the recovery of Cd are listed in Table 2. As it is seen, it is possible to determine Cd without being affected by the mentioned ions.

Analysis of the water samples

The prepared sorbent was used for analysis of real samples. In order to do this, the amounts of Cd were determined in different water samples namely: distilled water, tap water of Tehran (Tehran, taken after 10 min operation of the tap), rain water (Tehran, 25 January, 2013), snow water (Tehran, 7 February, 2013), and two synthetic samples containing different cations. The results are tabulated in Table 3. As it is seen, the amounts of Cd added to the water samples are extracted and determined quantitatively which indicates accuracy and precision of the present method.

Separation and speciation of cations by three-column system is possible to preconcentrate and at the same time separate the neutral metal complexes of CCG, anionic complexes and free ions from each other by this

Table 3. Recovery of Cd contents of water samples.

Water sample		Amount added (μg)	Found (μg)	% Recovery
		-	-	-
Sample distilled water (100 ml)	Cd	0.050	0.043(2.40)[a]	96
		0.100	0.094(2.60)	97
Tao water (100 ml)	Cd	-	0.015(3.0)	-
		0.050	0.068(2.42)	96
Snow water (50 ml)	Cd	-	0.048(2.25)	-
		0.100	0.155(2.30)	98.0
Rain water (100 ml)	Cd	-	0.045(2.25)	-
		0.100	0.143(2.40)	98
Synthetic sample 1 Na^+, Ca^{2+}, Fe^{3+}, $Co^{2+}Cr^{3+}$, Hg^{2+}, 1 mg L^{-1}	Cd	-	-	-
		0.100	0.104(2.40)	98
Synthetic sample 2 K^+ Ba^{2+}, Mn^{2+}, $Cd^{2+}Cd^{2+}$, Zn^{2+}, 1 mg L^{-1} of each cation	Cd	-	-	-
		0.100	0.105(2.70)	99

[a], Values in parenthesis are CVs based on three individual replicate measurements.

Table 4. Results of speciation of Cd in different samples by three-column system.

Column	Tap water (1000 ml)	Water sample (1000 ml)[a]	River water (50 ml)
		Cd(μg)	
Dowex 1X8	-	-	-
Silica C-18	-	-	-
Chelex-100	0.012(4.0)[b]	0.104(2.9)	0.103(2.8)

[a], This was a solution containing 0.1 μg of each cation in 1000 ml of distilled water; [b], values in parenthesis are CVs based on three replicate analysis. The samples are the same as those mentioned in Table 4.

method (Wang et al., 1997). Water samples were passed through the three connected columns: anoin exchanger, C_{18}-silica adsorber and chelating cation exchanger. Each species of Cd is retained in one of the columns; anionic complexes in the first column, neutral complexes of CCG in the second, and the free ions in the third. The results of passing certain volumes of different water samples through the columns are listed in Table 4. According to the results, it is indicated that Cd is present only as cations. On the other hand, the t-test comparing the obtained mean values of the present work with those published indicate no significant difference between them. We have proposed a method for determination and preconcentration of Cd in water samples using surfactant- coated C_{18} impregnated with a Schiff's base. The proposed method offers simple, highly sensitive, accurate and selective method for determination of trace amounts of Cd(II) in water samples.

ACKNOWLEDGMENT

The authors wish to thank the Chemistry Department of Varamin branch Islamic Azad University for financial support.

REFERENCES

Ahmad PH, Abdouss M, Ghiabi F, Moniri E, Mousavi SA (2011). Modification and characterization of poly (ethylene terephthalate)-grafted-acrylic acid/acryl amide fiber for removal of lead from human plasma and environmental sam J. Appl. Polym. Sci. DOI: 10.1002/app.34224.

Arca M, Blake AJ, Casab J, Demartin F, Devillanova FA, Garau A, Isaia F, Lippolis V, Kivekas R, Muns V, Schroder M, Verani G (2001). Conformationally locked pentadentate macrocycles containing the 1,10-phenanthroline unit. Synthesis and crystal structure of 5-oxa-2,8-dithia[9](2,9)-1,10-phenanthrolinophane (L) and its coordination properties to Ni II, Pd II, Pt II, RhIII and Ru II. J. Chem. Soc. Dalton Trans. pp. 1180-1189.

Blake AJ, Demartin F, Deviallonova FA, Garau A, Isaia F, Lippolis V, Schroder M, Verani G (1996). Synthesis of dinuclear group 6 metal carbonyl complexes bridged by 4,4′-bipyrazole ligands. J. Chem. Soc. Dalton Trans. pp. 3705-3712.

Boudreau SP, Cooper WT (1989). Analysis of thermally and chemically modified silica gels by heterogeneous gas-solid chromatography and infrared spectroscopy. Anal. Chem. 61:41-47.

Bruening ML, Mitchell DM, Bradshaw JS, Izatt RM, Bruening RL (1991). Removal of cesium from alkaline waste solution: Part II – Column ion exchange study. Anal. Chem. 63:21-27.

Dadler V, Lindoy LF, Sallin D, Schlaepfer CW (1987). Solid Phase Extraction of Phenolic from Natural Water by Modified Polyacrylonitrile Fiber. Aust. J. Chem. 40:1557-1563.

Ghoulipour V, Husain SW (2002). Studies on titanium(iv) tungstosilicate and titanium(iv) tungstophosphate. ii. separation and estimation of heavy metals from aquatic environments. Acta Chromatogr. 12:170-174.

Gomes-Gomes MM, Hidalgo Garcia MM, Palacio Corvillo MA (1995). On-line preconcentration of silver on a sulfhydryl cotton microcolumn and determination by flow injection atomic absorption spectrometry. Analyst 120:1911-1916.

Hashemi OR, Kargar-Razi M, Raoufi F, Moghimi A, Aghabozorg H, Ganjali MR (2001). Ultra-trace monitoring of copper in environmental and biological samples by inductively coupled plasma atomic emission spectrometry after separation and preconcentration by using octadecyl silica membrane disks modified by a new schiff's base. Microchem. J. 69:1-8.

Izatt RM, Bradshaw JS, Nielsen SA, Lamb JD, Christensen JJ (1985). This Week's Citation Classic. D. Sen. Chem. Rev. 85:271-277.

Izatt RM, Pawlak K, Bradshaw JS, Bruening RL (1991). trans-Dichlorido(1,4,8,11-tetraazacyclotetradecane)manganese(III) tetra-fluoridoborate. Chem. Rev. 91:1721-1726.

Izatt RM, Pawlak K, Bradshaw JS, Bruening RL (1995). Self-Assembled Ionophores. An Isoguanosine-K^+ Octamer. Chem. Rev. 95:2529-2532.

Leyden DE, Luttrell GH, Nonidez WK, Werho DB (1976). Adsorption of Co(II) and Cu(II) on silica gel surface modified with pyridinium ion from acetone and ethanol solutions. Anal. Chem. 48:67-72.

Liu QP, Liu JC, Tong Y, Cheng JK (1992). Development of a cloud-point extraction method for copper and nickel determination in food samples. Anal. Chim. Acta 269:223-228.

Liu QP, Zhao T, Liu JC, Cheng JK (1996). E-T based statistical modeling and compact statistical circuit simulation methodologies. Microchim. Acta 122:27.

Mahmoud ME (1997a). Silica gel-immobilized Eriochrome black-T as a potential solid phase extractor for zinc (II) and magnesium (II) from calcium (II). Talanta 45:309-314.

Mahmoud ME (1997b) Study of the selective extraction of iron (III) by silica-immobilized 5-formyl-3-arylazo-salicylic acid derivatives Talanta.44:15-21.

Mahmoud ME, Soliman EM (1997). Study of the selective extraction of iron(III) by silica-immobilized 5-formyl-3-arylazo-salicylic acid derivatives. Talanta 44:1063-1071.

Mahmoud ME (1999). Selective solid phase extraction of mercury(II) by silica gel-immobilized-dithiocarbamate derivatives. Anal. Chim. Acta 398:297-302.

Mahmoud ME (1998). In: Proceeding of the 25th FACSS Conference, Austin, TX, USA, 11–15 October.

Mishenina IV, Shapovalova EN, Bolshova TA, Smirnov PV, Shpigun OA (1996). Comparison of the liquid and gas chromatography of five classes of metal complexes), J. Anal. Chem. 51:270-276.

Moghimi A, Ghiasi R, Abedin AR, Ghammamy S (2009). Solid phase extraction of Cd(II) using mesoporous organosilicas and determination by FAAS. Afr. J. Pure Appl. Chem. 3(3):051-059.

Moghimi A (2007). Preconcentration and Determination of Trace Amounts of Heavy Metals in Water Samples Using Membrane Disk and Flame Atomic Absorption Spectrometry. Chin. J. Chem. 25(10):640-645.

Moghimi A, Tehrani MS, Waqif HS (2006). Preconcentration and Determination of Copper(II) Using Octadecyl Silica Membrane Disks Modified by 1,5-Diphenylcarhazide and Flame Atomic Absorption Spectrometry. Mater. Sci. Res. India 3(1a):27-32.

Nayebi P, Moghimi A (2006). Perconcentration and Determination of copper(II) by 1-(2-Pyridyl Azo)2-Naphtol(PAN) modified Octadecyl Silica. Oriental J. Chem. 22(3):507-512.

Smith MB (2009). Synthesis and exfoliation of isocyanate-treated graphene oxide nanoplatelets" March J. March's advanced organic chemistry: reactions, mechanisms, and structure. New York: John Wiley & Sons Inc. pp.1182-1183.

Unger K (1979). Porous Silica, Elsevier, Amsterdam

Shcherbinina NI, Myasoedova GV, Khabazova TA, Shroeva EY, Ishmiyarova GR, Nikitina IE, Bannykh LN (1990). Preconcentration of trace amounts of silver ion in aqueous samples on octadecyl silica membrane disks modified with hexathia-18-crown-6 and its determination by atomic absorption spectrometry. Zh. Anal. Khim. 45:2137-2141.

Shin DH, Ko YG, Choi US, Kim WN (2004). Fluorescence Enhancement of Coumarin Thiourea Derivatives by Hg^{2+}, Ag^+, and Silver Nanoparticles. Ind. Eng. Res. 43:2060-2065.

Tabarzadi M, Abdouss M, Hasani SA, Shoushtary AM, Mat.-wiss. U (2010). Ion adsorption studies of micro and nano acrylic fibers modified by ethanolamine. Ionen Adsorption von mit Ethanolamin modifizierten Mikro- und Nano-Acylfasern .Werkstofftech. 41(4):221-228.

Tong A, Akama Y, Tanaka S (1990). Sorption and preconcentration of some heavy metals by 2-mercaptobenzothiazole-clay. Anal. Chim. Acta 230:179-185.

Wang H, Zhang HS, Cheng JK (1999). Studies on 2-(2-thiazolylazo)-5-diethylaminophenol as a precolumn derivatizing reagent in the separation of platinum group metals by high performance liquid chromatography. Talanta 48:1-8.

Wang H, Zhang HS, Cheng JK, Qiu PH (1997). Application of hyphenated techniques in speciation analysis of arsenic, antimony and thallium. Microchem. J. 332-338.

Yang H, Shan C, Li F, Han D, Zhang Q, Niu L (2009). Covalent functionalization of polydisperse chemically-converted graphene sheets with amine-terminated ionic liquid. Chem. Commun. pp. 3880-3882.

Zargaran M, Shoushtari AM, Abdouss M (2008). Ion adsorption studies of micro and nano acrylic fibers modified by ethanolamine. Ionen Adsorption von mit Ethanolamin modifizierten Mikro- und Nano-Acylfasern. J. Appl. Polym. Sci. 110:3843-3849.

Zhang CP, Qi DY, Zhou TZ (1982). Sensitive spectrophotometric determination of traces of zirconium with 2-(6-bromo-2-benzothiazolylazo)-5-diethylaminophenol in the presence of sodium lauryl sulphate. Talanta 29:1119-1123.

Zhou TZ, Qi DY, Zhang CP (1983). Development of a cloud-point extraction method for copper and nickeldetermination in food samples. Acta Chim. Sin. 41:237-242.

Spectrophotometric studies of charge transfer complex of 8-hydroxyquinoline with 1,4-benzoquinone

Asmaa A. Ibrahim

Department of Chemistry, Faculty of Science, South Valley University, Aswan 81528, Egypt. E-mail: doc_somaia@yahoo.com.

The interaction between donor 8-hydroxyquinoline (HQ) and π - acceptor 1, 4-benzoqinone (BQ) has been studied in the solution as well as in the solid state. Infrared spectroscopic data indicate that a charge transfer interaction between the donor and acceptor are due to π → π*/ n → π* transition by the formation of radical ion pairs. In solution state, the formation constant, the extinction coefficient, the free energy and transition energy of the resulting charge transfer complex (CTC) were determined. The optimum experimental conditions, such as time, temperature and solvents for the CTC formation were established.

Key words: 8–Hydroxyquinoline, 1, 4–benzoqinone, the formation constant for the charge transfer complex and infrared spectroscopic.

INTRODUCTION

Charge transfer complexes have been exclusively studied (Yoshida and Kobayashi, 1970; Ayad, 1997; El-Mossalamy and Ibrahim, 2004; Sharma et al., 2008; Aljaber and Nour, 2008; Al-Qaradawi et al., 2008; Bazzi, 2008; Refat et al., 2010; Khan et al., 2010; Frag et al., 2011) due to their wide application as ion sensors in the field of environmental science and in the determination of drugs based on the charge-transfer (CT) complexes formed with electron acceptors (Gutmann et al., 1968; Karuna et al., 2006). They also can be used as organic semiconductors (Fortner et al., 1992), photocatalysts (Wu et al., 2011), dendrimers (Thomas and Thayumanavan, 2004) and in the studying of redox processes (Krysiñk et al., 1999). More recently, attention has been given to the isolation and investigation of physical properties of some CT-complexes in the solid states. Some of these complexes show interesting electrical conductivity properties and have found applications in many form of electronics and solar cells (Gogulamurali et al., 1992; Parisi et al., 2002; Trotter and White, 1978). Charge transfer interactions formed between 8–hydroxyquinoline and acceptors in different solvents has been studied spectrophoto metrically (Khan et al., 2009; Khan and Ahmed, 2009).

The reactions of 8-hydroxyquinoline (HQ) and 1, 4-benzoqinone (BQ) with number of organic compounds in solid states also have been studied spectroscopyically (Singh et al., 2000; Barreto and Zaia, 1994; Singh et al., 1987; Singh and Singh, 1988). On the other hand, BQ is the parent molecule of a class of a heterocyclic aromatic compounds which play a relevant role in biological systems (Morton, 1965) and in the formation of charge-transfer salts (Zhou et al., 1994; Kuroda et al., 2008), most of the organic conductors based on charge-transfer complexes contain a BQ and its derivatives as the acceptor component (Mansour, 2005). HQ is used as analytical colorimetric agent (Windheuser and Chu, 1967), chelating agent (Chang et al., 2008) and in drugs (Ferrero and Torre, 1986). The present paper is concerned with the results obtained from electronic and absorption measurements on the chemical product formed in the reaction of HQ, as an electron donor and BQ as an electron acceptor. Physical parameters and structure of the reaction product is determined. The aim of this paper is to study the optimum experimental conditions (time, temperature and solvents) for the charge transfer complex formation.

MATERIALS AND METHODS

HQ and solvents (Aldrich Chemicals Company) were of the highest purity and was used without further purification. HQ: Molecular

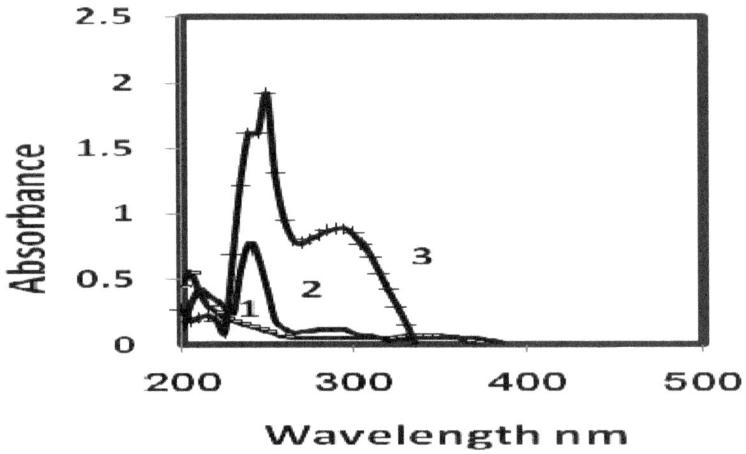

Figure 1. Electronic absorption spectra of 1- HQ, 2- BQ and 3- CT complex.

Formula: C_9H_7NO, Molecular weight: 145.16, Melting point: 76°C and Appearance: white crystalline needles. BQ: Molecular formula: 1,4-$C_6H_4O_2$, Molecular weight: 108.09, Melting point: 112.0 to 116.0°C and Appearance: yellow crystal. 1, 4–Benzoqinone was purified by sublimation.

Preparation of standard solutions

The concentration of HQ in the reaction mixtures was kept fixed at 0.4×10^{-4} M in methylene chloride. Solutions of BQ were changed over a wide range from 0.1×10^{-4} M to 1.2×10^{-4} M in the same solvent.

Synthesis of the solid CT complex from solution

The solid CT complex [(HQ) (BQ)] was synthesized by mixing the saturated solution of HQ (1 mmol) in methylene chloride (10 ml) and a saturated solution of BQ (1 mmol) in methylene chloride (10 ml). The mixture was stirred at room temperature for 3 h, a brown solid was formed, and the precipitate was filtered off, washed several times with small amounts (5 ml) of hexane and dried over $CaCl_2$. The product was identified as [(HQ)(BQ)] ($C_{15}H_{11}NO_3$; MW = 253.26 and MP = 210°C), analysis: Calculated (Found), %C: 71.14 (70.94), %H: 4.38 (4.12), %N: 5.53 (5.78).

Synthesis of the solid CT complex from solid

HQ (solid) and BQ (solid) were grinding together in the stoichiometric ratio 1:1.The change of the mixture color (brown) has indicated the beginning of the reaction. For a week, it was occasionally mixed during this period. The reaction product washed several times with hexane to remove unreacted components. The product was dried and kept in desiccators.

Micro structural studies

A clean glass slide was kept at melting point of HQ in an oven and a small amount of HQ was placed on the slide and immediately melted. The temperature of the oven was lowered and the molten HQ allowed crystallizing in one direction with the help of a cover slip (a very thin layer crystallized). This was examined under microscope and photographed (Singh et al., 2001). The photograph showed a very thin layer of crystallized HQ which represented the microstructure of HQ. In a small beaker BQ was taken and heated so that the beaker was filled with its vapors. A slide of HQ was kept on top of the beaker filled with BQ vapors and covered with a bigger beaker and kept at 70°C for 2 min. HQ reacted with the vapors of BQ and the change in microstructure was photographed.

Instruments

Infrared (IR) measurements (KBr discs) were carried out on a Unicam-Mattson 1000 Fourier Transform Infrared (FT-IR) spectrometer. Elemental analyses of Carbon, Hydrogen and Nitrogen (CHN) were measured. All analysis was carried out at the Micro analytical unit in Cairo University. Electronic spectra were recorded at room temperature via both Shimadzu Ultraviolet (UV) Spectrophotometer model 1601 PC with quartz cell of 1 cm path length and Jenway 6405 Spectrophotometer each for certain reactions.

RESULTS AND DISCUSSION

The reaction products obtained from the solution and the solid state reaction were brown colour. The elemental analysis, IR and the electronic spectra of the products are almost superimposable on each other. Charge-transfer interactions of HQ with BQ were investigated spectrophotometrically and found to form stable complex of [(HQ)(BQ)]. The donor site participated in CT complexation is the lone pair of electron on O atom of BQ. These complexes were easily synthesized from the reaction of HQ with BQ. IR, the elemental analyses, and Ultraviolet/Visible (UV-Vis) techniques characterize the [(HQ) (BQ)] charge transfer complex. The electronic spectra of the donor HQ, acceptor BQ and the resulting complex [(HQ)(BQ)] in methylene chloride were recorded in the visible range 200 to 500 nm (Figure 1). [(HQ)(BQ)] system has definite band at 295 nm due to the formation of CT complex. The nature of the binding of HQ and BQ

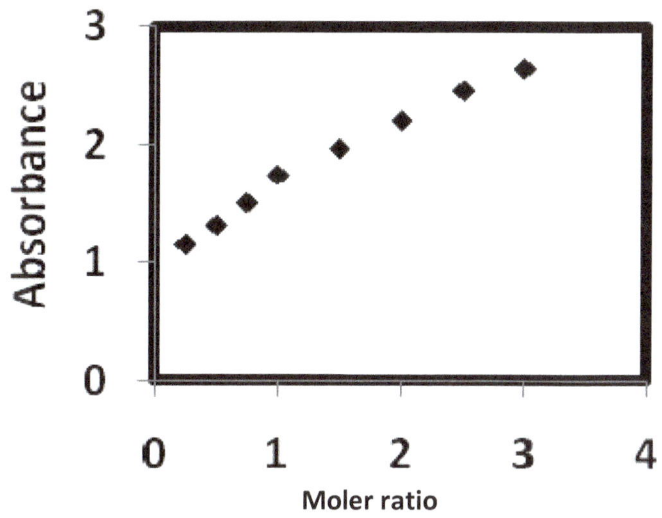

Figure 2. Molar ratio for the CT complex.

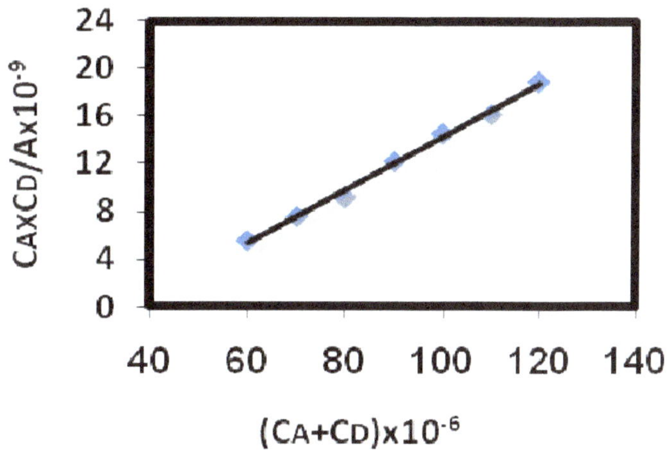

Figure 3. Relation between (C_A x C_D /A) and (C_A+ C_D) for the complex.

was determined by the molar ratio method (Yoe and Jones, 1944) indicated that a 1:1 complexation ratio has been occurred. The absorbance of the formed CT-complexes were measured and plotted as a function of the ratio C_D: C_A, according to the known method in ratios extending along the range from 1: 0.25 to 1: 3.00 (Figure 2).

The value of the formation constant, K_{CT} (mol^{-1}) and the extinction coefficient, ε_{CT} ($mol^{-1}cm^{-1}$) for the complex formed in solution was determine by using modified Benesi- Hildebrand equation (Benesi and Hildebrand, 1949). This equation can be written as follows:

$$(C_A \times C_D) / A = (C_A + C_D) / \varepsilon_{CT} + 1/ K_{CT}\, \varepsilon_{CT}$$

Where C_A and C_D are the initial concentrations of acceptor and donor, respectively, while A was the absorbance at the mentioned CT bands. On the plotting values of (C_A × C_D/A) values against the (C_A + C_D) values, straight lines were obtained with a slope of 1/ ε_{CT} and intercept of 1/Kε_{CT} (Figure 3). The data obtained throughout this calculation were given in Table 1. The standard free energy of complex in Table 1 can be determined from the formation constant by the following equation (Martin et al., 1969).

$$\Delta G^o = - RT \ln K_{CT}$$

Where ΔG^o was the free energy of the complex, R is gas constant, T is the temperature in Kelvin degree and K_{CT} was the formation constant of complex. The complex were found to have small value of - ΔG^o value (K_{CT} < 1,

$$D + A \leftrightarrow [D \rightarrow A] \leftrightarrow D^{+} + A^{-}$$

CT complex Radical ions

Scheme 1. Donor-acceptor mechanism in charge-transfer process.

Table 1. Physical parameters data of the complex.

Complex	λ_{CT} (nm)	K_{CT} (mol^{-1})	ε_{CT} (mol^{-1} cm^{-1})	E_{CT} (K cal mol^{-1})	ΔG^* (K cal mol^{-1})	Ω (s cm^{-1})
[(HQ)(BQ)]	295	27.723×10^{3}	4.484×10^{3}	97.14	4.99	0.72

Table 2. Effect of solvents on the position of the CT band of the complex in different organic solvents at 25°C.

Solvent	Dielectric constant	λ_{CT} (nm)	E_{CT} (K cal mol^{-1})
Chloroform	4.73	290	98.61
Methylene chloride	9.10	295	96.94
Acetone	20.70	330	86.66
Methanol	24.00	310	92.25

$\Delta G^{o} < 1$), indicating a weak nature of the complex and spontaneous reaction between D^{+} and A^{-} (Scheme 1).

The values transition energy reported in Table 1 was calculated from CT band of the electronic applying the following equation (Briegleb, 1964).

$$E_{CT} = 1243.667 / \lambda_{CT} \text{ nm}$$

Conductivity

The resulting [(HQ)(BQ)] solution exhibit appreciable conductance (Table 1) which may be explained by the possible formation of charge transfer complex between the reaction partners in solution (Bhowmik and Bhattacharyya, 1988; Ramadan et al., 1991; Yakuphanoglu and Arslan, 2004; Al-Shuja'a et al., 2011).

Effect of solvent

The wavelength of the CT band of the complex between HQ and BQ was measured in a number of solvents (Table 2). In the case of the oxygen containing solvents (methanol and acetone), no correlation appears between the polarity of the solvents and the transition energy. In free oxygen solvents (methylene chloride and chloroform), the transition energy of the molecular complex decreases as the polarity of the solvent increases. Accordingly, the transition energy of the molecular complex is inversely proportional to the polarity of the free oxygen solvents. The role of the polarity of the solvents arises from the stabilization of the excited state of the molecular complex through dipole - dipole or dipole - induced dipole interactions. Accordingly, the transition energy of the molecular complex is inversely proportional to the polarity of the no oxygenated solvent (Nour-El-Din and Mourad, 1983). The studies show that methylene chloride was the best solvent for the highest absorption intensity.

Effect of reagent concentration, temperature and time

Under the optimum conditions, a correlation was obtained between absorbance (A) and the concentration (C) over the range 0.1 to 1.5×10^{-4} M (Figure 4). It has been found that Beer's law was obeyed up to 0.6×10^{-4} M. In order to select the suitable temperature for CT complex, it has been made at different temperatures ranged from 5 to 45°C. The absorbance of CT complexes is measured at the suitable wavelength. The effect of temperature on these CT complexes is shown in Figure 5. It has been observed from the given results that the absorbance attains a maximum absorbance at 25°C, thus the room temperature was selected as the optimum temperature. At temperature 25°C, the optimum reaction time is determined by following the color development spectrophotometrically. It is attained after 30, 35 and 45 min for HQ to BQ complex (Figure 6). The results obtained indicated that the developed colours remained stable at room temperature for at least 30 min.

Micro structural studies

When a molten material is allowed to crystallize on a glass slide (only a very thin layer) in one direction, the structure observed under an optical microscopy is known

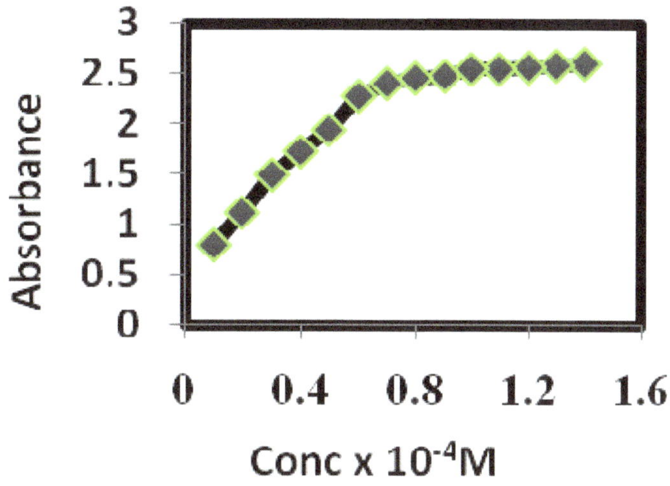

Figure 4. Effect of concentration.

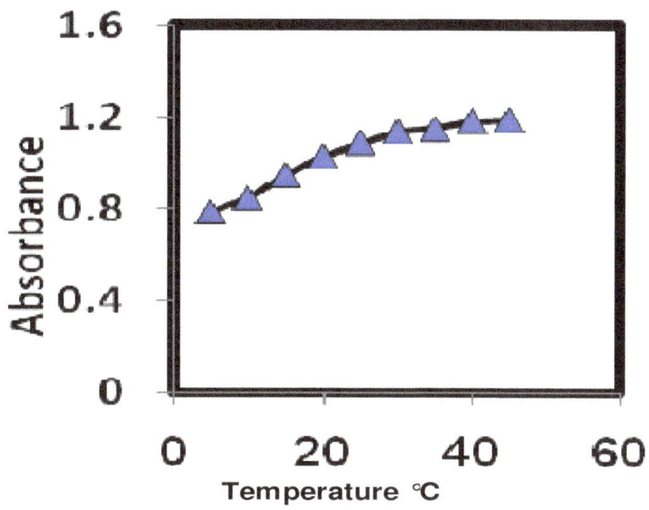

Figure 5. Effect of temperature.

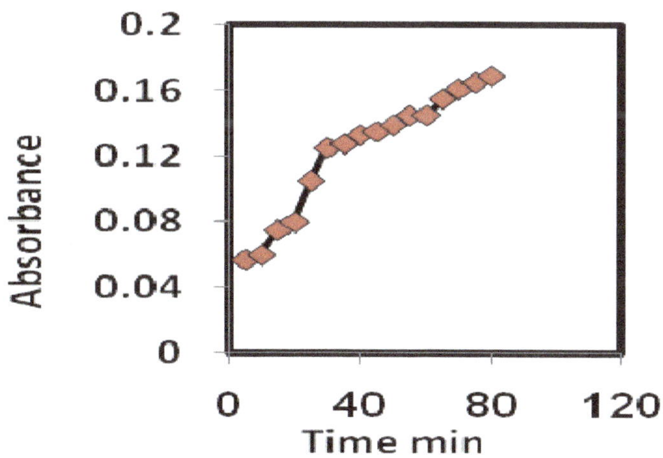

Figure 6. Effect of time on complex.

Figure 7. Microstructure of HQ exposed to vapors of BQ.

Table 3. Infrared frequencies and tentative assignments for HQ, BQ and the CT complex.

Compound	υ (OH)	υ (C=O) + υ (C=C)	Hydrogen bond
BQ		1680 vs 1410 vs 1311 ms 1220 vs	
HQ	3042 s, br	1577 ms 1500 vs 1473 w 1376 ms 1274 vs	
HQ - BQ	3410 s, br	1645 ms 1499 ms 1455 w 1411 w 1375 w	2363 v

as the microstructure of the material. The slide of HQ was made and exposed to vapors of BQ. The structural changes are given in Figure 7, shows that HQ crystallizes from one point and spreads over in all the directions. When the slide covered with HQ was allowed to react with the vapors of BQ, reaction started at the central point from which crystallization started and spread in other directions (Rastogi et al., 1977; Singh et al., 1994).

Infrared spectra

IR spectra of the characterized bands of HQ and BQ and [(HQ)(BQ)] were represented in Table 3. The formation of CT complex were associated by hydrogen bonding between one O atom of BQ and H atom of the OH group of HQ. However the majority of evidence suggested that the oxygen atom is preferred as a donation (n-π* interaction) (Gölcü et al., 2000; Singh et al., 2003; Refat et al., 2006). The shift of the bands of the acceptor to lower wavelength and those of the donor part to higher values reflects a donor to acceptor charge transfer of π-π*interaction, that is, $D_{HOMO} \rightarrow A_{LUMO}$ transition. On the basis of the results of complex obtained by elemental analysis, UV visible and IR analysis (Figure 8), the possible structure of the synthesized CT complex may

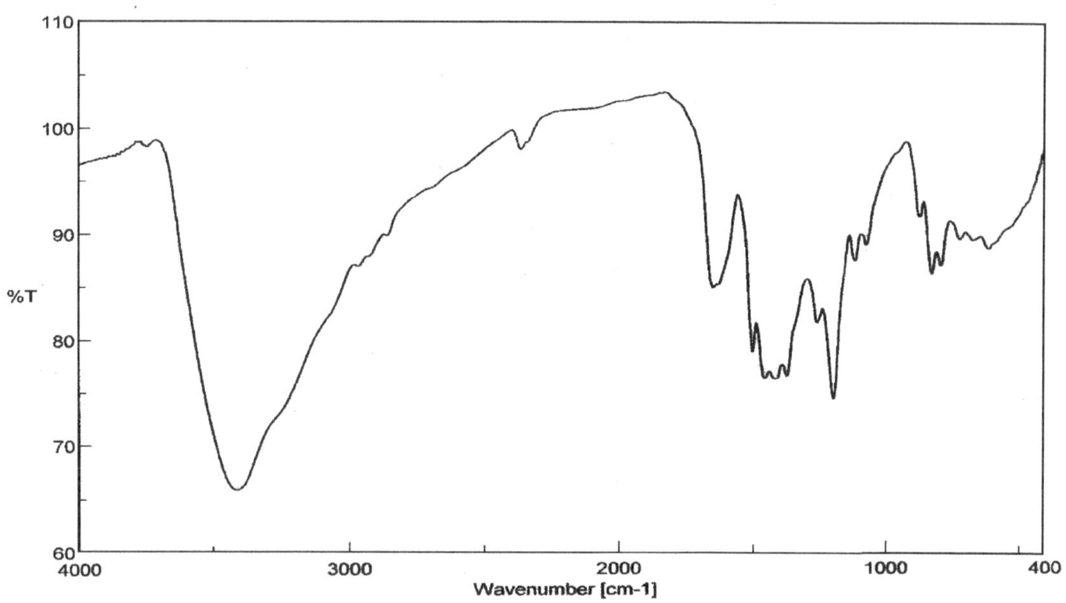

Figure 8. Infrared spectra of [(HQ)(BQ)] complex.

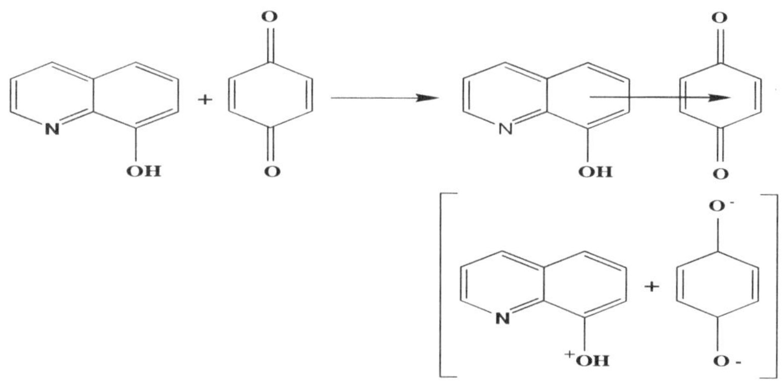

Scheme 2. Charge transfer transition between donor and acceptor.

be represented in Scheme 2

Conclusion

The interactions between the electron donor HQ and the acceptor BQ was studied spectrophotometrically in methylene chloride. A new charge transfer complex was isolated and characterized through elemental analysis, infrared and electronic spectra. The stoichiometry of the product was found to be 1:1. Accordingly, the formed CT-complex have the formula [(HQ)(BQ)]. Benesi-Hildebrand and its modification methods were applied to the determination of formation constant K_{CT}, and molar extinction coefficient (ε_{CT}). The free energy and transition energy of the resulting CTC were determined. The best solvent, time and temperature for the complex formation were established. In solid state, reaction proceeds by the diffusion of BQ into the crystal lattice of HQ.

REFERENCES

Aljaber AS, Nou EM (2008). " Spectroscopic investigation of the donor-Acceptor interaction between o-phenylenediamine and tetracyanoethylene in dichloromethane", Spectrochem. Acta A, 70(5): 997-1000.

Al-Qaradawi SY, Bazzi SS, Mostafa A, Nour E (2008). "Synthesis, spectroscopic and thermal investigations of solid charge-transfer complexes of 1,4,7-trimethyl-1,4,7-triazacyclononane and the acceptors iodine, TCNE, TCNQ and chloranil", Spectrochim. Acta A, 71(4): 1594-1598.

Al-Shuja'a OM, Obeid AO, El-Shekeil AG (2011). "Spectral, thermal and DC electrical conductivity of charge transfer complex formed between 5,7-dimethyl-1-oxo-2-phenyl-1H-pyrazolo [1,2-α]pyrazol-4-ium-3-olate and iodine", J. Macromol. Sci., A, 48, 5, 355-364.

Ayad MM (1997). "Charge-transfer complexes of benzo[b]thiophene with σ- and π-electron acceptors", Bull. Chem. Soc. Jpn., 70: 2369-2373.

Barreto WJ, Zaia DAM (1994), "The interaction of p-benzoqinone with amino acids in the solid state: A spectroscopic investigation", Spectrochem. Acta A, 50(6): 1031-1037.

Bazzi HS, Siham Y, AlQaradawi SY, Mostafa A, Nour EM (2008). "Synthesis, spectroscopic and thermal studies of the reactions of the donors piperazine and N,N′-dimethylpiperazine with σ- and π-acceptors", J. Mol. Struct., 879(1-3): 60-71.

Benesi H, Hildebrand J (1949). "A Spectrophotometric investigation of the interaction of iodine with aromatic hydrocarbons", J. Am. Chem. Soc., 71(8): 2703-2707.

Bhowmik BB, Bhattacharyya A (1988). "Solvent effects on charge-transfer intensities and heats of formation of chloranil complexes with aromatic hydrocarbons", Spectrochim. Acta A, 44: 1147-1151.

Briegleb G (1964). Electron affinities of organic molecules. Angew. Chem., 76(8): 326-341.

Chang F, Park S, Kim H (2008). "Highly CO_2-soluble 5-Amido-8-hydroxyquinoline Chelating Agents for Extraction of Metals in Sc-CO_2" Bull. Korean Chem. Soc., 29: 7.

El-Mossalamy EH, Ibrahim AA (2004). "Molecular charge transfer complexes of metal ions thiazolylazo compounds with iodine", Pigment and Reain Technol., 33(1): 32-38.

Ferrero ML, Torre CD (1986),"Cell Proliferation in ALLIUM CEPA I. meristems under 8-hydroxyquinoline, a chelating agent that affects DNA and RNA polymerases", J. Cell Set., 80: 171-180.

Fortner J, Saboungi ML, Enderby JE (1992). "Charge transfer in liquid semiconductors", Phys. Rev. Lett., 69: 1415-1418.

Frag EY, Mohamed GG, Farag AB, Yussof EB (2011). "Utility of π and σ–acceptor Reagents for the spectrophotometric determination of cefotaxime sodium antibacterial drug via charge transfer complex formation", Insight Pharm. Sci., 1(4): 47-54.

Gogulamurali N, Suthanthiraraj SA , Maruthamuthu P (1992)."Electrical conductivity of aromatic hydrocarbon-iodine charge transfer complexes", J. Mater. Sci., 27(1): 98-100.

Gölcü A, Kurtoğlu M, Serìn S (2000). "UV-Vis Spectrophoto-metric investigations of charge-transfer complexes of some compounds containing nitrogen atom and chloranil", Fen Ve Muhendislik Dergisi, 3(1): 23-27.

Gutmann F, Hermann AM, Rembaum A (1968), "Environmental and reaction studies on electrochemical cells based on solid charge-transfer complexes", J. Electrochem. Soc., 115(4): 359-362.

Karuna T, Neelima K, Venkateshwarlu G, Swamy PY (2006). "Spectrophotometric determination of drugs with iodine", J. Sci. Ind. Res., 65: 809-811.

Khan IM, Ahmad A (2009). "Spectrophotometric and spectroscopic studies of complexation of 8-hydroxyquinoline with π acceptor meta-dinitrobenzene in different polar solvents", Spectrochim. Acta A, 73(5):
966-971.

Khan IM, Naeem A, Ahmad A (2010)." Semiquantitative determination of some nitrogen compounds by the formation of charge-transfer complexes of diphenylamine with p-dimethylaminobenzaldehyde ", Chinese Chem. Lett., 21(6): 720-724.

Khan IM, Singh N, Ahmad A (2009). "Spectroscopic studies of multiple charge transfer complexes of 8-hydroxyquinoline with π-acceptor p-nitrophenol in different solvents at room temperature", Can. J. Anal. Sci. Spectrosc., 54(1): 31-37.

Krysiñk P, Tien HT, Ottova A (1999). "Charge-transfer processe s and redox reactions", Biotechnology Progress, 15(6): 974-990.

Kuroda R, Sato T, Imai Y (2008). "Varied charge-transfer complex crystals formed between diols and benzoquinone in the solid and solution States", Cryst. Eng. Comm., 10: 1881-1890.

Mansour IA (2005)." Spectral and electrical conductance studies of charge-transfer complexes derived from 6-aminoindazole Schiff bases and chloro-p-benzoquinone derivatives", Chem. Pap. 59(3): 174-177.

Martin AN, Swarbrick J, Cammarata A (1969). In: Lee, Febidger (Eds.), Physical Pharmacy, 3rd ed., (1969), Philadelphia, PA. p. 344

Morton RA (1965)."Biochemistry of Quinones", Academic Press: New York.

Nour El-Din AM, Mourad AE (1983). "Nitrones, II. Complexation Between Heterocyclic Nitrones and Tetracyanoethylene as well as 1,4-Benzoquinones", Monatshefte fiir Chemie, 114: 211-217.

Parisi J, Dyakonov V, Pientka M, Riedel I, Deibel C , Hummelenc JC (2002), "Charge transfer and transport in polymer-fullerene solar cells", Z. Naturforsch, 57 a: 995-1000.

Ramadan RM, El-Atrash AM, Ibrahim AMA, Etaiw SEH (1991). "The direct current electrical conductivity of the charge transfer complexes of some thiazoles and benzothiazoles with certain di- and trinitrobenzene derivatives", Thermochim. Acta, 178, 331-338.

Rastogi RP, Singh NB, Singh RP (1977). "Organic solid-state reactions", J. solid sate Chem., 20(2): 191-200.

Refat MS, Al- Didamony H, Abou El-Nour KM (2010). "Spectroscopic characterizations on the N, N0-bis-alkyl derivatives of 1,4,6,8-naphthalenediimide charge-transfer complexes", Arabian J. Chem., 4: 83-97.

Refat MS, Sadeek SA, Khater HM (2006)." Electronic, infrared, and 1HNMR spectral studies of charge-transfer complexes of o-toulidine and p-toluidine with alternation pi-acceptors", Spectrochim. Acta A, 64(3): 778-88.

Sharma R, Paliwal M, Singh S, Ameta R, Ameta SC (2008). "Synthesis, characterization and electronical conductivity of charge transfer complex of p-toludine and chloranil", Ind. J. Chem. Technol., 15: 613-616.

Singh NB, Singh RJ (1988), "Solid-state reaction between p-phenylenediamine and p-benzoquinone", J. Solid state Chem., 76(2): 375-390.

Singh NB, Pathak A, Fröhlich R (2003). "Solid-state complexation Reaction between p-benzoqinone and p-nitroaniline", Aust. J. Chem., 56(4): 329-333.

Singh NB, Singh NN, Laidlaw RK (1987),"Solid state reaction between α-naphthol and p-benzoquinone", J. Solid state Chem., 71(2): 530-539.

Singh NB, Singh NP, Kumar VA, Nethaji M (1994). "Study of the reaction between 1,2,3-trihydroxybenzene and 8-hydroxyquinoline in the solid state" Perkin Transactions, 2(2): 361-366.

Singh NB, Singh R, Singh K (2000). "Reaction between 8-hydroxyquinoline and salicylic acid in solid state and solution", Z. Phys. Chem., 214(2): 179-184.

Singh NB, Srivastava A, FrÖhlich R (2001). "Single crystal formation by solid state reaction between 8-hydroxyquinoline and 2,4,5-trichlorophenol", J. Chem. Soc. Perkin Trans., 2: 838-842.

Thomas KRJ, Thayumanavan S (2004). "Energy and charge transfer in dendrimers", Poly. Mat. Sci. Eng., 91: 418-419.

Trotter PJ, White PA (1978)."Resonance Raman determination of the triiodide structure in bis(tetrathiotetracene)triiodide organic conductor compared with the poly(vinyl alcohol)-iodine complex", Appl. Spectrosc., 32: 323-324.

Windheuser JJ, Chu DY (1967). "Colorimetric determination of iodochlorhydroxyquin and diiodohydroxyquin", J. Pharm. Sci., 56(4): 519-521.

Wu DY, Zhao LB, Liu XM, Huang R, Huang YF, Ren B, Tian ZQ (2011)."Photon-driven charge transfer and photocatalysis of p-aminothiophenol in metal nanogaps", Chem. Commun., 47(9): 2520-2522.

Yakuphanoglu F, Arslan M (2004). "Determination of electrical conduction mechanism and optical band gap of a new charge transfer complex", Solid State Commun., 132(3-4): 229-234.

Yoe JH, Jones AL (1944). "Colorimetric determination of iron with disodium-1,2-dihydroxybenzene-3,5-disulfonate", Ind. Eng. Chem. Anal. Ed., 16(2): 111-115.

Yoshida Z, Kobayashi T (1970). "Charge-transfer complexes of some 5-membered heteroaromatics with tetracyanoethylene, chloranil and maleic anhydride", Tetrahedron, 26(1): 267-271.

Zhou X, GZhang N, Zhang YQ (1994). "Charge-transfer reaction of sulfonamide drugs with p-benzoqinone", Fenxi Huaxue, 22(2): 190-192.

Pre-concentration and spectrophotometric determination of Co(II) by adsorption of its (E)-N[1]-((1-thiophen- 2- yl) ethylidene) benzene-1,2-diamine complex on an alumina column

Ali Moghimi

Department of Chemistry, Varamin (Pishva) Branch, Islamic Azad University, Varamin, Iran.
E-mail: alimoghimi@iauvaramin.ac.ir.

A simple method was developed for the pre-concentration of Co(II) based on the adsorption of its (E)-N[1]-((1-thiophen- 2- yl) ethylidene) benzene-1,2-diamine complex on a neutral alumina column. The influence of acidity, eluting agents, stability of the column, sample volume and interfering ions was investigated in details. The adsorbed complex could be eluted using environmentally benign polyethylene glycol (PEG 400) and the concentration of Co(II) was determined by visible spectrophotometry at a wavelength maximum of 636 nm. A detection limit of 6 µg/l could be achieved and the developed procedure was successfully applied for the determination of Co(II) in spiked water samples. The pre-concentration factor attainable for quantitative recovery (>96%) of Co(II) was 100 for a 1000 ml sample volume.

Key words: Co(II), pre-concentration, alumina column, polyethylene glycol, (E)-N[1]-((1-thiophen-2-yl) ethylidene) benzene-1,2-diamines.

INTRODUCTION

The direct determination of trace metals, especially toxic metal ions, such as Co, tin, arsenic, lead, antimony and selenium from various samples requires mostly an initial and efficient preconcentration step (Leyden et al., 1976). This preconcentration is required to meet the detection limits as well as to determine the lower concentration levels of the analyte of interest (Jones et al., 1983). This can be performed simply in many ways, including liquid and solid phase extraction techniques (Nambiar et al., 1998; Caroli et al., 1991). The application of solid phase extraction technique for preconcentration of trace metals from different samples results in several advantages, such as the minimal waste generation, reduction of sample matrix effects as well as sorption of the target species on the solid surface in a more stable chemical form (Alexandrova et al., 1993).

The normal and selective solid phase extractors are those derived from the immobilization of the organic compounds on the surface of solid supports which are mainly polyurethane foams (Arpadjan et al., 1997), filter paper (Leyden et al., 1975), cellulose (Gennaro et al., 1983) and ion exchange resins (Grote et al., 1985). Silica gel, alumina, magnesia and zirconia are the major inorganic solid matrices used to immobilize the target organic modifiers on their surfaces (Unger, 1979) of which silica gel is the most widely used solid support due to the well documented thermal, chemical and mechanical stability properties as compared to other organic and inorganic solid supports (Boudreau and Copper, 1989). The surface of silica gel is characterized by the presence of silanol groups, which are known as weak ion exchangers, causing low interaction, binding and extraction of the target analytes (Kvitek et al., 1982). For this reason, modification of the silica gel surface with certain functional groups has successfully been employed to produce the solid phase with certain selectivity characters (Bruening et al., 1991). Two approaches are known for loading the surface of solid phases with certain organic compounds, and they are defined as the chemical immobilization which is based on chemical bond formation between the silica gel surface groups and those of the organic modifier, and the other approach is known as the physical adsorption in which direct adsorption of the organic modifier with the active silanol groups takes place (Unger, 1979). Selective solid phase extractors and pre-concentrators are mainly based on impregnation of the solid surface with certain donor atoms, such as oxygen, nitrogen and sulfur containing compounds (Mahmoud, 1997; Mahmoud et al., 1997; Tong et al., 1990; Dadler et al., 1987). The most successful selective solid phases for soft metal ions are sulfur-containing compounds, which are widely used in different analytical fields. Amongst these sulfur-containing compounds are dithiocarbamate derivatives for selective extraction of Co(II) (Mahmoud, 1998, 1999) and preconcentration of various cations (Leyden et al., 1976; Narin et al., 2000; Akama et al., 2000; Ohta et al., 2001; Cuculic et al., 1997; Moghimi et al., 2006; Tehrani et al., 2005; Thurman, 1998; Pawliszyn, 1997; Izatt et al., 1996; Hagen et al., 1990; Krueger, 1995; Yamini et al., 1994; Shamsipur et al., 1999; Shamsipur et al., 2002; Brunner et al., 2003; Zelder et al., 2004; Boll et al., 2005; Nayebi et al., 2006; Moghimi et al., 2007; Moghimi 2008; Moghimi, 2007) and 2-mercaptobenzothiazol-modified silica gel for on-line preconcentration and separation of silver for atomic absorption spectrometric determinations (Qiaosheng et al., 1998; Taylor et al., 1995). Ammonium hexa-hydroazepin-1-dithiocarboxylate (HMDC)-loaded on silica gel as solid phase preconcentration column for atomic absorption spectrometry (AAS) and inductively coupled plasma atomic emission spectrometry (ICP-AES) was reported (Alexandrova et al., 1993). Mercapto-modified silica gel phase was used in pre-concentration of some trace metals from seawater (Moghimi et al., 2010). Sorption of Co(II) by some sulfur containing complexing agents loaded on various solid supports (Tajodini et al., 2010) was also reported. 2-Amino-1-cyclopentene-1-dithiocaboxylic acid (ACDA) for the extraction of silver(I), Co(II) and palladium(II) (Moghimi et al., 2006), 2-[2-triethoxysilyl-ethylthio] aniline for the selective extraction and separation of palladium from other interfering metal ions (Narin et al., 2000) as well as thiosemicarbazide for sorption of different metal ions (Campderros et al., 1998) and thioanilide loaded on silica gel for preconcentration of palladium(II) from water (Narin et al., 2000) are also sulfur containing silica gel phases. This paper describes the applications of modified (E)-N1-((1-thiophen-2-yl) ethylidene) benzene-1, 2-diamine complex on alumina column for selective extraction and solid phase pre-concentration of Co(II) from aqueous and natural water samples.

MATERIALS AND METHODS

Instrumentation

A Jasco V-576 (Japan) model double beam UV-Vis spectrophotometer fitted with tungsten lamp as the source was used for absorbance measurements. The 1 cm matched quartz cells were used for measuring the absorbance. The pH measurements were carried out by an ATC pH meter (EDT instruments, GP 353). The pH measurements were carried out by an ATC pH meter (EDT instruments, GP 353).

Chemicals and reagents

ACDA was prepared according to the procedure described by Takeshima et al. (1969). The solutions were prepared using analytical grade reagents. Triple distilled and deionized water was used for the preparation of solutions. A stock solution of 1000 µg/ml Co(II) was prepared by dissolving 0.1354 g of Co(II) chloride (Qualigens Fine Chemicals, Mumbai, India) in 100 ml water. A working solution of 10 µg/ml was prepared by suitable dilution. About 0.01 g of (E)-N^1-((1-thiophen- 2-yl) ethylidene)benzene-1,2-diamine was dissolved in minimum amount of acetone and diluted to 100 ml. Polyethylene glycol-400 (Qualigens Fine Chemicals, Mumbai, India) solution was prepared in the ratio of 7:3 by dissolving 7 ml of polyethylene glycol with 3 ml of deionized water. Neutral alumina (Himedia Research Laboratories, Mumbai, India) of particle size 150 mesh was used as the adsorbent. Sulfuric acid (Qualigens Fine Chemicals, Mumbai, India) of concentration of 1 mol/l was prepared by diluting 55.5 ml of concentrated sulfuric acid with 1 L of de-ionized water. Water samples (Pilani, India) were collected, acidified and stored in polythene bottles.

Preparation of alumina column

A glass column of 1.5 cm in diameter and 15 cm in length was used for the preconcentration of Co(II). About 0.8 g of neutral alumina was mixed with 25 ml of triple distilled water to form slurry and then loaded on to the column. Cotton was placed at the bottom for allowing alumina to settle properly. The column was packed up to a height of 3 cm.

Procedure for pre-concentration

A 1 ml volume of 10 µg/ml Co(II) solution was mixed with 5 ml of 0.5 mol/l sulfuric acid followed by the addition of 5 ml of (E)-N^1-((1-thiophen-2-yl) ethylidene) benzene-1,2-diamine solution and the resulting volume was maintained at 100 ml. The sample solution was loaded onto the column of neutral alumina maintaining a flow rate of 2 ml/min. The complex was adsorbed as a narrow band on the top of the column. The adsorbed complex was eluted using 10 ml of polyethylene glycol at a flow rate of 2 ml/min and the concentration of Co(II) was determined by visible spectrophotometry at 636 nm.

RESULTS AND DISCUSSION

Effect of acidity

The effect of acidity plays a significant role in the preconcentration studies. With pH meter, the volume of 0.5 mol/l sulfuric acid varied from 1 to 6 ml in 100 ml sample volume. The results are presented in Figure 1. Quantitative recovery (>96%) was obtained in the range of 3.5 to 6.0 ml of 0.5 mol/l sulfuric acid. Beyond 6 ml, there was no change in the recovery of Co(II).

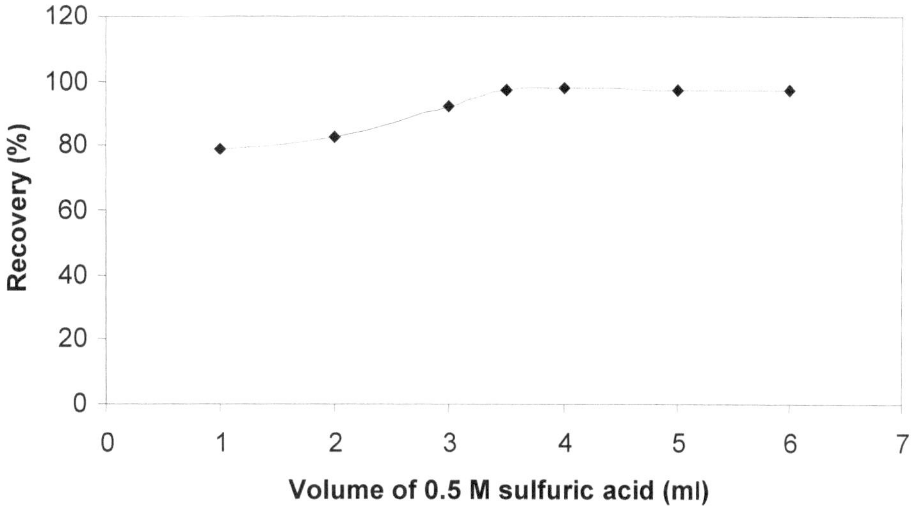

Figure 1. Effect of variation of volume of 0.5 mol/l sulfuric acid on the recovery of Co(II).

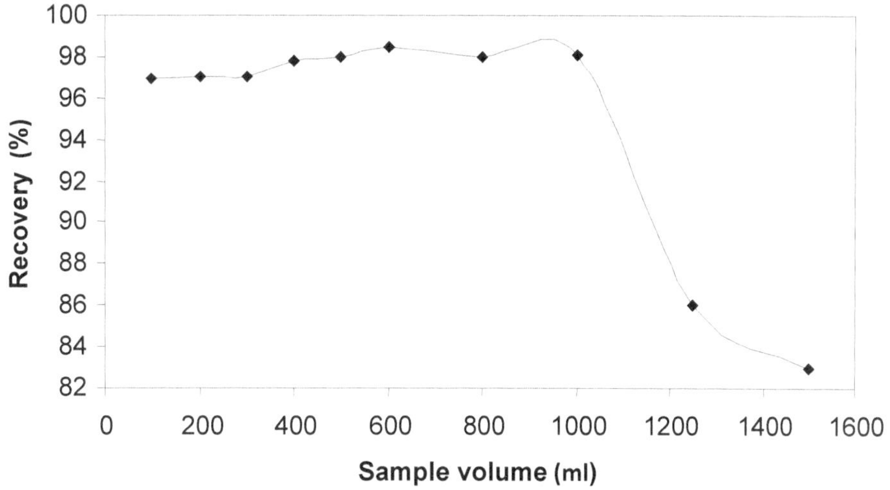

Figure 2. Effect of sample volume on the recovery of Co.

Choice of the eluent

A variety of reagents were tested in order to elute the adsorbed complex from the column. In order to choose the most effective eluent for the quantitative recovery of Co(II), methylisobutylketone, chloroform, acetone, sodium hydroxide, polyethylene glycol and ethanol were studied. The adsorption studies were carried out maintaining an overall Co(II) concentration of 10 µg in 100 ml sample volume. The percentage of recovery Co(II) was found to be quantitative with ethanol and polyethylene glycol as eluting agents. However, polyethylene glycol was preferred owing to its non-inflammability and less toxicity (Mahmoud et al., 1997). It was observed that when the ratio of polyethylene glycol-water mixture is 7:3, a recovery of 99.7% could be attained.

Effect of sample volume

The effect of sample volume on the recovery of the analyte was investigated in the range of 100 to 1500 ml maintaining an overall concentration of 0.025 mol/l sulfuric acid. The resulting complex was eluted using 10 ml of polyethylene glycol. The results are as shown in Figure 2. As can be seen from the figure, it is evident that the recovery of Co(II) is quantitative (>96%) up to 1000 ml sample volume. A preconcentration factor of 100 could be attained for quantitative recovery (>96%) of Co(II) when the sample volume was 1000 ml.

Effect of flow rate

The flow rate of 1 to 5 ml/min was found to be suitable for

optimum loading of Co(II) (E)-N[1]-((1-thiophen-2-yl) ethylidene) benzene-1,2-diamine complex on the neutral alumina column. At higher flow rates, there was a reduction in the percentage adsorption of Co(II). This could be probably due to the insufficient contact time between the sample solution and alumina. A flow rate of 2 ml/min was maintained for the elution of Co(II).

Effect of the amount of alumina

The amount of alumina loaded varied from 0.25 to 2.0 g and the preconcentration studies were carried out using the previous procedure. Quantitative recovery of Co(II) could be obtained in the range 0.75 to 2.0 g of alumina. For amounts less than 0.075 g/l of sample, there was a significant in the recovery.

Precision studies and limit of detection

Precision studies were carried out at 10 µg level of Co(II) by carrying out 10 separate determinations using the aforementioned procedure. The sample volume was maintained at 100 ml. The relative standard deviation of the method was found to be 3%. The sensitivity of the developed method is reflected by the limit of detection studies, defined as the lowest concentration of Co(II) below which quantitative recovery of the metal ion by alumina is not perceptibly seen. The limit of detection was found to be 6 µg/l.

Stability of the column

The stability of the column was tested using 10 µg Co(II) maintaining a sample volume of 100 ml. The adsorbed Co(II) (E)-N[1]-((1-thiophen-2- yl) ethylidene) benzene-1,2-diamine complex was eluted using 10 ml of polyethylene glycol-water mixture. The column could be used with good precision and quantitative recovery (>96%) for at least 10 cycles. Beyond 10 cycles, there was a significant reduction in the recovery of Co(II).

Effect of other ions

The interfering effect of diverse ions was studied at varying concentrations. The preconcentration studies were carried out as mentioned earlier using 10 µg Co(II) maintaining a sample volume of 100 ml. The studies indicated that Na^+, Mg^{2+}, Ca^{2+}, Hg^{2+}, $SO4^{2-}$, Cl^-, NO_3^-, Zn^{2+}, Co^{2+} and Fe^{2+} did not cause any significant reduction in the recovery of Co(II). The results are presented in Table 1 showing the recovery of Co(II) with varying concentrations of metal ions. The recovery was found to be quantitative in the concentration range of the metal ions that was investigated. Since, the ions that are commonly present in water samples did not interfere significantly, the method was applied to study the recovery of Co(II) in water samples.

Table 1. Effect of diverse ions on the recovery of 10 µg Co(II) in a sample volume of 100 ml.

Recovery of Co (%)	Amount (µg)	Ion
98.0	100	
98.8	1000	Ca^{2+}
97.9	500	
98.0	100	
98.8	500	Mg^{2+}
97.8	1000	
98.6	5	
98.6	10	Zn^{2+}
98.7	25	
98.7	100	
98.7	500	Cl^-
98.4	1000	
98.6	100	
98.1	50	NO_3^-
98.6	1000	
98.3	100	
97.8	500	SO_4^{2-}
98.3	1000	
98.3	5	
98.3	10	Co^{2+}
98.2	25	
98.7	5	
98.4	10	Hg^{2+}
98.5	25	
98.5	100	Fe^{2+}
98.0	250	

Recovery studies in tap water and well water samples

The validity of the proposed method was tested by spiking known concentrations of Co(II) to tap water and well water samples. The water samples were filtered and stored in polythene bottles. The recovery of Co(II) was found to be satisfactory with a relative standard deviation of 2% for five replicate measurements and the results

Table 2. Analytical results for the recovery of Co(II) in water samples.

Recovery (%)	Sample volume (ml)	Co(II) found (μg)	Co(II) added	Sample
98.0	100	99.6	100	-
98.6	250	99.8	100	Rain
97.8	1000	9.96	10	Water
98.6	100	9.94	10	-
98.0	100	9.96	10	-
98.8	200	9.95	10	Tap
98.7	500	9.94	10	
98.4	750	9.96	10	Water
98.6	1000	9.94	10	-

Table 3. Analysis of Co(II) content in city waste incineration ash sample.

Recovery (%)	Relative standard deviation (%)	Amount of Co(II) found (μg)	Amount of Co(II) added (μg)	Sample
–	2.3	31.7	0	City waste incineration ash (CRM176) 1 g/100 m
98.5	2.6	36.6	5.0	
97.9	2.8	41.5	10.0	

Table 4. Analytical results for the recovery of Co (II) in water samples.

Sample	Co (II) added (μg)	Found of Co (II) with FAAS (μg)	GF-AAS (μg)
Tap water	0.00	N.D[a]	N.D
	30.00	30.06 (0.4)[b]	29.85 (1.9)
Ground water	0.00	35.00 (1.3)	38.18 (2.2)
	30.00	52.3 (1.3)	51.52 (2.5)
Waste water	0.00	44.07 (1.0)	44.30 (1.3)
	30.00	64.09 (1.5)	64.18 (2.1)
Synthesis sample	0.00	N.D	N.D
0.2 mg of Co^{2+}, Cd^{2+}, Na^{+}, Al^{3+}, Ba^{2+}, Ca^{2+}	30.00	30.48 (1.1)	30.42 (1.5)

are as shown in Table 2.

Analysis of Co(II) content in real sample

The validity of the proposed method was tested by spiking known concentrations of Co(II) to tap water (Tehran, taken after 10 min operation of the tap) and rain water (Tehran, 2006) samples were analyzed (Table 3). As shown in Table 4, the added Co(II) ions can be quantitatively recovered from the water samples used.

A known amount of the sample was igested with $KMnO_4$–H_2SO_4–H_2O_2 mixture and diluted to a particular volume. Known aliquots were taken and subjected to the preconcentration procedure as mentioned earlier. The recovery of Co(II) was found to be quantitative and the results are presented in Table 3.

Comparison with other solid phase adsorbents

The proposed methodology was compared to a variety of solid adsorbents reported recently in the literature. Also, the proposed method was free of interference as compared to conventional procedures to determine Co (Moghimi, 2006; Choi et al., 2003; Saber et al., 2005). As can be seen from the references, it is evident that the preconcentration factor obtained with alumina is comparable to or even better than most of the other chelating matrices. The other significant feature of the proposed method is the use of environmentally benign polyethylene glycol for the elution of the complex.

Conclusion

The proposed spectrophotometric method for Co(II) is simple, sensitive and exhibits good selectivity. The elution of the complex does not involve strong acids or toxic organic solvents. The advantage of using polyethylene glycol as the eluent lies in the fact that it is non-inflammable, inexpensive and non-toxic. The conventional solvent extraction procedure associated with metal dithizonates is avoided in this methodology. The highest preconcentration factor attainable was 100 for a 1000 ml sample volume. The method showed minimum interferences with commonly found ions in water sample and the recovery of Co(II) was quantitative. The important features of the proposed method are its higher adsorption capacity with good preconcentration factor. The developed method is sensitive in detecting Co(II) at parts per billion (ppb) levels. The column could be used with good precision and quantitative recovery for at least 10 cycles. The quantitative recovery of Co(II) with a low relative standard deviation of 3% reflects the validity and accuracy of the proposed method when applied to real samples. The method developed was simple, reliable and precise for determining Co in water. Also, the proposed method was free of interference as compared to conventional procedures to determine Co (Moghimi, 2006; Choi et al., 2003; Saber et al., 2005).

ACKNOWLEDGEMENT

We gratefully acknowledge the Department of Chemistry, Islamic Azad University, Varamin (Pishva) branch for financial support.

REFERENCES

Akama Y, Ito M, Tanaka S (2000). Selective separation of cadmium from cobalt, copper, iron (III) and zinc by water-based two-phase system of tetrabutylammonium bromide. Talanta, 52: 645-651.

Alexandrova A, Arpadjan S (1993). Elimination of sulfide interference by sodium hypochlorite solution in the cold vapor atomic absorption spectrometric determination of mercury using tin(II) reduction in alkaline medium. Analyst, 118: 1309-1313.

Arpadjan S, Vuchkova L, Kostadinova E (1997). Study of the adsorption behavior of heavy metal ions on nanometer-size titanium dioxide with ICP-AES. Analyst, 122: 243-250.

Boudreau SP, Cooper WT (1989). Analysis of thermally and chemically modified silica gels by heterogeneous gas-solid chromatography and infrared spectroscopy. Anal. Chem., 61: 41-48.

Boll I, Kramer R, Brunner J, Mokhir A (2005). Oligonucleotide-templated reactions for sensing nucleic acids. J. Am. Chem. Soc., 27: 7849-7855.

Brunner J, Mokhir A, Kramer R (2003). Copper(II)-quenched oligonucleotide probes for fluorescent DNA sensing. J. Am. Chem. Soc., 125: 12410-12415.

Bruening ML, Mitchell DM, Bradshaw JS, Izatt RM, Bruening RL (1991). Removal of cesium from alkaline waste solution: Part II – Column ion exchange study. Anal. Chem., 63: 21-27.

Campderros ME, Acosta A, Marchese J (1998). Selective separation of copper with Lix 864 in a hollow fiber module. Talanta, 47:19-23.

Caroli C, Alimanti A, Petrucci F, Horvath Z (1991). Selective pre-concentration and solid phase extraction of mercury(II) from natural water by silica gel-loaded dithizone phases. Anal. Chim. Acta, 248: 241-247.

Choi YS, Choi HS (2003). Studies on solvent sublation of trace heavy metals by continuous flow system as ternary complexes of 1,10-phenanthroline and thiocyanate ion. Bull. Korean Chem. Soc., 24: 222-228.

Cuculic V, Mlakar M, Branica M (1997). Synergetic adsorption of copper(I1) mixed ligand complexes onto the SEP-PAK Cl8 column. Anal. Chim. Acta, 339: 181-189.

Dadler V, Lindoy LF, Sallin D, Schlaepfer CW (1987). Selective pre-concentration and solid phase extraction of mercury(II) from natural water by silica gel-loaded dithizone phases. Aust. J. Chem., 40: 1557-1563.

Gennaro MC, Baiocchi C, Campi E, Mentasti E, Aruga R (1983). Undesirable and harmful metals in wines-determination and removal. Anal. Chim. Acta, 151: 339-344.

Grote M, Kettrup A (1985). Liquid-liquid extraction of noble metals by formazans: Analytical applications of silver extraction by ortho-substituted formazans. Anal. Chim. Acta, 175: 239-244.

Izatt RM, Bradshaw JS, Bruening RL (1996). Non-destructive separation of metal ions from wastewater containing excess aminopolycarboxylate chelant in solution with an ion-selective immobilized macrocyclic. Pure Appl. Chem., 68: 1237-1242.

Jones JS, Harrington DE, Leone BA, Bramdstedt WR (1983). Determination of trace elements in analytical-reagent grade sodium salts by atomic absorption spectrometry and inductively coupled plasma atomic emission spectrometry after preconcentration by column solid phase extraction. Atom. Spectrosc., 4: 49-57.

Hagen DF, Markell CG, Schmitt GA (1990). Enhanced biosensor performance for on-site field analysis of explosives in water using solid-phase extraction membranes. Anal. Chim. Acta., 236:157-163.

Krueger CJ, Fild JA (1995). Method for the analysis of triadimefon and ethofumesate from dislodgeable foliar residues on turfgrass by solid-phase extraction and in-vial elution. Anal.Chem., 67: 3363-3369.

Kvitek RJ, Evans JF, Carr PW (1982). Denaturation of purple membranes at the air/water interface studied by SEM. Anal. Chem. Acta, 144: 93-98.

Leyden DE, Luttrell GH, Nonidez WK, Werho DB (1976). Adsorption of Co(II) and Cu(II) on silica gel surface modified with pyridinium ion from acetone and ethanol solutions. Anal. Chem., 48: 67-72.

Leyden DE, Luttrell GH, Sloan AE, DeAngelis NJ (1976). Automated separation and preconcentration of copper(II) from natural waters using a column treatment. Anal. Chem. Acta, 84: 97-102.

Leyden DE, Luttrell GH (1975). Comparison of chelating agents immobilized on glass with chelex 100 for removal and preconcentration of trace copper(II). Anal. Chem., 47: 1612-1616.

Mahmoud ME (1997). Silica gel-immobilized Eriochrome black-T as a potential solid phase extractor for zinc(II) and magnesium(II) from calcium(II). Talanta, 45: 309-314.

Mahmoud ME (1997). Silica-immobilized formylsalicylic acid as a selective phase for the extraction of iron(III). Talanta, 44: 15-21.

Mahmoud ME, Soliman EM (1997). Study of the selective extraction of iron(III) by silica-immobilized 5-formyl-3-arylazo-salicylic acid derivatives. Talanta, 44: 1063-1069.

Mahmoud ME (1998). Proceeding of the 25th FACSS Conference, Austin, TX, USA, 11–15 October.

Mahmoud ME (1999). Selective solid phase extraction of mercury(II) by silica gel-immobilized-dithiocarbamate derivatives. Anal. Chem.

Acta, 398: 297-302.
Moghimi A, Tehrani MS, Waqif HS (2006). Preconcentration and determination of chromium species using octadecyl silica membrane disks and flame atomic absorption spectrometry. Mater. Sci. Res. India, 3(1a): 27-32.
Moghimi A,Tajodini N (2010). Preconcentration of copper(ii) in water samples using polyurethane foam/2-(6'-ethyl-2'-benzothiazolylazo) chromotropic acid. Asian J. Chem., 22(5): 3325-3334.
Moghimi A (2006). Solid phase extraction of lead on octadecyl bonded silica membrane disk modified with Isopropyl 2- [(isopropoxy carbothioyl)disulfanyl]ethane thioate and determination by flame atomic absorption spectrometry. Oriental J. Chem., 22(3): 527-532.
Moghimi A (2007). Preconcentration and determination of trace amounts of heavy metals in water samples using membrane disk and flame atomic absorption spectrometry. Chinese J. Chem., 25(10): 640-645.
Moghimi A (2007). Solid phase extraction of trace copper(ii) using octadecyl silica membrane disks modified with *N,N'*-disalicylideneethylenediamine. Chinese J. Chem., 25(10): 1536-1542.
Moghimi, A. (2008). Preconcentration of copper(ii) using mesoporous organo-silicas and determination by flame atomic absorption spectrometry. Chinese J. Chem., J. Korean Chem. Society, 52(2): 155-163.
Moghimi A, Ghammamy S (2007). Preconcentration of copper(ii) in water samples using polyurethane foam/2-(6'-Ethyl-2'-benzothiazolylazo) chromotropic acid. Environ. Chem. Indian J., 2(3).
Moghimi A, Tehrani MS, Waqif HS (2006). Preconcentration and determination of copper(ii) using octadecyl silica membrane disks modified by 1,5-diphenylcarhazide and flame atomic absorption spectrometry. Material Sci. Res. India, 3(1a): 27-34.
Nambiar DC, Patil NN, Shinde VM (1998). Effects of mercuric chloride on the immunological, hematological, biochemical parameters and diseases resistance of nile tilapia challenged with *Aeromnas hydrophila. fresenius* J. Anal. Chem., 360: 205-212.
Narin I, Soylak M, Elic L, Dogan M (2000). Solid phase extraction of copper(ii) from aqueous solutions by adsorption of its 2-propylpiperidine-1-carbodithioate complex on alumina column. Talanta, 52: 1041-1047.
Nayebi P, Moghimi A (2006). Preconcentration of mercury by adsorption of its diphenylthiocarbazone complex on an silica gel-immobilized Schiff base diphenylthiocarbazone (DPC) column and spectrophotometric determination. Oriental J. Chem., 22(3): 507-512.
Ohta K, Tanahasi H, Suzuki T, Kaneco S (2001). Degradation of reactive yellow 86 with photo-Fenton process driven by solar light. Talanta, 53: 715-722.
Pawliszyn J (1997). Solid-phase microextraction, theory and practice, Wiley-VCH, New York.
Qiaosheng P, Qiaoyu S, Zhide H, Zhixing S (1998). Application of 2-mercaptobenzothiazole self-assembled monolayer on polycrystalline gold electrode as a nanosensor for determination of Ag(I). Analyst, 123: 239-243.
Shamsipur M, Ghiasvand AR, Yamini Y (1999). Preconcentration, determination and speciation of chromium(iii) using solid phase extraction and flame atomic absorption spectrometry. Anal.Chem., 71: 4892-4897.
Shamsipur M, Ghiasvand AR, Sharghi H (2002). Preconcentration of Ultra Trace Hg(II) in Aqueous Samples on Octadecyl Silica Membrane Disks Modified by Dibenzodiazathia-18-crown-6-dione and Its Determination by Cold Vapor Atomic Absorption Spectrometry Int. J. Environ. Anal. Chem., 82: 23-29.
Saber TM, Rastegar F, Parchehbaf A, Rezvani Z (2005). Preconcentration and determination of copper(ii)and zinc(ii) using octadecyl silica membrane disks modified by 1,5-diphenylcarhazide and flame atomic absorption spectrometry. Chinese J. Chem., 23: 1437-1442.
Takeshima M, Yokoyama T, Imamoto M, Asaba H (1969). Electrochemical oxidation of the Ni(II) complex of 2-amino cyclopentene-1-dithiocarboxylate at a Pt electrode. J. Org. Chem., 34: 730-735.
Taylor KZ, Waddell DS. Reiner EJ (1995). Direct elution of solid phase extraction disks for the determination of polychlorinated dibenzo-p-dioxins and polychlorinated dibenzofurans in effluent samples. Anal. Chem., 67: 1186-1192.
Tajodini N, Moghimi A (2010). Preconcentration of mercury by adsorption from natural water by silica gel -loaded schiff's base diphenylthiocarbazone (dpc) phases. Asian J. Chem., 22(7): 4994-5000.
Tehrani MS, Moghimi A, Waqif HS. (2005). Preconcentration and determination of trace amounts of heavy metals in water samples using membrane disk and flame atomic absorption spectrometry. Mater. Sci. Res. India, 3(2): 135-142.
Thurman EM, Mills MS (1998). Solid-phase extraction, principles and practice, Wiley, New York.
Tong A, Akama Y, Tanaka S (1990). Selective preconcentration of Au(III), Pt(IV) and Pd(II) on silica gel modified with γ-aminopropyltriethoxysilane. Anal. Chim. Acta, 230: 179-185.
Unger K, Porous S (1979). Comparison of an ordered mesoporous aluminosilicate, silica, alumina, titania and zirconia in normal-phase high-performance liquid chromatography. Elsevier, Amsterdam.
Yamini Y, Ashraf-Khorassani M. (1994). Solid-phase extraction and determination of trace amounts of lead(ii) using octadecyl silica membrane disks modified with a recently synthesized anthraquinone derivative and atomic absorption spectrometry. High Resolut J. Chromatogr., 17: 634-639.
Zelder F H, Brunner J, Kramer (2005). A fluorogenic and chromogenic probe that detects the esterase activity of trace copper (II). R. Chem. Commun., 902-911.

25

Structural study of complex lipids present in soils: Characterization of monocarboxylic acids resulting from polar lipids fraction alkaline hydrolysis

Okome Mintsa Madeleine[1]*, Nguele Jean-Calvin[1], Eba François[2], Amblès André[3] and Joffre Janie[3]

[1]Department of Chemistry-Biochemistry, Faculty of Medicine, Health Sciences University, Owendo, Gabon.
[2]Laboratory of Chemistry, Department of Sciences, École Normale Supérieure, Libreville, Gabon.
[3]Laboratory of Synthesis and Reactivity of Natural Substances, UMR–CNRS 6514, Faculty of Sciences, University of Poitiers, 40, avenue du Recteur Pineau, 86022 Poitiers-Cedex, France.

In acid soils such as turf, the biological activity is reduced, so lipids tend to accumulate. Those lipids are made of two fractions: one fraction, named "simple lipids", directly analyzable, and the other, named "polar lipids" or "complex lipids". The structure and origin of these "polar lipids" had been investigated using chemical degradation methods. The degradation procedure consisted in an alkaline hydrolysis, where 18-crown-6 ether was used as a phase transfer catalyst. This operation yielded a large amount of acids and hydroxyl functions. Gas chromatography (GC) displayed similar distributions of monocarboxylic acids in simple lipids and in hydrolyzed polar lipids, except the eicosanoic acid which was abundant in polar lipids. Gas chromatography-mass spectrometry (GC-MS) analysis of monocarboxylic acids showed large quantities of aliphatic acids. Some hopanoic acids were also found. It seems that monocarboxylic acids were chemically linked to the polar matrix by ester group.

Key words: Polar lipids, alkaline hydrolysis, monocarboxylic acids, 18-crown-6.

INTRODUCTION

The soil is a vital environment for numerous living beings. However, its natural balance is often disturbed by various dumping such as industrial wastes or excessive pesticides, which bring about toxic effects for environment. In order to avoid those harmful outcomes, it is necessary to have a good knowledge of physical, chemical and biological properties of the soil. These properties are widely conditioned by its organic components.

Lipids are part of organic material of soils. They are from microbial (Stevenson, 1965) or plant origins (Fustec-Mathon et al., 1979; Kolattukudy, 1980; Amblès et al., 1989), and are resulting from the transformation of plant residues by microorganisms (Amblès et al., 1993). Generally, lipids rates vary in accordance with the type of soil (Albrecht, 1970; Arpino, 1973; Jambu et al., 1978), climate, vegetation and biological activity or soil characteristics. They are extracted by organic solvents and are made of simple lipids directly hydrolysable and polar lipids with complex structure, which seemingly correspond to a set of organic macromolecules resulting from simple lipids polymerization (Maggi-Churin, 1986).

We undertook this work with the aim of characterizing the structure and the origin of those polar lipids. Considering that some lipids are not directly analyzable, we resorted to using reactions of degradation. Such methods were used for kerogen structural characterization (Kribii et al., 2001). In order to identify their natural precursors, we disrupted the ester bonds using an alkaline hydrolysis reaction in the presence of crown ether, a phase transfer catalyst. Other hydrolysis catalysts were used, but the 18-crown-6 ether appeared to be the best material (Okome, 1991). This hydrolysis released carboxylic acids and alcohols. We have chosen to characterize carboxylic monoacids for they represent

*Corresponding author. E-mail: okommintsa@yahoo.fr.

Table 1. Soil sample analytical characteristics.

pH(H_2O)	Clays %	Alluvium %	Sand %	C org. %	N org. %	Fe_2O_3 free %	Al_2O_3 free %	P_2O_5 %
4.3	12.5	14.4	37.3	20.3	1.5	1.0	1.1	0.14

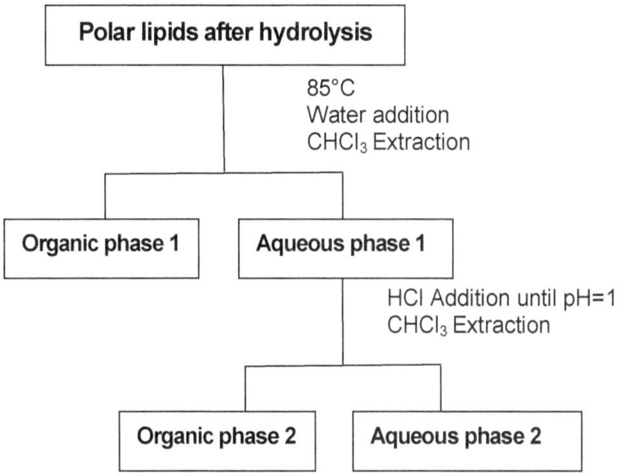

Figure 1. General Protocol of soil lipids extraction.

the most abundant family amongst hydrolysis products. Monoacids of simple lipids from same soil were analyzed for identification and comparison.

MATERIALS AND METHODS

Origins and characteristics of soil samples

Soil sample was obtained from the acid turf horizon of Millevaches plateau (Corrèze, France) between 20 and 40 cm deep. It is known to be a low-biological activity sodden soil (Table 1)

Equipment

Column chromatography

We used glass columns with two-way faucets (glass or Teflon key). For various fractionation purposes, we used 60G Merck Kieselgel silica (granules of 0.063 to 0.2 mm). For column obstruction, we used cotton and Fontainebleau sand. Silica, sand and cotton were thoroughly washed in soxhlet for 24 h, using bi-distilled chloroform and then they were vacuum-dried before being used.

Thin layer chromatography

Analyses were run on 0.25 mm thick commercial Merck plates, or on lab-made ones on which we spread 0.5 mm thick 60 G Merck silica.

Gas chromatography

Chromatography apparatuses were Packard model 427 and 438. They were equipped with a FID detector, along with an entry-divider ROS injector, a 25 mm Crompack CP^{tm} Sil5 melted silica capillary column, and a Merck D 2000 Shimadzu integrator. For analyses, temperature was selectively programmed: detector and injector (300°C); oven initial temperature (70 or 120°C); final temperature (300°C); temperature gradient (2°C/min).

Mass spectrometry (MS)

Mass spectra presented in this work were registered by two types of apparatuses: Kratos MS 25 mass spectrometer coupled with a Perkin-Elmer Sigma 3 gas chromatograph and a Data General Nova 4C computer, and a Finnigan Incos 500 spectrometer coupled with a Varian 3300 gas chromatograph and Data General computer.

Spectroscopic analyses (NMR and IR)

Proton and carbon 13 NMR spectra were measured on a Bruker Spectrospin 200 MHz WP 200 SY apparatus. IR spectra were seized thank to a Beckman Acculab 2 spectrometer.

Analytic protocol

Soil samples were dried in the open air, homogenized, ground, and then sifted at 200 µm before extraction.

Lipids extraction

Soil samples were introduced in cellulose cartridge, and twice extracted by bi-distilled chloroform through soxhlet, in order to avoid any sort of potential contamination. Chloroform was our organic solvent of reference for lipids extraction because previous experiments (Bilong, 1979) have shown that the use of chloroform would increase lipid rate and more particularly that of polar lipids, in comparison to the mixture ether of petrol (3/4) – ethyl acetate (1/4), which was more commonly used. After cooling off, the solvent was discarded by evaporation under partial vacuum (temperature lesser than 50°C) using a rotary evaporator. After extraction, we obtained directly fractionable "free" lipids. "Associated" lipids were extracted after alkaline treatment in order to free humic and fulvic acids and then acid treatment for organic and mineral matrix destruction to which they were covalently linked. Soil lipids extraction protocol was described in Figure 1.

Lipid extracts fractionation

After solvent evaporation and drying, each of both lipid sets (free and associated) was pre-fractionated using silica potash impregnated chromatography, according the Mac Carthy and Dutie Method (McCarthy and Duthie, 1962). We obtained three fractions: neutral, acid and polar. The neutral fraction was dissolved in a small volume of ether of petrol and then fractionated in silica column. The fractionation was followed by thin layer chromatography. The acidic fraction was methylated using diazomethane before being fractionated within the column. Amongst neutral and acidic fractions, we found numerous

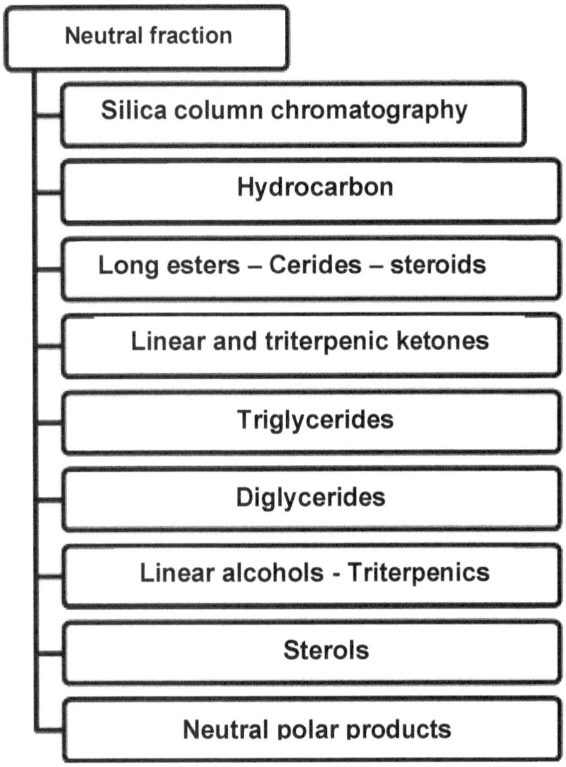

Figure 2. Neutral fraction's chromatographic analysis.

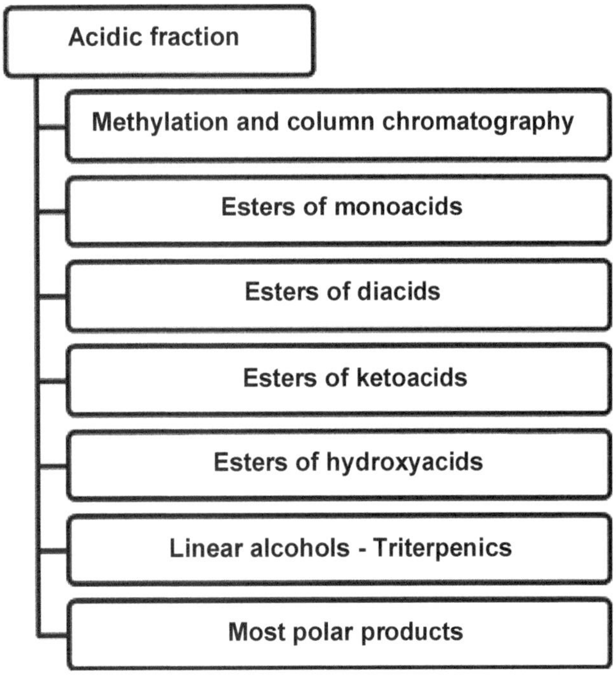

Figure 3. Acidic fraction's chromatographic analysis after methylation.

analyzable families as shown by silica column chromatography results displayed in Figures 2 and 3.

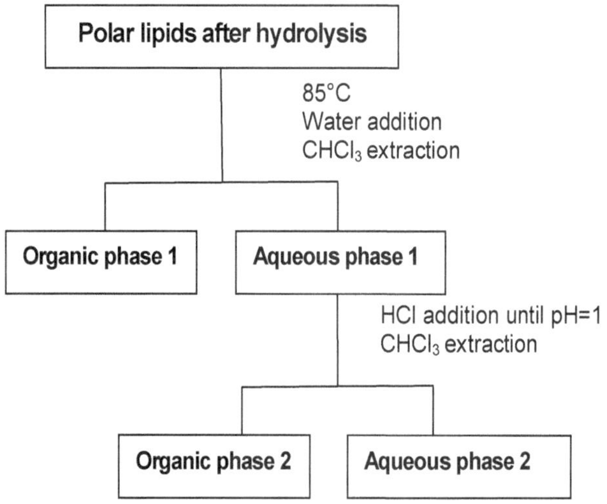

Figure 4. Hydrolysis products extraction protocol.

Complex lipids fraction obtaining

The neutral and acidic fractions' most polar lipids were those that did not migrate on the chromatographic plate in an ethyl acetate/cyclohexane mixture (40/60). They were added to the polar fraction. We then obtained the "polar lipids" fraction or "complex lipids" for which we would like to determine the structure.

Alkaline hydrolysis

Since the polar lipids structure was complex, we resorted to chain breaking reactions. We have performed an alkaline hydrolysis in the presence of a phase transfer catalyst: 1,4,7,10,13,16-hexaoxacyclooctadecane crown ether or 18-crown-6 ether. Crown ether role was to bring potassium hydroxide (KOH) in the organic phase, forming a stable complex in toluene (Dehmlow, 1983). Reactive and phase transfer catalyst were therefore within the same organic phase, then the reaction would more easily take place. The hydrolysis was performed in inert atmosphere (nitrogen), with reflux heating under agitation during 72 h. After cooling, hydrolysis products were extracted according the protocol described in Figure 4.

The organic phase 1, containing neutral and alkaline products, was obtained after alkaline extraction. After aqueous phase acidification, the extraction led to the organic phase 2 containing acidic products. First step products were separated in column and analyzed. Polar residues were hydrolyzed a second time according the same protocol.

Products analysis

After diazomethane methylation, acids resulting from the hydrolysis were fractionated using silica column chromatography and analyzed through GC-MS-coupled capillary gas chromatography.

RESULTS

Lipids extraction

After drying under dessicator and weighing lipid extracts,

Table 2. Weight results of lipids extracted and obtained polar lipids.

Lipid pools	Lipid extracts	Polar lipids
Free lipids (mg/kg of dry soil)	18 704	12 840 (68.6%)
Associated lipids (mg/kg of dry soil)	7 380	5 807 (78.7%)

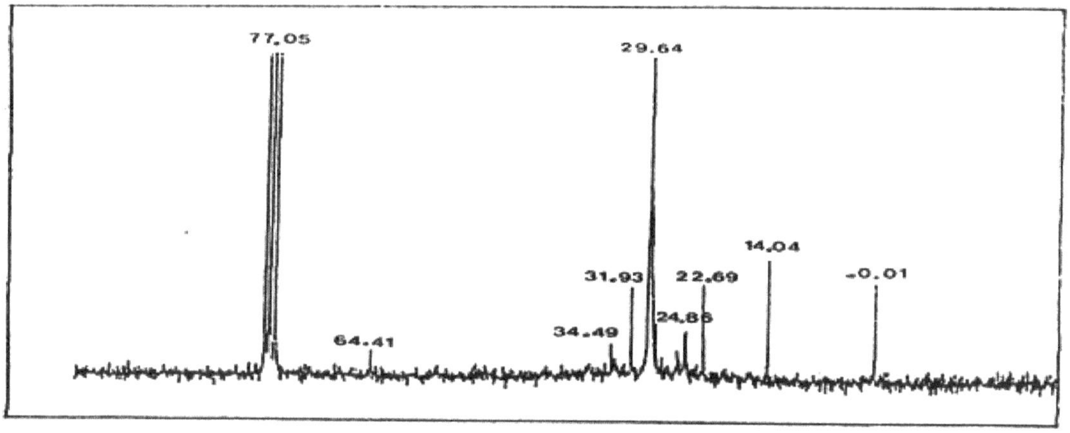

Figure 5. ^{13}C NMR Spectrum of free lipids polar fraction.

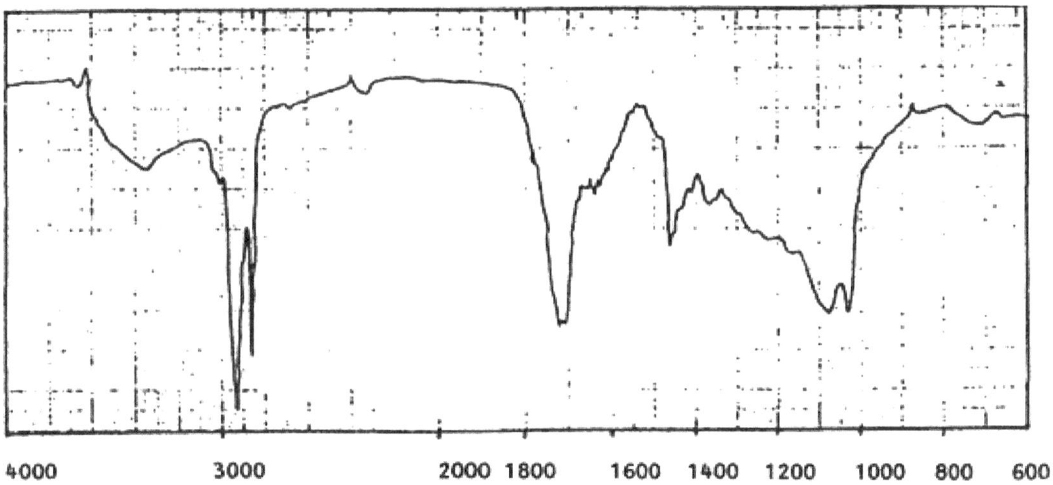

Figure 6. IR spectrum of free lipids polar fraction.

we obtained the following results. We noticed that the turf contained a higher concentration of polar lipids within "associated" lipids (78.7%) than within "free" lipids (68.6%), both lipidic fractions (free and associated) were studied separately (Table 2).

Complex lipids characterization tryout

Obtained polar lipids were under the appearance of a dark brown solid, not very soluble in methylene chloride or ether (even under heat), but they were easily soluble in chloroform or methanol. They did not migrate on silica plate chromatography with polar elution liquid containing 50% of cyclohexane and 50% of ethyl acetate. The elemental analysis has given the following results.

Carbon and hydrogen rates were less important in associated polar lipids than free ones, but on the other hand, their oxygen and nitrogen rates were lower. Free polar fractions were studied under spectroscopy: from both illustrated spectra in Figures 5 and 6, we inferred that we had non-aromatic material since nothing was disclosed under NMR ^{13}C.

The ^{13}C-NMR analyses of free lipids polar fraction

Table 3. Results of elemental polar lipids analyses.

Samples	C %	H %	N %	O %
Free polar lipids	72.39	10.16	0.27	16.96
Associated polar lipids	58.87	7.51	0.78	22.17

Table 4. Rate of different products after « free » complex lipids hydrolysis.

Products	1st step (%)	2nd step (%)	Total (%)
Monoacids	7.6	1.1	8.7
Diacids	2.7	1.4	4.1
Hydroxyacids	7.9	0.6	8.5
Alcohols	1.3	0.3	1.6
Other products	14.0	2.8	16.8
Polar residue	60.5	35.8	35.8
Total of identified products*	19.5	3.4	22.9
Mass default	6.0	58.0	24.5

*Identified products were monoacids, diacids, hydroxyacids and alcohols.

Table 5. Results of the « Associated » polar lipids hydrolysis.

Products	1st step (%)	2nd step (%)	Total (%)
Monoacids	2.3	0.5	2.8
Diacids	0.5	0.7	1.2
Hydroxyacids	3.7	0.5	4.2
Alcohols	1.4	0.1	1.5
Other products	3.8	1.2	5.0
Analyzed products	7.9	1.8	9.7

yielded the spectrum shown in Figure 5. We did the analysis of the same fraction in IR and we obtained the spectrum shown in Figure 6.

Alkaline hydrolysis

"Free" polar lipids hydrolysis

Organic phase 1 products were fractionated on the column, while those of the Organic phase 2 were fractionated after diazomethane methylation. Results of the hydrolysis were displayed in Table 3.

At the first step of hydrolysis, we obtained most of the products. The yield of second step was poor along with an important mass default probably because lighter products were freed in the course of the reaction and were lost during solvent evaporation; meanwhile others could have been dissolved in the aqueous phase. Hydroxyacids and monoacids were the most abundant products while alcohols were very much in the minority (Table 4).

Associated polar lipids hydrolysis

The yield of "associated" polar lipids was clearly lower than that of "free" ones. In the majority, we obtained hydroxyacids and monoacids, diacids and alcohols remaining a minority. We have chosen to study monoacids because they represented one of the most predominant families amongst hydrolysis products (Table 5).

Monoacids analysis

In order to better apprehend the existing relations between complex and simple lipids, we have concomitantly monoacids and simple lipids.

"Free" lipids

They represent the most abundant family amongst analyzed products resulting from « free » polar lipids

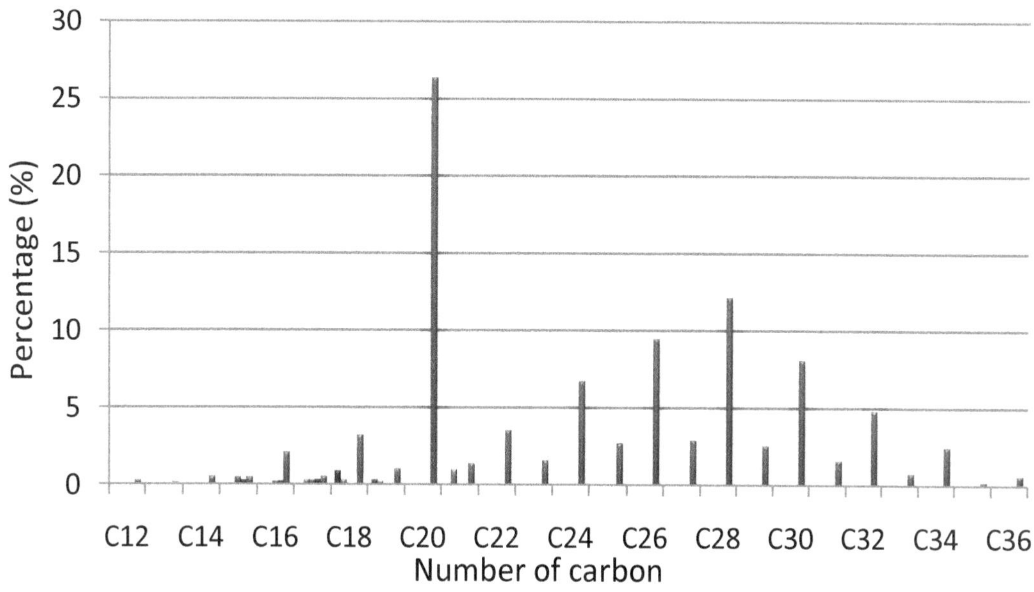

Figure 7. Histogram distribution of monoacids resulting of the first step of « free » polar lipids hydrolysis. (We separated the two histogramms).

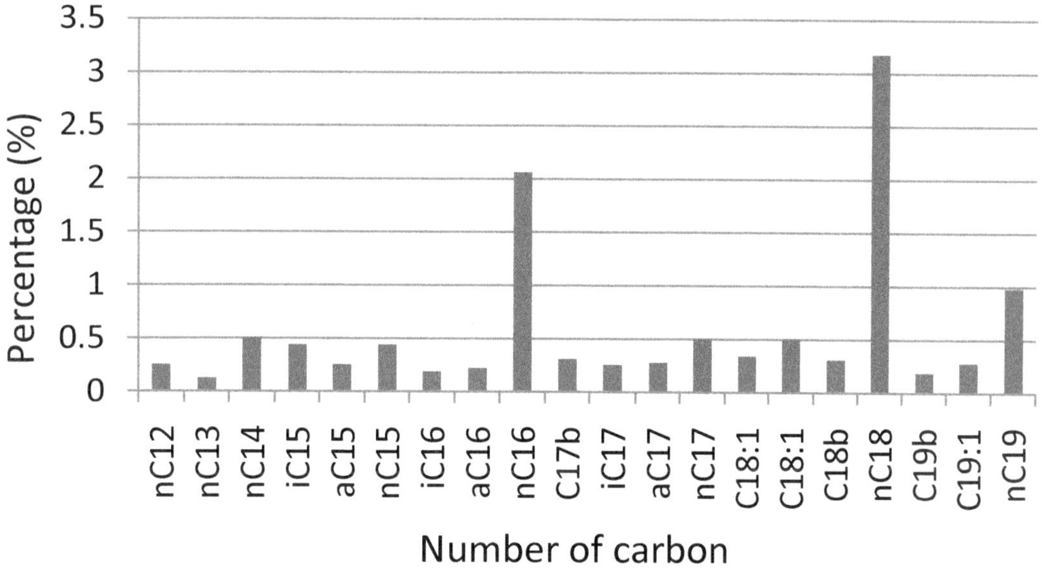

Figure 8. Histogram distribution of short-chain monoacids resulting of first step of « free » polar lipids hydrolysis.

hydrolysis. They were constituted of components whose chain length oscillate from C_{12} to C_{36}. The distribution of monoacids from the first step hydrolysis was shown in Figure 7.

As shown in Figure 8, we distinguished a set of short-chain monoacids (C_{12} to C_{19}) for which the principal ones were linear acids between C_{16} and C_{18}. We also noticed the presence of desaturated acids ($C_{18:1}$ and $C_{19:1}$) along with ramified acids such as iso- and anteiso-C_{15}.

A set of long-chain monoacids (C_{20} to C_{36}) essentially pair numbers of carbon such as nC_{26} and nC_{28} monoacid. We noticed a neat preponderance of eicosanoic acid (nC_{20}).

The outcome of the second step of hydrolysis was Analog to that of the first one, as shown on the histogram made of linear compounds, saturated with predominantly in Figure 9. Simple lipid monoacids distribution, shown on Figure 10, was different from polar ones by a larger proportion of long paired-linear carbon chain, and a lower eicosanoic acid concentration.

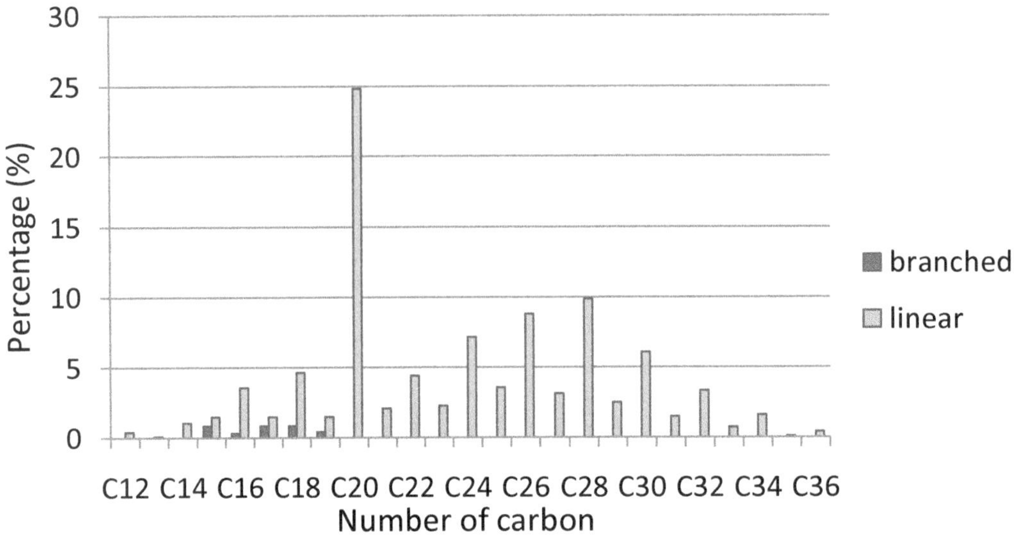

Figure 9. Histogram distribution of monoacids resulting of the second step of « free » polar lipids hydrolysis.

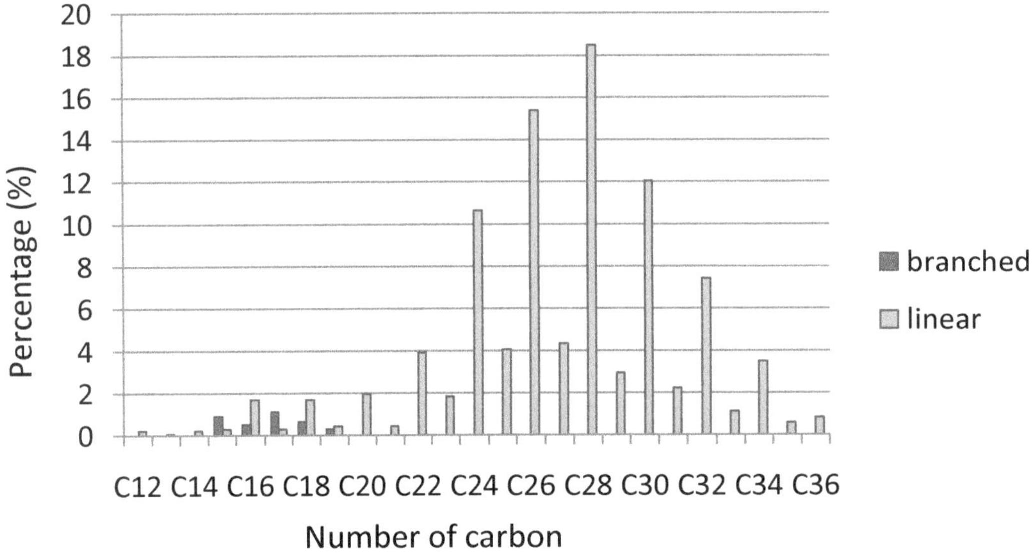

Figure 10. Histogram distribution of simple lipids monoacids.

"Associated" lipids

Monoacids represented 2.8% of hydrolysis products. After hydroxyacids, they were the most abundant. Their chromatogram, which was identical for both steps, was revealed on Figure 11.

The distribution was different from that of free lipids. On the other hand, we recorded numerous branched iso- and anteiso-compounds. We also identified branched monoacids whose mass spectra revealed a methylated carbon-10 structure. Mass spectra of the compound C_{17} is shown in Figure 12.

We mostly obtained a set of long-chain, linear, saturated and paired-numbers of carbon monoacids stretching from C_{20} to C_{34}. We have identified, using GC-MS, four hopanoic acids: C_{30}, C_{31}, C_{32}, and C_{33} (respectively A, B, C and D in the chromatogram shown on Figure 11) and traces of a diterpenic acid: dehydroabietic acid. The mass spectra of their methylic ester are shown on Figure 13 (compound A); Figure 14 (compound B), and Figure 16 (dehydroabietic acid).

The chromatogram of "Associated" simple lipids monoacids, presented in Figure 17, displays a distribution resembling those resulting from complex lipids hydrolysis.

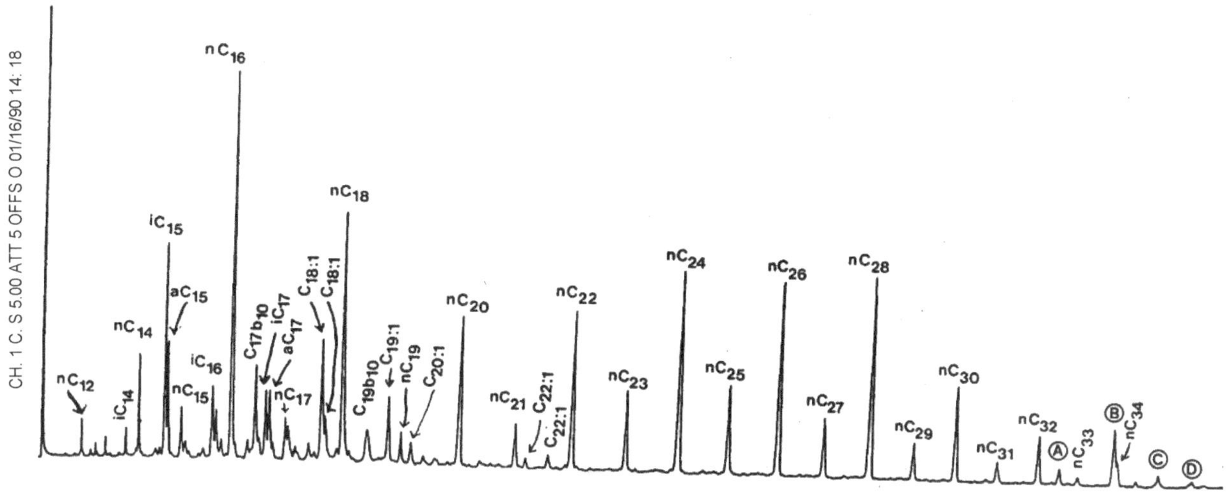

Figure 11. Chromatogram of monoacids resulting of the « associated » polar lipids hydrolysis.

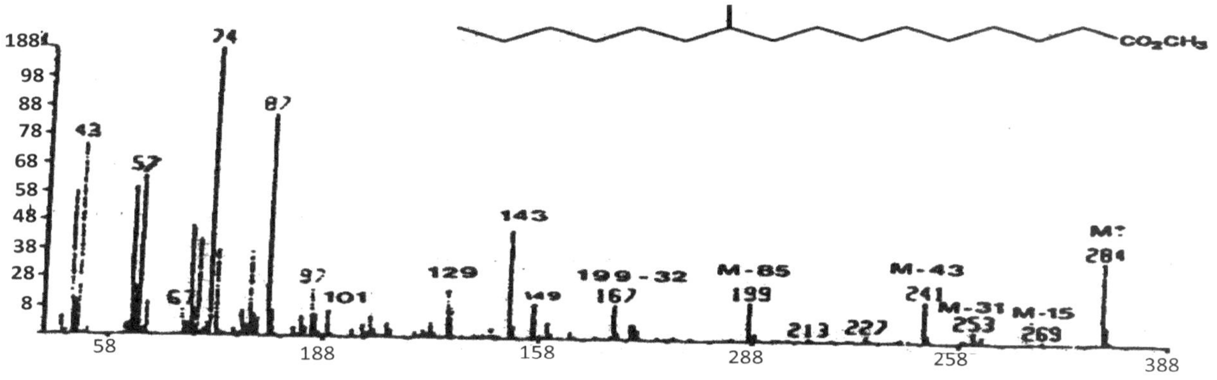

Figure 12. Mass spectrum of methyl-10-methylhexadecanoate (C_{17b10}).

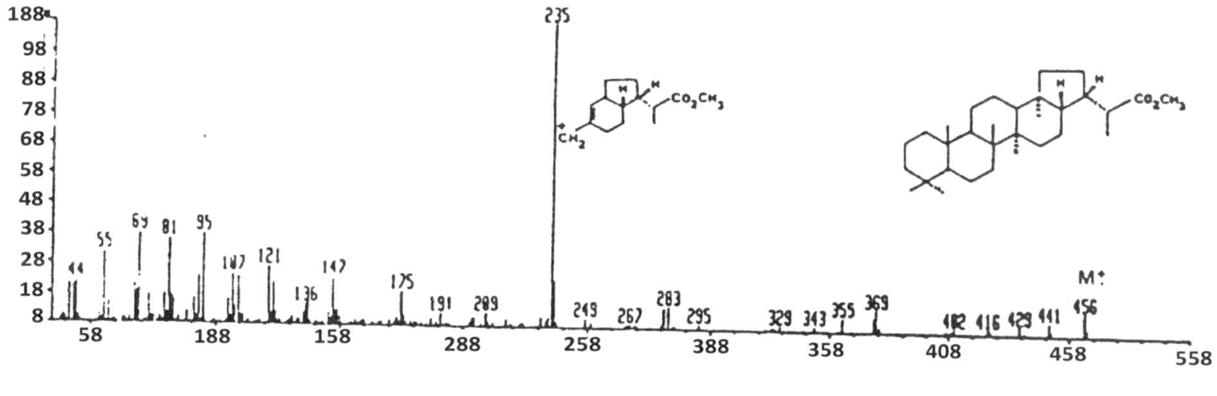

Figure 13. The mass spectrum of 17β-H-hopanoic acid (C_{30}) methylic ester.

Monoacids resulting from "associated" polar lipids hydrolysis were unlike those produced with « free » polar lipids by a higher concentration of short compounds (< C_{20}), a normal rate of eicosanoic acid and by the presence of hopanoic acids. Distribution of those monoacids is equivalent to their homologous within the corresponding simple lipids fraction. Like the "free" lipids, we would encompass that "associated" lipids could be

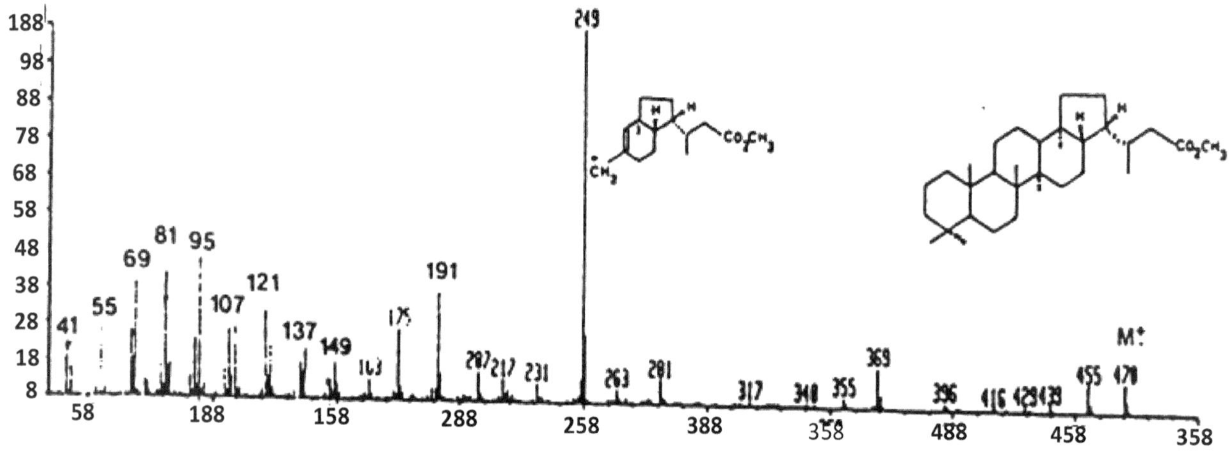

Figure 14. The mass spectrum of 17β-H-hopanoic acid (C31) methylic ester.

incorporated in the polar matrix.

DISCUSSION

The acid turf horizon of plateau de Millevaches, the soil of our study, contains about 70 to 80% of polar lipids (Table2). We have chosen that type of soil because it appeared that most of "free" and "associated" lipids were "complex lipids". An independent and parallel study realized on a podzol, under the same conditions, has shown a lower rate of extracted lipids (43.5% of free polar lipids) comparatively to that of the anmoor (68.6% of polar lipids) (Okome Mintsa, 1991).

Carbon and hydrogen rates were less important in associated polar lipids than free ones (Table 3), but on the other hand, their oxygen and nitrogen rates were lower. Free polar fractions were studied under spectroscopy: from both illustrated spectra in Figures 5 and 6, we inferred that we had non-aromatic material since nothing was disclosed under NMR ^{13}C. We noticed a 64.41 ppm peak which was probably resulting from carbon atoms with O-CH_2- groups (Hammon et al., 1998). On the other hand, we could disclose the non-aromatic nature of this complex mixture through the absence of a peak between 110 and 160 ppm, which was confirmed by IR analysis whose spectrum displayed around 3030, 1500, along with 1600 and at less than 900 cm^{-1} (Hammon et al., 1998; Allinger et al., 1982).

The large band seen around 3200 cm^{-1} had shown the presence of OH group (Allinger et al., 1982). Interestingly, previous data (Mahieu et al., 1999) supported our hypothesis by showing that ^{13}C CPMAS-NMR yielded different peaks according the nature of chemical groups among which the carbon was pertaining. They disclosed that depending chemical shift regions, alkyl groups had between 0 to 50 ppm peaks, whereas O-alkyls had 50 to 110 ppm; 110 to 160 for aromatics and 160 to 200 for carbonyls.

Following the same interest, other teams have also shown that ^{13}C-NMR spectroscopy could help to assess the interaction of soil organic matter with aluminium and allophane. They found a peak at 179 ppm assigned to carboxylic groups (Parfitt et al., 1999). Alkaline hydrolysis with crown-ether was performed according two steps; the second one could complete the first in case of inaccessibility of certain ester bounds.

At the first step of hydrolysis, we obtained most of the products. The yield of second step was poor along with an important mass default probably because lighter products were freed in the course of the reaction and were lost during solvent evaporation; meanwhile others could have been dissolved in the aqueous phase.

Hydroxyacids and monoacids were the most abundant products while alcohols were very much in the minority. The yield of "associated" polar lipids was clearly lower than that of "free" ones (Table 5). In the majority, we obtained hydroxyacids and monoacids, diacids and alcohols remaining a minority. We have chosen to study monoacids because they represented one of the most predominant families amongst hydrolysis products.

Monoacids distribution displayed a neat preponderance of eicosanoic acid (nC_{20}) (Figures 7 and 9). We also found that acid in this soil's cerides and triglycerides. The relevant acid could result from their enzymatic hydrolysis (Magnoux, 1982; Amblès et al., 1985), followed by a recombination amongst complex lipids.

Simple lipid monoacids distribution, shown in Figure 10, was different from polar ones by a larger proportion of long paired-linear carbon chain, and a lower eicosanoic acid concentration. Long-chain, linear, paired-carbon number and saturated monoacids are from plant origin (Martin and Jupiler, 1970; Tulloch, 1976; Stevenson, 1982), while short-chain, branched and desaturated were from microbial origin (Boon et al., 1977; Volkman et al., 1980). Surprisingly, apart from the eicosanoic acid (nC_{20}) we noticed that fatty monoacids freed by hydrolysis such as "free" complex lipids were close to simple monoacids.

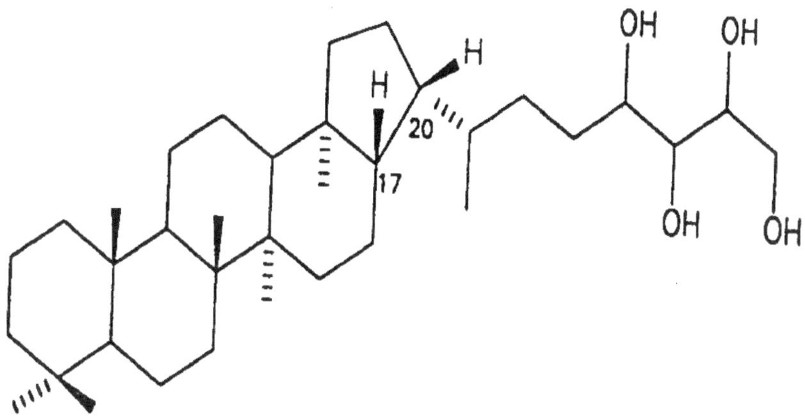

Figure 15. Chemical structure of bacteriohopanetetrol.

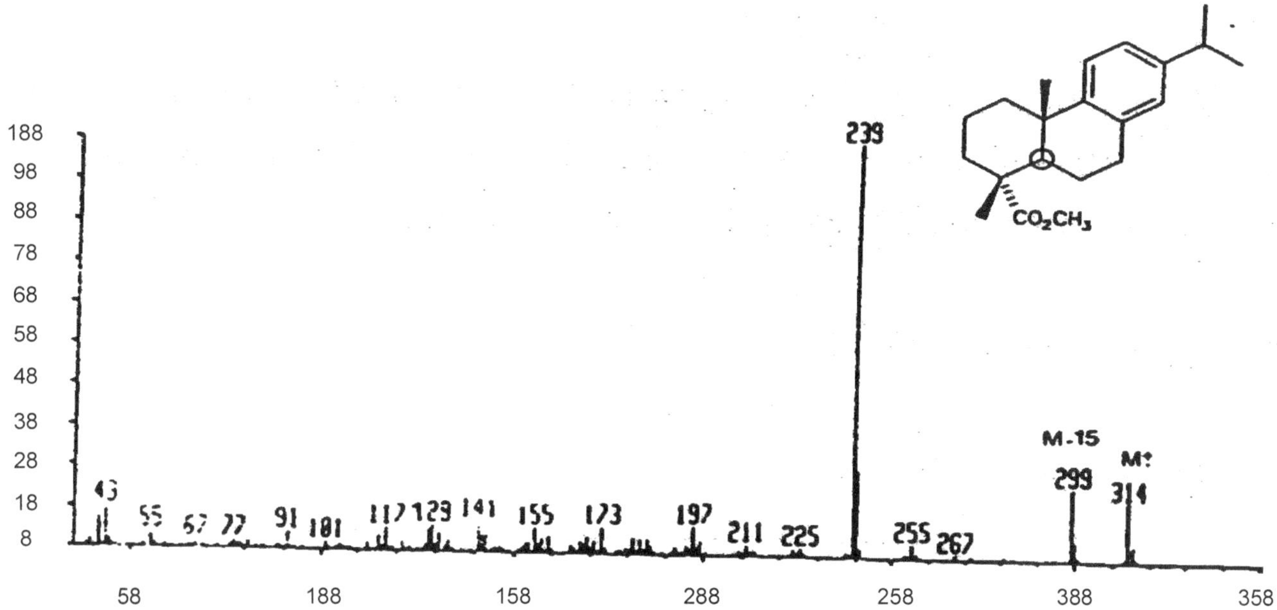

Figure 16. Mass spectrum of methyle dehydroabietate.

It seemed that simple lipids would have incorporated complex soil organic lipids.

The distribution was different from that of free lipids. Nevertheless, we distinguished two modes: a short mode, comporting in most cases: linear, saturated and predominantly paired-numbers of carbon monoacids stretching from 12 to 19 (nC_{16} was the major one). They were from microbial origin (Parker, 1969; Boon, 1977). On the other hand, we recorded numerous branched iso- and anteiso- compounds. We also identified branched monoacids whose mass spectra revealed a methylated carbon-10 structure (Figure 12).

Hopanoic acids were triterpenic acids often found in the form of acids or hydrocarbons within sedimentary organic material (Ensminger et al., 1973; Tulloch, 1976; Ourisson et al., 1979; Stevenson, 1982). In conformity with previous studies, we hypothesized that microorganisms were the main source of hopanes (Förster et al., 1973; Rohmer, 1975; Ourisson et al., 1984; Rohmer and Ourisson, 1984), with bacteriohopanetetrol as a putative precursor (Ensminger et al., 1973; Förster et al., 1973; Rohmer and Ourisson, 1976, 1977; Dastillung et al., 1980). The chemical structure of the bacteriohopanetetrol is shown in Figure 15.

Dehydroabietic acid is a diterpenic acid which is abundantly found in forest soils and in recent sea sediments (Simoneit, 1977, 1979). The chromatogram of "Associated" simple lipids monoacids, presented in Figure 17, displays a distribution resembling those resulting from complex lipids hydrolysis. Distribution of monoacids resulting from "associated" polar lipids hydrolysis is equivalent to their homologous within the

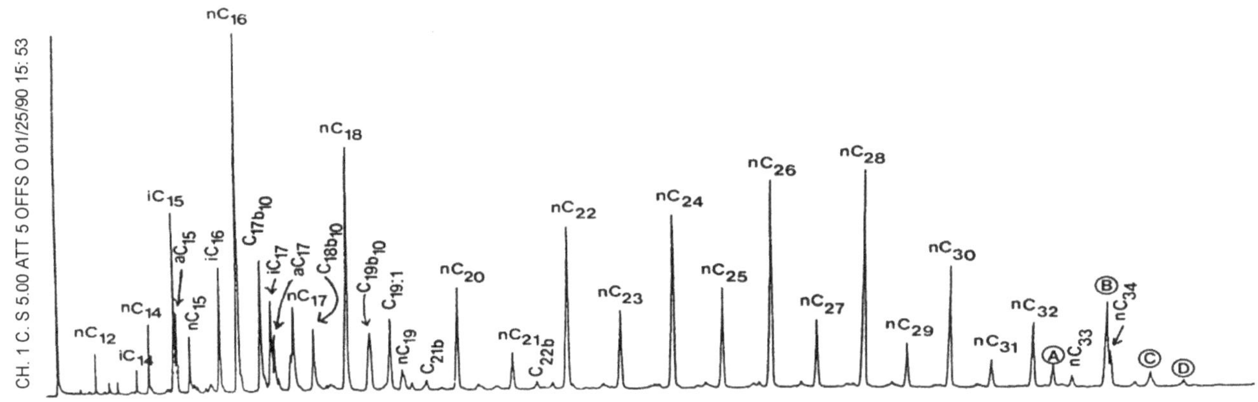

Figure 17. Chromatogram of Monoacids from "Associated" simple lipids.

corresponding simple lipids fraction (Figure 17). Like the "free" lipids, we would encompass that "associated" lipids could be incorporated in the polar matrix.

Conclusion

Polar lipids represented an important fraction of soil organic material. It had been shown that those lipids could never be directly analyzed (Amblès et al., 1985). Alkaline hydrolysis in the presence of a phase transfer catalyst had helped to free numerous molecules bearing carboxyl and hydroxyl functions, which indicated thepresence of an important number of ester bonds in the polar lipidic fraction of this soil. Some analogous methods of structure characterization have been used for kerogen characterization (Kribii et al., 2001).

"Free" complex lipids hydrolysis had yielded more products than that of "associated" ones. It seemed that in the anmoor, hydrolysable functions of "free" polar lipid were more important than the "associated" lipids. Nevertheless, the same classes of lipids were found but under different proportions.

The analysis of monoacids resulting from alkaline hydrolysis in the presence of crown ether revealed the presence of aliphatic chains, linked to the macromolecular network by ester bonds. This study has shown that simple molecules, through the oxidation process, could have been incorporated into the soil's polar matrix. This strong statement had also been demonstrated in a complementary study (Amblès et al., 1993). There must be a subtle equilibrium between simple and complex lipids. Supplementary investigation methods could allow determining the nature of other possible bonds existing between the free lipids and the soil's polar matrix.

REFERENCES

Albrecht P (1970). Study of the organic components of the sedimentary sequences of Logbaba and Messel. Digénétic transformations. Memory of the geological map of Alsace and Lorraine, No. 32.

Allinger NL, Cava MP, De Jongh DC, Johnson CR, Lebel NA, Stevens CL (1983). Organic chemistry. Applications. McGraw-Hill Ed., 3: 990-991.

Amblès A, Magnoux P, Jambu P, Jacquesy R, Fustec E (1989). Effects of the addition of bentonite on the hydrocarbon fraction of a podzol soil (A1) horizon. J. Soil Sci., 40: 685-694.

Amblès A, Jacquesy R, Jambu P, Joffre J, Fustec E (1985). Influence of iron on the dynamics of lipids in hydromorphic acidic environment. Agrochimica, 24(2, 3, 4): 199-209.

Amblès A, Jacquesy JC, Jambu P, Mayoungou P, Okome-Mintsa M, Hita C, Parlanti E (1993). High molecular weight lipids present in soil. Nature and origin. Org. Geochem. (K. Øygard, Ed.): 668-671.

Amblès. A, Magnoux P, Jambu P, Jacquesy R, Fustec E (1989). Effects of addition of bentonite on the hydrocarbon fraction of a podzol soil (A1 horizon). J. Soil Sci., 40: 685-694.

Arpino P (1973). Lipids from the Eocene lakes sediments. Geological science thesis, Strasbourg.

Bilong P (1979). Evolution of the compounds humic and lipid of the soils of the moors of the Plateau de Millevaches. Doctorate thesis, University of Poitiers, p. 113.

Boon JJ, De Leeuw JWD, Hoek JH (1977). Significiance and taxonomic value of iso and anteiso monoenoic fatty acids and branched β-hydroxyacids in Desulfovibrio Desulfuricans. J. Bacteriol., 129(3): 1183-1191.

Dastillung M, Albrecht P, Ourisson G (1980). Aliphatic and polycyclic alcohols in sediments. Alcohols derived from the hopane and 3-methyl hopane. J. Chem. Res., (M): 2353-2374.

Dastillung M, Albrecht P, Ourisson G (1980). Aliphatic and polycyclic ketones in sediments C_{27} - C_{35} ketones and aldehydes of hopane series. J. Chem. Res., (S): 166-167, (M): 2325-2352.

Dehmlow EV, Dehmlow SS (1983). Phase transfert catalysis, Verlag Chimie, p. 386.

Ensminger A, Van Dorsselaer A, Spyckerelle C, Albrecht P, Ourisson G (1973). Pentacyclic triterpenes of the hopane type as ubisquistous geochemical markers: Origin and significance. Advances in organic geochemistry. Edition Tissot B and Biehner F. Technip 1974, pp. 245-260.

Förster HJ, Biemann K, Haig WG, Tattrie NH, Colvin JR (1973). The structure of novel C_{35} pentacyclic terpenes from Acetobacter-xylium. Biochem. J., 135: 133-143.

Fustec-Mathon E, Jambu P, Bilong P, Amblès A, Jacquesy R (1979). Inter-relationships between lipids and iron in soils. Influence of the addition of iron A1 podzol horizon. International symposia of the CNRS n° 303. Organo-mineral migration in temperate soils.

Hamon M, Pellerin F, Guernet M, Mahuzier G (1998). Analytical chemistry. Spectral methods and organic analysis. Masson Ed, 3: 179.

Jambu P, Fustec E, Jacquesy R (1978). Lipids of soil: nature, origin, evolution, properties. Soil science. Bulletin de l'A.F.E.S., 4: 229-240.

Kolattukudy PE (1980) cutin, suberin and waxes. The biochemistry in plants. Lipids, structures and functions. P.K. stumpf Ed., 4: 571-645.

Kribii A, Lemée B, Chaouch A, Amblès A (2001). Structural study of the Moroccan Timahdit (Y-layer) oil shale kerogen using chemical degradations. Fuel, 80: 681-691.

Maggi-Churin R (1986). Organic matter of soils. Structural study of the polar lipid fraction. Doctorate thesis, University of Poitiers, p. 63.

Magnoux P (1982). Study of the influence of inflow of clay on the composition of the lipid fraction of a deficient soil. Doctorate thesis, University of Poitiers, 115 p.

Mahieu N, Powlson DS, Randall EW (1999). Statistical analysis of published carbon-13 CPMAS NMR spectra of soil organic matter. Soil Sci. Soc. Am. J., 63: 307-319.

Martin JT, Juniper BE (1970). The cuticles of plants. Edward Arnold, London, 347 p.

McCarthy RD, Duthie AH (1962). A rapid method for the separation of free fatty acids from other lipids. J. lipids Res., 3: 177-179.

Okome Mintsa M (1991). Study of complex lipids in acidic soils: structure and origin. Doctorate thesis, University of Poitiers, 189 p.

Ourisson G, Albrecht P, Rohmer M (1979). The hopanoids. Paleochemistry and biochemistry of a group of natural product. Pure Appl. Chem., 51: 709-729.

Ourisson G, Albrecht P, Rohmer M (1984). The microbial origin of fossil fuels. Pour la Science. 84: 55-56.

Parfitt RL, Yuan G, Theng BKG (1999). A ^{13}C-NMR study of the interactions of soil organic matter with aluminium and allophone in podzols. Eur. J. Soil Sci., 50: 695-700.

Parker PL (1969). Fatty acids and alcohols. Org. Geochem. Methods and results (Eglinton G. and Murphy M. T. Ed.). Spring Verlag.

Rohmer M (1975). Triterpenoids of prokariots. Science doctorate thesis, University Louis Pasteur of Strasbourg.

Rohmer M, Ourisson G (1976). Structure of the bactériohopanetétrols of acetobacter-xylium. Tetrahedron lett., 40: 3633-3636.

Rohmer M, Ourisson G (1977). Derived from the bacteriohopane: structural variation and distribution. Tetrahedron lett., 40: 3637-3640.

Rohmer M. Bouvier-Nave P, Ourisson G (1984). Distribution of hopanoid triterpenes in prokaryotes. J. Gen. Microbiol., 130: 1137-1150.

Simoneit BRT (1979). Eolian particulates from oceanic and rural areas. Their lipids fulvic and humic acids and residual carbon. In Advances in Organic Geochemistry. Physics and chemistry of the Earth, 12: 343-352.

Simoneit BRT (1977). Diterpenoid compounds and other lipids in deep-sea sediments and their geochemical significance. Geochim. Cosmochim. Acta., 41: 463-476.

Sljivar Brosic S, Saban M, Amblès, Vitorovic D (1987). Newly discovered constituents of the bitumen from Aleskinac (Yougoslavia) shale. Isoprenoids and hopanoic acids. J. Serb. Chem. Soc., 52(4): 187-195.

Stevenson JF (1982). Soil lipids in humus chemistry. Genesis, composition, reactions. A. Wiley Interscience Publication John Wiley and Sons Inc. Stevenson JF (1966). Lipids in soils. Am. Oil Chem. Soc., 43(4): 203-210.

Tulloch AP (1976). Epicuticular wax of Agropyron Smithili leaves. Phytochemistry, 15(7): 1153-1156.

Van Dorsselaer A, Ensminger A, Spyckerelle C, Dastillung M, Sieskind O, Arpino P, Albrecht P, Ourisson G (1974). Degraded and extended hopane derivatives (C_{27} to C_{35}) as ubiquitous geochemical markers. Tetrahedron lett., 14: 1349-1352.

Volkman JK, John RB, Gillan FT, Perry GJ, Bavor HJJr (1980). Microbial lipids of an intertidal sediments I. Fatty acids and hydrocarbons. Geochim. Cosmochim. Acta., 44: 1133-1134.

Adsorption kinetics of 2-chloroacétophénone and 4-bromoacétophénone thiosemicarbazone on Oueme Valley clays (R. Benin)

Tadjou DO REGO[1,2], Serge Fredys Rodrigue O. SENOU[1,3], Fidèl DIMON[3], Salome KPOVIESSI[2], Alban HOUNGBEME[1], Coffi AZANDEGBE ENI[3], Georges C. ACCROMBESSI[2], Jacques POUPAERT[4] and Fernand A. GBAGUIDI[1,2*]

[1]Laboratoire National de Pharmacognosie, Centre Beninois de la Recherche Scientifique et Technique (CBRST), BP 06 Oganla, Porto-Novo, Benin.
[2]Laboratory of Physic and Synthesis Organic Chemistry (LaCOPS), University of Abomey-Calavi, Faculty of Sciences and Technics (FAST),BP: 4521 Cotonou, Benin.
[3]Laboratory of Physical Chemistry, University of Abomey-Calavi (UAC), Faculty of Sciences and Technics (FAST), Cotonou, Benin.
[4]Louvain Drug Research Institute (LDRI), School of Pharmacy, Université catholique de Louvain, B1 7203 Avenue Emmanuel Mounier 72, B-1200 Brussels, Belgium.

2-chloroacetophenone-thiosemicarbazone and 4-bromoacétophénone-thiosemicarbazone are compounds endowed with numerous pharmacological properties. In order to produce therapeutic agents using clay as excipient, this paper deals with the absorption kinetics of a clay originating from Oueme valley in Benin. A prospection in this area gave rise to the collection of 4 samples classified according to their color (yellow, white, black, and red). Toxicity studies performed on larvae gave no evidence of toxicity. These clays are highly hydrophilic and can accept a high load of organic material. The white clay, which is the least acidic, shows the best adsorption capacity. Additional experiments were carried out with this clay in ethanol. Best adsorption was obtained after 2 h. Absorption isotherms indicate low affinity between the clay and the thiosemicarbazones and can be modelized using Langmuir and Freundlich's models.

Key words: Clay, thiosemicarbazones, kinetic, absorption, larval toxicity, desorption.

INTRODUCTION

2-chloroacetophenone thiosemicarbazone and 4-bromoacetophenone thiosemicarbazone are interesting pharmacomolecules whose usefulnes has been proved in microbiology (Fatondji 2011). On the other hand, clay is a

Tabe 1. Physicochemical characteristics of synthesized 2-CAT and 4-BAT. Eluent CH_2Cl_2 / EtOAc 2/1 (Fatondji, 2011).

Adsorbats	2-CAT	4-BAT
Molecular formula	$C_9H_{10}ClN_3S$	$C_9H_{10}BrN_3S$
Molecular weight (g/mol)	227.72	272.16
Melting point	157-158	198-199
Front report	0.88	0.80

natural material with various therapeutic applications (antibacterial, antiseptic).

In Benin, this natural resource still awaiting exploitation is available in the Oueme valley and is scarcely used by local population to cure ulcers, muscular lesions, and persistent diarrhea. In order to incorporate these thiosemicarbazones into clays as excipients to improve their therapeutic efficacy, the clay must be endowed with good capacities of sorption and desorption of these organic compounds. The aim of this paper is to follow up the evolution of the clay adsorption of 2-chloroacetophenone thiosemicarbazone (2-CAT) and 4-bromoacetophénone thiosemicarbazone (4-BAT) in ethanol.

EXPERIMENTAL

Materials

Young shrimp larvae (*Artemia salina*) were used as biological material. 2-chloroacetophenone-thiosemicarbazone (2-CAT) and 4-bromoacetophenone-thiosemicarbazone (4-BAT) used as adsorbates were previously synthesized, purified and characterized in the Laboratory of Organic Chemistry and Physical Synthesis (LACOPS / UAC). Their physicochemical characteristics are summarized in Table 1.

White, yellow, black and red consumables clay, were taken from the Oueme valley, in the township Dangbo and Aguegues (southern of Benin). A portion of each sample was dried in an oven at 105 ± 2°C, powdered, sieved to 180 µm and stored for the sorption study. The second parts were kept in plastic boxes at room temperature for toxicity studies, because freezing may reduce their sensitivity (Caquet., 1994).

Characterization of clays

Water and organic material content, acid-base nature, specific surface area and cation exchange capacity of each sample of clays were determined.

Water content

Water content was determined by Audigié et al. (1978) method.

Organic material content

Clay's organic material content is measured according to NF EN 12879 norm.

pH measurements

The pH was measured following the method proposed by Aïssata (2009).

Cation-exchange capacity (C.E.C.)

Cation-exchange capacity (CEC) is the maximum quantity of total cations, of any class, that a soil is capable of holding, at a given pH value, available for exchange with the soil solution. It is expressed as milliequivalent of hydrogen per 100 g of dry soil (m_{eq}/100 g), C.E.C was assessed according to Metson (1956) for clays below 180 µm and following the AFNOR NF X 31-130 norm.

Specific surface

The specific area in m^2 /g is assessed by methylene blue colorimetry according to Hicham El (2006).

Cytotoxicity test on shrimp larvae

The test is performed against *Artemia salina* Leach by the method of Michael et al. (1956) resumed by Vanhaecke et al. (1981) and Sleet and Brendel (1983) and proposed in the literature as a simple bio - assay method for assessment of preliminary toxicity of natural active products (Solis et al., 2001). The eggs of *A. salina* were incubated in sea water until hatching of young larvae (48 h). Then, series of solutions of each tested clay sample at varying and progressive concentrations were prepared. A defined number of larvae, (sixteen) were introduced into each solution. All solutions and control solution containing no active substance were left under stirring for 24 h. Counting under a microscope the number of dead larvae in each solution was used to evaluate the toxicity of the solution. In the case where there was death in the control medium, the data was corrected by Abbott's formula:

% death = [(test - control) / control)] × 100 (Abbott, 1925).

Data (dose-response) are transformed by logarithm and the LC_{50} were determined by linear regression (Hafner et al., 1977). Tests were carried out in triplicate.

Adsorption kinetics of 2-CAT and 4-BAT in ethanol

Sorption technique

The initial concentrations of thiosemicarbazone were: 0.0625, 0.125, 0.25 and 0.5 mg/ml. 0.02 g of clay (< 180 µm) is mixed with 2 ml of an ethanolic solution of thiosemicarbazone. The suspensions are stirred for 10 h at room temperature. The supernatant is assayed by high performance thin layer chromatography (HPTLC) after centrifugation to determine the

Table 2. Physicochemical characteristics of consumables clays.

Variables		Yellow clay	Black clay	Red clay	White clay
Te(%)		36.70	54.73	33.38	59.95
M. O. (%)		14.34	23.84	18.12	13.76
C E C (m_{eq} /100 g)		23.79	20.13	22.21	16.01
S.S. (m^2/g)		3.32	2.29	2.46	4.82
pH medium	KCl	3.54±0.01	3.29±0.01	4.18±0.03	5.02±0.03
à 25°C	H_2O	4.65±0.00	4.17±0.01	5.12±0.01	6.04±0.02

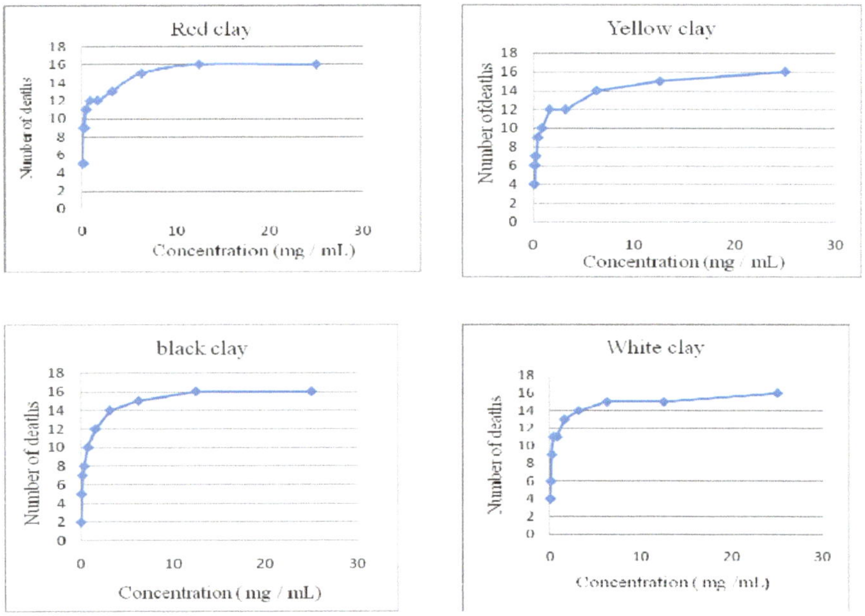

Figure 1. Dose-response curves illustrating the sensitivity of shrimp larvae to clay species.

residual thiosemicarbazone concentration.

Analyses of the thiosemicarbazones

Thiosemicarbazones are analyzed by CCMHP densitometer (Camag TLC Scanner III) at 300 nm. A mother solution (1 $mg.mL^{-1}$) is prepared in ethanol and further diluted to get concentrations of 0.0625, 0.125, 0.25 and 0.5 mg/ml used for standardization. HPTLC plates (Silica gel 60 F254S Merck®) are developed using a toluene: acetonitrile mixture (6 : 7, v/v).

RESULTS AND DISCUSSION

Physicochemical parameters of the clays

As can be seen in Table 2, the studied clays are rather hydrophilic with a content of water ranging between 33.38 and 59.95%. The white (59.95%) and black clays (54.73%) are notably more hydrophilic than the red (33.38%) and yellow clays (36.70). Clay's CEC values are found in the range 16.01 and 23.79 m_{eq} / 100 g. These figures are satisfactory since Morel (1996) found for kaolinite values between 5 and 15 m_{eq} /100 g, for illite and chlorite values between 10 and 40 m_{eq} /100 g. These clays exhibit a weak adsorption capacity in accordance with their low specific area (2.29 – 4.82 $m^2.g^{-1}$). The white clay exhibits the highest specific surface. Clay's pH figures are all below 7, attesting to their acidic character. Yellow, red and black clays are found clearly more acidic ($3.5 \le pH \le 5$), than the white one (6.04).

Larval toxicity

The shrimp larvae are sensitive to clays. We noticed a gradual increase in the number of dead larvae gradually as the extract concentration increased (Figure 1). The LC50 (Table 3) compared to the values of table drawn by

Table 3. LC50 value of the various supplies clays.

	Yellow clay	Black clay	Red clay	White clay
CL_{50} value (mg/mL)	0.17	0.10	0.17	0.29

Table 4. Correspondence between LC50 and toxicity.

CL_{50}	Toxicity
CL_{50} > 100 µg /mL ou 0.1 mg /mL	-
100 µg /mL > CL_{50} ≥ 50 ou 0.05 mg /mL	++
50 µg /mL > CL_{50} ≥ 10 g /mL ou 0.01 mg /mL	++
CL_{50} < 10 µg /mL	+++

- No toxic, + Low toxicity,++ : Moderate toxicity, +++ : High toxicity.

Table 5. Isotherm parameters studied 2-CAT and 4-BAT for white clay.

Models		Settings			
		q_m	b	R_L	R^2
Langmuir	2-CAT	$-1.35.10^{-5}$	-17.386	0.999 – 0.996	0.99
	4-BAT	$-2.03.10^{-5}$	-3.640	0. 556 – 0.843	0.97
Freundlich	2-CAT	$1/n$ = 4.20		k_f = 4.426	0.96
	4-BAT	$1/n$ = 1.86		$k_f = 5.14.10^{-2}$	0.88

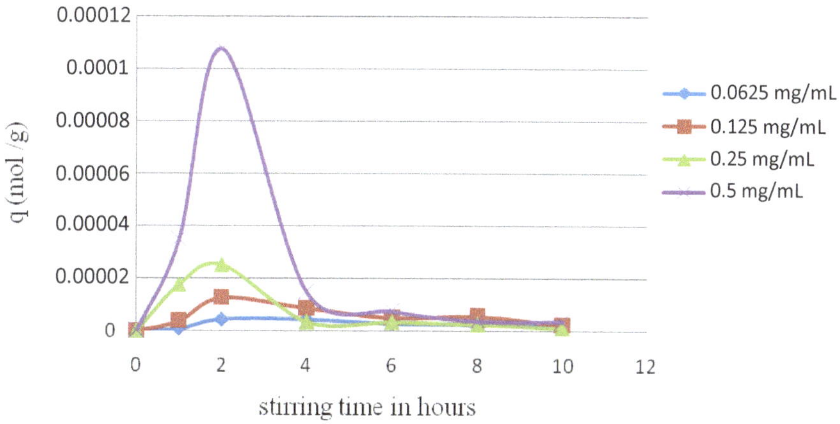

Figure 2. Kinetics test of adsorption 4-BAT on the white clay in ethanol medium.

Mireille (1995) sparkling (Table 4) shows that these clays are not toxic. Thus, consumables clays of the valley can be used without toxicological risk. In view of these results of LC_{50} figures (Table 5) compared with those established by Mireille (1995) demonstrate the lack of toxicity of these clays.

Adsorption kinetics of 2-CAT and 4-BAT in ethanol

According to Figures 2 and 3, adsorption kinetics of 2-CAT and 4-BAT on white clay shows that there is no apparent equilibration time for adsorption of thiosemicarbazones by this clay and that adsorption is

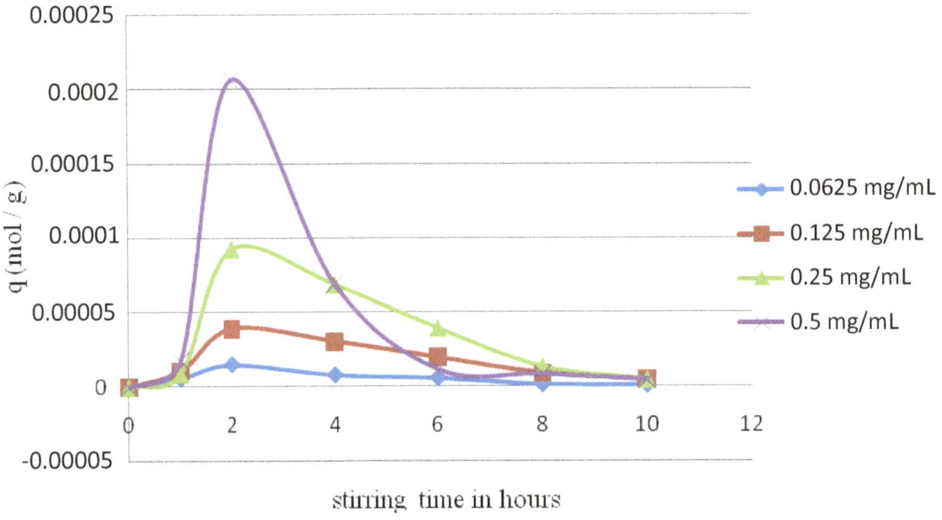

Figure 3. Kinetics test of adsorption 2-CAT on the white clay in ethanol medium.

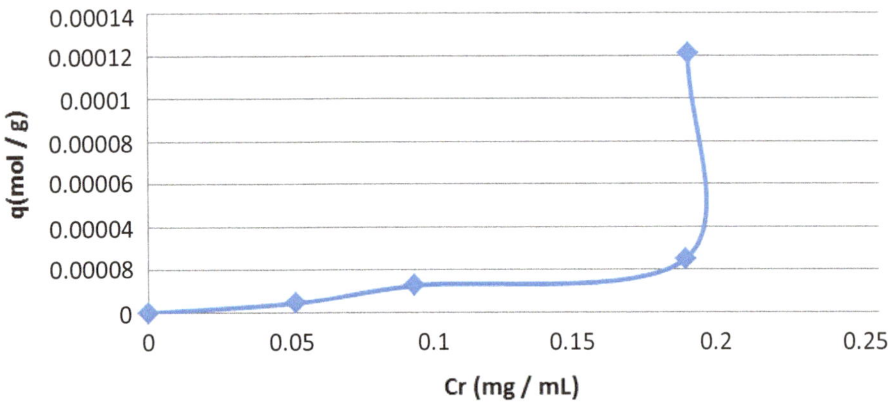

Figure 4. Adsorption isotherms of 4-BAT on white clay in ethanol medium at 25 ± 2°C.

reversible. After a fast initial phase, adsorption is maximal after 2 h. Desorption follows the same kinetic behaviour and reaches a plateau after 2 h. Thiosemicarbazone's adsorption is considerably influenced by its initial concentration in ethanol. Maximal sorption goes up to 63% for 4-BAT and 89% for 2-CAT after 2 h equilibration.

Sorption isotherm of 2-CAT and 4-BAT

A sorption isotherm (also adsorption isotherm) describes the equilibrium of the sorption of a material at a surface (more general at a surface boundary) at constant temperature. It represents the amount of material bound at the surface (called the sorbate) as a function of the material present in the solution. In this study, sorption isotherms were determined at room temperature after 2 h equilibration and were found of the concave S-shape type, suggesting a low affinity of the white clay (Figures 4 and 5 and Table 5) for thiosemicarbazone derivatives according to Giles et al. (1974). Modeling of these isotherms give intensity parameters which fit with those computed following to the models of Langmuir and Freundlich ($R^2 > 0.8$).

Conclusion

Selected clays originating from Oueme valley of Benin are not toxic. They are highly hydrophilic, rich in organic materials and are of acidic nature. They exhibit low specific area and CEC comparative analysis of their characteristics shows that the white clay, less acidic and hydrophilic shows the best adsorption capacity.

Adsorption isotherms suggest a low affinity between the white clays and thiosemicarbazones and can be

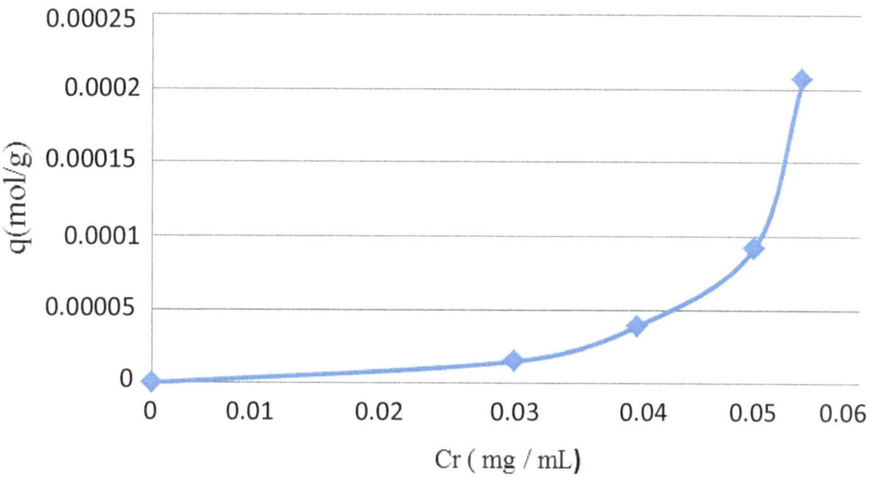

Figure 5. Adsorption isotherms of 2-CAT on the white clay in ethanol medium at 25 ± 2°C.

rationalized following the models of Freundlich and Langmuir. Adsorption kinetics demonstrates that adsorption is reversible and optimal after 2 h.

Conflict of Interest

The author(s) have not declared any conflict of interests.

REFERENCES

Abbott WS (1925). A method of computing the effectiveness of an insecticide. J. Econ. Entomol. 18:265.

Aïssata C (2009). Lutte contre *sitophilus oryzae* l. (coleoptera: curculionidae) et *tribolium castaneumherbst* (coleoptera: tenebrionidae) dans les stocks de riz par la technique d'étuvage traditionnelle pratiquée en basse-guinée et l'utilisation des huiles essentielles végétales. Thèse de doctorat en sciences de l'environnement de l'université du québec à montréal P.173.

Audigié D, Dupont G, Zonszain T (1978). Manipulation d'analyse biochimique. Ed. Doin. Paris. Pp.27-74.

Caquet T (1994). Quelles substances ? Quels effets ? Comment agissent-elles ? Où les chercher et comment les mesurer ? Conférences Professionnelles de l'Agence de l'Eau Artois-Picardie (12/04/94) : Maîtriser les substances toxiques dans le milieu naturel et les Rejets ; un nouveau défi en Artois-Picardie. Pp.5-51.

Fatondji RH (2011). Synthesis, Characterization and antimicrobial activities of antiparasitic semicarbazones, thiosemicarbazones and their derivatives Thiadiazolines. PhD thesis: University of Abomey. Pp.45-135.

Giles CH, Smith D, Huitson A (1974). A general treatment and classification of the solute adsorption isotherm I Theoretical. Colloid Interface Sci. (47):755-765.

Hafner E, Heiner E, Noack E (1977). Arzn eim-Forsch. 27:1871.

Hicham El B, Abdalhamid O (2006). Développement de nouvelles techniques de détermination des pesticites et contribution à la réduction de leur impact sur les eaux par utilisation des substances organiques naturelles. Thèse de doctorale (université Abdelmalek Essaâdi) P.193.

Metson AJ (1956). Methods of chemical analysis for soil survey samples. N Z Soil Bur Bull n° P.12.

Michael AS, Thompson CG, Abramovitz M (1956). Artemia Salina as a test organism for a bioassay. Science 123:464.

Mireille M (1995). Test de toxicité sur larves d'*Artemia salina* entretien d'un élevage de balanes . Rapport de stage de 2ème année. Deust aquaculture. Centre universitaire de nouvelle Calédonie., France.

Morel R (1996). Les sols cultivés. Lavoisier, Paris.

Sleet RB, Brendel K (1983). Improved methods for harvesting and Counting synchronous populations of *Artemia nauplii* for use in developmental toxicology. Ecotoxicol. Env. Sefty. 7:435-446.

Solis PN, Wright CW, Anderson MM, Gupta MP, Phillipson JD (2001). A microwell cytotoxicity assay using *Artemia salina*. Plant Med,1993, 59: 250-252. Thèse de doctorat. Université Abdelmalek Essaâdi, Tetouan,. P.180.

Vanhaecke P, Persoone G, Claus C, Sorgeloos P (1981). Proposal for a Short-term toxicity test *with Artemia nauplii*. Ecotoxicol. Env. Safety. 5:382-387.

Substituent effects of 2-aryl-*trans*-decahydroquinolin-4-ones in CTADC oxidation: Spectrophotometric approach

Sharmila N.[1], Rao R. S. C. [2], Hari Babu B.[1], Trimurtulu G.[2] and Satyanarayana P. V. V.[1]

[1]Department of Chemistry, Acharya Nagarjuna University, Nagarjunanagar – 522 510, A.P., India.
[2]Laila Impex R & D Centre, Jawahar Auto Nagar, Vijayawada, AP-India.

A kinetic study was carried out using spectrophotometric approach on the oxidation of nine differently substituted 2-aryl-*trans*-decahydroquinolin-4-ones by Cetyltrimethylammonium dichromate (CTADC) as oxidant at 30°C and in aqueous acetic acid (50%) containing catalytic amount of sulphuric acid (6N). The reactions were studied at 350 nm. The reaction was found to follow second order kinetics. The substitution of methyl group at position three of the decahydroquinoline ring and the presence of electron releasing groups in the aryl group increased the rate of oxidation of the substrates. The change in the rate of oxidation with temperature and solvent composition was also studied. The products formed during the oxidation were also confirmed by using high-performance liquid chromatography (HPLC) and spectral analysis.

Key words: 2-aryl-*trans*-decahydroquinolin-4-ones, kinetics, oxidation, Cetyltrimethylammonium dichromate (CTADC).

INTRODUCTION

The kinetic studies on the oxidation of different aliphatic (Singh et al., 1982; Mahadevappa and Swamy 1988; Singh et al., 1978a; Chiba et al., 1995), alicyclic (Mushran et al., 1976; singh et al., 1978b; Karnojitzky 1981), aromatic (Ogata and Sawaki 1972; Manivannan and Maruthamuthu 1986; Khandual et al., 1973; Annapoorna et al., 1998; Devries et al., 1995) ketones and a series of 2,6-diaryl-4-piperidones (Kumabe et al., 2001; Selvaraj et al., 1979) have been studied extensively by various workers and suitable mechanisms have been proposed. Recently the kinetics of oxidation of present ketones were reported in aqueous acetic acid using thallium(III), cerium(IV) and lead(IV) (Satyanarayana et al., 2010, 2013a, b) as oxidizing agents. In the present study CTADC was selected as oxidizing agent, as it is found to be selective, mild phase transferring and chemoselective oxidant (Patel and Mishra, 2007a, 2004b; Sahu et al., 2005; Patel and Mishra, 2006c, d). It was found to

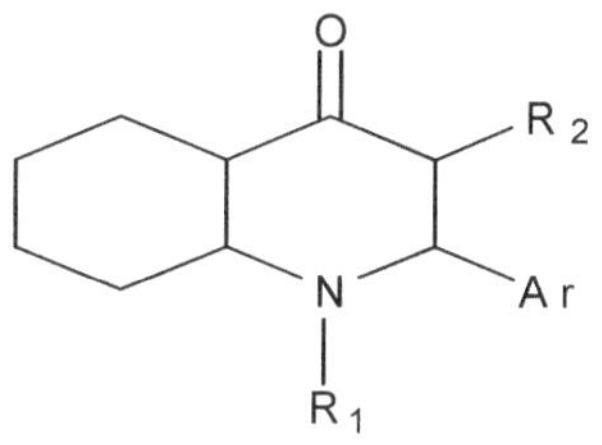

S/No.	Ar	R_1	R_2
1.	C_6H_5	H	H
2.	*p*-Me-C_6H_4	H	H
3.	*p*-OMe-C_6H_4	H	H
4.	*o*-Cl-C_6H_4	H	H
5.	C_6H_5	H	Me
6.	*p*-Me-C_6H_4	H	Me
7.	*p*-OMe-C_6H_4	H	Me
8.	*p*-Cl-C_6H_4	H	Me
9.	C_6H_5	Me	Me

Figure 1. 2-aryl-*trans*-decahydroquinolin-4-ones.

oxidize aromatic amines and thiols to the corresponding coupled dehydrogenated products (Patel and Mishra, 2004b), aldoximes to the nitriles (Sahu et al., 2005), cholesterol to 7-dehydrocholesterol (Patel and Mishra, 2006c), arylthiourea to corresponding urea (Sahoo et al., 2010), alcohols to ketones (Vimala et al., 2009), oxidation of diols (Patel et al., 2008f) selectively. However, the study of the kinetics of oxidation of 2-aryl-*trans*-decahydroquinolin-4-ones with CTADC was not reported so far. The present investigation was taken up with a view to identify whether the complex reagent CTADC useful for oxidation of the said ketones and also to identify the possible products during oxidation process. Hence, it is considered to investigate the kinetics of oxidation of 2-aryl-*trans*-decahydroquinolin-4-ones and also to study the effect of substituent(s) in aryl group and in the heterocyclic ring. This prompted the authors to carry out the investigation on a total nine compounds (1-9) (Figure 1).

EXPERIMENTAL

Reagents and apparatus

Preparation of ketones

The ketones were prepared by the method developed by Baliah and Natarajan (1981), and purified by recrystallization from suitable solvents to constant melting points. All the samples were dried in vacuum before use. Cetyltrimethylammonium dichromate (CTADC) was prepared by the known method (Patel et al., 2005f) and its purity was checked by estimating Cr(VI) iodometrically (Vogel, 1961).

Acetic acid

Acetic acid glacial (Excelar), supplied by 'Qualigens Fine Chemicals', was refluxed with chromium trioxide for 6 h and fractionally distilled. The fraction boiling at 390-391K was collected and was used.

Other reagents

Cetyltrimethylammonium bromides, Potassium dichromate, Potassium iodide, Sulphuric acid used were all A.R grade. Doubly distilled water was used for all purposes.

Instrument

The instrument used was analytical technologies UV/visible spectrophotometer of Model Spectro2080.

Kinetic procedure

The oxidation kinetics of 2-aryl-*trans*-decahydroquinolin-4-ones by CTADC in the presence of aqueous acetic acid using a UV-Vis spectrophotometric method and all the kinetic measurements were carried out at 350 nm (Figure 2). The measurements were performed in 50:50(v/v) acetic acid: water containing 6N H_2SO_4 at 30°C. The temperature was controlled by using thermostat of accuracy ±0.1°C. The required amount of CTADC solution was prepared by dissolving the necessary amount of CTADC in the solvent medium. The solutions of the substrates (ketones) were

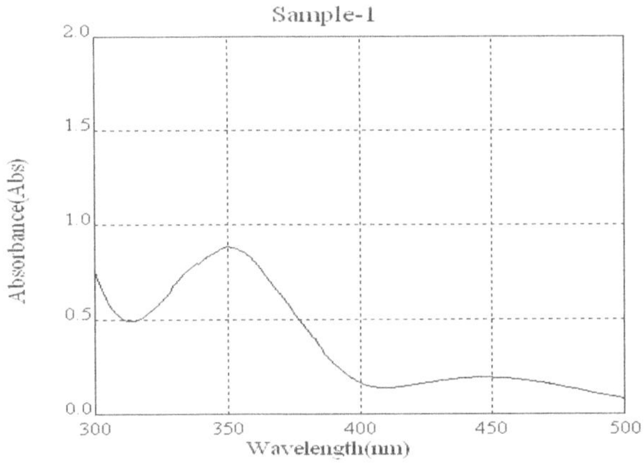

Figure 2. UV-absorption maxima of the CTADC reagent.

Table 1. Second order rate constants for the oxidation of substituted 2-aryl-*trans*-decahydroquinolin-4-ones by CTADC.

S/No.	Compound	$k_2 \times 10^2$ Mole^{-1} Sec^{-1}
1	2-ph-4-one	07.53 + 0.12
2	2-(*p*-tolyl)-4-one	10.48 + 0.15
3	2-(*p*-OMe)-4-one	14.12 + 0.11
4	2-*o*-Cl-Ph-4-one	20.81 + 0.25
5	3-Me-2-ph-4-one	29.08 + 0.32
6	3-Me-2-*p*-tolyl-4-one	39.23 + 0.20
7	3-Me-2-(*p*-OMe)-4-one	16.27 + 0.02
8	3-Me-2-(*p*-Cl)-4-one	20.03 + 0.06
9	1,3-di-Me-2-ph-4-one	22.31 + 0.21

Temp = 30°C, H^+] = 6.0 N, Solvent = 50% AcOH.

prepared by dissolving the appropriate quantity of the compounds in the same solvent, so that the concentration of the ketones was maintained always higher than the concentration of CTADC. The reaction was initiated by mixing CTADC to ketones and the progress of the reaction was followed spectrophotometrically by monitoring the decrease in absorbance at 350 nm. The second order conditions were followed for determining the rates of the reactions.

Product analysis

A mixture of ketone and CTADC was allowed to react in aqueous acetic acid (50%, v/v) in presence of sulphuric acid (6.0 N). The concentration of CTADC was maintained slightly excess than the concentration of ketone. The resulting mixture was kept aside at 30°C temperature for 2 to 3 days. The mixture turned reddish green, indicating the formation of reduced Cr (III). After that the reaction mixture was neutralized with saturated solution of sodium carbonate, extracted with ether and the combined ether extract separated and the crude product was obtained after distillation of the ether layer.

Stoichiometry

The stoichiometry of the reaction was determined by allowing a known excess of the oxidant CTADC to react with the substrate in solvent medium at 30°C and the un-reacted CTADC was estimated. The stoichiometry was found to be in the mole ratio of 1:1 for oxidant to substrate.

$$R_2C{=}O + \text{CTADC} \longrightarrow \text{Product} + \text{Cr(III)}$$

Calculation of the rate constants

The rate constants were calculated using the second order rate equation:

$$k_2 = \frac{2.303}{t\,(a-b)} \log \frac{b(a-x)}{a(b-x)}$$

Where, a=Initial concentration of substrate in moles/lit; b=Initial concentration of CTADC in moles/lit.

x= Amount of CTADC reacted in time t (seconds); t= Reaction time in seconds.

RESULTS AND DISCUSSION

A kinetic study was made on the oxidation of 2-aryl-*trans*-decahydroquinolin-4-ones (1 to 9) by using CTADC as oxidizing agent in aqueous acetic acid (50%) medium. The reaction was found to follow over all second order kinetics. The rate constants obtained for the selected substrates were given in Table 1.

An examination of the rate constants for the oxidation of 2-aryl-*trans*-decahydroquinolin-4-ones 1 to 3 (where there is no methyl group at 3rd position) and those of corresponding 3-Methyl-2-aryl-*trans*-decahydroquinolin-4-ones 5, 6 and 7 (where there is methyl group at 3rd position) revealed that the latter series of compounds 5, 6, and 7 were oxidized at a faster rate than former series of compounds 1 to 3. This observation is as expected as the electron-releasing inductive effect of methyl group in heterocyclic ring enhances the rate of oxidation of these series of compounds. However, the observation made in the present investigation was found to be contrary to the observations made by us in earlier studies for same compounds with metal ions Tl(III), Pb(IV) and Ce(IV) (Satyanarayana et al., 2010, 2013a, 2013b). The results of the study were given in Table 2.

The introduction of electron releasing and electron withdrawing substituent's on the phenyl group at *ortho*- and *para*- positions had provided useful information to know the structure-activity relationship of these compounds. Substitution at *ortho* and *para* positions of aryl ring is expected to produce negligible effect on the rate of oxidation, because aryl ring is far away from the reaction centre. But the substituents at *ortho* and *para* positions played a vital role on rate of oxidation. This may be mainly due to the polar effect exerted by the substituents. The electron releasing groups increased the rate of oxidation. For example for compounds 1 to 3, the rate of oxidation

Table 2. Second order rate constants for the oxidation of 2-aryl-*trans*-decahydroquinolin-4-ones by Pb(IV), Tl(III) and Ce(IV).

S/No.	Compound	$k_2 \times 10^2$ Mole^{-1} Sec^{-1}	$k_2 \times 10^3$ Mole^{-1} Sec^{-1}	
		LTA 98% AcOH	Tl(III) 80% AcOH	Ce (IV) 50% AcOH
1	2-ph-4-one	11.42 ± 0.05	8.44 ± 0.09	1.19 ± 0.04
2	2-(*p*-tolyl)-4-one	13.48 ±0.15	8.47 ± 0.13	1.07 ± 0.02
3	2-(*p*-OMe)-4-one	15.12 ± 0.11	11.22 ± 0.01	1.23 ± 0.01
4	2-*o*-Cl-Ph-4-one	17.40± 0.16	24.98± 0.04	--
5	3-Me-2-ph-4-one	03.78 ±0.15	1.39 ± 0.04	0.25 ± 0.12
6	3-Me-2-*p*-tolyl-4-one	03.66 ± 0.07	1.44 ± 0.01	0.03 ± 0.02
7	3-Me-2-(*p*-OMe)-4-one	00.79 ± 0.02	0.93± 0.04	--
8	1,3-di-Me-2-ph-4-one	03.55 ± 0.21	30.40± 0.03	--

Table 3. Effect of solvent composition on the oxidation of 2-Phenyl-4-one by CTADC.

Name of the compound	Percentage of acetic acid ($10^2 \times k_2$ dm^3 mol^{-1}sec^{-1})		
	40%	50%	60%
2-Phenyl-4-one	3.88±0.23	07.53 ± 0.12	16.27 ±0.24

[H_2SO_4] = 6N; Temp = 30°C.

Table 4. Effect of acid concentration on the oxidation of 2-Phenyl-4-one by CTADC.

Name of the compound	Variation in [H^+] ($10^2 \times k_2$ dm^3 mol^{-1}sec^{-1})		
	5N	6N	7N
2-Phenyl-4-one	3.87±0.15	07.53 ± 0.12	14.86±0.19

Solvent: AcOH/H_2O : 50/50(v/v), Temp = 30°C.

was found to be increased with increase in electron releasing nature of the substituent at *p*-position of phenyl ring. But, the trend was not followed for compound 7 that is, for *p*-methoxy substituted compounds in case of 3-methyl substituted compounds 5 to 7. This may be because highly bulky methoxy substituent on phenyl ring in addition to 3-methyl substituent on decalin ring.

The introduction of a methyl group in the 3rd position in decalin ring increased the rate of oxidation of ketones, because of the combined effects of inductive as well as steric and was discussed above. But, further introduction of methyl group on 1st position that is, on N atom of heterocyclic ring has showed some influence on the rate of oxidation. The rate of oxidation of 1, 3-dimethyl-2-phenyl-*trans*-decahydroquinolin-4-one (9) (presence of methyl group at both 1 and 3 positions) was found to be high when compared with the rate of oxidation of compound (1), having no methyl group in position 3 but was showed less rate of oxidation compared to compound 5 (having methyl group at position 3). Further the rate of oxidation of 2-*o*-chlorophenyl-*trans*-decahydroquinolin-4-one was expected to be low, but was found to be high when compared with the rates of oxidation of compounds 1 to 3 and is almost similar to the rate constant of compound 8. This type of higher rate constant of *o*-chloro compound was already reported in oxidation of the same compounds with thallium (III) and lead (IV) metal ions (Table 2) but has not gained proper reason.

The oxidation kinetics of 2-aryl-*trans*-decahydroquinolin-4-ones has also been carried out in different compositions of acetic acid and the relevant data was given in Table 3. The concentration of sulphuric acid was maintained constant at 6.0 N in all these studies and the temperature was maintained at 30°C. Increase in the percentage of acetic acid of the medium increases the rate of oxidation. Further increase in concentration of sulphuric acid also increased the rate of oxidation of the said compounds and the data was given in Table 4. This observation was quite consistent with Cr(VI) reagents where the oxidation potential of this type of reagents increases with increase in concentration of acid in the reaction medium. The oxidation studies were also carried out at different temperatures (Table 5).

Table 5. Effect of temperature on the oxidation of 2-Phenyl-4-one by CTADC.

Name of the Compound	Variation in temperature ($10^2 \times k_2$ dm^3 $mol^{-1}sec^{-1}$)		
	30°C	40°C	50°C
2-Phenyl-4-one	07.53 ± 0.12	20.70 ± 0.35	51.35±0.04

$[H^+]$ = 4.0 N, Solvent = 50 % AcOH.

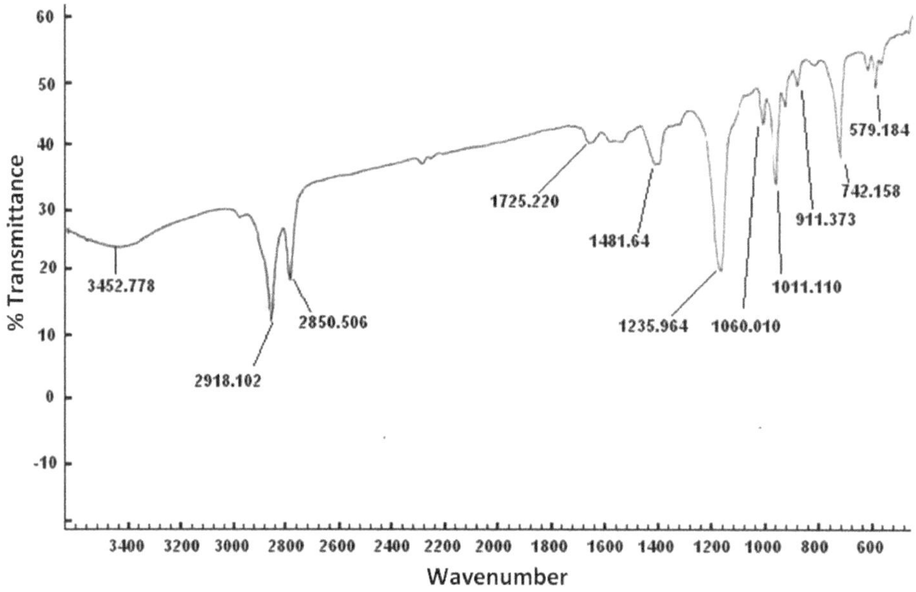

Figure 3. IR spectrum of the crude product of compound **3.**

In product analysis, the completion of the reaction was confirmed by comparison of Thin layer chromatography (TLC) of starting material with reaction mixture. A preliminary examination was carried out in order to identify the product. A spectral study was carried out to establish the compound present in the crude product in the workup of the reaction mixture. An examination of the IR spectrum (Figure 3) of the product mixture revealed that they are quite different from the spectrum of the original compounds taken.

A broad peak (3452 cm^{-1}) in the region of 3300 to 3460 cm^{-1} indicated the presence of –OH of a carboxylic group or an -N-H group in the mixture. The sharp peaks at 2918 and 2850 cm^{-1} indicated the presence of aromatic system. Further, on spraying ninhydrin reagent on TLC of the product mixture gave spots of purple color. These spots might be due to presence of nitrogen atom in the crude product.

The earlier reports in respect of oxidation of the present type of systems indicated that the product is an amino acid (Meenal and Roopakalyani 1988a, Dhar and Vardarajan 1991; Satyanarayana et al., 2010, 2013b). However, the earlier authors have not conclusively confirmed the formation of amino acid. This prompted the present authors to isolate the product from the oxidation of compound (3) through reverse phase HPLC and was successfully isolated the products. The products were analyzed through ^{1}H NMR (Figure 4) and mass spectroscopy (Figure 5), which gave surprising results.

The ^{1}H NMR spectrum of this compound revealed that one of the products was *p*-methoxy benzamide and this was further confirmed by the mass spectrum of the product obtained (m/z=152.2). These observations prompted us to suggest that probably the carbon-nitrogen bond of alicyclic ring and piperidone ring might have been broken to form an amide, when the alicyclic ring was found to be completely broken from the heterocyclic ring. The other product is likely to be cyclohexane derivative which could not be identified and it needs further investigation. A reasonable scheme for the formation of the product is given in Scheme 1.

Conclusions

A kinetic study was carried out for the substituted 2-aryl-

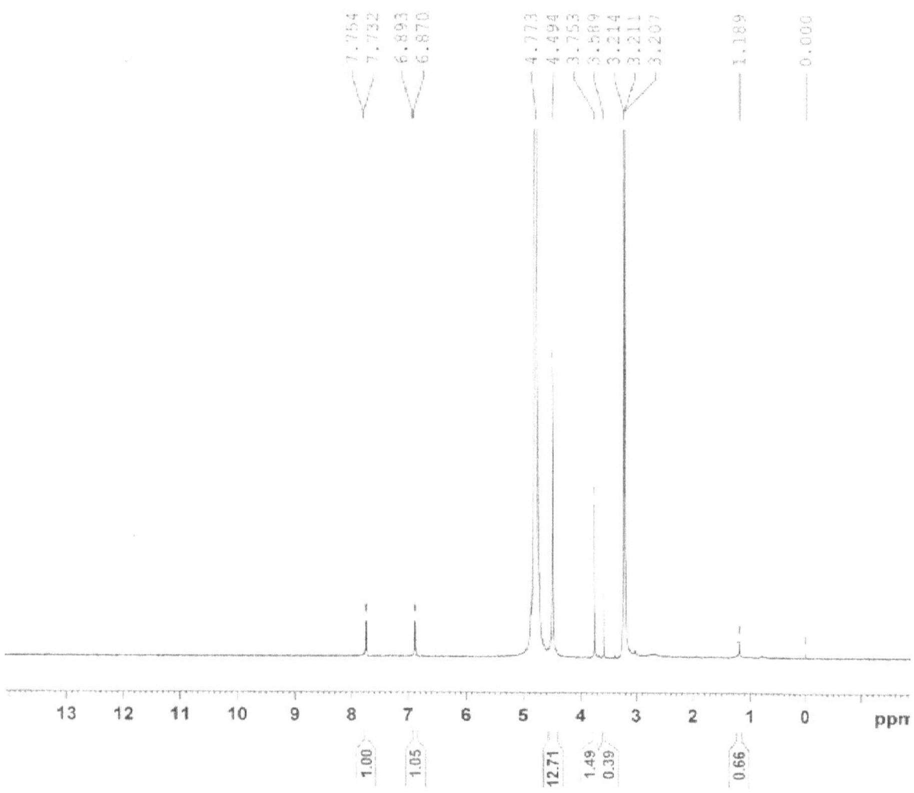

Figure 4. ^{1}H NMR spectrum of the product of compound **3.**

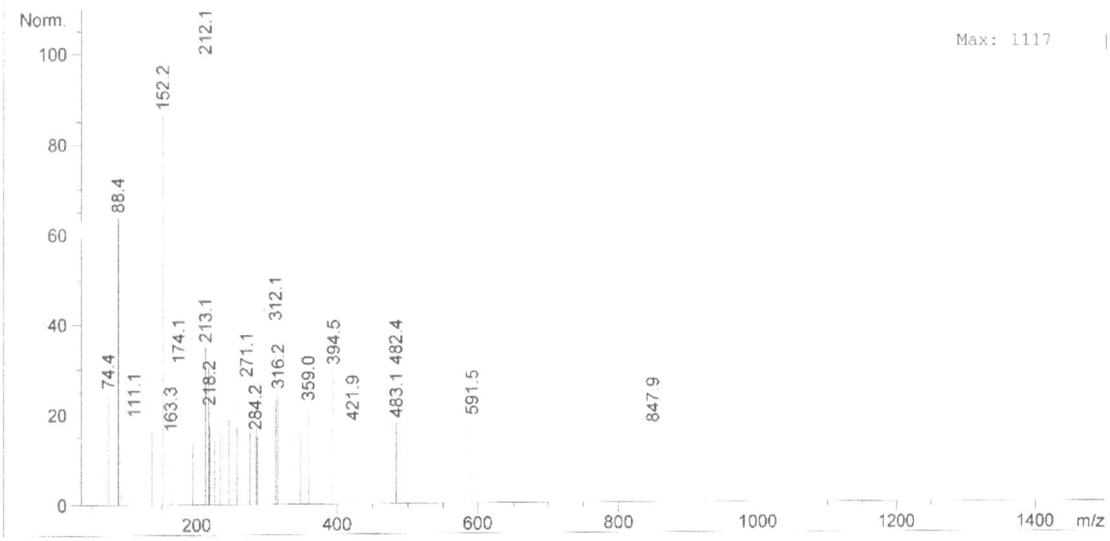

Figure 5. Mass spectrum of the product of compound 3.

trans-decahydroquinolin-4-ones system using spectrophotometric approach. Substituent effects on the rate of oxidation of 2-aryl-*trans*-decahydroquinolin-4-ones were also studied. The presence of electron releasing groups on aryl group at C-2 position and methyl group at C-3 position increased the rate of oxidation of the

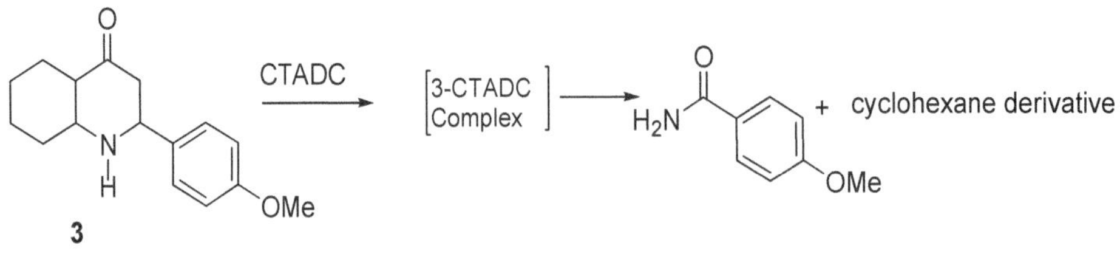

Scheme 1. Oxidation of 2-*p*-methoxyphenyl-*trans*-decahydroquinolin-4-one with CTADC.

compounds with CTADC. The products obtained during the oxidation were established.

Conflict of Interest

The author(s) have not declared any conflict of interests.

ACKNOWLEDGEMENTS

The authors greatly acknowledge the Acharya Nagarjuna University for constant encouragement and also thankful to UGC, New Delhi, India for financial support through UGC-BSR fellowship (F. No.11-67/2008 dt. 06.03.2012). Further, the authors are also thankful to Laila Impex R&D centre, Vijayawada for providing instrumental support in respect of product analysis.

REFERENCES

Annapoorna SR, Prasad Rao M, Sethuram M (1998). Effect of multiple substituents on oxidation of phenyl styryl ketone and its substituted analogs by V(V) in acid medium: Akinetic study. Indian J. Chem. 37B:874.

Baliah V, Natarajan A (1981). Synthesis of 2-aryl and 2-aryl-3-methyl-decahydroquinolin-4-ones. Indian J. Chem. 20B:830.

Chiba T, Sakagami H, Murata M, Okimoto M (1995). Electrolytic oxidation of ketones in ammonical methnol in the presence of catalytic amounts of KI. J. Org. Chem. 60:6764-6770.

Devries EFJ, Ploeg L, Colao M, Brussee J, Vander Gen A (1995). Enantio selective oxidation of aromatic ketones by molecular oxygen, catalysed by chiral monoaza-crown ethers. Tetrahedron Asymmetry. 6(5):1123-1132.

Dhar RK, Vardarajan R (1991). Kinetics and mechanism of oxidation some substituted 4-piperidones by pyridinium fluorochromate. Inidan J. Chem. 30A:936.

Karnojitzky V (1981). Oxidation of ketones by molecular oxygen. Russ. Chem. Rev. 50:883.

Khandual NC, Satpathy KK, Nayak PL (1973). Kinetics of thallium (III) oxidation of aromatic ketones. Proc. Math. Sci. 77(4):163.

Kumabe R, Nishino H, Yasutake M, Nguyen VH, Sawa KK (2001). A study of conformational effects on oxidation kinetics of N-H and N-methyl-4-piperidones with N-bromosuccinimide: A Comparative analysis. Tetrahedron Lett. 42:691-72.

Mahadevappa DS, Swamy P (1988). Kinetics of oxidation of aliphatic ketones with BromamineT in acid medium. Bull. Chem. Soc. Japan. 61(2):543.

Manivannan G, Maruthamuthu P (1986). Kinetics and mechanism of oxidation of aliphatic and aromatic ketones by peroxomonosulphate. J. Chem. Soc. Perkin Trans. 2:565.

Meenal KR, Roopakalyani P (1988a). Kinetic behavior of some substituted 4-piperidones towards oxidation by aqueous acidic CrO_3. J. Indian Chem. Soc. 65:624.

Mushran SP, Bose AK, Tiwari JN (1976). Kinetic study of oxidation of cycloheptanone by N-bromosuccinimide. Monatsch Chem. 107:1021.

Ogata Y, Sawaki Y (1972). Kinetics of the Baeyer-Villiger reaction of aromatic ketones with perbenzoic acid. J. Org. Chem. 37(19):2953.

Patel S, Minati K, Biswa B, Mishra BK (2005). Cetyltrimethylammonium dichromate: A phase transferring oxidant. Synth. Commun. 35:1033-1037.

Patel S, Mishra BK (2004b). Cetyltrimethylammoniumdichromate: A mild oxidant for Coupling amines and thiols. Tetrahedron Lett. 45:1371.

Patel S, Mishra BK (2006c). Oxidation of cholesterol by a biomimetic oxidant CTADC. J. Org. Chem. 71:3522.

Patel S, Mishra BK (2006d). Oxidation of alcohol by lipopathic Cr(VI): A mechanistic Study. J. Org. Chem. 71:6759.

Patel S, Mishra BK (2007a). Chromium (VI) oxidants having quaternary ammonium ions: Studies on synthetic applications and oxidation kinetics. Tetrahedron 63:4367.

Patel S, Sung DD, Mishra BK (2008e). Oxidation of diols by cetyltrimethylammonium dichromate. Indian. J. Chem. 47A:1218.

Sahoo P, Mayeesahoo S, Patel S, Mishra BK (2010). Oxidation of arylthiourea by CTADC. Indian. J. Chem. 49A:1483-1487.

Sahu S, Patel S, Mishra BK (2005). Oxidation of oximes by CTADC: A facile conversion of oximes to nitriles. Synth. Commun. 35:3123.

Satyanarayana PVV, Sharmila N, Hari Babu B, (2013a). A kinetic study on oxidation of 2-aryl-trans-decahydroquinolin-4-ones by cerium(IV). Oxidation Commun. 36(1):71-77.

Satyanarayana PVV, Udaya LK, Sharmila N, Hari Babu B, Raman GV, Nageswara RB (2013b). Chemical Technology: An Indian J. 8(2). (In press).

Satyanarayana PVV, Vijayasaradhi CH, Rao RSC, Hari Babu B (2010). Kinetics of oxidation of 2-aryl-*trans*-decahydroquinolin-4-ones by thallium(III). Oxid. Commun. 33(2):371.

Selvaraj K, Senthilnathan VP, Ramalingam K (1979). Kinetics of oxidation of some substituted 4-piperidones y vanadium (V). Indian J. Chem. 17A:589.

Singh B, Pandey L, Sharma J, Panday SM (1982). Mechanism of oxidation of some aliphatic ketones by N-Bromosuccinimide in Acidic media. Tetrahedron 38:167.

Singh K, Tiwari JN, Mushran SP (1978a). Kinetics and mechanism of oxidation of diethyl ketone by N-bromosuccinimide. Int. J. Chem Kinetic 10:995.

Singh K, Pandey L, Mushran SP (1978b). Kinetics of N-bromosuccinimide oxidation of cyclooctanone in acid media. Indian J. Chem. 16A:875.

Vimala BC, Nagendrappa G (2009). Solvent free reactions using CTAP and CTADC- cis-1,2-dihydroxylation of alkenes, oxidation of alcohols and regeneration of aldehydes and ketones from oximes. J. Saudi Chem. Soc. 13:169.

Vogel AIA (1961). Longmans Textbook of Quantitative Inorganic Analysis, 3rd ed. London.

Effects of solvent on the UV-visible absorption spectra of acenaphtho(1,2-b) quinoxaline and acenaphtho(1,2-b)benzo(g)quinoxaline

S. A. Ahmed[1]*, N. O. Obi-Egbedi[2], N. W. Odozi[2], I. Iweibo[2] and M. D. Adeoye[3]

[1]Department of Chemistry, University of Agriculture, Abeokuta, Nigeria.
[2]Department of Chemistry, University of Ibadan, Ibadan, Nigeria.
[3]Department of Chemical Sciences, Fountain University, Osogbo, Nigeria.

The effects of solvent on the electronic absorption spectra has been used to investigate the structural and molecular properties of acenaphtho(1,2-b)quinoxaline and acenaphtho(1,2-b) benzo(g)quinoxaline. The structures of the compounds have been established by UV-visible spectrophotometer. The spectra behaviours in terms of dipole moments in the ground and excited states have been interpreted and related to the transition energies, intensities and the molecular structures of the compounds.

Key words: Solvent effects, dipole moments, solvent polarizability, transition energies.

INTRODUCTION

Various aspects on the spectra of some quinoxaline compounds have been studied previously by several authors, such as, electronic absorption bands of quinoxalines in cyclohexane by Mason (1962), of 2,3-diphenylquinoxalines by Bohlmann (1951), spectra properties of some derivatives of 6H-indolo{2,3-b}quinoxaline by Sarki and Al-Badri (1980), transient absorption fluorescence spectra in n-hexane by Greene et al. (1978), the fluorescence quantum yield of quinoxaline in iso-octane by Yamatoto et al. (1978), fluorescence decay of apparent $S_n \leftarrow S_i$ absorption in n-hexane by Halliday and Topp (1977), excited state absorption spectra of quinoxaline in benzene, iso-octane and cis-1,3-pentadiene by Anderson et al. (1979), the electronic transition dipole moments of charge transfer absorption and fluorescence of some quinoxaline derivatives by Czerwienice et al. (2000). However, no attempts have been made neither to report the absorption spectra nor assign the observed band for these quinoxaline derivatives.

In this work, besides the assignment of the spectra bands, the spectra behaviours in terms of dipole moments in the ground and excited states vis-à-vis the transition energies and intensities are interpreted and related to the molecular structures of these two compounds.

*Corresponding author. E-mail: ahmedakinyeye@yahoo.co.uk.

MATERIALS AND METHODS

Spectroscopic grade of acenaphtho(1,2-b)quinoxaline and acenaphtho(1,2-b)benzo(g) quinoxaline were obtained from Tokyo Kasei (Japan). Methanol, ether and chloroform are products of British Drug House Ltd and were redistilled several times.

Electronic spectra of dilute solution of each compound in the concentration range of 10^{-5} to 10^{-6} M and 200 to 500 nm respectively, were determined with a computerized, double beam Schimadzu UV-1650 coupled with UV probe 2.3 version software.

Molar absorptivities and integrated absorption coefficient were determined by deconvolution of the spectra. A plot of molar extinction coefficient against wave number was used to determine the transition energy. The oscillator strength (f) in solution from states i to j was calculated using equation described in Iweibo et al. (1990) and given as Equation 1:

$$f = \frac{2.303 m_e c_4^2 \varepsilon_0}{N_o e^2 n} \int \varepsilon_{(\bar{v})} dv \tag{1}$$

where e is the electronic charge, m is mass of the electron, c is speed of light, ε vacuum permittivity, n the refractive index of solvent and $\varepsilon_{(\bar{v})}$ the molar absorptivity of the absorbing solute at the wave number ṽ. The integral $\int \varepsilon_{(\bar{v})} d\bar{v}$ was calculated using

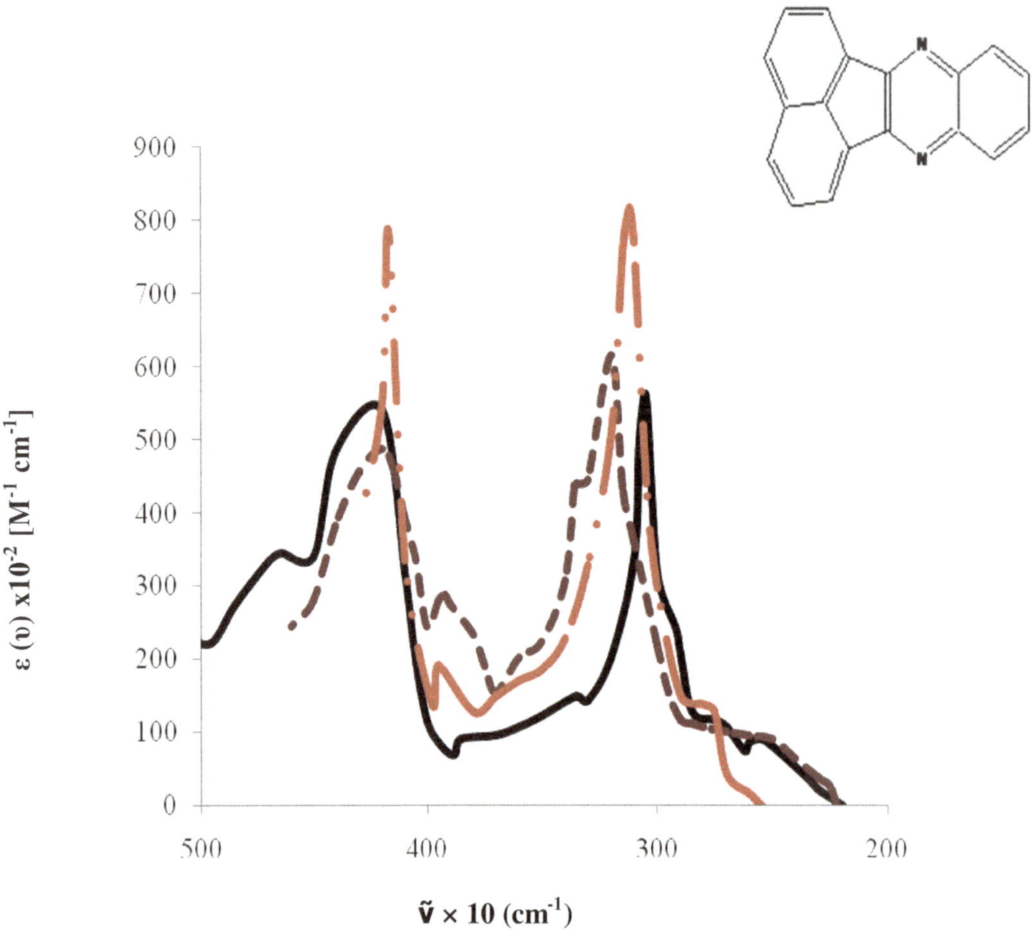

Figure 1. Electronic absorption spectra of acenaphtho(1,2-b)quinoxaline in methanol (———), ether (– – – –) and chloroform (—··— ··—).

assumptions of symmetrical bands or conformity with Gaussian distributions as described by Iweibo et al. (1982).

RESULTS AND DISCUSSION

Presented in Figures 1 and 2 are the electronic absorption spectra of acenaphtho(1,2-b) quinoxaline and acenaphtho(1,2-b)benzo(g)quinoxaline respectively in the representative solvents (methanol, ether and chloroform). Also, detailed assignment of the spectra properties in few solvents have been displayed for each compound in Table 1 to avoid overcrowding of symbols and masking of salient spectra features.

Acenaphtho(1,2-b) quinoxaline

The absorption spectra of acenaphtho(1,2-b)quinoxaline showed four distinct bands which were designated as $S_o \rightarrow S_1, S_o \rightarrow S_2, S_o \rightarrow S_3$ and $S_o \rightarrow S_4$ in order of increasing energy.

The $S_o \rightarrow S_1$ transition is a broad band with low intensity in the three representative solvents. The bands show shoulders towards red at 26,074 cm^{-1} in methanol, at 27,454 cm^{-1} in chloroform. Because of their low intensities, the bands are assigned as ^1L_b band (using Platt's notations) which implies that they are both quantum mechanically and overlap forbidden (that is, totally forbidden).

The $S_o \rightarrow S_2$ transition was of concentrated intensities in the three solvents, they are characterized with shoulders towards blue at 31,738 cm^{-1} in methanol, 32,972 cm^{-1} in ether, and 32,509 cm^{-1} in chloroform. $S_o \rightarrow S_2$ peak in methanol is exceptionally sharp and submerges the $S_o \rightarrow S_3$ peaks. Two likely causes of this observation from theoretical explanation are:

i. The possibility of fusion of two bands, the solvent has little or no effect on their separation or decomposition,
ii. The formation of charge transfer due to

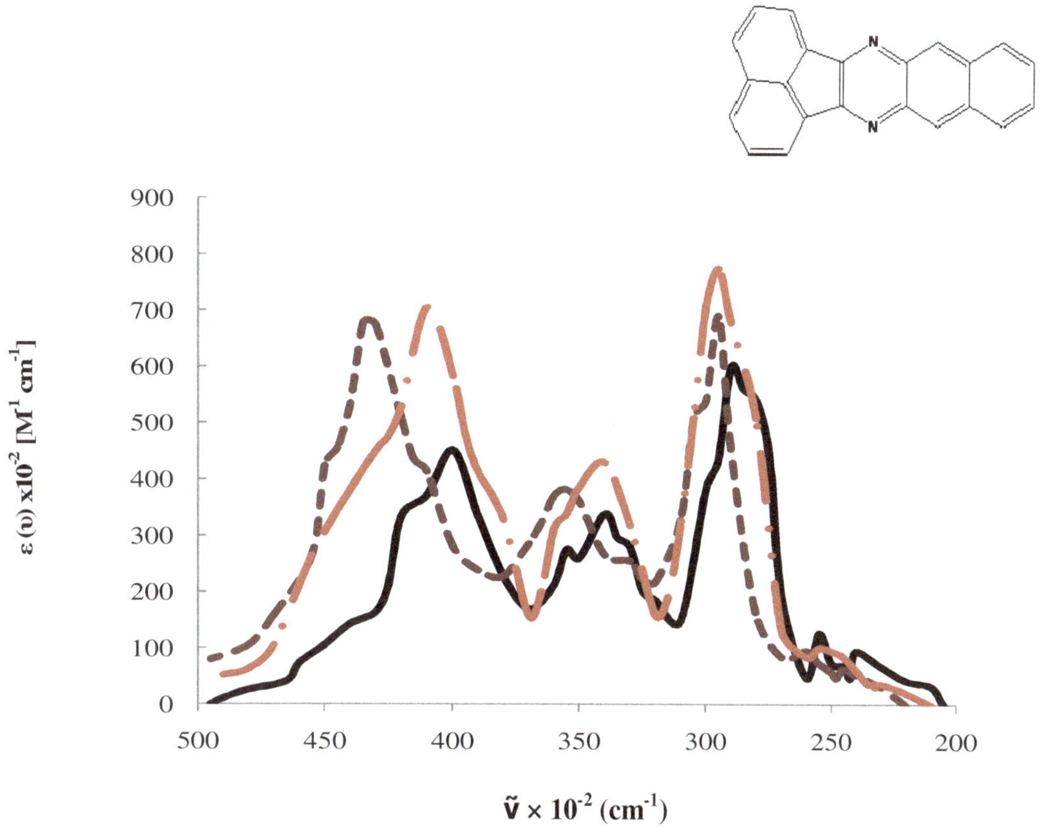

Figure 2. Electronic absorption spectra of acenaphtho(1,2-b)benzo(g)quinoxaline in methanol (———), ether (– – – –) and chloroform (—··—··—).

donor–acceptor interaction as suggested by Czerwierniee et al. (2000).

The $S_o \rightarrow S_3$ transition had broad bands in all the solvents, it is overlapping with the $S_o \rightarrow S_2$ bands and can be assigned as ^1B_b which implies that they are both quantum mechanically and overlap forbidden. The $S_o \rightarrow S_4$ transitions in the three solvents had intense bands and are assigned as ^1B_a bands which mean that they are both quantum mechanically and overlap allowed.

Generally, it was observed that the vibrational fine structure in all the transitions were completely lost in the solvents (much more pronounced in methanol). This suggests a specific solute interaction.

Acenaphtho(1,2-b)benzo(g)quinoxaline

Acenaphtho(1,2-b) benzo(g) quinoxaline shows five absorption band peaks in methanol, ether and four in chloroform. The fifth band peak suggests that the additional ring in acenaphtho(1,2-b)benzo(g)quinoxaline extends the radius of the π-electron of the ring. This is in line with free electron molecular orbital {FEMO} theory, leading to relative red shift of the transition energies. The obscurity of the fifth band in chloroform agrees with the fact that chlorinated hydrocarbon suffer the disadvantage of non-transparency in the near UV region (where the transition occurred). $S_o \rightarrow S_1$ transition was a broad band in the representative solvents. The bands were of low intensities but considerable distortion due to instrumental response was recorded for the peak in chloroform. These bands are labeled as ^1L_a band indicating that the transition is totally forbidden. $S_o \rightarrow S_2$ transition in ether and methanol is broad, red-shifted and of low intensity with little distortion owning to instrumental response. However, the band in chloroform is intense and the transition is both quantum mechanically allowed and overlap allowed.

The $S_o \rightarrow S_3$ transition was intense in all the solvents and the transition is allowed. However, it was observed that the $S_o \rightarrow S_3$ transition in chloroform derived its intensity from the more intense $S_o \rightarrow S_2$ transition. This phenomenon is known as intensity borrowing. The

Table 1. Summary of observed absorption bands of compounds in different solvents.

Compounds and transitions	Methanol			Ether			Chloroform		
	$\bar{\nu}_{max}\ (cm^{-1})$	$\varepsilon_{max}\ (M^{-1}\ cm^{-1})$	f	$\bar{\nu}_{max}\ (cm^{-1})$	$\varepsilon_{max}\ (M^{-1}\ cm^{-1})$	f	$\bar{\nu}_{max}\ (cm^{-1})$	$\varepsilon_{max}\ (M^{-1}\ cm^{-1})$	f
Acenaphtho(1,2-b)quinoxaline									
$S_0 \rightarrow S_1$	27693	11483	0.28	28992	17287	0.18	28867	17841	0.22
$S_0 \rightarrow S_2$	30157	56926	0.91	31481	59230	0.84	31196	85463	0.87
$S_0 \rightarrow S_3$	40545	8978	0.37	39415	26258	0.21	39082	14097	0.13
$S_0 \rightarrow S_4$	42582	55216	0.89	42043	54923	0.76	41675	77093	0.59
Acenaphtho(1,2-b)benzo(g) quinoxaline									
$S_0 \rightarrow S_1$	24050	4286	0.03	24518	3712	0.04	25530	8197	0.07
$S_0 \rightarrow S_2$	25437	6388	0.05	25936	5568	0.05	29546	75956	0.73
$S_0 \rightarrow S_3$	28960	62120	0.52	29501	69142	0.61	35746	30055	0.40
$S_0 \rightarrow S_4$	33892	24562	0.26	35781	23666	0.34	42024	59290	0.52
$S_0 \rightarrow S_5$	41760	68084	0.57	43495	71928	0.76	-	-	-

$S_o \rightarrow S_4$ transition was of low intensity in methanol and ether and was submerged under the intensity of the $S_o \rightarrow S_3$ transition, it is designated as ^1B_b (that is, quantum mechanically allowed but symmetry forbidden). $S_o \rightarrow S_4$ band peaks in chloroform was intense with a shoulder at 41,162cm^{-1} and is designated as ^1B_a band (that is, both quantum mechanically and overlap allowed).

Band intensity

The transition intensities recorded in the form of oscillator strengths (*f*), in different solvents as shown in Table 1 reveal that the trend of the oscillator strength is $f(S_o \rightarrow S_1) < f(S_o \rightarrow S_3) < f(S_o \rightarrow S_2) < f(S_o \rightarrow S_4)$ for the transitions in acenaphtho(1,2-b)quinoxaline.

For acenaphtho(1,2-b)benzo(g)quinoxaline, the oscillator strength follow the trend $f(S_o \rightarrow S_1) < f(S_o \rightarrow S_2) < f(S_o \rightarrow S_4) < f(S_o \rightarrow S_3) < f(S_o \rightarrow S_5)$ for methanol and ether. However, in chloroform, the trend is $f(S_o \rightarrow S_1) < f(S_o \rightarrow S_3) < f(S_o \rightarrow S_2) < f(S_o \rightarrow S_4)$. Table 1 shows that the $S_o \rightarrow S_2$ transition is far more sensitive to solvent perturbation than $S_o \rightarrow S_1$ or $S_o \rightarrow S_3$ in terms of changes in intensity and energy of the transition.

Conclusion

From the structure and symmetry of the acenaphtho(1,2-b)quinoxaline and acenaphtho (1,2-b)benzo(g)quinoxaline (both having C_2V symmetry); it is obvious that the molecules possess permanent dipole μ (nitrogen being more electronegative than carbon) in the ground state.

The blue shift observed in the most intense bands in polar solvents relative to non-polar solvents implies that the dipole moment in the ground state is greater than those in the excited states of acenaphtho(1,2-b)quinoxaline and acenaphtho(1,2-b)benzo(g)quinoxaline.

The earlier suggestion of the existence of $n \rightarrow \pi^*$ transition in acenaphtho(1,2-b)quinoxaline and acenaphtho(1,2-b)benzo(g)quinoxaline is not likely because the non-bonding, lone-pair electrons on the N-atom are orthogonal to the π-electron system which gives rise to the observed absorption in the compounds.

REFERENCES

Anderson RW, Damschen DE, Scott GW, Talley LD (1979). Picosecond relaxation processes in excited state quinoxaline. J. Chem. Phys., 71(3): 1134.

Bohlmann F (1951). Constitution and light absorption: Dicarbonyl derivatives. Chem. Ber., 84: 860-865.

Czerwieniee R, Herbich J, Kapturkiewiez A, Nowacki J (2000). Radiative electron transfer in planar donor–acceptor quinoxaline derivatives. Chem. Phys. Lett., 325(5-6): 589-598.

Greene BI, Hochstrasser RM, Weisman RB (1978). Picosecond transient spectroscopy of molecules in solution. J. Chem. Phys., 70(3): 1247-1252.

Halliday LA, Topp MR (1977). Picosecond luminescence detection using types II phase matching frequency conversion. Chem. Phys. Lett., 46: 8-14.

Iweibo I, Obi-Egbedi NO, Chongwain PT, Lesi AF, Abe T (1990). Theory of electronic intensity in solution or condensed media. J. Chem. Phys., 93(4): 2238-2245.

Iweibo I, Oderinde RA, Faniran JA (1982). Electronic absorption spectra and structures of aniline and its 4-chloro, pentafluoro and pentachloro derivatives. Spectrochim. Acta, 38A(1): 1-7.

Mason SF (1962). The electronic spectra of *N*-heteroaromatic systems - *n*→π-Transitions of polycyclic azines. J. Chem. Soc., part IX: 493-494.

Sarki GY, Al-Badri HT (1980). Synthesis and spectroscopic studies of some new substituted 6*H*-indolo[2,3-*b*]quinoxaline. J. Heterocyclic Chem., 17(4): 813-815.

Yamatoto K, Takemura T, Baba H (1978). Fluorescence and dual phosphorescence spectra of Quinoxaline in fluid solution. Bull. Chem. Soc. Jpn., 51: 729-732.

Theoretical study of phthalane oxidation and effect of substituents by using Hyperchem program

Mohammed Oday Ezzat and Mohd Nizam Mordi*

Centre for Drug Research, Universiti Sains Malaysia, 11800 Minden, Penang, Malaysia.

The effect of substituents on the relative stabilization energy (E_{RS}) of phthalane-X carbocations has been studied with PM3 semi-empirical molecular orbital theory. The results show that the amino group gives E_{RS} less than zero for very big amount and that will increase stability of carbocation by π-electron donation, whereas the nitro group gives height amount of E_{RS} more than zero and that will lead to difficulties of an intermediate carbocation formation because of its strong σ- and π-accepting nature. E_{RS} decreases with increase in the phenyl rings attached to cation. When electron withdraws substituents attached, it will destabilize the cation, because they act as π-electron donors.

Key words: Relative stabilization energy (E_{RS}), PM3 semi-empirical, isodesmic reactions.

INTRODUCTION

Phthalane (1,3-dihydroisobenzofuran or 1,3-dihydro-2-benzofuran) also known as isocoumaran, is a bicyclic aromatic organic compound and it can be oxidized to phthalic acid (Schulz, 1959). Phthalic acid is an isomer of isophthalic and terephthalic acid (Christensen et al., 1977). In inhibiting the amine pump in adrenergic neurons, phthalane is approximately equipotent with tricyclic compounds like antidepressants (Maxwell et al., 1980). For full potency to be expressed, it requires the presence of a phenyl ring on the bicyclic nucleus (Heudorf et al., 2007; Kolarik et al., 2008). Phthalane contained methyl groups on their bicyclic ring system which are involved in binding to the receptor (Sathyanarayana et al., 2008). Derivatives are found in the drug candidate Lu 10 to 171 and drug citalopram (Hernández-Díaz, 2009).

Thermochemistry is the study of nature of energy, energy change in chemical reactions, enthalpy, calorimetry, standard enthalpy of formation and reaction and thermodynamics-thermochemistry (Lomenick et al., 2010).

In chemical reaction, the energy could be release or absorb (Getman, 1918), also, a phase could be changed as melting or boiling. Thermochemistry could be used to predict if the reaction is going to forward direction or back direction (Laidler and Meiser, 1982). The system in thermochemistry is the part that has been studied from the universe (Perrot, 1998). Heat is absorbed in endothermic reactions and it release in exothermic reactions (Atkins and Paula, 2006).

Hehre et al. (1970) defined isodesmic reactions as "chemical changes in which there is retention of the number of bonds of a given formal type, but with a change in their relation to one another". In isodesmic reactions, reactants and products have the same number of electron pairs (George et al., 1976). In a thermochemical sense, it can be used to quantify the stability of reactive intermediates (Rose and Williams, 2002). By hypothetical reaction, the effect of a particular structural feature could get and the best way to cancel out errors in computations (Matthias et al., 2006). Homodesmotic reactions are a subclass of isodesmic reactions where reactants and products share even larger similarity (Grimme et al., 2007). Various definitions have been given for this class in the past and the most consistent one (Wheeler et al., 2009).

The object of this study is to theoretically study the effect of adding substitution on phthalane at para position to find the relative stabilization energy (E_{RS}) according to isodesmic reactions by comparing the energies (which equals ΔH) of the difference adding groups.

*Corresponding author. E-mail: mnizam@usm.my.

COMPUTATIONAL DETAILS

All structures, were optimized in the gas phase using parameterized model number 3 (PM3) (Stewart, 1998), a semi-empirical method for the quantum calculation of molecular electronic structure in computational chemistry (Stewart, 2004), implemented in the HyperChem8.0 package (Hypercube, 2007). Adding substituted and conjugate acids flowing in the order as shown in Figure 1 and Table 1.

The heat of formation and total energy was found, calculation of E_{RS} was by calculating the differences in heats of formation close to difference in binding energies (which we often use just for the sake of convenience); however, in this case, we follow the practice of most chemistry journal authors (El-Nahas and Clark, 1995).

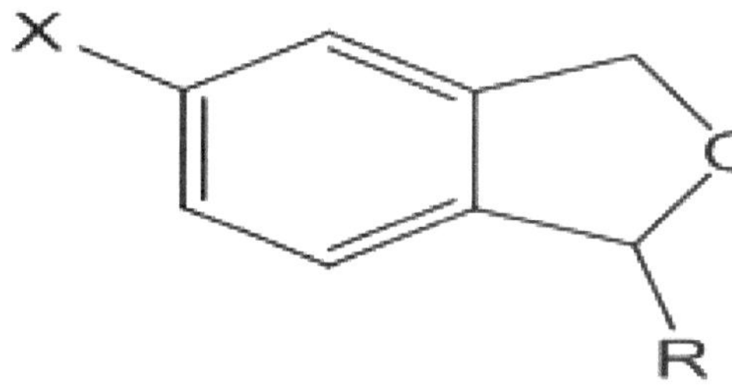

Figure 1. Adding substituted and conjugate acid to phthalane.

RESULTS AND DISCUSSION

The study of acid catalyzed hydrolysis of an acetal related to a 4-substituted phthalane and the effect different 4-substituents might have on the rate of hydrolysis. We tested the conclusion by investigating the stabilities of similar carbocations using a computational chemistry tool called an isodesmic equation, as shown in Figure 2. The rates of these reactions were measured experimentally by Rose and Williams (2002).

The objective of this study is to computationally determine the effect of different groups X on the formation of the carbocations. It is difficult to just compare the energies of the different carbocations, because the groups X differ. The difference in energy between the aforementioned reaction for X= H and X = another group, as shown in Figure 3. This is called the "Relative Stabilization Energy", which equals ΔH for the following reaction.

Figure 3 is an isodesmic reaction, defined as a chemical reaction in which the type of chemical bonds broken in the reactant is the same as the type of bonds formed in the reaction product. This is often used to calculate substituent effects. From Table 2, after calculation of E_{RS}, it shows for primary carbocation when substituted group (R) is hydrogen (H) at compounds (1 to 12), and the E_{RS} increases at the sequence of 4<3<6<2<7<5<8<1<10<11<9<12.

Negative value of E_{RS} less than zero (E_{RS} < 0) refers to the substituted group which will increase the ability to form the carbocation more than when the substituted group is hydrogen. At compound 4, when the substituted group is (NH_2), there will be increased stability of carbocation in very big amount according to E_{RS} less than zero (E_{RS} < 0). This will increase the stability of an intermediate carbocation formation, and will increase reaction speed. This is because the amino group has the highest stabilization by π-electron donation. Also, for compound 12 when the substituted group is (NO_2), the E_{RS} has the highest amount and that will lead to difficulties of an intermediate carbocation formation. This is because nitro group is most destabilizing because of its strong σ- and π-accepting nature, thereby leading to decrease in the stability of an intermediate carbocation

Table 1. Adding substituted and conjugate acid to phthalane.

Compound	X	R
Original compound	H	H
1	F	H
2	OH	H
3	OCH_3	H
4	NH_2	H
5	CH_3	H
6	C_6H_5	H
7	$CH=CH_2$	H
8	C≡CH	H
9	CN	H
10	CHO	H
11	COOH	H
12	NO_2	H
13	F	CH_3
14	OH	CH_3
15	OCH_3	CH_3
16	NH_2	CH_3
17	CH_3	CH_3
18	C_6H_5	CH_3
19	$CH=CH_2$	CH_3
20	C≡CH	CH_3
21	CN	CH_3
22	CHO	CH_3
23	COOH	CH_3
24	NO_2	CH_3

formation. When substituted group (R) is methyl (CH_3) at compounds (13 to 24), it will form secondary carbocation, and E_{RS} will increase at the sequence of 16<15<18<14<19 <17<20<13<22<23 <21<24. Also, at the same substituents at compound 16 when substituted group is NH_2, the lower amount of E_{RS} is less than zero (E_{RS} < 0), and at compound 24 when the substituted group is NO_2, E_{RS} is in the highest amount. This methyl group increases all the amount of E_{RS} more than how hydrogen do, by decreasing the stability of an intermediate carbocation formation and will lead to decrease

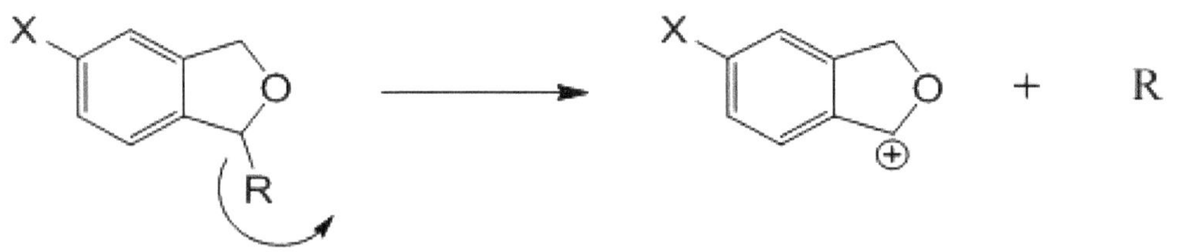

Figure 2. Acid catalyzed hydrolysis of an acetal.

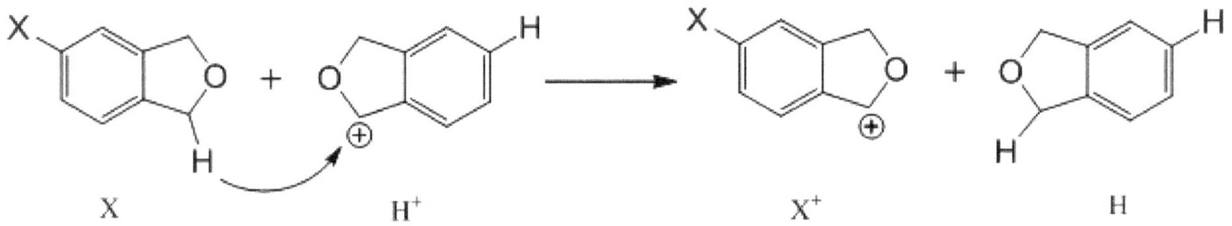

Figure 3. Formation of the carbocation.

Table 2. Theoretical value of calculated relative stabilization energy (ERS) (kcal/mol).

Compound	Heat of formation for H (R=H, X=H)	Heat of formation for H^+ (R=H, X=H)	Heat of formation for X	Heat of formation for X^+	E_{RS} = ΔH(X) - ΔH(H)
1	-13.338	177.659	-56.805	137.25	3.058
2	-13.338	177.659	-58.403	127.864	-4.73
3	-13.338	177.659	-51.323	133.567	-6.107
4	-13.338	177.659	-15.557	163.434	-12.006
5	-13.338	177.659	-22.701	165.579	-2.717
6	-13.338	177.659	10.775	196.561	-5.211
7	-13.338	177.659	2.327	189.72	-3.604
8	-13.338	177.659	37.885	226.849	-2.033
9	-13.338	177.659	22.042	220.317	7.278
10	-13.338	177.659	-47.245	148.919	5.167
11	-13.338	177.659	-102.726	94.287	6.016
12	-13.338	177.659	-21.711	183.937	14.651
13	-19.153	163.798	-62.648	123.383	3.08
14	-19.153	163.798	-64.246	114.673	-4.032
15	-19.153	163.798	-57.163	120.495	-5.293
16	-19.153	163.798	-21.389	151.214	-10.348
17	-19.153	163.798	-28.54	152.01	-2.401
18	-19.153	163.798	4.961	183.477	-4.435
19	-19.153	163.798	-3.482	176.399	-3.07
20	-19.153	163.798	32.071	213.384	-1.638
21	-19.153	163.798	16.207	206.208	7.05
22	-19.153	163.798	-53.058	134.802	4.909
23	-19.153	163.798	-108.542	80.118	5.709
24	-19.153	163.798	-27.541	169.247	13.837

reaction speed, except when the substitutes are CHO, COOH and NO_2 and will increase E_{RS} as compared to when the substituted group (R) is hydrogen (H). The effect for substituents on E_{RS} at phthalane will follow the same order for primary and secondary carbocation.

We can classify substituents into four categories: (1)

lone pair electron donors (F, OH, OCH_3 and NH_2, (2) conjugated systems (C_6H_5, $CH=CH_2$ and $C\equiv CH$), (3) hyperconjugative groups (CH_3) and (4) σ- and π-acceptors (CN, CHO, COOH and NO_2). If we look to F, OH, OCH_3 and NH_2 as lone pair electron donors, despite of their σ-withdrawing nature, F, OH, OCH_3 and NH_2 will give large amount of E_{RS}, with the amino group being the most effective. The stability of these carbocations decrease with increasing substituent electro negativities.

For conjugated systems (C_6H_5, $CH=CH_2$ and $C\equiv CH$), these groups have double bonds which can resonate with the adjacent cation center. The E_{RS} order is C_6H_5 < $CH=CH_2$ < $C\equiv CH$ in line with the ability of these substituents to donate their π-electrons. As a hyperconjugative group, methyl group shows a moderate effect on E_{RS} due to its π-electron donation through hyperconjugation, while σ- and π-acceptors, (CN, CHO and COOH) shows significant π-interaction between these groups and the vacant p orbital at the cation center. The E_{RS} order CN > COOH > CHO indicating the degree of π-contribution. The strongest electron-withdrawing group, NO_2, give the most destabilized carbocations as the highest E_{RS} in the whole series.

In all compounds, the ability of reaction increases with the increase of stability for an intermediate carbocation formation when E_{RS} < 0 and the ability of reaction decreases with decrease of stability for an intermediate carbocation formation when E_{RS} > 0.

Conclusions

In this study, E_{RS} of phthalane oxidations and effect of substituents on the original compound has been calculated to investigate the stabilities of similar carbocations by using computational chemistry called an isodesmic equation. For primary and secondary carbocations, the effect for substituents on E_{RS} at phthalane will follow the same order. The amino group will increase the stability of carbocation in very big amount according to the E_{RS} (that is, less than zero), and the nitro group is the most destabilizing.

ACKNOWLEDGEMENTS

The author (Mohammed Oday Ezzat) thank the Academy of Sciences for the Developing World (TWAS) and Universiti Sains Malaysia (USM) for financial support under TWAS-USM fellowship.

REFERENCES

Atkins P, Paula DJ (2006). Atkins' Physical Chemistry. WH Freeman, pp. 122-128.

Christensen AV, Fjalland B, Pedersen V, Danneskiold-Samsoe P, Svendsen O (1977). Pharmacology of a new phthalane (Lu 10-171) with specific 5-HT uptake inhibiting properties. Eur. J. Pharmacol., 41(2): 153-162.

El-Nahas AM, Clark T (1995). AM1 and PM3 Calculations on the Effect of Substituents on the Stabilities of Carbocations in the Gas Phase and in Solution. J. Org. Chem., 60(24): 8023-8027.

George P, Trachtman M, Bock CW, Brett AM (1976). Homodesmotic reactions for the assessment of stabilization energies in benzenoid and other conjugated cyclic hydrocarbons. J. Chem. Soc. Perkin Trans., 2(11): 1222–1227.

Getman FH (1918). Outlines of Theoretical Chemistry. John Wiley & Sons. Hoboken, 155-161.

Grimme S, Antony J, Schwabe T, Mueck-Lichtenfeld C (2007). Density Functional Theory with dispersion corrections for supramolecular structures, aggregates, and complexes of (Bio)organic molecules. Org. Biomol. Chem., 5, 741-758.

Hehre WJ, Ditchfield R, Radom L, Pople JA (1970). Molecular orbital theory of the electronic structure of organic compounds. V. Molecular theory of bond separation. J. Am. Chem. Soc., 92(16): 4796-4801.

Hernández-Díaz S, Mitchell AA, Kelley KE, Calafat AM, Hauser R (2009). Medications as a Potential Source of Exposure to Phthalates in the U.S. Population Environ. Health Perspect., 117: 185–189.

Heudorf U, Mersch-Sundermann V, Angerer J (2007). Phthalates: Toxicology and exposure. Int. J. Hyg. Environ. Health, 21(5): 623–634.

Hypercube (2007). Inc. HyperChem 8.0 Package. Gainesville Florida. USA. http://www.hyper.com/.

Kolarik B, Bornehag C, Naydenov K, Sundell J, Stavova P, Nielsen O (2008). The concentrations of phthalates in settled dust in Bulgarian homes in relation to building characteristic and cleaning habits in the family. Atmos. Environ., 42(37): 8553–8559.

Laidler KJ, Meiser JH (1982). Physical Chemistry. John Wiley & Sons. 175-182.

Lomenick JP, Calafat AM, Melguizo Castro MS, Stenger P, Foster MB, Wintergerst KA (2010). Phthalate Exposure and Precocious Puberty in Females. J. Pediatr., 156(2): 221-225.

Matthias F B, Holger L H, Gernot B (2006). Stabilization of Carbanions: Fluorine is more effective than the Heavier Halogens. Angewandte Chemie International Edition. 45(5): 823- 826.

Maxwell RA, Ferris RM, Woodward EC, Tang F, Eckhardt S (1980). Structure-Activity Relationships for Potent Phthalane and Thiophthalane Inhibitors of Norepinephrine Uptake A Comparison with Desipramine and Related Compounds. Molecular Pharmacology. 17(3): 321-327.

Perrot P (1998). A to Z of Thermodynamics. Oxford University Press. 92-96.

Rose P, Williams A (2002). Nucleophilic additions to oxides intermediates. J. Chem. Soc., 56(2): 1589-1595.

Sathyanarayana S, Karr CJ, Lozano P (2008). Possible sources of infant phthalate exposure. Pediatrics, 121(2): 260–268.

Schulz M (1959). Autoxydation der Phthalane Papierchromatographie organ. Peroxyverbindgn. Math.-naturwiss. F. 6:152.

Stewart JJP (1998). "PM3" in Encyclopedia of Computational Chemistry, Wiley, 95-104.

Stewart JJP (2004). Optimization of parameters for semiempirical methods IV: extension of MNDO, AM1, and PM3 to more main group elements. J. Mol. Model., 10(2): 155-164.

Wheeler G, Houk KN, Schleyer PVR, Allen WD (2009). A Hierarchy of Homodesmotic Reactions for Thermochemistry. J. Am. Chem. Soc., 131(7): 2547 – 2560.

Adsorption of o-nitrophenol onto nano iron oxide and alginate microspheres: Batch and column studies

Anjali Soni[1], Alka Tiwari[1]* and A. K. Bajpai[2]

[1]Department of Chemistry, Government V.Y.T.PG., Autonomous College Durg (C.G.), 491001, India.
[2]Bose Memorial Research Laboratory, Department of Chemistry, Government Autonomous Science College, Jabalpur (M.P.), 482001, India.

In the present study, the adsorption behavior of o-nitrophenol from aqueous medium, using nano iron oxide loaded calcium alginate beads were studied using equilibrium batch and column flow techniques. The effect of pH, adsorbent dose, contact time and temperature on adsorption capacity of the magnetic beads was investigated. The optimum pH value was defined to be 2 at temperature 30°C. The equilibrium data were fitted to Langmuir and Freundlich isotherm models. Batch adsorption data was better described by Freundlich isotherm model than Langmuir model. In column experiments the effect of the initial concentration, bed height and flow rate on o- nitrophenol adsorption were studied. The bed depth service time (BDST) model was used to analyze the experimental data. The sorption capacity per unit bed volume (N_0) and rate constant (Ka) were calculated to be 578.4 mg/L and 1.18 L/mg/min, respectively. The characterization of the prepared magnetic beads has been accomplished by Fourier transform infrared spectroscopy (FTIR), scanning electron microscopy (SEM) and X-ray diffraction (XRD) measurements.

Key words: Adsorption isotherm, magnetic beads, o- nitrophenol.

INTRODUCTION

Industrial development is associated with the disposal of a large number of toxic pollutants that are difficult to degrade by natural means. Many industries discharge wastewater containing hazardous substances such as phenols, dyes and heavy metals. It is well known that the contamination of water from the industries provides many problems to human health. In our world, organic pollutants are still a problem for the environment and drinking water. A typical group of organic chemicals that are considered as both unwanted by-products as well as raw material in several industries is the phenol. Phenol and its corresponding substituted compounds are important environmental pollutants because of their toxic effects towards life in the aquatic environment.

There are many methods such as oxidation, ion exchange, electrochemical oxidation, reverse osmosis (Goncharuk et al., 2002) and photo-catalytic degradation, which have been used for the removal of phenol and its derivatives.

Adsorption is a well established and powerful technique for the removal of organic pollutants from wastewater. Adsorption has been found to be superior choice than other techniques for water re-use because of low initial cost, simplicity of design, ease of operation and insensitivity to toxic substances (Juny et al., 2001). Nitrophenol compounds are common toxic, industrial and persistent pollutants. It is desirable to reduce their concentrations to less than 20 ppb. They are contained in

*Corresponding author. E-mail: alkatiwari18@yahoo.co.in, alkatiwari18@gmail.com.

many industrial effluents such as those from oil refineries, plastics, textile, rubber processing plants and pesticide production (Yaneva and Koumanova, 2006). O-nitrophenol is widely used in manufacturing fuel, vegetation protectant and drugs. This organic pollutant can enter our body through lungs and its toxicity lies in the chemical reaction with protein in the bioplasm of organism cell, producing insoluble protein and resulting in the loss of activity of the cell. It also could restrain microorganism growth, thereby decreasing the self purifying ability of water. Therefore, this compound must be removed from drinking water.

A number of adsorption materials have been studied for their capacity to remove pollutants including ion exchange resins, bentonite (Banat et al., 2000), silica gel, activated carbon (Roostaei and Tezel, 2004) etc. Some of these materials such as ion exchange resins are quite effective but expensive and others like coal and straw are inexpensive but not very effective. Activated carbon is the most widely used adsorbent with high adsorption capacity for organic compounds, but its use is usually limited due to its high cost.

The most widely studied polysaccharide based biosorbents are alginate, gelatin, chitosan, and starch. Alginate is a natural high molecular weight biopolymer, extracted mainly from brown seaweed (algae) (Pandey et al., 2007). It is a linear polysaccharide of (1-4)-linked α – L guluronate (G) and β - D mannuronate (M) residues arranged in a non-regular, block-wise pattern along the linear chain (Zouboulis and Katsoyiannis, 2002). Gelation of alginate is possible by interaction of carboxylate groups with divalent ions. Calcium cations are commonly used to cross-link sodium alginate (Pathak and Se-Jony, 2009). Alginates are very much important, due to their biodegradability, immunogenicity and ability to form gel with a variety of cross- linking agents.

The use of metal nanoparticles for environmental remediation is a relatively new area of research in which, only limited progress has happened so for (Zhany, 2003). Magnetic separation technique using magnetic polymeric particles is a quick and easy method for sensitive and reliable capture of inorganic or organic pollutants. The magnetic sorbents behave similar or even better than various commercial adsorbents (Ngah et al., 2006). Nanoscale particles are promising in this area because of their unique properties such as small particle sizes, large surface to volume ratio. After the usage, the magnetic sorbent can be easily separated from the solution by simple magnetic force. In this study, iron oxides have been found to be successfully used as composite materials with host materials in fabricating magnetic sorbent (Oliveira et al., 2003, 2004; Chang and Cheng, 2006).

The aim this study is to prepare nano iron oxide loaded calcium alginate beads to determine the adsorption capacity of o-nitrophenol from aqueous solution under batch equilibrium and column flow experimental conditions and to study the effect of various experimental parameters such as aqueous phase pH, adsorbent dose, contact time and initial adsorbate concentration on uptake capacity, predict the kinetic and isotherm models, deduce the adsorption mechanism and accumulate data for their potential industrial application. The characterizations of prepared magnetic beads were carried out by Fourier transform infrared spectroscopy (FTIR), scanning electron microscopy (SEM) and X-ray diffraction (XRD) measurements. The important column design parameters such as flow rate of fluid, bed height and initial concentration of o- nitrophenol solutions have been investigated. The breakthrough curves for the adsorption of o-nitrophenol were analyzed using bed depth service time (BDST) model.

MATERIALS AND METHODS

O-nitro phenol and alginate were obtained from Loba- Chemie Mumbai, India and used as received. Other chemicals such as ferrous chloride, ferric chloride and calcium chloride (cross linking agent for alginate) (Molychem, Mumbai, India) were used of AR grade. Double distilled water was used throughout the experiments.

Preparation of magnetic particles

To prepare the nano particles of iron oxide, to the mixture of ferrous chloride and ferric chloride (0.5 M) 25% ammonia solution was added at 80°C followed by the addition of concentrated ammonia solution drop wise under drastic stirring. After 15 min, black magnetic particles dispersed in liquid, which were then separated by an external magnetic field, washed several times with double distilled water and finally stored as a stock solution by adding 50 ml distilled water for the subsequent experiments (Chen and Wang, 2007).

Preparation of magnetic beads

The sodium alginate gel was prepared by dissolving 1 g sodium alginate in double distilled water and was mixed with magnetic particle solution in 1:1 ratio. This blended gel was stirred for a period of 30 min for homogeneity and kept aside for 30 min to obtain a bubble free solution. Spherical shaped beads were then prepared by drop- wise addition of this mixture into a 0.05 M $CaCl_2$ solution with constant stirring. The beads so produced were allowed to harden by leaving them in solution for 24 h and thereafter filtered and washed with double distilled water to remove excess calcium. A photograph of dry and swollen magnetic beads is shown in Figure 1.

Adsorption experiments

The adsorption of o-nitrophenol was carried out by the batch contact method and column flow method. In batch method, the swollen magnetic beads (0.1 g) were added to a 10 ml known solution of o-nitro phenol of 0.2 mg/ml concentrations at constant pH 2 and temperature 30°C. The suspension was shaken in a wrist action shaking machine for 4 h, which was found to be a sufficient

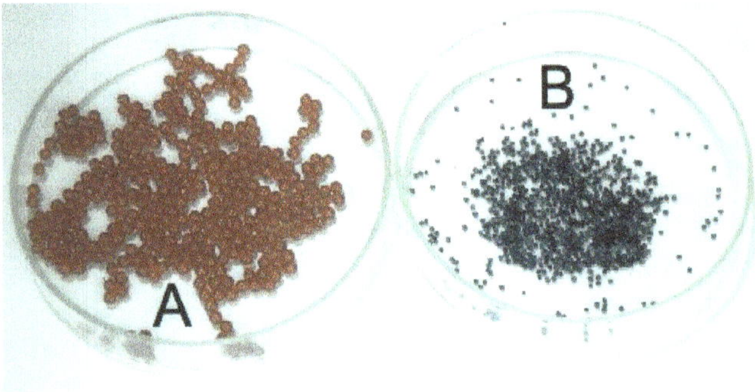

Figure 1. A Photograph showing (A) dry, (B) swollen magnetic beads.

time to attain equilibrium adsorption. After shaking, the suspension was filtered through whatman filter paper and the concentration of o-nitrophenol was determined by measuring absorbance using UV/Visible spectrophotometer at 273 nm. The amount of o-nitrophenol was calculated by the following mass balance equation.

$$\text{Adsorbed amount (mg/g)} = \frac{(C_i - C_f)V}{m} \quad (1)$$

where C_i and C_f being the initial and final concentrations (mg/ml) of o-nitrophenol solution respectively, V being the volume of adsorbate and m is the weight of swollen beads taken as adsorbent.

Desorption

In order to desorb magnetic adsorbent the beads loaded with o-nitrophenol were exposed to 2 ml of 0.1 N HCl for 15 min, filtered and washed with double distilled water. The beads were then again exposed to HCl to strip any remaining o-nitrophenol and then used again.

Column adsorption experiments

Column flow adsorption experiments were conducted in a glass column of about 1.0 cm internal diameter and 25 cm length. The column was filled with a known weight of the adsorbent by tapping so that the column was filled without gaps. The inlet solution of known concentration of o-nitrophenol was allowed to pass through the bed at constant flow rate, 1 ml/min in down flow manner. The complete cycle of operation of each column experiment includes three steps: pH precondition of the adsorbate solution, solution flow and adsorption of solute until column exhaustion occurs. The outlet solution was collected at different time intervals and the concentration of adsorbate in the outlet solution was determined by UV- Visible spectrophotometer. The column experiments were carried out to study the effects of process parameters on adsorption of o-nitrophenol.

Modeling and analysis of column data

The performance of a fixed- bed column has been described through the concept of the breakthrough curve. The time for breakthrough appearance and the shape of the breakthrough curve are very important characteristics for determining the operation and the dynamic response of an adsorption column because they directly affect the feasibility and economics of the sorption phenomena. Experimental determination of these parameters is dependent on column operating conditions such as feed inlet concentration and flow rate. The breakthrough curve is usually expressed in terms of adsorbed o-nitrophenol concentration (C) or normalized concentration which is defined as the ratio between the adsorbate concentration in the liquid at the outlet and inlet of the column (C/C_0), in a function of time or volume of the inlet, for a bed depth fixed (Aksu and Gonen, 2004). The volume of the outlet, V_{eff} can be calculated through the following equation.

$$V_{eff} = Qt_{total} \quad (2)$$

where t_{total} is the total time in minutes, Q is the flow rate which circulated through the column in ml/min.

The total amount of o- nitrophenol passed through the column in mg, can be calculated from the following Equation (3).

$$m_{total} = C_0Qt_{total}/1000 \quad (3)$$

The total (%) of o-nitrophenol removal can be calculated from the ratio of o- nitrophenol adsorbed (q_{total}) to the total amount of o-nitrophenol sent to the column (m_{total}) as:

$$\%R = \frac{q_{total}}{m_{total}} \times 100 \quad (4)$$

The amount of o-nitrophenol adsorbed at equilibrium or adsorption capacity q_{eq} (mg of adsorbed o-nitrophenol/g of adsorbent) can be determined using the following equation.

$$q_{eq} = \frac{q_{total}}{X} \quad (5)$$

where X is the mass of adsorbent in g.

The time at which the break point is attained is called breakthrough time (t_b) and the portion of the influent concentration curve as a function of time is called breakthrough curve. The time at which o-nitrophenol concentration in the influent exceeded 99% of the inlet concentration is called bed exhaustion time (t_e).

The mass transfer zone (Δt) is given by:

$$\Delta t = t_e - t_b \quad (6)$$

FTIR Spectra

The FTIR spectra of magnetic beads of nano iron oxide and alginate (before adsorption and after adsorption) were recorded on a Perkin- Elmer spectrophotometer [CDRI. Lucknow].

SEM analysis

To examine the morphological characteristics of the magnetic beads before and after adsorption, samples were viewed using a scanning electron microscope (IUC, DAE, Indore).

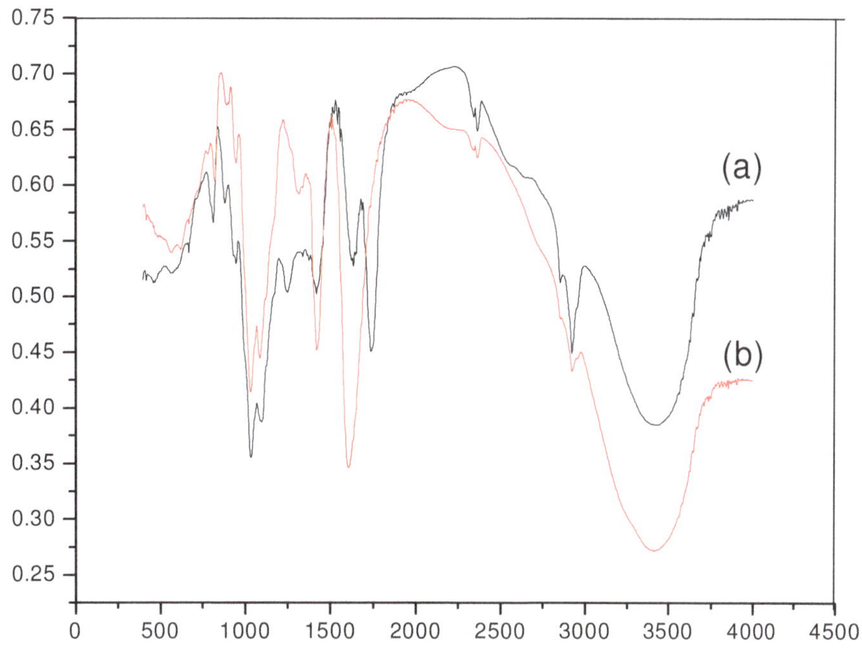

Figure 2. FTIR Spectra of (a) magnetic beads and (b) o-nitrophenol adsorbed magnetic beads.

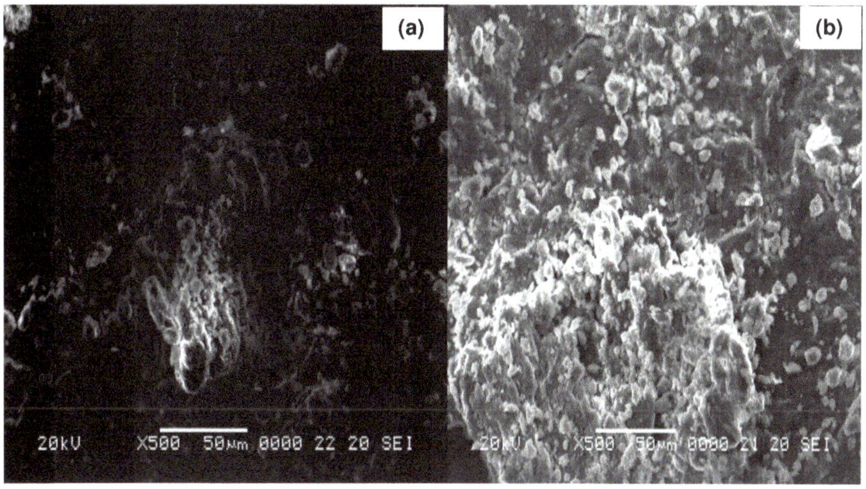

Figure 3. SEM image of bare (a) magnetic beads and (b) SEM image of magnetic beads after adsorption.

XRD analysis

The crystalline nature of the bare alginate beads and iron oxide loaded alginate beads was studied on a rotating X- ray diffractometer (IUC, DAE, Indore).

RESULTS

The FTIR spectra of bare beads and o-nitrophenol loaded magnetic alginate beads are illustrated in Figure 2a and b. In the FTIR spectra the major bands were obtained at 3418, 1607, 1421 and 500 to 700 cm^{-1}.

The SEM micrographs of the nano iron- oxide loaded alginate beads are shown in Figure 3a and b respectively.

The XRD patterns of the prepared alginate beads and iron oxide loaded alginate beads are shown in the Figure 4a and b, respectively. The crystallinity of iron oxide

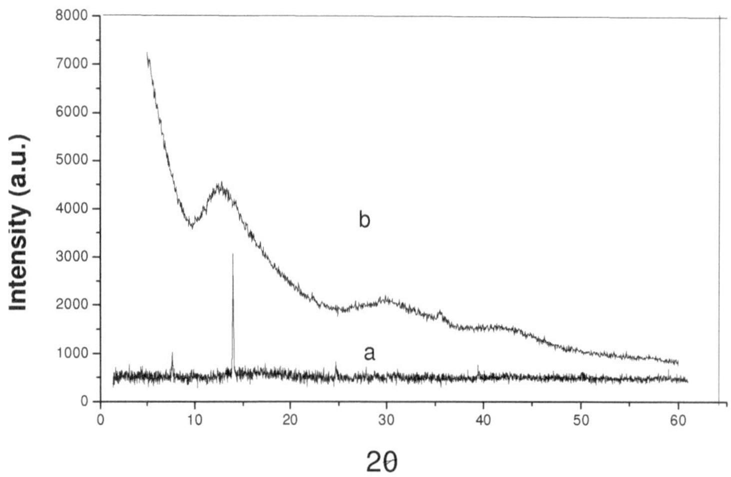

Figure 4. XRD Pattern of (a) alginate beads and (b) nano iron oxide loaded alginate beads.

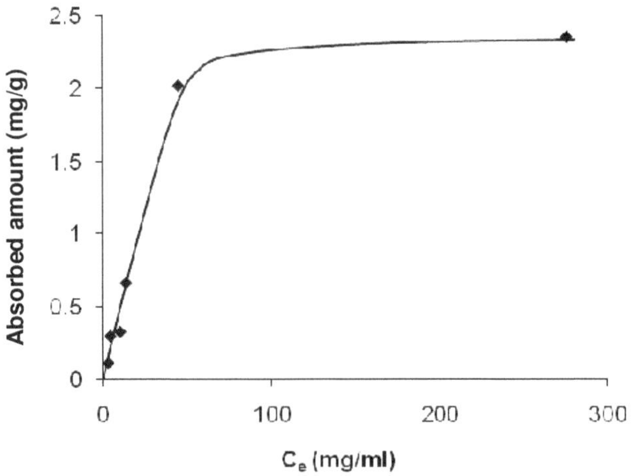

Figure 5. Langmuir adsorption isotherm drawn between adsorbed amount of o- nitrophenol and equilibrium concentration of o-nitrophenol solution showing the adsorption of o-nitrophenol onto magnetic beads of iron oxide and alginate = 0.1 g, pH = 2, temp. = 30°C.

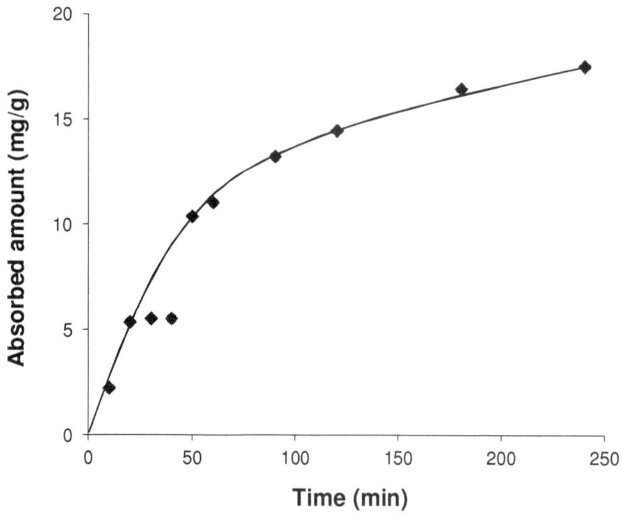

Figure 6. Variation in the adsorption amount of o-nitrophenol with time for fixed amount of magnetic beads = 0.1g, pH = 2, temp. = 30°C.

loaded alginate beads were determined by degree of crystallinity. The numerical formula used to calculate the percent crystallinity of alginate and iron oxide loaded alginate beads are shown in the following equation:

$$Xc\ (\%) = (Ac/Aa+Ac) \times 100 \tag{7}$$

where Ac and Aa are the area of crystalline and amorphous phases, respectively (Gupta and Bajpai, 2010).

The batch experiment was performed at different pH levels, that is, 2.0 to 9.0 and the highest adsorption was obtained at pH 5 Figure 5.

The adsorption experiment was also conducted taking different adsorbent dosage (0.1 to 0.5 g) which was found to be 96% removal with 0.1 g Figure 6.

In order to understand the adsorbent-adsorbate interaction, adsorption experiments with different o-nitrophenol concentrations in the range (0.012 to 0.3 mg/ml) were carried out and it was found that with the increase of o-nitrophenol concentration, the amount of adsorption also increases.

The adsorption experiments were performed at different temperatures from (10 to 50°C). It has been

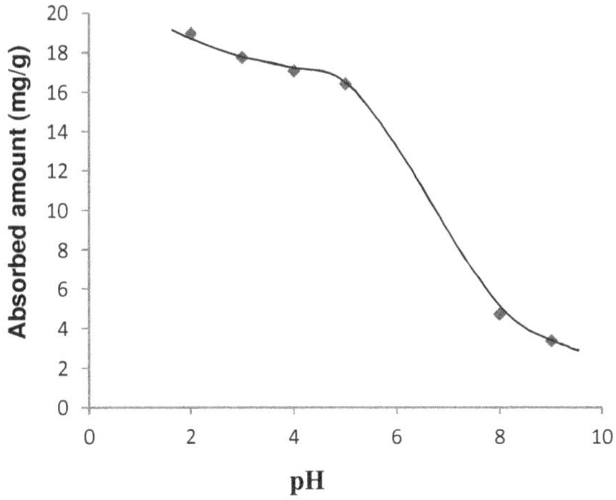

Figure 7. The effect of initial pH on the adsorption of o-nitrophenol onto magnetic beads = 0.1, temp. = 300°C, time = 180 min, o-nitrophenol solution 10 ml of initial conc. 200 mg/L.

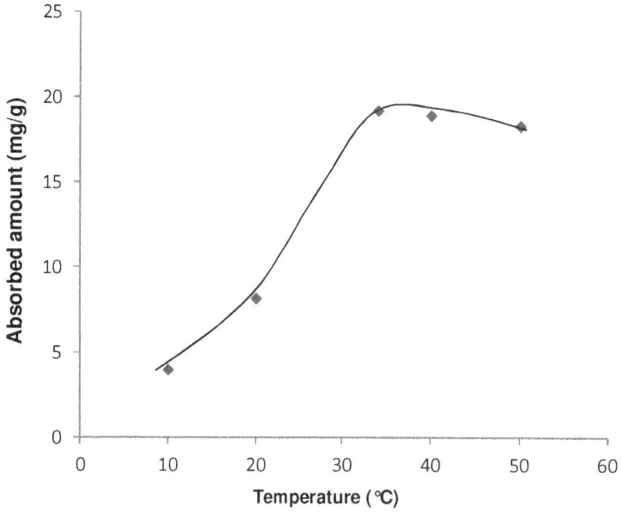

Figure 9. The effect of temperature on the adsorption of o-nitrophenol, magnetic beads = 0.1 g, pH = 2, time = 180 min, o-nitrophenol solution 10 ml of initial conc. 200 mg/L.

Figure 8. The effect of adsorbent dose on the adsorption of o-nitrophenol, pH = 2, temp. = 30°C, time = 180 min, o-nirtophenol solution 10 ml of initial conc. 200 mg/L.

observed that the adsorption of o-nitrophenol increases from 10 to 30°C, while beyond 30°C a sudden fall in adsorption is noticed as shown in Figure 7.

The chemical composition of beads has a significant effect on the adsorption of o-nitrophenol because upon variation in the amount of polymer components in the adsorbent beads, the number and nature of active sites will be largely affected which in turn influence the adsorption process. In the case of iron oxide nano particles, the amount was varied in the range from 0.05 to 0.2 ml. The results are shown in Figure 8.

Adsorption of o-nitrophenol is dependent on the quantity of adsorbent in the column. Breakthrough experiments were conducted at constant influent concentration of 0.2 mg/ml, flow rate of 1 ml/min and various bed heights of 2.5, 5 and 10 cm and breakthrough curves were plotted between ratio of final (outlet) and initial influent concentration (C/C_0) and time (min) as shown in Figure 9.

Solution flow rate is an important parameter, which also influences the overall performance and design of the adsorption column, as the contact time between the adsorbate and adsorbent gets affected with change in flow rate, and in turn it affects the mass transfer parameters and column capacity. It was studied at constant influent concentration of 0.2 mg/ml and bed height of 10 cm by varying flow rate from 1 to 5 ml/min. Breakthrough curves are shown in Figure 10.

The effect of concentration on breakthrough curves has been depicted in Figure 11. A high concentration difference provides the high driving force for the adsorption process and this may explain why higher adsorption capacities were achieved in the column fed with higher o-nitrophenol concentration than that with a lower concentration. It is clear that adsorbent gets saturated early at high concentration (0.2 mg/ml) compared to low concentration (0.05 mg/ml).

DISCUSSION

FTIR Spectra

According to Shem and Xu (2006), the band near 3418

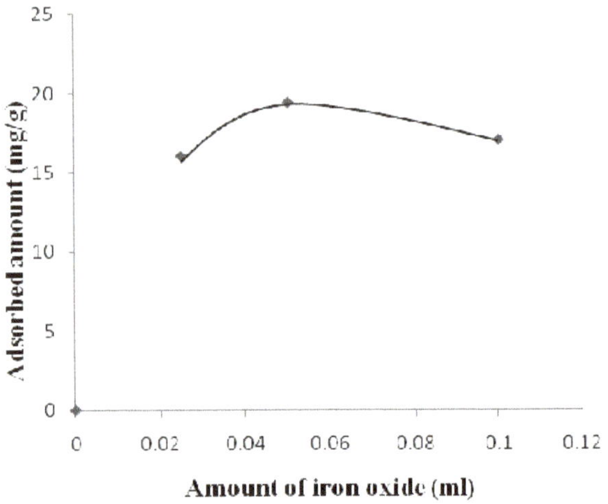

Figure 10. The effect of magnetic nanoparticles on the adsorption of o-nitrophenol, magnetic beads = 0.1 g, pH = 2, temp. = 30°C, time = 180 min, o-nitrophenol solution 10 ml of initial conc. 200 mg/L.

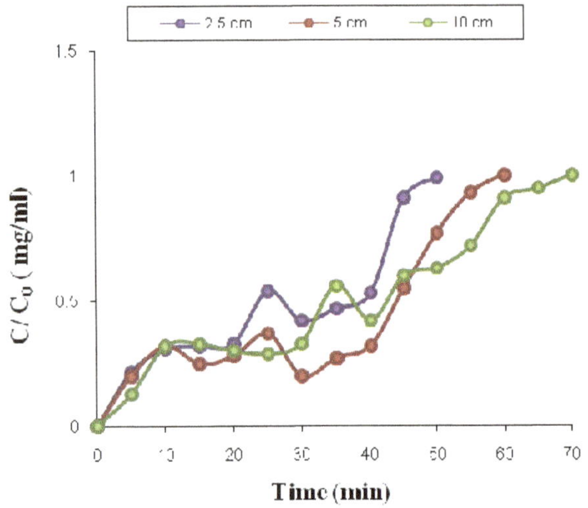

Figure 11. The effect of bed height on breakthrough curve at temp. = 30°C, pH = 2, bed height of magnetic beads = 2.5, 5 and 10 cm and flow rate = 1 ml/min and o-nitriphenol solution of initial conc. 200 mg/L.

cm^{-1} is due to –OH stretching. The band around 1607 cm^{-1} is assigned to the strong asymmetrical stretching vibration band (CO) in carboxyl group and the band near 1421 cm^{-1} is weak symmetrical stretching vibration band (C-O) in carboxyl group. The band at 1085 cm^{-1} is due to the C-O stretching of ether group and the band at 1029 cm^{-1} is attributed to the –C-O stretching of alcoholic group. There is an obvious absorption band at low frequency zone of 500 to 700 cm^{-1}, which is assigned to the stretching vibration of the Fe-O bond in iron oxide. A general explanation is that alginate is attached to iron oxide via the interaction between the carboxylic groups of alginate and the surface hydroxyl groups of iron oxide (Finotelli et al., 2004). By comparing the curves of magnetic beads before adsorption and after adsorption the following conclusions can be drawn. The –OH stretching band shifting from 3418 to 3422 cm^{-1} suggests the attachment of o-nitrophenol on the adsorbent surface. The shifting of CO stretching band to the higher frequency (from 1607 to 1741 cm^{-1}), may result from low electron density induced by the sorption of o-nitrophenol onto the adjacent hydroxyl group. The C-O band shifting to the higher frequency (from 1421 to 1629 cm^{-1}) can be attributed to the associations of the hydroxyl group with o-nitrophenol. A slight shift of Fe-O band from 563 to 571 cm^{-1} displays that some of the o-nitrophenol molecules are adsorbed onto the nano iron oxide surface.

SEM analysis

SEM is one of the most widely used surface diagnostic tools. The SEM micrographs of the nano iron oxide loaded alginate beads and o-nitrophenol adsorbed beads are shown in Figure 3a and b respectively. There are clear gaps which indicate that the surface of sorbent is rough with abundant protuberance, which might be the magnetite agglomeration. Figure 3b of o-nitrophenol adsorbed nano iron oxide loaded polymeric beads reveals the progressive changes and the presence of new uneven bulky particles over the surface which are absent before loading o-nitrophenol. Many pores are also present on the surface of adsorbent beads, which favor the adsorption of o-nitrophenol.

XRD analysis

In the XRD nature of alginate beads amorphous nature of alginate could not be observed due to the multivalent cations that is, Crosslinker, hence Figure 4a of cross linked alginate beads show low crystallinity and in Figure 4b the appearance of new sharp peak at (12.5) suggests the impregnation of iron oxide into alginate beads. The present crystallinity of iron oxide loaded alginate beads was calculated to be approximately (26.44%).

Adsorption isotherms

Adsorption isotherms are essential for the description of interaction of solutes with adsorbents, and are critical in optimizing the use of adsorbents. The equilibrium adsorption isotherms are one of the most important data to understand sorption mechanism. Several isotherm equations are available and two important isotherms

Table 1. Static and kinetic parameters for adsorption of o-nitrophenol onto magnetic beads of iron oxide and alginate.

Parameter	Value
Adsorption coefficient, K (g^{-1} dm^3)	65.7
Adsorption capacity, a_s (mg g^{-1})	2.924
Rate constant for adsorption, k_1 (S^{-1})	3.33×10^{-1}
Rate constant for desorption, k_2 (dm^{-3} S^{-1})	1.13×10^{-1}
Predicted sorption, K_f (mg g^{-1})	1.720
Freundlich constant, n	1.152
Separation factor, RL	0.829-0.048
Kad (mg g^{-1})	1.006
Kp (mg g S-$^{1/2}$)	1.781
β	1.003
A	0.555

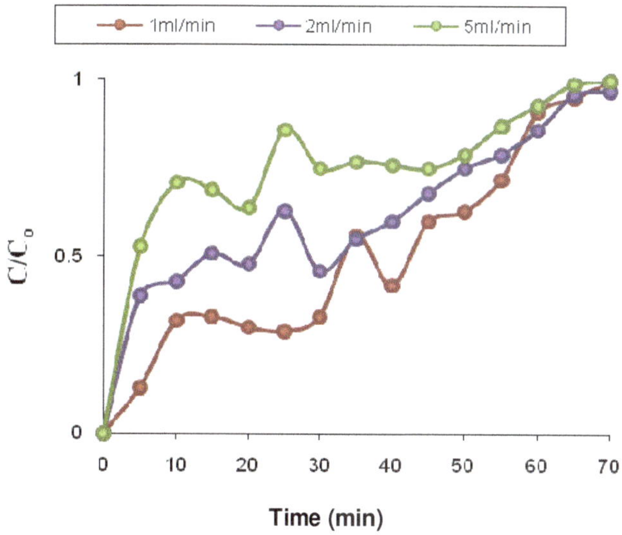

Figure 12. The effect of flow rate on breakthrough curve at temperature = 30°C, pH = 2, flow rate = 1, 2 and 5 ml/min, bed height = 10 cm and o-nitrophenol of initial conc. 200 mg/L.

selected for this study are the Langmuir and Freundlich models for the evaluation of experimental results.

Langmuir isotherm

The Langmuir isotherm is valid for monolayer adsorption on a surface containing a finite number of identical sites. The model assumes uniform adsorption on the surface and no transmigration on the plane of the surface. The linear form of the langmuir isotherm can be represented by the following equation.

$$\frac{C_e}{a} = \frac{1}{a_s} K + \frac{C_e}{a_s} \quad (8)$$

where C_e (mg/L) is the equilibrium concentration of the adsorbate, a (mg/g) is the amount of adsorbate adsorbed per unit mass of adsorbent. In the present study, a plot of C_e (equilibrium concentration) versus a (adsorbed amount of o-nitrophenol) resulted in a linear graphical relation indicating the application of the previous model. The values of Langmuir isotherm constants are given in Table 1. It has been found that the removal of o-nitrophenol increases with the increase in concentration of (0.012 to 0.3 mg/ml) and ultimately decreases showing the saturation in the adsorbed amount. The results are shown in Figure 12. The langmuir isotherm can be expressed by a separation factor (RL), which is given by the Equation (9).

$$RL = \frac{1}{1 + K\,C_0} \quad (9)$$

where C_0 is the initial o-nitrophenol concentration and K is the langmuir constant.

The RL value of in all the systems studied was observed between 0 and 1, indicating favorable adsorption of o-nitrophenol onto magnetic alginate beads.

Freundlich isotherm

The freundlich adsorption isotherm model is based on multilayer adsorption. The linear form of the freundlich isotherm model is given by the following relation (Rengaraj et al., 2002; Banat et al., 2000; Aksu and Yener, 2001; Khalid et al., 2000).

$$a = K_f\, C_e^{1/n}$$

$$\text{or} \quad \log a = K_f + (1/n) \log C_e \quad (10)$$

where a is the adsorbed amount of o-introphenol at equilibrium (mg/g), C_e is the equilibrium concentration of the adsorbate (mg/ml) and K_f and 1/n are the freundlich constants related to adsorption capacity and adsorption intensity respectively of the adsorbent. The higher value of k_f, shows easy uptake of o-nitrophenol from aqueous solution. 1/n is the measure of the nature and strength of the adsorption process and the distribution of active sites. If (1/n) < 1, the bond energies increase with the surface density, if (1/n)>1, the bond energies decrease with the surface density and when n = 0 all surface sites are equivalent. Alternatively, it has been shown using mathematical calculation that value of n between 1 and

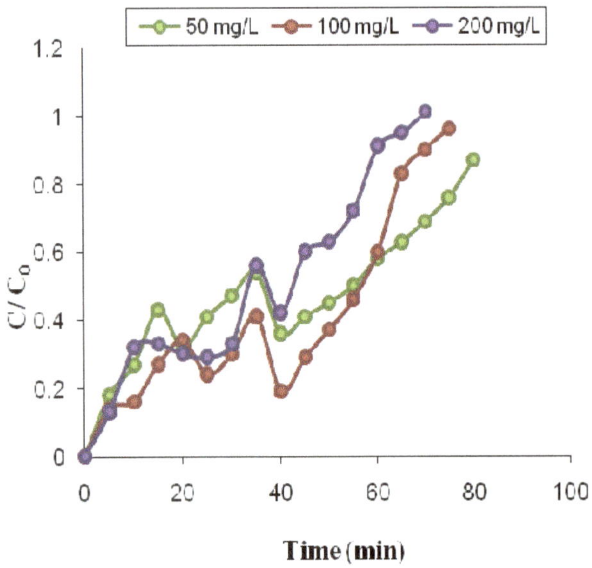

Figure 13. The effect of inlet adsorbate concentration on breakthrough curve at temp. = 30°C, pH = 2, initial concentration = 50, 100 and 200 mg/L, bed height = 10 cm and flow rate = 1 ml/min.

10, represent beneficial adsorption. The values of freundlich constants are given in Table 1.

The higher regression values showed that the equilibrium data for o-nitrophenol fitted well to both Langmuir and Freundlich isotherms in the studied concentration ranges. Based on the correlation coefficients (R^2), the equilibrium data was slightly better fitted in the freundlich isotherm than the Langmuir isotherm.

Modified freundlich isotherm

Freundlich isotherm may be reduced to the modified freundlich isotherm as given:

$$\log \frac{a}{(a_s - a)} = \beta \log C_e + \log \frac{A}{a_s} \qquad (11)$$

where β and A are empirical constants which can be determined through linear regression analysis. The values of β and A were calculated and are given in Table 1.

Adsorption kinetics

In the adsorption system contact time plays a vital role, irrespective of the other experimental parameters that affect the adsorption kinetics. The effect of contact time on the removal of o-nitrophenol is presented in Figure 13. It was found that the removal of o-nitrophenol increases with the increase in contact time to some extent, then reaches up to the maximum value and then decreases, which may be due to the desorption process. The relative increase in the extent of removal of o-nitrophenol after 4 h is negligible and hence it is fixed as the optimum contact time. The kinetics parameter, which is helpful for the prediction of adsorption rate, gives important information for designing and modeling the process. The kinetics scheme is given as follows:

$$\frac{1}{C} = k_1 \frac{t}{C_0} + \frac{1}{C_0} \qquad (12)$$

where C and C_o being the concentrations of o-nitrophenol at any time t and at zero time respectively (that is, initial concentration) and k_1 is the rate constant for adsorption. The value of k_1 was calculated, as summarized in Table 1. Once a_s and k_1 are the known, the value of k_2 (rate constant for desorption) may also be calculated, which is given in Table 1.

The lagergreen model assumes a first-order kinetics rate based on surface reaction (Mondal and Lalvani, 2000). It was found that the model represented the experimental data reasonably well as could be confirmed from the R^2 values. The lagergreen rate equation is given as:

$$\text{Log } (a_s - a) = \log a_s - (K_{ad}/2.303)\, t \qquad (13)$$

where k_{ad} is the lagergreen rate constant for adsorption, given in Table 1.

Intraparticle diffusion

In adsorption study it is necessary to determine the rate limiting step. Hence, the results obtained from the previous experiment were used in the adsorption process. The rate constant for intraparitcle diffusion (kp) is calculated by the following equation:

$$a = kp.t^{1/2} \qquad (14)$$

where a is the amount of o-nitrophenol adsorbed (mg/g) at time (t) and kp (mg/g $min^{0.5}$) is the rate constant for intraparticle diffusion. Value of kp was calculated as shown in Table 1. The kp value was obtained from the slope of the linear portion of the curve. The initial curve portion is attributed to the boundary layer diffusion, while the linear portion is responsible for the intraparticle diffusion.

Table 2. Results of breakthrough curve at different bed height for adsorption of o-nitrophenol on magnetic beads.

Bed height (cm)	Flow rate (ml/min)	t Total (min)	m Total (mg)	q Total (mg)	Total removal (%)	q_{eq} (mg/g)
2.5	1	50	11.25	5.25	46.6	3.5
5	1	60	13.5	6	44.4	2
10	1	70	15.75	9.8	62.2	1.63

Effect of pH

The pH value is an important parameter in adsorption taking place on the surface of adsorbent. The o-nitrophenol is an ionized organic pollutant which makes much difference in sorption ability of adsorbent beads. Unionized molecule is more easily adsorbed on the surface of polymeric beads, as it has higher hydrophobicity than ionized molecule. Acidity of medium can affect its sorption behavior through altering its existing form. At more acidic pH values, the adsorbent surface is positively charged, while at pH values above 6.5, it is negatively charged. Therefore, the pH value will have a significant effect on the adsorption desorption properties at the adsorbent surface.

Effect of adsorbent dosage

Increase in the amount of adsorbent, the site for o-nitrophenol adsorption increases. The maximum removal of o-nitrophenol (96%) from the aqueous solution could be achieved at 0.1 g adsorbent dose, which on further increase in adsorbent dose, the concentration of o-nitrophenol in the solution did not show any marked variation.

Effect of initial concentration

The increase of o-nitrophenol concentration, the amount of adsorption also increases. It means that the adsorption is highly dependent on the initial concentration of adsorbate. This is because at lower concentration, the ratio of the initial number of o-nitrophenol molecules to the available surface area is low; subsequently the fractional adsorption becomes independent on the initial concentration. However, at higher concentrations the available sites for adsorption becomes fewer and hence, the percentage removal of o-nitrophenol is noticed to be dependent upon the initial concentration.

Effect of temperature

The observed increase in adsorption may be attributed to the fact that by increasing temperature, greater number of active sites may be generated on the magnetic beads because of enhanced rates of protonation / deprotonation of functional groups on beads. This will clearly bring about an increase in the adsorption of o-nitrophenol. The increased adsorption at higher temperatures also suggests for the possibility of formation of some co- ordinate type of bonds between o-nitrophenol and electron rich donor atoms of adsorbent beads. The observed decrease in the adsorption at much higher temperature could be attributed to the weaking of the binding forces between o-nitrophenol and active sites on the magnetic beads.

Effect of beads composition

The adsorption of o-nitrophenol increases with the increase in nano iron oxide concentration up to 0.1 ml and then further increase in iron concentration results a decrease in the adsorption. The increased adsorption in initial range (0.05 to 0.1 ml) may be due to the increase in the number of active sites on beads while beyond 0.1 ml nano iron concentration the decrease in adsorption was observed which may be due to the reason that at higher concentration, beads become morphologically compact and consequently the diffusion of o-nitrophenol may be hindered.

Effect of bed height

The results indicate that the breakthrough time increases with increasing bed height, due to the availability of more number of binding sites (that is, the total surface area increases) (Zulfadhly et al., 2011; Vijayaraghavan et al., 2004). The o-nitrophenol adsorption capacity was also increased with bed height and was found to be maximum at 10 cm, which was selected as the optimized bed height for further experiments. The results showing the effects of bed height on the adsorption capacity of o-nitrophenol are shown in Table 2. In the continuous flow method the probability of contact between the adsorbate and the adsorbent is less when compared to batch mode, which ultimately results in lesser equilibrium sorption capacity in column mode.

Table 3. Results of breakthrough curve at different flow rate for adsorption of o-nitrophenol on magnetic beads.

Bed height (cm)	Flow rate (ml/min)	t Total (min)	m Total (mg)	q Total (mg)	Total removal (%)	q_{eq} (mg/g)
10	1	70	15.57	9.8	62.2	1.63
10	2	42	18.9	10.08	53.3	1.8
10	5	14	15.75	3.85	24.4	0.64

Table 4. Results of breakthrough curve at different inlet adsorbate concentration for adsorption of o-nitrophenol on magnetic beads.

O nitrophenol conc. (mg/ml)	Bed height (cm)	Flow rate (ml/min)	t Total (min)	m Total (mg)	q Total (mg)	Total removal (%)	q_{eq} (mg/g)
0.05	10	1	85	2.75	1.5	54.5	0.25
0.1	10	1	80	9.3	5.7	61.2	0.95
0.2	10	1	70	15.75	9.8	62.2	1.63

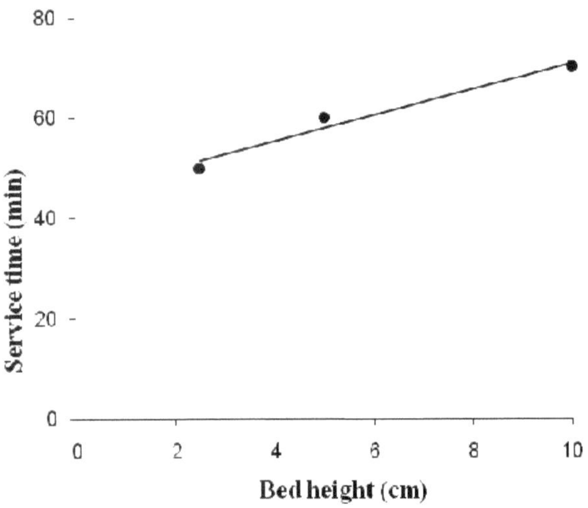

Figure 14. BDST model for o-nitrophenol adsorption by magnetic beads, flow rate = 1 ml/min, initial o-nitrophenol concentration 200 mg/L.

Effect of flow rate

Breakthrough curves indicating clearly that there is rapid uptake of o-nitrophenol in the initial stages and rate decreases thereafter and finally reaches saturation. The breakthrough time decreases when flow rate increase from 1 to 5 ml/min, and the breakthrough curve becomes flatter at higher flow rate. This is due to decreased contact time between the adsorbate and the adsorbent at higher flow rate. The effects of flow rate on the adsorption capacity of o-nitrophenol are shown in Table 3. The effect of flow rate is helpful for the large-scale treatment systems in order to utilize the bed for its maximum capacity with minimal flow rate (Aksu et al., 2007).

Effect of concentration

The breakthrough time decreased with increase in initial o-nitrophenol concentration. This is because the adsorbent near the top of the column comes into contact with o-nitrophenol before adsorbent near the bottom of the column is saturated. So at high concentrations the adsorbent becomes more quickly saturated, thereby decreasing the breakthrough time. The major effect is the increase in adsorption capacity with increasing o-nitrophenol concentration as shown in Table 4.

Modeling of column adsorption by the bed depth service time (BDST) model

BDST is a simple model which states that bed height (Z) and service time (t) of a column bears a linear relationship. Service time is referring to the time utilized by the adsorption column to reach the breakthrough point with specific saturation percentage, and it was related to adsorption operating conditions:

$$t = N0Z/C_0Q - 1/K_aC_0 \ln (C_0/C_t - 1) \quad (15)$$

where N_0 is the sorption capacity of bed (mg/L), Q is flow rate (ml/min) and K_a is rate constant (L/mg/min).

The plot of service time against bed height at a flow rate of 1ml/min as shown in Figure 14 was linear (R^2 = 0.964) indicating the validity of BDST model for the present system. N_0 (the sorption capacity of the bed per unit bed volume) was calculated from the slope of BDST plot, assuming initial concentration (C_0), and flow rate (Q) as constant during the column adsorption. The rate constant (K_a), calculated from the intercept of BDST plot,

Table 5. Calculated constants of Bed-Depth Service Time equation for adsorption of o-nitrophenol on magnetic beads.

Adsorbent	N_0	k_a	R^2
Iron oxide loaded alginate beads	578.4	1.18	0.964

characterizes the rate of solute transfer from the fluid phase to the solid phase (Ko et al., 2000).

The calculated N_0 and K_a values were 578.4 mg/L and 1.18 L/mg/min respectively. If the value of K_a is large then even a short bed will avoid breakthrough but if K_a, decreases a progressively longer bed is required to avoid breakthrough. It is shown in Table 5.

Thus, the BDST model parameters can be useful to scale up the process for other flow rates without further experiment run.

Conclusion

In this study, nano iron oxide loaded calcium alginate beads were prepared to study the removal of o-nitrophenol from aqueous solution by batch and column flow techniques. This study identified magnetic alginate beads as an appropriate adsorbent to be utilized for continuous removal of o-nitrophenol from aqueous solution. The FTIR analysis confirmed that o-nitrophenol can be adsorbed onto the surface of magnetic beads through carboxyl and hydroxyl functional groups. The SEM image reveals the rough and heterogeneous, relatively spherical shape and narrow size distribution of the iron oxide in alginate. In the XRD studies, percentage (%) crystallinity results indicate the amorphous nature of alginate beads and crystalline nature of iron oxide loaded alginate beads. The pH studies indicate that the optimum o-nitrophenol sorption can be achieved at pH 2 and the maximum removal (96%) was obtained in 4 h. A series of column experiments revealed that the continuous adsorption system represented by the breakthrough curves was dependent on the initial o-nitrophenol concentration, bed height and the solution flow rate. The BDST model constants were determined and proposed for the use in column design. Consequently, the experimental results suggest that, as a kind of potential adsorbent, the nano iron oxide loaded alginate beads can be used for efficient removal of o-nitrophenol from wastewater using the technology of magnetic separation. Higher adsorption capacity was obtained in batch experiments than in column experiments.

ACKNOWLEDGMENTS

The authors are thankful to UGC-DAE, Consortium for Scientific Research, Indore, for FTIR, XRD and SEM analysis.

REFERENCES

Aksu Z, Cagatay SS, Ganen F (2007). Cantinuous fixed bed biosorption of reactive dyes dried Rhizopus arrhizus: Determination of column capacity. J. Hazard. Mater 143: 362-371.

Aksu Z, Gonen F (2004). Biosorption of phenol by immobilized activated sludge in a continuous packed bed: Prediction of breakthrough curve. Process Biochem. 39: 599-613.

Aksu Z, Yener J (2001). A comparative adsorption/biosorption study of mono-chlorinated phenols onto various sorbent. Waste Manage. 21: 695-702.

Banat AF, Al-Bashir, Bal-Asheh S, Hayajineh O (2000). Adsorption opf phenol by Bintonit. Environ. Pollut. 107: 391-398.

Chang YC, Chen DH (2006). Recovery of gold(III) ions by a chitosan-coated magnetic nano adsorbent. Gold Bull. 39: 98-102.

Chen C, Wang JL (2007). Influence of metal ionic characteristics on their biosorption capacity by saccharomyces cerevisiae. Appl. Microbiol. Biotechnol. 74(4): 911-917.

Finotelli PC, Morales MA, Russi AM (2004). Magnetic studies of iron (III) nanoparticles in alginate polymer for drug delivery application. Mater. Sci. Eng. 24: 624-629.

Goncharuk VV, Kucheruk DD, Kochkodan VM, Badekha VP (2002). Removal of organic substances from aqueous solution by reagent enhanced reverse osmosis. Desalination 143: 45-51.

Gupta R, Bajpai AK (2010). Magnetically guided release of ciprofloxacin from super paramagnetic polymer nanocomposites. J. Biomater. Sci. 1-26.

Juny MW, Ahn KH, Lee Y (2001). Adsorption characteristics of phenol and chlorophenols on granular activated carbons (GAC). Microchem. J. 70: 123-131.

Khalid NS, Ahmad A, Ahmad J (2000). Potential of rice husks for antimony removal. Appl. Radiat. Isotopes. 52: 30-38.

Ko DCK, Porter JF, Mckay G (2000). Optomized correlation for the fixed bed adsorption of metal ions on bone char. Chem. Eng. Sci. 55: 5819-5829.

Mondal K, Lalvani SB (2000). Modeling of mass transfer controlled adsorption rate based on Langmuir adsorption isotherm. Sep. Sci. Technol. 35: 2583-2599.

Ngah IW, Hwang KV, Jeon C (2006). Removal of Pb ion from water by magneticticlly modified Zeclite. Miner. Eng. 19: 1452-1455.

Oliveira LCA, Fabris A, Sapag JD (2003). Clay-iron oxide magnetic composites for adsorption of contaminants in water. Appl. Clay. Sci. 22: 169-177.

Oliveira LCA, Petkowicz D, Smaniotto IA (2004). Magnetic zeolites: a new adsorption for removal of metallic contaminants from water. Water Res. 38: 3699-3704.

Pandey A, Shukla D, Ray L (2007). Studies on Cr(VI), Pb(II) and Cu(II) adsorption/desorption using calcium alginate as biopolymer. Chem. Sep. Bioavailab. 19(1): 17-24.

Pathak TS, Se-Jony (2009). Effect of cross-linker and cross-linker concentration on porosity, surface morphology and thermal behavior of metal alginates prepared from algae. Carbohydr. Polym. 78: 717-724.

Rengaraj S, Moon SH, Sivabalan R (2002). Agricultural solid waste for the removal of organics: adsorption of phenol from water and waste water by palm seed coat activated carbon. Waste Manage. 22: 543-548.

Roostaei N, Tezel FH (2004). Removal of phenol from aqueous solutions by adsorption. J. Environ. Manage. 70(2): 157-164.

Shem H, Xu JR (2006). The colloidal stability and core shell structure of magnetic nanoparticles coated with alginate. Appl. Surf. Sci. 253: 2158-2164.

Vijayaraghavan K, Jegan J, Palanivelu K (2004). Removal of nickel (II) ions from aqueous solution using crab shell particles in a packed bed up- flow column. J. Hazard. Mater. 113B(1-3): 223-230.

Yaneva Z, Koumanova B (2006). Comparative modeling of mono and dinitrophenols sorption on yellow bentonite from aqueous solutions. J. Colloid Interface Sci. 293: 303-311.

Zhany WX (2003). Nanoscale iron particles for environmental remediation: An overview. J. Nanopart. Res. 5: 323-332.

Zouboulis AI, Katsoyiannis IA (2002). Arsenic removal using Iron Oxide Loaded Alginate Beads. Ind. Eng. Chem. Res. 41: 6149-6155.

Zulfadhly Z, Mashitah MD, Bhatia S (2011). Heavy metals removals in fixed bed column by the macrofungus pycnoporus sanguineus. Environ. Pollut. 112(3): 463-470.

Comparing the conducting properties of benzobispyrrole-derived and benzobisthiazoles-derived squaraines

I. A. ADEJORO and Adebayo A. ADENIYI*

Department of Chemistry, University of Ibadan, Ibadan, Nigeria.

Smaller or vanishing value of semiconductor energy band-gap (Eg) is a good factor that determines the conductivity of semiconductor polymers and it has high correlation with the intrinsic electronic, optical and magnetic properties of materials. Squaraine dyes because of their unique optical properties have been shown to be an ideal candidate as one of the building blocks organic dyes with physical property of low HOMO (Highest occupied molecular orbital)-LUMO (lowest unoccupied molecular orbital) separation. This research dealt with the correlation of the change in theoretical Eg values computed for different models using BLYP, SVWN, B3LYP and ROHF functional and the excited properties with the available experimental Eg. The study showed that increasing the strength of an electron releasing group or use of stronger electronic donor that is bonded to squaric unit led to increasing HOMO width and consequently reduced Eg and Increased UV absorption maximal.

Key words: HOMO (highest occupied molecular orbital), LUMO (lowest unoccupied molecular orbital), band gap, UV absorption, DFT (density functional theory) and HF (Hybrid functional) methods, excited state property.

INTRODUCTION

Polysquaraines are defined as the organic dyes of squaraines and a class of π-conjugated polymers which can be synthesized by condensing 3,4-dihydroxy-3-cyclobutene 1,2-dione (squaric acid) with electron-rich aromatic compounds through a nucleophilic attack on the electron deficient squaric acid {Maahs and Hegenberg, 1966; Ziegenbein, and Sprenger, 1966) The advantages of the Polysquaraines were shown in terms of their favourable optical properties (Arunkumar, et al., 2006; Jyothish, et al., 2007) and extremely large flexibility for synthetic manipulation and their suitability for the design of conducting polymers with low optical Eg (Ajayaghost, 2005; Lu, et al., 2004). They are classed as organic dyes with very intense fluorescence spectra, typically in the red and near infra-red region with absorption maxima between 630 and 670 nm and their emission maxima between 650 to 700 nm. Their major characteristic feature is the unique aromatic four member ring system derived from squaric acid and what leads to the increasing success of designing varieties of symmetrical and unsymmetrical squaraines with favourable optical properties is the possibility of synthesizing variety of electron-rich aromatic and heterocyclic systems which react with squaric acid (Ajayaghost, 2005)

Polysquaraines have several technological applications (Jyothish et al., 2007; Ajayaghost, 2005; Paterson et al., 2002) as in photodynamic therapeutics, organic solar cells, xerographic sensitizers, optical recording media, photoreceptor, photolabeling, DNA sequencing and laser printing due to their suitable optical properties. It was further revealed that such organic dyes are extensively used as the signaling units in chemosensor design which are integrated to a chromophore, for the specific detection of analytes in different fields such as chemistry, biology, medicine, and environmental studies (Ajayaghost, 2005). This application is traceable to their intense absorption and emission properties that are sensitive to external inputs (Bissellet al., 1992; de Silva et al., 1997) and their peculiar zwitterionic structure that can be perturbed with the polarity of the medium, temperature, pH, and other additives (Ajayaghost, 2005). These polysquaraines are also revealed to photosensitize large band gap semiconductors such as TiO_2 through a

*Corresponding author. E-mail: a.aadeniyi@yahoo.com

charge-injection process from a singlet excited state (Paterson et al., 2002). The mechanism of photo-sensitization is based on the fact that squaraines dyes are one of the rare π-conjugated polymers that absorb in the near-IR wavelength (Cho et al., 2002; Tsuda and Osuka, 2001), therefore as sensitizers they are excited at a longer wavelength than that of the semiconductor (TiO_2) band gap which resulted to charge transfer from the excited dyes to the semiconductor. The extension of conjugation in squaraine dyes through vinylaromatic end groups imparts interesting optical and redox properties (Buschel et al. (2003).

The difficulty of the early attempts of synthesizing polysquaraines has been traced to the insoluble and intractable nature of the resulted materials ((Chen and Hall, 1986; Treibs and Jacob 1965). The discovery of the condensation of benzobisthiazoles (BT) and benzobispyroles (BP) with squaric acid to form polysquaraines with low optical band gaps brings about a great progress in the synthesis of polysquaraines (Ajayaghosh, 2003b, c, 2005; Havinga, et al., 1992).

The structural modification through the control of HOMO-LUMO gap is very important in designing of low Eg polymers with intrinsic conductivity (conductivity without doping) yet without disregarding the important properties like solubility and processability (Ajayaghosh, 2003b). Therefore in this research the structures of the two forms of the squaraines dyes were modify by adding different chain of methyl to the N group of the models as shown in Figures 1 and 2 and the effect of this methylation on the intrinsic conductive properties were determined. Also the significant effect of the two electron-rich groups (that is, benzobisthiazoles and benzobispyroles) on the conductivity of the squaraines dyes was considered.

The trivial name given to Benzobisthiazoles derived Polysquaraines is BT1 when $R=CH_3$, $R'=CH_3$ and BT2 when $R=C_{12}H_{25}$, $R'=CH$.

The trivial name given to Benzobispyrrole derived Polysquaraines is BP1 when R=H, BP2 when $R=CH_3$, BP3 when $R=C_4H_9$, BP4 when $R=C_7H_{15}$, BP5 $R=C_{12}H_{25}$. In all these cases R=R'.

COMPUTATIONAL DETAILS

We built the Monomers as Donor-Acceptor (D-A) and D-A-D-A as Dimmers using Gabedit package (Allouche 2002-2007). The monomers were allow to bond to extra carbon outside the normal unit to represent the continuity of the structure. Also all the Gaussian input files were prepared using gabedit.

Gaussian 03 packages (Frisch et al., 2004) was used for all the calculations and all the thermodynamic properties were computed at Temperature of 298.15K and a pressure of 1 atm.

The optimization and single point calculations were computed using DFT functional BLYP and SVWN, hybrid Functional B3LYP and Hatree-Fock ROHF. A basis set of 6-311+g was used. The purpose of using many functions is to discover which one will best explain the trend of change in the experimentally available Eg values and also give relatively close values.

The band width of HOMO and LUMO of the models were computed through the population analysis. The band-gap (Eg) was obtained by subtracting the first orbital energy of the LUMO from the last orbital energy of HOMO.

$$Eg = E^{l}_{HOMO} - E^{f}_{LUMO}$$

Where E^{l}_{HOMO} is the last orbital energy of HOMO and E^{f}_{LUMO} is the first orbital energy of LUMO.

The excited state properties were computed with CIS functional and a bases set of 6-311+g to obtain the UV absorption for all the models. In this excited state properties calculation diffusion function was included because it has been proved to be essential in order to obtain a good result in the excited state calculations (Foresman, and Leen, 1996). For the larger models with a basis set more than 200, an equation that uses AO integrals and recomputed them as they are needed was used through the keyword CIS=Direct.

The purpose of considering the monomers and the dimers is to see if just D-A of the monomers will give a Eg and UV values that are close and good enough to explain the trend of change in the experimental Eg values in order to reduce the computational cost of using D-A-D or D-A-D-A. More so to see if the same trend of change can be reproduced when consider the dimers.

RESULTS

The schematic representation of the geometry for the five forms of Benzobispyrrole (BP1 to BP2) and the two forms of Benzobisthiazoles (BT1 and BT2) of squaraines dyes are shown in Figures 1 and 2 with numbering of the common atoms in the models.

The optimized geometries of the Monomers and the Dimers of the all the models of Squaraines were obtained using the *ab-initio* ROHF and DFT BLYP,SWVN and Hybrid Functional B3LYP. The common and the significant bond lengths for the BP1 and BT1 are shown together for better comparison. The bond length of BP1 are given without the parenthesis while that of BT1 are given in parenthesis as follows: $C_1=C_2$ 1.34 (1.34), C_8-C_{14} 1.41 (1.4), $C_{14}=C_{15}$ 1.37(1.38), $C_{16}=O_{21}$ 1.27(1.26), $C_{18}=O_{22}$ 1.25(1.25), $C_2=N_{13}$ 1.41(1.41), $C_{12}=N_{13}$ 1.39(1.4).

Considering also the Mulliken Charge Distribution of BP1 and BT1, the values for BP1 are given outside the parenthesis while that of BT1 are given in parenthesis. C_2 0.6(0.09), C_{12} 0.37(0.53), C_4 -0.09(-0.48), N_{13} -0.89(-0.76), O_{21} -0.49(-0.43), O_{22} -0.38(-0.4), C_{16} 0.15(0.17), C_{18} 0.15(0.13), C_{15} -0.15(-0.09), C_{17} -0.09(-0.09), C_8 0.45(0.06).

The electronic properties are in Table 1 showing the computed band-gap (Ev) of the monomers and the dimers using different functional and the experimental band-gap (Ev) extracted from the supplementary work of Ajayaghosh (Ajayaghosh, 2003b). The excited state properties are in Table 2 shows the calculated excitation energy (Ev), the oscillator strength and UV in nm for the Monomers only.

DISCUSSION

The bond lengths which are common features in the BP1

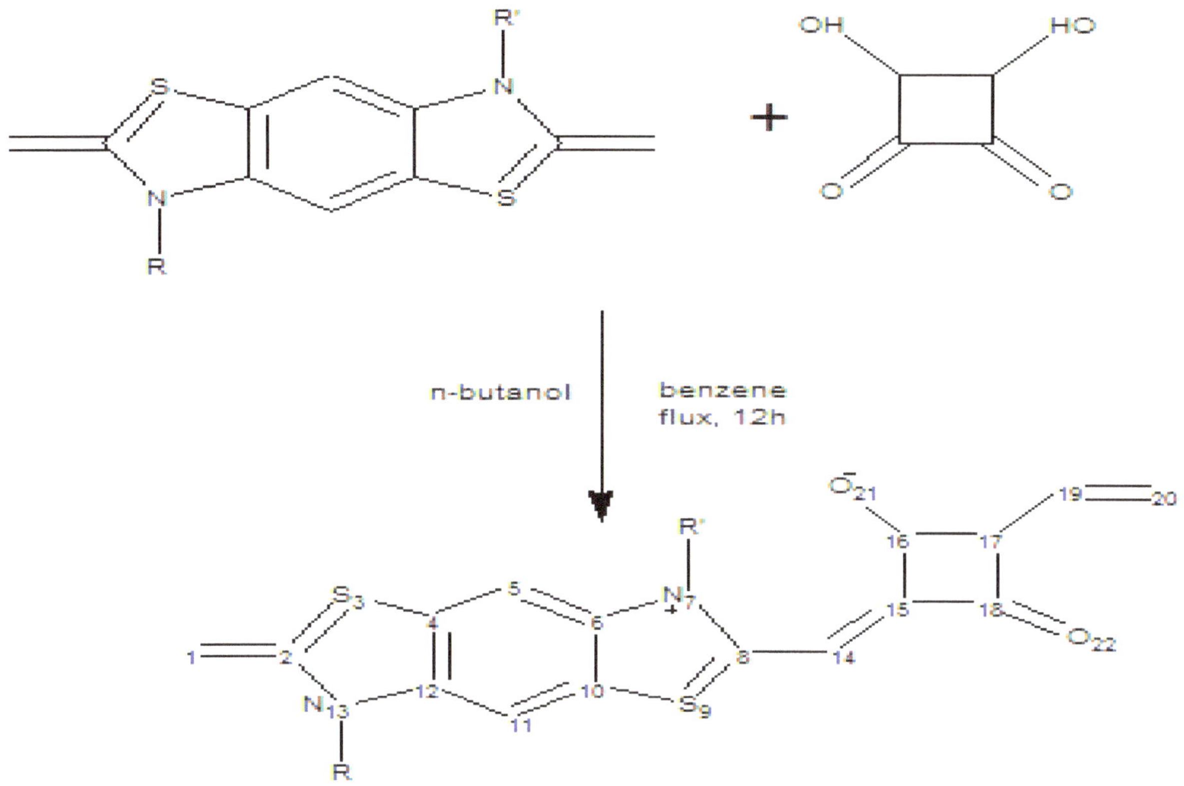

Figure 1. Benzobisthiazoles derived Polysquaraines showing the common parts of its two models BT1 and BT2 (Ajayaghosh, 2003).

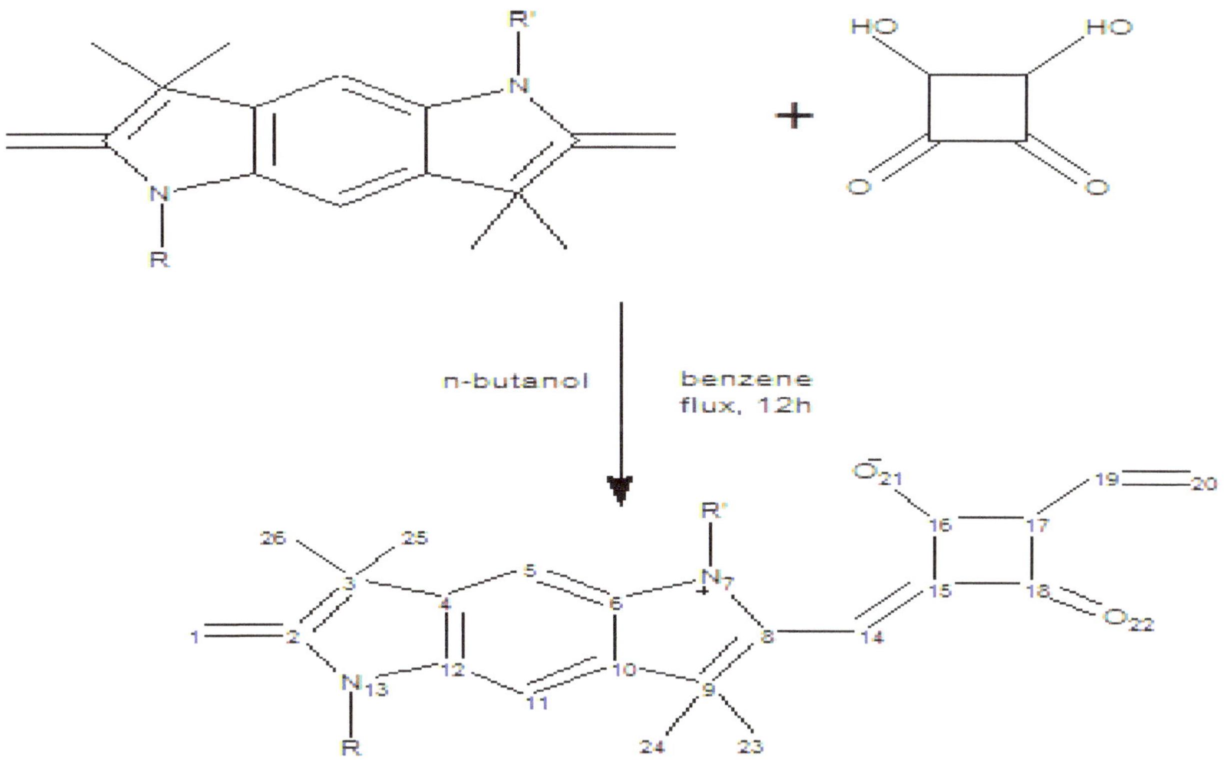

Figure 2. Benzobispyrrole derived Polysquaraines showing the common part of its five models BP1 to BP5 (Ajayaghosh, 2003).

Table 1. Energy band gap of various squaraines dyes considered.

Squaraines type	Functional / basis set	Computation band-gap (Ev)		Experimental band-gap (Ev)
		Monomer	dimer	
BP1	B3LYP/ 6-311G	2.35	1.28	1.15
	BLYP/ 6-311G	1.26	0.47	
	SWVN/ 6-311G	1.33		
	ROHF/ 6-311G	7.47	2.73	
BP2	B3LYP/ 6-311G	2.31	1.26	1.1
	BLYP/ 6-311G	1.25	0.46	
	SWVN/ 6-311G	1.32		
	ROHF/ 6-311G	7.44		
BP3	B3LYP/ 6-311G	2.29	1.24	1.2
	BLYP/ 6-311G	1.24	0.44	
	SWVN/ 6-311G	1.29		
	ROHF/ 6-311G	7.32	2.63	
BP4	B3LYP/ 6-311G	2.30	1.24	1.2
	BLYP/ 6-311G	1.25	0.44	
	SWVN/ 6-311G	1.31		
	ROHF/ 6-311G	7.39	2.64	
BP5	B3LYP/ 6-311G	2.30	1.24	1.15
	BLYP/ 6-311G	1.25	0.44	
	SWVN/ 6-311G	1.31		
	ROHF/ 6-311G	7.39	2.65	
BT1	B3LYP/ 6-311G	2.30	1.23	0.8
	BLYP/ 6-311G	1.27	0.47	
	SWVN/ 6-311G	1.36		
	ROHF/ 6-311G	7.25	2.44	
BT2	B3LYP/ 6-311G	2.31	1.27	0.8
	BLYP/ 6-311G	1.27	0.95	
	SWVN/ 6-311G	1.36		
	ROHF/ 6-311G	6.68	2.41	

The experimental energy band gap values are from supplementary work of Ajayaghosh (Ziegenbein, and Sprenger, 1966).

and BT1 were measured and the result showed that they are still relatively the same from one form of the model to another. However, in the terms of charge distribution there is a significant different in the value of the atomic Mulliken charges from BT1 to BP1 which can be explained as substituting lone pair atom S in BT1 with -$C(CH_3)_2$ to give BP1. The lone pair atom S resulted to a higher negative charge on adjacent carbon C_4 towards the squaraines unit. This effect is also obvious on carbon C_8 which shows that higher number of electrons was pushed towards the squaraines unit in BT1 than in BP1 which consequentially resulted to increase in intrinsic conductivity that is associated with decrease Eg values of BT forms compare to BP forms (Table 1).

There is a wide deviation of calculated Eg values of the monomers from the experimental Eg as shown in Table 1 which is an indication that D-A units of monomers is not sufficient to find correct Eg. Therefore, to obtain a better result in agreement with experimental results at least D-A-D unit is required. Dimers which are D-A-D-A units give calculated Eg values that are very closed to experimental Eg values. But the trend of the calculated Eg values obtained even for the monomers is very interesting and informative for it is correlated with trend observed in the experimental Eg values. Considering the band gap calculated for monomers and Dimers by B3LYP functional, band gap in BP forms gradually reduces as alkylation increases from BP1 to BP3. But from BP4 to

Table 2. Excited state properties of various monomers of squaraines dyes considered using functional/basis set of CIS/6-311+G.

Molecule	Symmetry	Calculated excitation energy (Ev)	Oscillator strength	UV (Monomer) (nm)
BP1	Cs	3.15	1.39	394.05
BP2	C1	3.11	1.57	398.47
BP3	C1	3.10	1.6	400.54
BP4	C1	3.12	1.58	397.65
BP5	C1	3.21	1.57	386.27
BT1	C1	3.06	1.63	405.68
BT2	C1	3.06	1.65	405.55

BP5 there is increase in Eg values which can be explained to be the prominent effect of steric hindrance. For all the computational methods, B3LYP functional gives a Eg values of the dimers that are very closed to experimental Eg values. The Eg calculated with ROHF functional as usual is larger than expected values because of its lack of correlation function. These *ab initio* Hartree Fock calculations, despite its overestimating the band gap values, correctly reproduced the trend of changes observed in the experimental values and calculated UV absorption. The DFT, BLYP and SVWN is about 40 to 50% lower than experimental value. This is understood since DFT band plot is a representation of the DFT Kohn-Sham energies. The Kohn-Sham electronic structure is not quasiparticle electronic structure of a system, and there is no Koopman theorem holding for Kohn-Sham energies, as there is for Hartree-Fock energies, which can be truly considered as an approximation for quasiparticle energies that can give the correct band energies of the system. This is why the band structure calculated by Hartree-Fock, ROHF, and hybrid functional, B3LYP which is a combination of Hartree-Fock and Density Functional theory give a better representation of the band energies.

The increase in the width of HOMO and LUMO through polymerization which consequentially resulted in lower Eg value is due to the increase in molecular orbitals. The calculated UV values for the monomers in Table 2 are lower than the expectation based on the available literature on squaraines ((Arunkumar et al., 2006; Jyothish, et al., 2007; Paterson et al., 2002; Ajayaghosh, 2003a; Law et al., 1986; Thomas et al., 1997) and the reason is that the form of the Monomers used which are just D-A unit is not enough to give the complete charge distribution in a unit structure of squaraines dyes. However, the trend of change in the UV values is very significant to describe the change in the intrinsic conductivity for a higher UV value is associated with lower Eg value.

The increase in the UV maximum absorption of the monomers from BP1 to BP3 is in good agreement with decreased calculated Eg values (Figure 3). The significant effect of steric hindrance which resulted in increased Eg values from BP4 to BP5 is also reflected through the reduction in the UV maximum absorption. A significant change in the UV absorption is notice by substituting pyrole unit in BP with thiazole unit to form BT.

The trend of UV absorption calculated adds meaning to the change in the values of band-gaps for different models of Squaraines. The squaraines models with the highest UV absorption is BT1 and BT2 which is consequentially characterized with lowest values of calculated and experimental Eg (Figure 3). The Eg calculated from B3LYP Functional shows that BT1 as the lowest band-gap even though it over estimated the band-gap of BT2. A better trend is obtained from the plot of band-gap of the Dimers calculated by ROHF Functional which gave relatively the same lowest band-gap to BT1 and BT2 which is line with UV absorption calculated and the experimental band-gap. This is the positive effect of introduction of lone pair atom, S, to replace -$C(CH_3)_2$. This can be further explained that the presence of lone pair atom S in BT form widens the HOMO width since it is an electron releasing group which is referred to as alpha effect.

As can be seen, the gradual drop in the UV absorption from BP4 to BP5 is also reflected from the calculated band-gap from ROHF functional which shows a gradual increase in band-gap from BP4 to BP5. This is a reflection of the steric hindrance constituted by larger substituted alkyl on their nitrogen atoms (Figure 2). This steric effect is even more pronounced from the experimental band-gap as it start increasing from BP3 and even though BP5 is a bit lower than BP4 it is still higher than BP2.

The inference from this explanation is that increasing alkylation which is an electron releasing group will widen the width of HOMO and thereby reduces the band-gap but at a certain level of alkylation, the steric hindrance will over rule the positive effect of alkylation and therefore results in increasing band-gap.

Conclusion

In conclusion, the calculated Eg values through ROHF best reproduced the trend of changes in experimental Eg values and the next to it is B3LYP. But B3LYP gave values that are much closer to the experimental values than the rest of the functionals while ROHF

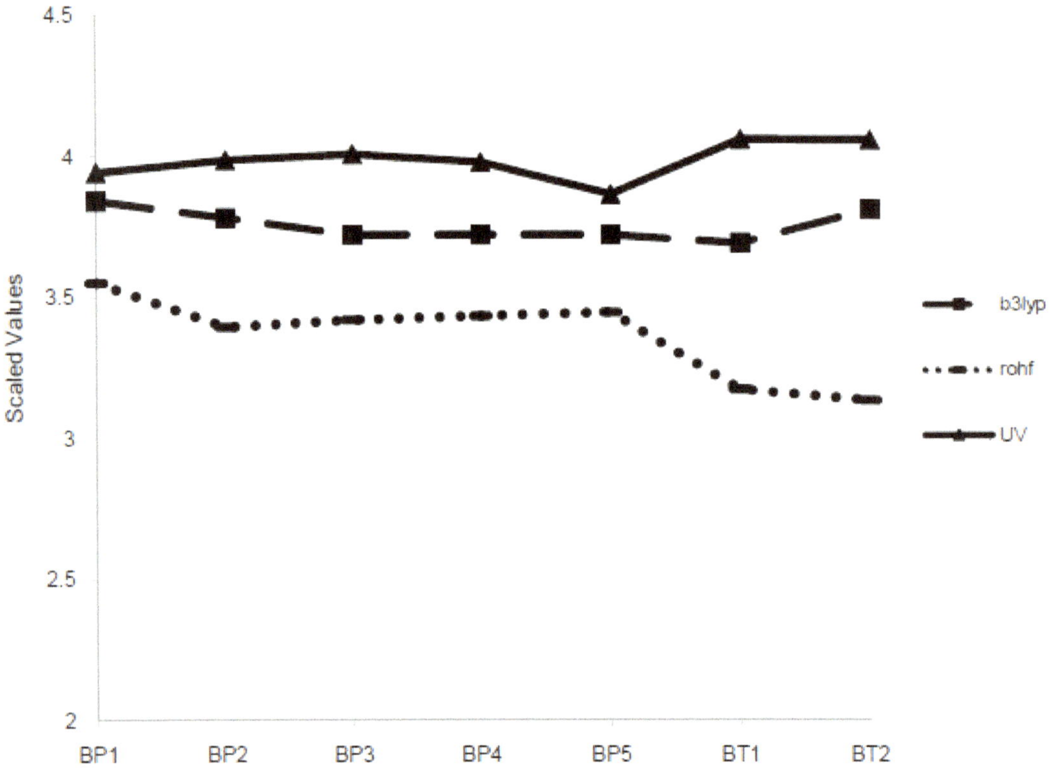

Figure 3. The plots to compare the trend of change in calculated band-gap by B3LYP (scaled by 3) and ROHF (scaled by 1.3) functionals on the dimers of all the models with calculated UV absorption (scaled by 1/100) by CIS functional of the monomers with the experimental band-gap

overestimated the values and BLYP and SVWN underestimated the Eg values. The substitution of pyrole unit in BP with thiazole unit to form BT significantly improved the intrinsic conductivity of the polymers and resulted to lower band-gap energy and increased UV absorption values. This is reasonable when taking into account the considerably greater nucleophilic nature of the thiazole ring as compared with that of the pyrole ring. However, the study shows that the quality of Squaraines dyes intrinsic conductivity can be improved by increasing the strength of electron releasing group that is bonded to the squaric unit and by avoiding introducing too big side chain molecules that can constitute a steric hindrance. Also, to obtain a calculated Eg values and excited state energy values that are closed to experimental value, there is need to consider at least D-A-D unit.

AKNOWLEDGEMENT

We like to acknowledge the PSMN Computational Unit of École Normale Supérieure de Lyon which provided us the computational facility to carry out this research and the professional effort of Associate Prof. Paul Fleurat-Lessard of Laboratoire de Chimie, École Normale Supérieure de Lyon, France, Professor Emmanuel I. Iwuoha of UWC, South Africa; who introduced me into the field of conjugated polymers.

REFERENCES

Ajayaghosh A (2003b). Supplementary material for chemical society. Re-views:http://www.rsc.org/suppdata/cs/b2/.
Ajayaghosh A (2003a). Donor-acceptor type low band gap polymers: polysquaraines and related systems. Chem. Soc. Rev., 32: 181-191.
Ajayaghost A (2005). Chemistry of squaraine-derived materials: Near-ir dyes, low band gap systems, and cation sensors. Acc. Chem. Res., 38: 449-459.
Allouche AR (Copyright 2002-2007). Gabedit 2.1.8. Graphical User Interface to GAMESS-US, Gaussian, Molcas, Molpro, MPQC, PCGamess and Q-Chem computational chemistry packages.
Arunkumar E, Fu N, Smith BD (2006). Squaraine-derived rotaxanes: Highly stable, fluorescent near-ir dyes. Chem. Eur. J., 12: 4684 4690.b204251g.
Bissell RA, de Silva AP, Gunaratne HQN, Lynch PLM, Maguire GEM, Sandanayake KRAS (1992). Molecular fluorescent signaling with fluor-spacer-receptor systems: Approaches to sensing and switching devices via supramolecular photophysics. Chem. Soc. Rev., 21: 187-195.
Buschel M, Ajayaghosh A, Arunkumar E, Daub J (2003). Redox-switchable squaraines with extended conjugation. Org. Lett., 5: 2957-2978.
Chen YY, Hall HKJ (1986). Polycondensation of squaric acid with n-alkylcarbazoles. Polym. Bull., 16: 419-425.
Cho HS, Jeong DH, Cho S, Kim D, Matsuzaki Y, Tanaka K, Tsuda A, Osuka A (2002). Photophysical properties of porphyrin tapes. J. Am. Chem. Soc., 124: 14642-14654.
De Silva AP, Gunaratne HQN, Gunnlaugsson T, Huxley AJM, McCoy CP, Rademacher JT, Rice TE (1997). Signaling recognition events

with fluorescent sensors and switches. Chem. Rev., 97(5): 1515-1566.

Foresman JB, leen Frisch (1996). Exploring Chemistry with Electronic Structure Methods, 2nd Edition. Gaussian, Inc.

Frisch MJ, Trucks GW, Schlegel HB, Scuseria GE, Robb MA, Cheeseman JR, Montgomery JA, Jr. TV, Kudin KN, Burant JC, Millam JM, Iyengar SS, Tomasi J, Barone V, Mennucci B, Cossi M, Scalmani G, Rega N, Petersson GA, Nakatsuji H, Hada M, Ehara M, Toyota K, Fukuda R, Hasegawa J, Ishida M, Nakajima T, Honda Y, Kitao O, Nakai H, Klene M, Li X, Knox JE, Hratchian HP, Cross JB, Bakken V, Adamo C, Jaramillo J, Gomperts R, Stratmann RE, Yazyev O, Austin AJ, Cammi R, Pomelli C, Ochterski JW, Ayala PY, Morokuma K, Voth GA, Salvador P, Dannenberg JJ, Zakrzewski VG, Dapprich S, Daniels AD, Strain MC, Farkas O, Malick DK, Rabuck AD, Raghavachari K, Foresman JB, Ortiz JV, Cui Q, Baboul AG, Clifford S, Cioslowski J, Stefanov BB, Liu G, Liashenko A, Piskorz P, Komaromi I, Martin RL, Fox DJ, Keith T, Al-Laham MA, Peng CY, Nanayakkara A, Challacombe M, Gill PMW, Johnson B, Chen W, Wong MW, Gonzalez C, Pople JA (2004). Gaussian 03, Revision E.01. Gaussian, Inc., Wallingford CT.

Havinga EE, Ten Hoeve W, Wynberg HA (1992). New class of small band gap organic polymer conductors. Polym. Bull., 29: 119-126.

Jyothish K, Avirah RR, Ramaiah D (2007). Development of squaraine dyes for photodynamic therapeutical applications: Synthesis and study of electronic factors in the dye formation reaction. ARKIVOC, 8: 296-310.

Law KY, Bailey FC, Bluett LJ (1986). Squaraine chemistry, on the anomalous mass spectra of bis(4-dimethylaminopheny1)squaraine and its derivatives. Can. J. Chem., 64: 1607-1619.

Lu HC, Whang WT, Cheng BM (2004). Thermal stability of polysquaraine studied by infrared spectroscopy. 14A1 BM - IR Microscopy.

Maahs G, Hegenberg P (1966). Syntheses and derivatives of squaric acid. Angew. Chem. Int. Ed. Engl., 5: 888-893.

Paterson MJ, Blancafort L, Wilsey S, Robb MA (2002). Photoinduced electron transfer in squaraine dyes: Sensitization of large band gap semiconductors. J. Phys. Chem., 106: 11431-11439.

Thomas KG, Thomas KJ, Das S, George MV (1997). A squaraine-based near-infrared absorbing sensor for the selective detection of transition and other metal ions in aqueous media. Chem. Commun., 6: 597-598.

Treibs A, Jacob K (1965). Cyclotrimethine dyes derived from squaric acid. Angew. Chem., Int. Ed. Engl., 4: 694-695.

Tsuda A, Osuka A (2001). Fully conjugated porphyrin tapes with electronic absorption bands that reach into infrared. Science, 293: 79-82.

Ziegenbein W, Sprenger HE (1966). Condensation products of squaric acid and azulenic hydrocarbons. Angew. Chem, Int. Ed. Engl., 5: 893-894.

Synthesis of benzothiazoyl salicylate moieties

Diepreye Ere* and Allen T. Ekubo

Department of Chemical Sciences, Niger Delta University, Wilberforce Island, Bayelsa State, Nigeria.

Aminosalicylates are derivatives of the highly functionalized compound aminosalicylic acid. Benzothiazoyl salicylates were obtained in this work from the highly functionalized aminosalicylic acid by reacting ammonium thioureas as a nucleophile to aminosalicylates that were derived from aminosalicylic acid and bromine was used as the cyclising agent. Results obtained indicated that it was the brominated cyclised product that was formed.

Key words: Aminosalicylates, aminosalicylic acid, benzothiazoles, thioureas.

INTRODUCTION

Benzothiazoles are heterocyles in which benzene ring is fused with thiazole ring. Benzothiazoles are usually known as precursors (Sum et al., 2003; Gupta and Rawat, 2010) to natural products, pharmaceutical agents and other compounds that exhibit a wide spectrum of biological activity. Literature search indicates that benzothiazoles are very important. Leong et al. (2003) reported that benzothiazoles represent a potent and highly selective class of antitumour agents. Para-aminosalicylic acid is an old agent used for the treatment of tuberculosis. It is envisaged that benzothiazoles derived from *para*-aminosalicylic acid moieties would be useful for multidrug resistant tuberculosis which is becoming prevalent by the day (Gandhi et al., 2006). Therefore "benzothiazoles salicylates" was synthesized which could also become biological active compounds targeted at multidrug drug resistant tuberculosis (MDRTB). Jacobson synthesis (Jacobson, 1903) of benzothiazoles, via oxidative cyclisation of an arylthioamide on an unsubstituted ortho position using potassium fericyanide in a basic medium is one of the established methods used for the synthesis of substituted benzothiazoles. Downer and Jackson (2004) attributed the success of preparing their benzothiazoles to the formation of the intermediate arylthioanilides (Scheme 1).

Jordan et al. (2003) amongst others, have also reported an efficient conversion of substituted aryl thioureas to 2-aminobenzothiazoles using either bromine or benzyl trimethyl ammonium tribromide. A variation on this theme has also been used for making 2-amino benzothiazoles directly in one pot using ammonium thiocyanate and aniline or *p*-toluidine to form the thiourea which was cyclised *in situ*.

Following the aforementioned methods, we have used para-aminosalicylic alkyl esters and reacted them with ammonium thiocyanate and benzyltrimethyl ammonium tribromide in dichloromethane to give benzothiazole products in very good yields (92 to 97%).

The aim of this work is to make benzothiazoles centered on para-aminoaminosalicylic acid that will be active against multi-drug resistant tuberculosis strands (Schemes 2 and 3).

MATERIALS AND METHODS

Reagents and reaction solvents

Reagents used in this study were without further purification unless otherwise stated.

Evaluation of materials

Proton nuclear magnetic resonance (^{1}H-NMR) spectroscopy. The structures of the products were confirmed by ^{1}H-NMR and ^{13}C-NMR spectroscopy using JOEL Lambda 400 spectrometer. An internal standard of trimethylsilane (TMS) was used. The following abbreviations are examples used to describe the splitting patterns: s-singlet, d-doublet, t-triplet, q-quartet, quint-quintet, dd-double-doublet, m-multiplet, b-broad.

1. Mass spectrometry (MS): Mass spectra were recorded using a Finnigan-MAT 1020 GC/MS and a QP505A Schimadzu GC/MS

*Corresponding author. E-mail: diepreyeere@yahoo.co.uk.

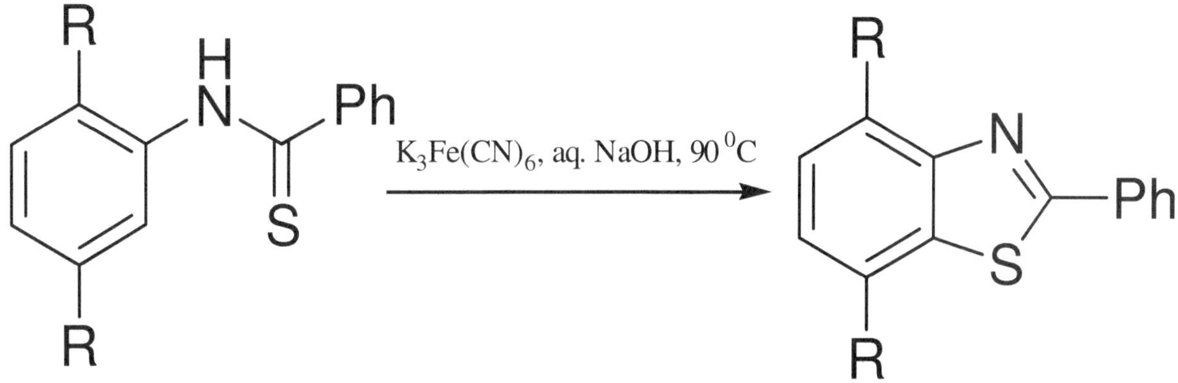

Scheme 1. General synthesis of benzothiazoles.

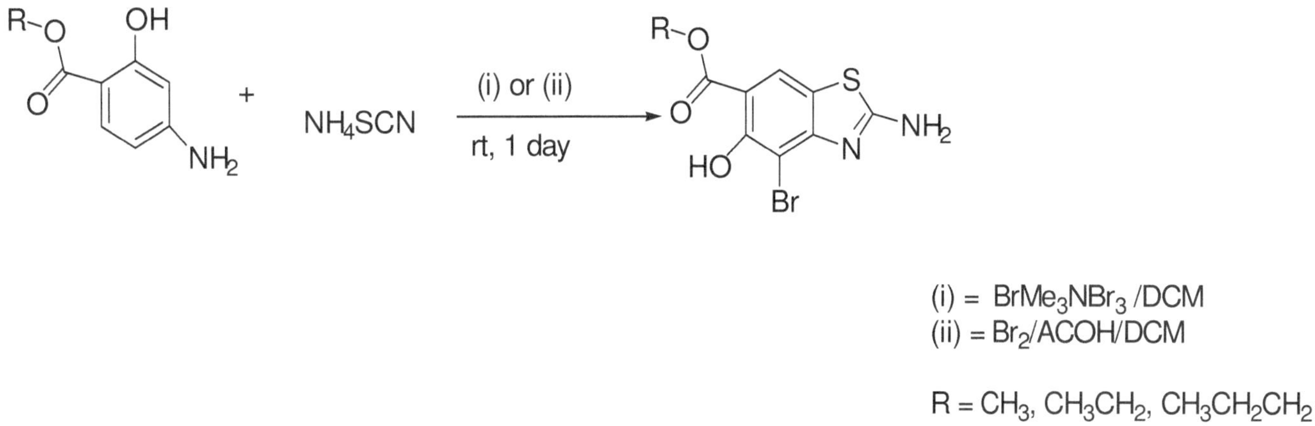

Scheme 2. Experimental synthesis of 2-amino-4-bromo-5-hydroxybenzothiazoles.

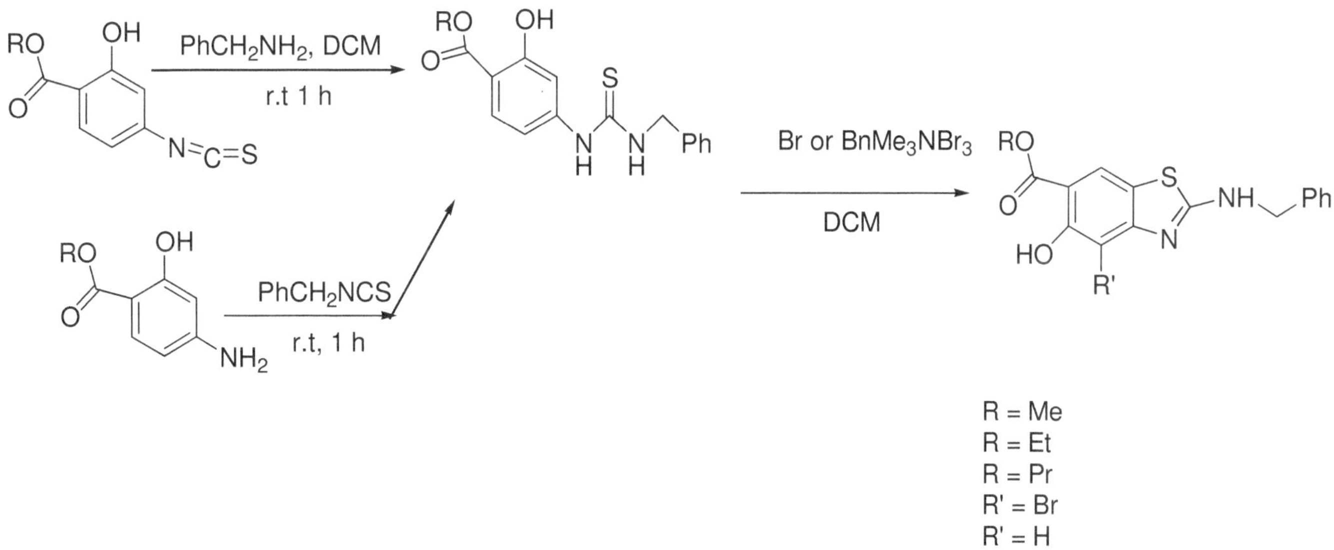

Scheme 3. Experimental synthesis of 2-benzylamino benzothiazoles.

spectrometer. Where present the mass ion of the material is identified by M^+.

2. Chromatography: The progress of reactions was frequently monitored by thin layer chromatography (TLC). Glass or aluminium backed TLC plates coated with silica gel (60 F254 Merck) were utilized.

Purification of intermediates and final products were mainly achieved by column chromatography using silica gel 60 (230-400 Mesh). The purity of compounds in this work was confirmed by ^1H and ^{13}C NMR spectroscopic and elemental analysis.

3. Melting points (mp) were measured in open capillaries for each synthesized compound using Gallenkamp melting point apparatus and were uncorrected.

Methodology

2-Amino-4-bromo-5-hydroxybenzothiazole-6-carboxylic acid alkyl esters (A-D)

To a stirred mixture of alkyl-4-aminosalicylate (0.06 mol) and ammonium thiocyanate (0.13 mol) in glacial acetic acid (90 ml) at 0℃ was added drop wise bromine (0.06 mol) in glacial acetic acid (20 ml) over 45 min such that the temperature of the reaction mixture did not exceed 5℃. The reaction mixture was then allowed to stir for a further 1 h in ice-bath. The precipitates that were formed were collected and dissolved in hot water (100 ml) and then saturated sodium hydrogen carbonate added to neutralize the mixture. The precipitate that was formed were collected by filtration, washed with water and then dried in vacuo over calcium chloride to yield the compound subsequently stated.

Synthesis of 2-Benzylamino-4-bromo-5-hdroxybenzothiazole-6-carboxylic acid alkyl esters (E)

To a stirred solution of alkyl-4-(3-benzylthioureido)-2-hydroxybenzoate (3.4 mmol) in dichloromethane (10 ml) was added 10 (ten) drops of bromine from a Pasteur pipette and the resulting mixture stirred at room temperature for 2 h. Dichloromethane (10 ml) was added and the suspension that was formed was neutralized by the drop wise addition of saturated solution of sodium hydrogen carbonate. Dichloromethane (3 × 10 ml) was used to extract the solute and washed with water (2 × 10 ml) and then dried over magnesium sulphate and then filtered. Solvent was then removed in vacuo to obtain the crude product. The product was then purified by recrystallisation from methanol to yield the recorded products.

RESULTS

2-Amino-4-bromo-5-hydroxybenzothiazole-6-carboxylic acid methyl ester (A)

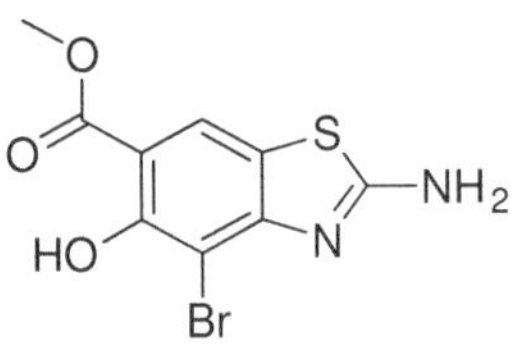

Yield (97%), mp 240 to 24℃, R_f = 0.9 (EtOAc); (Found: C, 35.86; H, 2.04; N, 9.27; S, 10.57% $C_9H_7N_2O_3SBr$ requires C, 35.66; H, 2.33; N, 9.24; S, 10.58 %); ^1H-NMR [d_6-DMSO, 400 MHz] δ_H 3.90(3H, s, OCH_3), 8.13 (1H, s, Ar-H), 8.36 (2H, bs, NH_2), 11.39(1H, s, Ar-OH); ^{13}C-NMR [d_6-DMSO, 100 MHz] δ_c 52.7, 106.1, 121.1, 121.7, 129.2, 156.2, 157.4, 169.6, 171.8, m/z 304/302 (M^+, 27 %), 272/270 (100), 244/242 (9), 135 (22), 93 (51).

2-Amino-4-bromo-5-hydroxybenzothiazole-6-carboxylic acid ethyl ester (B)

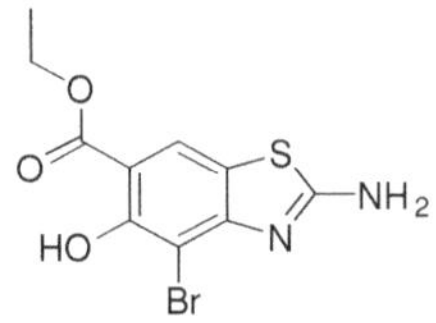

Yield (80%) mp 224 -226 ^oC, R_f = 0.7 (EtOAc); (Found: C, 37.85; H, 2.88; N, 8.83; S, 10.11 $C_{10}H_9BrN_2O_3S$ requires C, 37.87, H, 2.86; Br 25.19; N, 8.83; S, 10.11; O, 15.13); ^1H-NMR [d_6-DMSO, 400 MHz] δ_H 1.52 (3H, t, J =6, 2.4, 6 Hz -CH_3), 4.29 (2H, q, J = 2.4, 6, 6, 2.4 Hz = -CH_2-O-), 8.38 (2H, bs, NH_2), 8.60 (1H,s, Ar-H), 11.45 (1H, s, Ar-OH); ^{13}C-NMR [d_6 -DMSO, 100 MHz]δ_c 14.2, 59.1, 105.6, 117.2, 117.3, 125.2, 157.1, 158.4, 172.4, 166.9; m/z 317/315 (M^+, 32%).

2-Amino-4-bromo-5-hydroxybenzothiazole-6-carboxylic acid propyl ester (C)

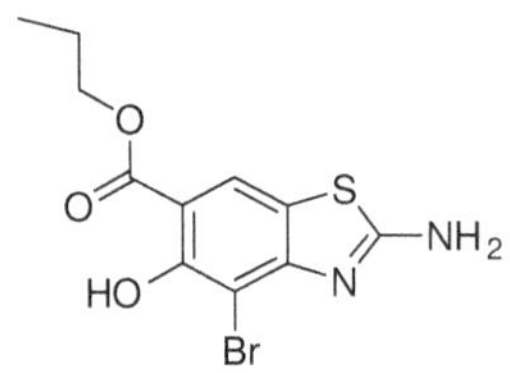

Yield (72%) mp 218 -219.5 ^0C, R_f = 0.75 (EtOAc); (Found: C = 39.80, H 3.32, N, 8.46, S 9.67, $C_{11}H_{11}BrN_2O_3S$ requires C 39.89, H 3.35, Br 24.13, N 8.46, O 14.49, S 9.68) ^1H-NMR [d_6-DMSO, 400 MHz] δ_H 0.98 (3H, t, J = 2.5, 0.6, 2.5 Hz, CH_3), 1.82 (2H, m, CH_2), 4.27 (2H, t, J = 2.5, 0.6, 2.5 Hz, CH_2), 8.27 (2H, bs, NH_2) 8.62 (1H, s, Ar-H), 11.40 (1H, s, Ar-OH); ^{13}C-NMR [d_6-DMSO, 100 MHz] δ_C 10.2, 22.7, 68.3,103.8, 117.2, 117.3, 124.5, 157.2, 158.4, 167.2, 174.8 m/z 331/329 (M^+, 43%)

2-Amino-4-bromo-5-hydroxybenzothiazole-6-carboxylic acid benzyl ester (D)

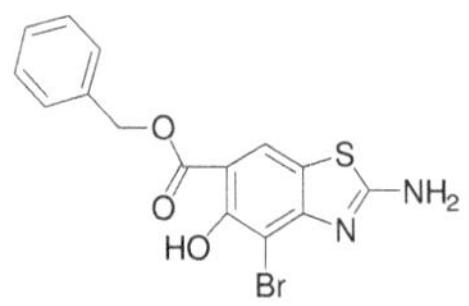

Yield (84%), mp 204 to 205°C, R_f = 0.58 (EtOAc); (Found: C 47.56, H 2.88, Br 21.07, N 7.39, S 8.44: $C_{15}H_{11}BrN_2O_3S$ requires: C 47.51, H 2.92, Br 21.07, N 7.39, O 12.66, S 8.46); ^{1}H-NMR [d_6-DMSO, 400 MHz] δ_H 5.52(2H, bs, NH_2), 10.22(1H, s, Ar-OH) 7.25 (4H, s, 4 x Ar-H), 8.60 (1H, S, Ar-H); ^{13}C-NMR [d_6-DMSO, 100 MHz] δc 71.8, 103.8, 117.2, 117.3, 124.3, 127.3 (2 X), 127.4, 128.7 (2 X), 140.9, 157.1, 158.4, 167.0, 174.5; m/z 379/377 (M^+, 47%).

2-Benzylamino-4-bromo-5-hydroxy-benzothiazole-6-carboxylic acid methyl ester (E)

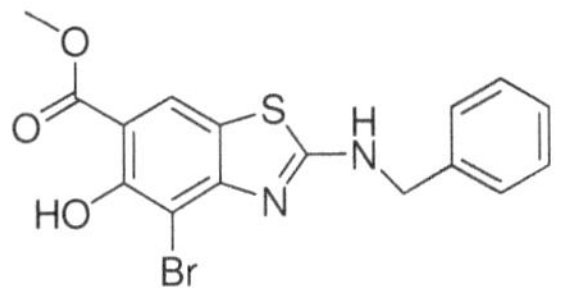

Yield (74%) mp 226 to 227°C, R_f = 0.62 (EtOAc); (Found: C 48.87, H 3.33, S 8.15, N 7.12, $C_{16}H_{13}BrN_2O_3S$ requires: C 48.87, H 3.33, Br 20.32, N 7.12, O 12.21, S 8.15), ^{1}HNMR [d_6-DMSO, 400 MHz]δ_H 3.87 (3H, s, OCH_3), 4.60 (2H, d, J – 5.5 Hz, CH_2-), 6.85 (1H, s, Ar-H), 7.32-7.35(5H, m, 5 x Ph-H), 8.09 (1H, s, Ar-H), 8.97 (1H, bt, J = 5.5 Hz, NH), 10.65 (1H, s, Ar-OH); ^{13}C-NMR [d_6-DMSO, 100 MHz] δ_c 49., 56.8, 103.6, 117.2, 117.3, 124.3, 126.5, 127.1, 128.3, 142.4, 157.1, 158.4, 167.0, 174.5, m/z 394 (M^+, 20%), 393 (17), 362 (17), 361 (14), 106 (17), 91 (100), 65 (21).

DISCUSSION

Para-aminosalicylic alkyl esters undergo nucleophilic addition with ammonium thiocyanate and benzyltrimethyl ammonium tribromide as the cylisation agent in dichloromethane to give benzothiazole products in very good yield (92 to 97%). However spectroscopic evidence showed that it was the bromine products rather than the earlier envisaged products that were formed.

Earlier one had envisaged that cylisation will occur para to the hydroxyl group, due to steric factors, but the products isolated were 2–amino-4-bromo-5-hydroxybenzothiazole-6-carboxylic acid alkyl esters (A-D) as shown in the Scheme 2. The possible explanation for this occurrence is that the electron rich nature of para-aminosalicylic acid (PAS) may have caused bromination to take place prior to cyclisation or more likely bromination pre- thiourea formation. Therefore, the position of the bromine atom could either be ortho or para to the hydroxyl group and the products are likely to be 2-amino-4-bromo-7-hydroxybenzothiazole-6-carboxylic acid alkyl esters. Bromination of para-aminosalicylic acid and N-acetyl para-aminosalicylic acid gave a mixture of isomers 4-amino-5-bromo-2-hydroxybenzoic acid and 4-amino-3-bromo-2-hydroxybenzoic acid as well as 4-acetylamino-5-bromo-2-hydroxybenzoic acid and 4-acetylamino-3-bromo-2-hydroxybenzoic acid. It was difficult to confirm which isomers of the bromobenzothiazoles that were formed. Whichever of the isomers formed would possess the same one aromatic hydrogen in ^{1}H-NMR. Due to the fact that the compounds are highly functionalized molecules, possessing many heteroatoms, neither the ^{1}H nor ^{13}C NMR chemical shifts would indicate which isomer is formed. Efforts to get crystalline material for X-ray crystallographic studies were unsuccessful; therefore the final confirmations for which isomers are formed are open to debate.

In order to obtain non-brominated benzothiazole carboxylic acid alkyl esters, several other routes were investigated. The reaction between the para-amino salicylic alkyl esters and ammonium thiocyanate were investigated in order to obtain the intermediate thioureas prior to cyclisation. The production of thioureas using this method was not very reliable as repeatability problem were encountered for all the alkyl esters. One method amongst several others that was tried in order to obtain intermediate thioureas was when para-aminosalicylic acid alkyl esters and 1,1'-thiocarbonyl di-2,2'-pyridone (TDP) were reacted smoothly to produce 2-hydroxy-4-isothiocyanato benzoic acid alkyl esters in good yield (73 to 93%). This method was reported by (Clarke and Boyle, 1999) and had been developed for the synthesis of isothiocyanatoporphrins from aminoporphyrins.

Once isothiocyanates were produced successfully efforts to synthesize the thiourea by addition of an amine was undertaken. Benzylamine was deliberately chosen because the benzyl group could be removed by hydrogenation at a later stage. The 2-hydroxy-4-isothiocyanatobenzoic acid alkyl esters underwent nucleophilic addition with benzylamine in dichloromethane at room temperature to form the cyclisation precursor, 2-hydroxy-4-(2-phenylaminothioacetylamino) benzoic acid alkyl esters in good yield (76 to 82 %). These thioureas were made in a “reverse way” using Para-amino salicylic alkyl ester and commercially available benzylisothiocyanate as the analogous products using methyl isothiocyanate.

Conclusion

Benzothiazole is an important class of heterocycles which are known to exhibit a wide range of biological properties. Several other methods have been reported for the synthesis and cyclisation of benzothiazoles; however this is the first method for the synthesis of highly functionalised benzothiazoles which may possibly become effective for the cure of tuberculosis.

REFERENCES

Clarke OJ, Boyle RW (1999) Isothiocyanatoporphyrins, useful intermediates for the conjugation of porphyrins with biomolecules and solid supports. Chem. Commun., pp. 2231-2232.

Downer NK, Jackson YA (2004). Synthesis of benzothiazoles via ipso substitution of ortho-methoxythiobenzamides. Org. Biomol. Chem., 20(2): 3039-3043.

Gandhi NR, Moll A, Sturm AW, Pawinski R, Govender T, Lollo U, Zeller K, Andrews J, Friedland G. (2006). Extensively drug-resistant tuberculosis as a cause of death in patients Co-infected with tuberculosis and HIV in a rural area of South Africa. The Lancet, 368: 1575-1580.

Gupta A, Rawat S. (2010). Synthesis and cyclisation of benzothiazoles: Review. J. curr. Pharm. Res., 3(1): 13-23.

Jacobson P (1903). Synthesis and cylisation of 2-substituted benzothiazoles Ber. Dtsch. Chem. Ges., 36: 3121 A.

Jordan D, Luo C, Reitz AB (2003). Efficient conversion of substituted aryl thioureas to 2-aminobenzothiazoles using benzyltrimethylammonium tribromide, J. Org. Chem., 68: 8693-8696.

Leong CO, Gaskell M, Martin EA, Heydon RT, Farmer PB, Bibby MC, Cooper PA, Double JA, Bradshaw TD, Stevens MFG (2003). Antitimour 2-(4-aminophenyl) benzothiazoles generate DNA adducts in sensitive tumour cells in vitro and in vivo. Br. J. Cancer., 88(3): 470-477.

Sum PE, How D, Torres N, Newman H, Petersen PJ, Mansour TS (2003). Novel benzoxazole derivatives of mannopeptimycin glycopeptides antibiotics. Bioorg. Med. Chem. Lett., 13(15): 2607-2610.

Kinetics and thermodynamic studies of adsorption of malachite green onto unmodified and EDTA-modified groundnut husk

Bamgbose Janet Titilayo[1]*, Bamigbade Akeem Adesina[1], Nkiko Mojisola Olubunmi[1], Ahmed Sikiru Akinyeye[1] and Ikotun Adebomi Ayodeji[2]

[1]Department of Chemistry, University of Agriculture, Abeokuta, Ogun State, Nigeria.
[2]Department of Chemistry, Bowen University, Iwo, Osun State, Nigeria.

A batch experiment was applied to study the adsorption of malachite green from aqueous solution by unmodified and ethylene diamine tetraacetic acid (EDTA)-modified groundnut husk. The effect of initial dye concentration, temperature, contact time and pH on the sorption process was investigated. The Langmuir and Freundlich adsorption isotherms were applied to test the experimental data. The chemically modified groundnut husk agreed well with both the Langmuir and Freundlich models at the temperature below the room temperature (25°C) and above the room temperature. Unmodified groundnut husk showed poor agreement with the isotherm at temperature below the room temperature but fits well into the model at higher temperature. Kinetics of the adsorption with respect to the initial dye concentration and temperature was also investigated. The pseudo-first-order, pseudo-second-order and intraparticle diffusion models were used to test the kinetic data. Obtained data were fitted properly in the pseudo-second-order kinetic model. The thermodynamic parameters such as change in free energy (ΔG°), enthalpy (ΔH°) and entropy (ΔS°) were also determined and the results obtained confirmed that the sorption process is feasible, spontaneous, and endothermic.

Key words: Adsorption capacity, ethylene diamine tetraacetic acid (EDTA)-modified, groundnut husk, diffusion, Kinetic modelling.

INTRODUCTION

There has been increasing scientific concern about the hazardous effects of coloured dyes which are widely used in industries such as textile, leather, paper, plastics, and food technology etc. to colour their final products (Mohamed, 2004). These industries are the important sources of contaminants responsible for the continuous pollution of the environment. Reactive dyes exhibit a wide range of different chemical structures, primarily based on substituted aromatic and heterocyclic groups. A large number of reactive dyes are azo compounds that are linked by an azo bridge (Change et al., 2001). Many reactive dyes are toxic to some organisms and cause direct destruction of aquatic creatures (Schnick, 1998). In addition, since reactive dyes are highly soluble in water, their removal from effluent is difficult by conventional physicochemical and biological treatment methods (Ozacar and Sengil, 2003).

Malachite green (4-[(4-dimethylaminophenyl)-phenylmethyl]-N,N-dimethylaniline) is a basic dye that is mostly used in industries (Khaltri et al, 1997 and Papinulti et al, 2006). Though the use of this dye has been banned in several countries and has not been approved by the US Food and Drug Administration (Raymound et al, 1984). Nevertheless, it is still being used in many parts of the world due to its lack of a proper alternative (Papic et al., 2004). When malachite green is discharged into receiving streams, it affects the aquatic life and causes detrimental effects in the liver, gill, kidney, intestine and the pituitary gland of living organism (Srivastava et al., 2004). Therefore, the treatment of effluent containing such dye is desirable due to its aesthetic impacts on receiving waters.

*Corresponding author. E-mail: titibam2003@yahoo.com.

Table 1. Some important properties of dye (malachite green) used.

Item	Detail
Name of dye	Malachite green (MG^+)
CI name	Basic green 4 (BG4)
Colour index number	42,000
Empirical formula	$C_{23}H_{25}N_2Cl$
Molecular weight	365
Dye contents	90%
Maximum wavelength of absorption	619 nm

Various techniques have been employed for the removal of dyes from wastewaters. These methods include adsorption, nano-filtration, electro-kinetic coagulation, coagulation and precipitation, advanced chemical oxidation, electrochemical oxidation, ozonisation, supported liquid membrane, liquid-liquid extraction, and biological process (Mahmoud et al., 2007). The adsorption process is one of the efficient methods to remove dyes from effluent due to its low initial cost, simplicity of design, ease of operation, and insensitivity to toxic substance (Robinson et al., 2001). Activated carbon is the most widely used adsorbent with great success due to its large surface area, micro-porous structure, high adsorption capacity etc. However, its use is limited because of its high cost. This has led to research for cheaper substitutes (Allen and Koumanova, 2005). Many investigators have studied the feasibility of using inexpensive alternative materials like pearl millet husk, date pits, saw dust, wool waste, coconut shell, rice to remove dyes and metal ions in effluents (Arivoli et al., 2007a, b; Sekaran et al., 1995; Selvarani, 2000).

Other materials including fly ash, tree bark, and human hair have been tested and are reported to give good results in several areas of application (Malik, 2003). However, the adsorption capacities of the above adsorbents are not very high. In order to improve the efficiency of the adsorption processes, it is necessary to develop cheap and easily available adsorbent with high adsorption capacities. Research for a low-cost and easily available adsorbent has led to the investigation of materials of biological origin as potential adsorbent.

Groundnut husk, an agricultural waste material, that is widely cultivated in both Northern and South-western Nigeria and indiscriminately litters around Nigeria urban areas has attracted the attention of many researchers as a potential adsorbent to concentrate and adsorb metal ions. Groundnut husk is cellulose that contains polysaccharide with carboxyl acid functional groups, hydroxyl groups, carboxyl groups of alginate and amino groups. It has been observed that biosorbent needs further modification to increase the active binding site and also make them available for sorption (Huang et al., 1996). Treatment of biomass removes the surface impurities on the biosorbent and exposed the available binding site for sorption (Cabuk et al., 2005). This research is targeted at using natural biomass (groundnut husk) both unmodified and ethylene diamine tetraacetic acid (EDTA)-modified groundnut husk to remove malachite green from aqueous solution.

MATERIALS AND METHODS

Groundnut husk was collected locally from popular Lafenwa market in Abeokuta, Ogun state, Nigeria. The groundnut husk was washed with distilled water to remove the attached dust and impurities and then sun-dried. The washed groundnut husk was ground in a mechanical grinder to form powder. The biomass was sieved to particles with size less than 0.1 mm diameter and soaked in 7% v/v aqueous H_2SO_4 for 4 h for hydrolysis. The mixture was filtered and the residue washed with distilled water several times to remove any acid content and dried at 50°C for 8 h. The dried biomass was stored in an air tight glass bottles to protect it from moisture.

Malachite green was obtained from Bektoh (Germany) and used without further purification. It was characterized by its visible spectrum, which gave a molar extinction coefficient which is in agreement with the literature value at λ_{max} of 619 nm (Green, 1990). The summary of the properties of malachite green used is presented in Table 1. Other chemicals and reagents used in this study were analar grade and used without further purification.

EDTA-modification of groundnut husk

A mixture of 17 g of the hydrolyzed husk, 0.3 dm^3 of pyridine and 56.7 g EDTA was heated under reflux for 3 h. The reaction mixture was filtered, diluted with 0.03 dm^3 of distilled water and filtered again. EDTA-modified husk was washed several times with distilled water until it is free of pyridine odour and then air-dried. The modified husk was stored in air tight glass bottles for use (Okieimen et al., 1991).

Batch kinetic studies

In experiment of batch kinetic adsorption, 0.2 g of EDTA-modified groundnut husk and unmodified groundnut husk of 0.1 mm diameter particles size and 50 cm^3 of dye solution was poured in a 100 ml flask and agitated with a maximum speed of 500 rpm using a shaker with a water bath to control temperature (Haake Model K-F3). At particular time intervals, the aqueous sample (5 cm^3) was taken and the concentration was analyzed. The concentrations of dyes were measured with ultraviolet (UV)/visible spectrophotometer (Hitachi Model U-2000). The experimental parameters included contact-time, initial dye concentration, temperature and pH. All of the experiments were carried out in duplicate and the average values were used in calculations. The maximum deviation was found to be ±2%. The amount of adsorption at time t, q_t (mg/g), equilibrium adsorption, (mg/g) and percentage dye removal at time t, and at equilibrium was obtained as follows:

$$q_t = \frac{(C_o - C_t)V}{W} \tag{1}$$

$$q_e = \frac{(C_o - C_e)V}{W} \tag{2}$$

$$\%\ Dye\ removal = \frac{(C_o - C_e)100}{W} \tag{3}$$

Where C_o, C_t and C_e (g/m^3) are the liquid-phase concentrations of solutes at initial, any time t, and at equilibrium, respectively. M is the dosage of adsorbent in the solution (kg/m^3) and V is the volume of the dye used in (cm^3).

Kinetic models of adsorption

Pseudo-first and second-order equations

In order to examine the controlling mechanism of adsorption processes such as mass transfer and chemical reaction, several kinetic models were used to test experimental data. The large number and array of different functional groups on the groundnut husk both modified and unmodified (-NH_2, -COOH, -OH) imply that there are many types of adsorbent-solute interactions. From a system design viewpoint, a lumped analysis of adsorption rate is thus sufficient to practical operation. A simple kinetic analysis of adsorption is the pseudo-first order equation in the form (HO and Mackay, 1998, 2000):

$$\ln(q_e - q_t) = \ln q_e - K_1 t \quad (4)$$

Where K_1 is the rate constant of pseudo-first order adsorption and q_e denotes the amount of adsorption at equilibrium.

In addition, a pseudo-second-order equation based on adsorption equilibrium capacity may be expressed in the form of (HO and Mckay 1998; HO and Mckay, 2000);

$$\frac{t}{q_t} = \frac{1}{k_2 q_e^2} + \frac{t}{q_e} \quad (5)$$

Where k_2 is the rate constant of pseudo-second order adsorption.

The intra-particle diffusion equation (Furusawa and Smith, 1997) can be written by the following equation:

$$q_t = K_{ipd} t^{1/2} + C \quad (6)$$

Where q_t is the amount of dye adsorbed onto adsorbent at various time $t^{1/2}$ (mg/g), C is the intercept and K_{ipd} is the intra-particle diffusion rate constant (mg/g/min).

Thermodynamic parameters such as change in free energy ($\Delta G°$) (kJ/mol), enthalpy ($\Delta H°$) (kJ/mol) and entropy ($\Delta S°$) (J/K/mol) for the adsorption process were determined using the following equation (Vadivelan and Vasanthakumar, 2005):

$$\Delta H^o = \frac{R[(T_2 T_1)]}{(T_2 - T_1)} \ln\left(\frac{K_2}{K_1}\right) \quad (7)$$

$$\Delta G^o = -RT \ln K_o \quad (8)$$

$$\Delta G^o = \Delta H^o - T\Delta S^o \quad (9)$$

$$K_o = \frac{C_{solid}}{C_{liquid}} \quad (10)$$

$$\log K_o = \frac{(\Delta S^o)}{2.303R} - \frac{(\Delta H^o)}{2.303RT} \quad (11)$$

Adsorption isotherm: The adsorption isotherm is an equation relating the amount of solute adsorbed onto the solid and equilibrium concentration of solute in solution at constant temperature. Several isotherm equations are available. In this study, the adsorption data correlated with the Freundlich (Freundlich, 1906) and Langmuir (Ozacar and Sengil, 2003) models. The following linearized Freundlich adsorption isotherm is employed.

$$\ln q_e = \ln K_f + \left(\frac{1}{n}\right) \ln C_e \quad (12)$$

Where q_e the amount is adsorbed at equilibrium (mg/g), K_f the Freundlich constant, $\left(\frac{1}{n}\right)$ is the heterogeneity factor which are related to the capacity and intensity of the adsorption, respectively and C_e is the equilibrium concentration (mg/L). The value of K_f and $\left(\frac{1}{n}\right)$ can be obtained from the slope and intercept of the plots of $\ln q_e$ against $\ln C_e$.

A well-known linear form of the Langmuir equation can be expressed as:

$$\frac{C_e}{q_e} = \frac{1}{(q_m x b)} + \frac{C_e}{q_m} \quad (13)$$

Where, C_e and q_e has usual meaning and q_m and *b* are the Langmuir constant related to the capacity and energy of adsorption, respectively. According to the Equation 13, a plot of $\frac{C_e}{q_e}$ versus C_e should be a straight line with a slope $\frac{1}{q_m}$ and intercept $\frac{1}{(q_m x b)}$ when the adsorption obeys the Langmuir equation.

The important characteristic of the Langmuir isotherm can be expressed in terms of a dimensionless factor, R_L (Ozacar and Sengil, 2003) which is defined as:

$$R_L = \left(\frac{1}{(1 + bC_o)}\right) \quad (14)$$

The R_L values indicate the type of the isotherm to be either unfavourable ($R_L>1$), linear ($R_L= 1$), favourable ($0<R_L<1$) or irreversible ($R_L= 0$).

RESULTS AND DISCUSSION

Kinetic and the thermodynamic studies of the adsorption of malachite green onto unmodified and EDTA-modified groundnut husk have been investigated and the results presented.

Effect of contact time

The contact time between the dye molecules and the sorbent is significant in the dye treatment by sorption process. The effect of contact time on the adsorption of malachite green (MG^+) was investigated at initial dye concentration of 50 mg/L, a sorbent mass of 0.2 g, a

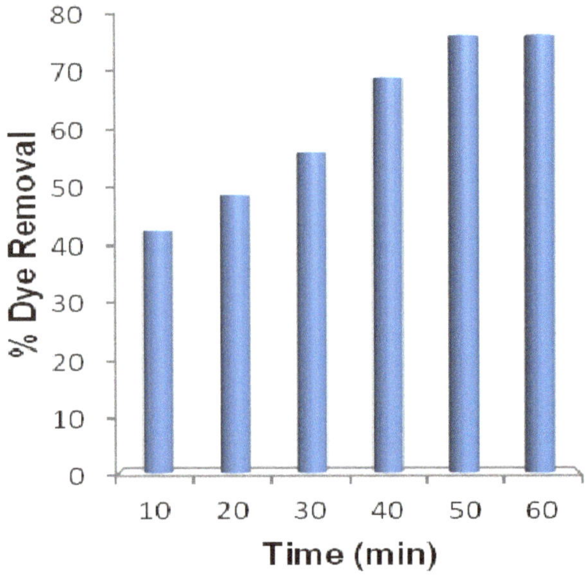

Figure 1a. Effect of contact time on adsorption of malachite green by unmodified groundnut husk.

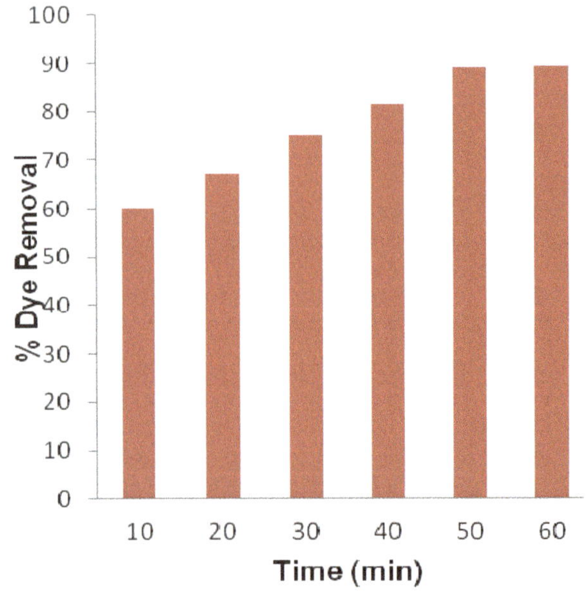

Figure 1b. Effect of contact time on adsorption of malachite green by EDTA-modified groundnut husk.

solution volume of 50 ml, an agitation speed of 250 rpm and temperature of 25°C. Figures 1a and b present the effect of contact time on the adsorption of malachite green (MG^+) onto unmodified groundnut husk and EDTA-modified groundnut husk, respectively. The figures reveal that the uptake of sorbate species is fast at initial stage of the contact time (30 min) and thereafter it becomes slower near the equilibrium. Between these two stages of the uptake, the rate of sorption is found to be nearly constant. This is clear from the fact that a large number of vacant binding sites are available for sorption during the initial stage and after a lapse of time, the remaining vacant surface site are difficult to be occupied due to repulsive forces between the solute molecule on the solid and bulk phases. Equilibrium was established at exactly 50 min for both adsorbents, the corresponding percentage dye removed at this time by each adsorbent stood at 78.6% for unmodified groundnut husk, but there was significant increase in the percentage dye removed by the EDTA-modified groundnut husk at equilibrium contact time which is 92.4%, suggesting an increased in the sorption efficiency of the modified adsorbent.

At this point, the amount of dye sorbed by the adsorbent is in dynamic equilibrium with the amount of dye desorbed from the sorbent. Additionally, the chart is single, smooth, and continuous leading to equilibrium which indicates the possible monolayer coverage of the surface of the adsorbent.

Effect of initial dye concentration

The experimental results of adsorption of malachite green (MG^+) onto both unmodified and the chemically modified groundnut husk at various initial concentrations of 50, 100, 150, 200 and 250 mg/L at different temperatures are presented in Tables 2a and b. These tables reveal that the percentage adsorption decreased with increasing initial dye concentration but adsorption capacity of the adsorbent increased with increase in dye concentration. This indicates that the initial dye concentration plays an important role in the adsorption capacity of malachite green on the groundnut husk that is, highly dependent on the initial concentration of the dye. At lower concentration, the ratio of initial number of dye molecules to the available surface area is low. Subsequently, the fractional adsorption becomes independent of initial dye concentration. However, at high concentration the available sites of adsorption become fewer and hence the percentage removal of dye is dependent upon initial concentration (Namasivayam et al., 1997; Namasivayam and Yamuna, 1995).

Effect of temperature

The effect of temperature on the adsorption of malachite green onto unmodified and EDTA-modified ground nut husk was studied at different temperatures of 20, 25, 30 and 35°C and the results shown in Tables 2a and b. The results show an increase in the adsorption capacity of malachite green onto groundnut husk with increasing temperatures. The chemically modified groundnut husk had the highest percentage dye removal across all the

Table 2a. Effect of initial concentration on adsorption of malachite green by unmodified groundnut husk.

C_o (mg/L)	C_e (mg/L)				Q_e (mg/L)				% Dye removal			
	20°C	25°C	30°C	35°C	20°C	25°C	30°C	35°C	20°C	25°C	30°C	35°C
50	10.70	7.950	4.25	2.75	9.83	10.5	11.4	11.8	78.6	84.1	91.9	94.5
100	29.20	23.70	19.5	13.8	17.7	19.1	20.1	21.6	70.8	76.3	80.5	86.2
150	84.38	44.25	37.1	29.4	16.4	26.4	28.2	30.2	62.5	70.5	75.3	80.4
200	87.20	75.40	57.0	46.4	28.2	31.2	35.8	38.4	56.4	62.3	71.5	76.8
250	124.8	106.5	89.8	71.3	31.3	35.9	40.1	44.7	50.1	57.4	64.1	71.5

Table 2b. Effect of initial concentration on adsorption of malachite green by EDTA-modified groundnut husk.

C_o (mg/L)	C_e (mg/L)				Q_e (mg/L)				% Dye removal			
	20°C	25°C	30°C	35°C	20°C	25°C	30°C	35°C	20°C	25°C	30°C	35°C
50	7.40	3.95	2.05	0.45	10.7	11.5	11.9	12.4	85.2	92.1	95.9	99.1
100	19.5	13.5	9.80	5.50	20.1	21.6	22.6	23.6	80.5	86.5	90.2	94.5
150	39.3	28.2	22.4	15.3	27.7	30.5	31.9	33.7	73.8	81.2	85.1	89.8
200	69.6	49.0	39.0	33.0	32.6	37.7	40.3	41.8	65.2	75.5	80.5	83.5
250	98.8	72.0	59.5	51.3	37.8	44.5	47.6	49.7	60.5	71.2	76.2	79.5

Table 3a. Equilibrium constant (K_o) and thermodynamic parameters of adsorption of malachite green by unmodified groundnut husk.

C_o (mg/L)	K_o				$\Delta G°$ (kJ/mol)				$\Delta H°$	$\Delta S°$
	20°C	25°C	30°C	35°C	20°C	25°C	30°C	35°C	(kJ/mol)	(J/molk^{1})
50	3.67	5.29	10.8	17.2	-2.98	-4.43	-5.89	-7.35	82.39	291.35
100	2.43	3.22	4.13	6.25	-2.04	-2.88	-3.72	-4.56	47.30	168.39
150	0.78	2.39	3.05	4.10	-0.03	-1.33	-2.70	-4.06	79.80	272.26
200	1.29	1.65	2.51	3.31	-0.53	-1.40	-2.26	-3.12	50.06	172.67
250	1.00	1.35	1.79	2.51	-0.04	-0.76	-1.56	-2.36	46.79	159.57

temperatures. Therefore, the sorption of malachite green onto groundnut husk (both modified and unmodified) is temperature-dependent. This may be due to the fact that increase temperature may produce swelling effect within the internal structure of the groundnut husk thereby enlarged its pore size.

Equilibrium and thermodynamics of adsorption of Malachite green by modified and unmodified groundnut husk

The values of K_o, $\Delta G°$ (kJ/mol), $\Delta H°$ (kJ/mol), and $\Delta S°$ (J/mol K^{-1}) were calculated from the slope and intercept of Van't Hoff plots of Equation 11 and the results are presented in the Tables 3a and b above. The results indicate that K_o decreased with increase in the concentration of dye but increased with increase in temperature. The positive values of ΔH show the endothermic nature of the adsorption and the possibility of physisorption. In the case of physical adsorption, while increasing the temperature of the system, the extent of dye adsorption increases, this ruled out the possibility of chemisorption.

The negative values of $\Delta G°$ show that the adsorption is highly favourable and spontaneous. The positive values of $\Delta S°$ indicate an increased disorder and randomness at the solid-solution interface of the adsorbent. The adsorbed water molecules, which are displaced by the adsorbate (dye) species, gain more translational entropy than it is lost by the adsorbate molecules, thus allowing the prevalence of randomness in the system (Namasivayam et al., 1997; Namasivayam and Yamuna, 1995). From the results, we could conclude that more efficient is the physisorption.

Langmuir and Freundlich isotherms

The isotherms in Figures 2a, b and 3a, b were fitted in the linear Langmuir and Freundlich equations.

Table 3b. Equilibrium constant (K_o) and thermodynamic parameters of adsorption of malachite green by EDTA-modified groundnut husk.

C_o (mg/L)	K_o				ΔG (kJ/mol)				ΔH	ΔS
	20°C	25°C	30°C	35°C	20°C	25°C	30°C	35°C	(kJ/mol)	(J/molk[1])
50	5.76	11.7	23.4	110	-3.79	-6.35	-8.91	-11.5	146.35	512.41
100	4.13	6.41	9.20	17.2	-3.35	-4.62	-5.89	-7.16	71.10	254.10
150	2.82	4.32	5.71	8.80	-2.53	-3.54	-4.55	-5.56	56.68	202.09
200	1.87	3.08	4.13	5.06	-1.69	-2.58	-3.47	-4.36	50.31	177.49
250	1.53	2.47	3.20	3.88	-1.21	-2.03	-2.85	-3.66	46.78	163.78

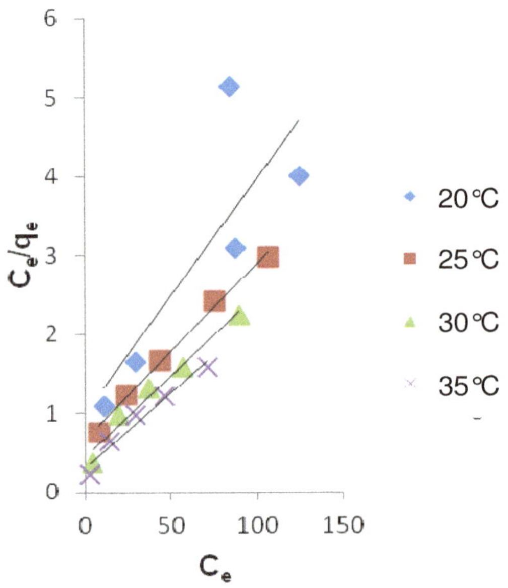

Figure 2a. Langmuir adsorption Isotherm for unmodified adsorbent.

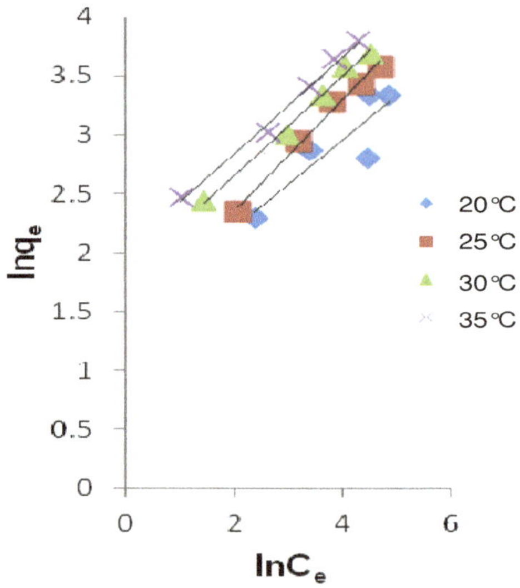

Figure 3a. Freundlich isotherm for adsorption of malachite green onto unmodified adsorbent.

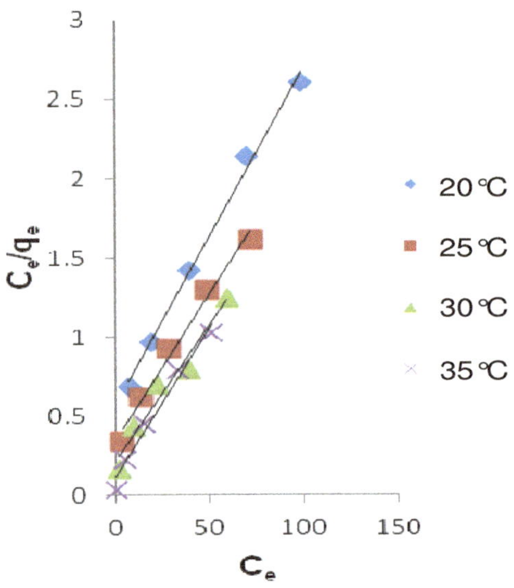

Figure 2b. Langmuir adsorption Isotherm for modified adsorbent.

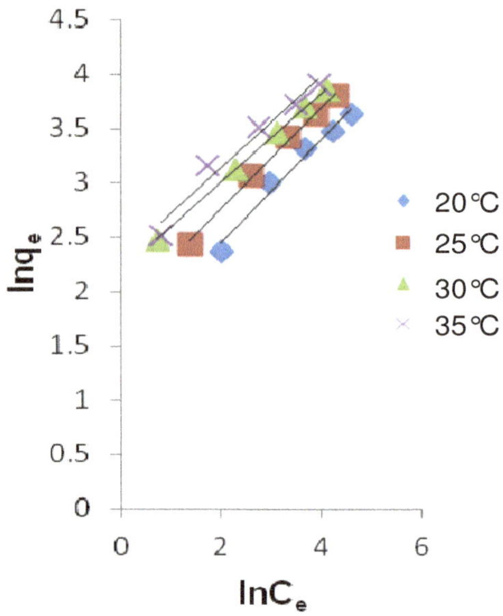

Figure 3b. Freundlich Isotherm for adsorption of malachite green onto modified adsorbent.

Table 4a. Langmuir isotherm constant for adsorption of malachite green by unmodified groundnut husk.

Temperature	20 °C	25 °C	30 °C	35 °C
q_m	33.90	44.84	48.08	52.91
b	0.029	0.034	0.047	0.061
R^2	0.6821	0.9938	0.9713	0.9631

Table 4b. Langmuir isotherm constant for adsorption of malachite green by EDTA-modified groundnut husk.

Temperature	20 °C	25 °C	30 °C	35 °C
q_m	46.73	54.05	58.14	52.36
b	0.038	0.054	0.081	0.181
R^2	0.9948	0.9845	0.9576	0.9778

Table 5a. Dimensionless separation factor (R_L) for unmodified groundnut husk.

	R_L			
Temperature/C_o (mg/L)	20 °C	25 °C	30 °C	35 °C
50	0.408	0.370	0.299	0.247

Table 5b. Dimensionless separation factor (R_L) for EDTA-modified groundnut husk.

	R_L			
Temperature/C_o (mg/L)	20 °C	25 °C	30 °C	35 °C
50	0.345	0.270	0.198	0.0995

Table 6a. Freundlich Isotherm constant for the adsorption of malachite green by unmodified groundnut husk.

Temperature	20 °C	25 °C	30 °C	35 °C
K_f	4.27	4.10	6.07	7.49
n	2.64	2.11	2.36	2.41
R^2	0.7698	0.9889	0.9913	0.9921

Table 6b. Freundlich Isotherm constant for the adsorption of malachite green by EDTA-modified groundnut husk.

Temperature	20 °C	25 °C	30 °C	35 °C
K_f	4.46	6.22	8.88	10.01
N	2.10	2.14	2.43	2.38
R^2	0.976	0.997	0.999	0.9649

Tables 4a, b and 6a, b show the computation results. The Langmuir and Freundlich fit quite well with the experimental data for the EDTA-modified groundnut husk both below the room temperature (20 °C) and above room temperature as its correlation coefficient R^2 >0.9523 but for the unmodified groundnut husk it shows poor agreement of both the Langmuir and Freundlich isotherm with the experimental data at lower temperature (below room temperature) as its correlation coefficient R^2 < 0.7898, but agreed well at the temperature above room temperature. The calculated n-values are greater than unity, suggesting favourable adsorption (Freundlich, 1906). Also, the dimensionless factor (R_L) values presented in Tables 5a and b lies in between 0 and 1. This suggested the favourability of the sorption process.

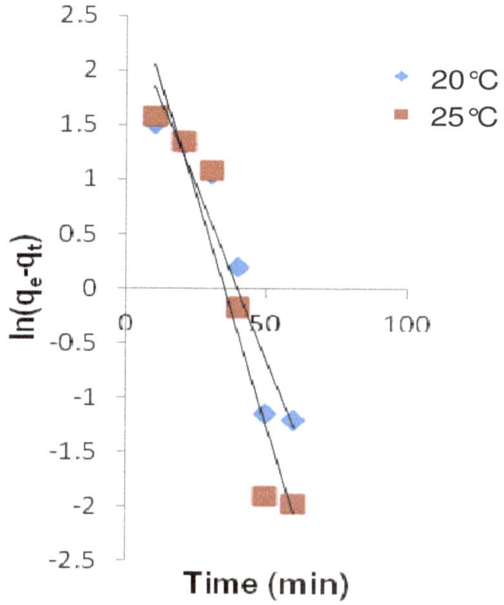

Figure 4a. Pseudo-first-order kinetics of unmodified groundnut husk.

Pseudo-first-order, pseudo-second-order and the intraparticle diffusion Models

The pseudo-first-order, pseudo-second-order and the intraparticle diffusion model of the adsorption process at 20 and 25 °C are plotted in Figures 4a, b, 5a, b and 6a, b, respectively.

Tables 7a and b presented the rate constant and correlation coefficient of the pseudo-first-order, pseudo-second-order and the intraparticle diffusion model at 20 and 25 °C. The correlation coefficient R^2 for the pseudo-second-order adsorption model has extremely high values (R^2 > 0.9614), and its calculated equilibrium adsorption capacities q_{ecal} fit well with the experimental data. These suggest that the pseudo-second-order adsorption mechanism is predominant and the overall rate of this sorption process appears to be controlled by the chemical process (Mckay, 1983).

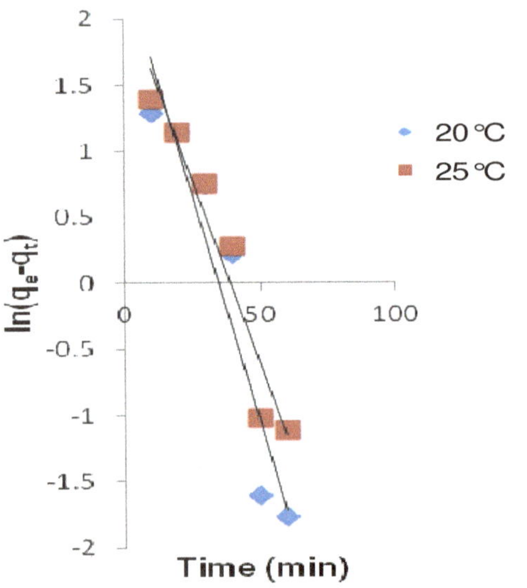

Figure 4b. Pseudo-first-order kinetics of EDTA-modified groundnut husk.

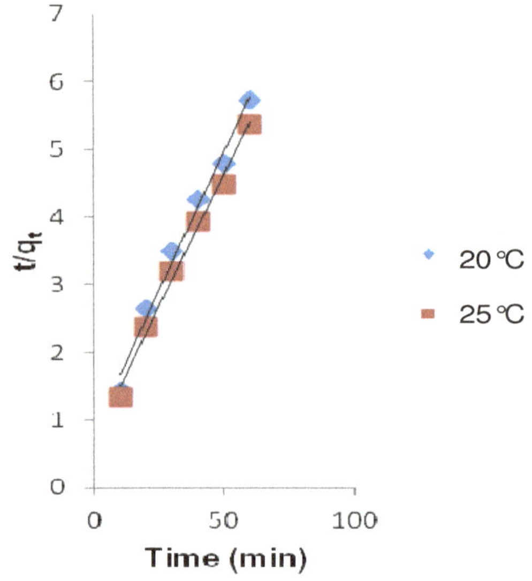

Figure 5b. Pseudo-second-order kinetics of EDTA-modified groundnut husk.

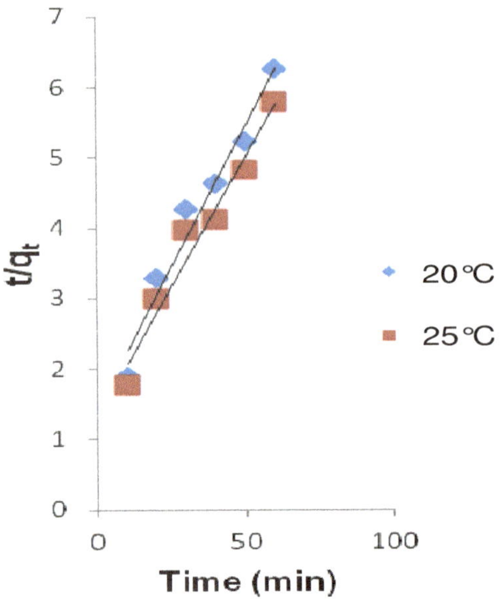

Figure 5a. Pseudo-second-order kinetics of unmodified groundnut husk.

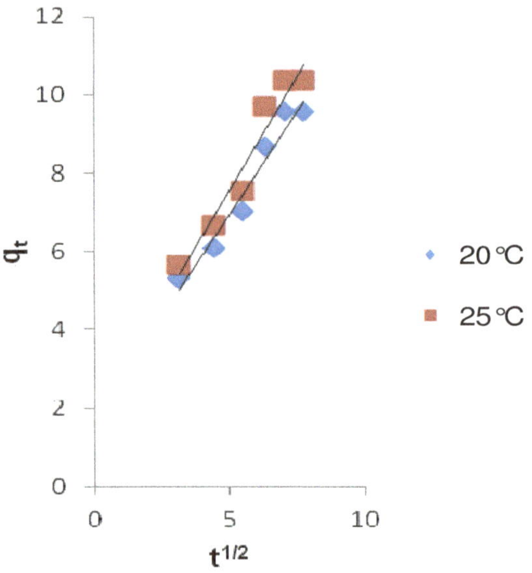

Figure 6a. Intraparticle diffusion of adsorption of malachite green onto unmodified groundnut husk.

The results in Tables 6a and 6b above describe the rate constant studied at two different temperatures calculated by the three models. For the pseudo-second-order, both the rate constant and the initial adsorption rate increased with increasing temperatures. This may be explained that higher temperature increases the reaction rate and also decreases the particle density thereby enlarging the pore sizes for dye molecules to diffuse. The correlation coefficient for the intraparticle diffusion is lower than pseudo-second-order kinetics, which indicates the present system may be followed by intraparticle diffusion.

Effect of pH

The effect of pH on adsorption of malachite green onto unmodified and EDTA-modified groundnut husk was investigated, while the initial concentration, shaking time, amount of adsorbent and temperature were fixed at 50 mg/L, 250 min, 0.2 g and 25°C, respectively.

pH affects not only the adsorption capacity but also the

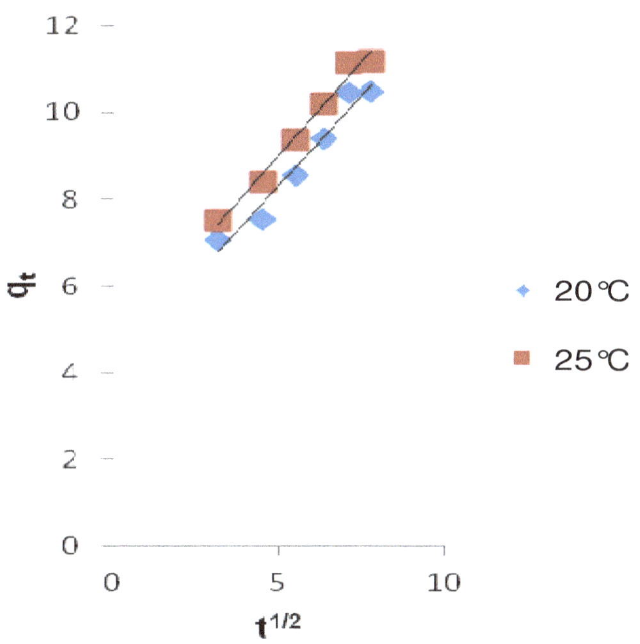

Figure 6b. Intraparticle diffusion of adsorption of malachite green onto EDTA-modified groundnut husk.

Table 7a. Pseudo-first-order, pseudo-second-order and the intraparticle diffusion models for unmodified adsorbent.

C_o (mg/L)	20℃			25℃		
	50	100	150	50	100	150
Pseudo-first-order						
k_1	0.0625	0.0891	0.0909	0.0823	0.0798	0.0923
q_e	11.93	35.28	51.32	17.80	30.51	57.24
R^2	0.9169	0.8972	0.8674	0.908	0.9038	0.9022
Pseudo-second-order						
k_2	0.0044	0.0025	0.0542	0.0040	0.0028	0.0556
q_e	12.48	22.73	30.12	13.62	23.53	34.13
R^2	0.965	0.9758	0.9582	0.9625	0.9831	0.9799
Intraparticle diffusion						
k_{ipd}	1.0512	1.9345	2.51	1.669	1.975	2.9263
R^2	0.9578	0.9732	0.9577	0.948	0.9832	0.9803

colour of the dye and the solubility of some dye. Therefore, the pH value of solution was an important controlling factor in the adsorption process (Waranusantigul et al, 2003). The effect of pH on malachite green adsorption by unmodified and chemically modified groundnut husk is shown in Figures (7a) and (7b), respectively.

Figure (7a) shows that the equilibrium sorption capacity is minimum at pH 2 and increases monotonically up to pH 5. A further increase in pH leads to an appreciable increase in percentage adsorption. The absence of sorption at low pH can be explained by the fact that at this acidic pH, H^+ may compete with dye ions for the adsorption sites of the adsorbent, thereby inhibiting the adsorption of dye. At higher pH, the groundnut husk may get negatively charged, which enhances the adsorption of positively charged dye cations through electrostatic forces of attraction. Moreover, a change of solution pH

Table 7b. Pseudo-first-order, pseudo-second-order and the intraparticle diffusion models for EDTA-modified adsorbent.

C_o (mg/L)	20°C			25°C		
	50	100	150	50	100	150
Pseudo-first-order						
k_1	0.0556	0.0907	0.0909	0.0686	0.0903	0.0923
q_e	11.05	35.66	58.67	8.88	39.40	68.17
R^2	0.8902	0.842	0.8674	0.9311	0.8289	0.9022
Pseudo-second-order						
k_2	0.0079	0.0825	0.0658	0.0084	0.0808	0.0687
q_e	12.18	23.75	33.78	12.85	25.45	37.59
R^2	0.9859	0.9814	0.9617	0.9921	0.9678	0.982
Intraparticle diffusion						
k_{ipd}	0.835	1.7633	2.659	0.8671	1.8262	3.0053
R^2	0.9622	0.9457	0.9468	0.9821	0.9367	0.9691

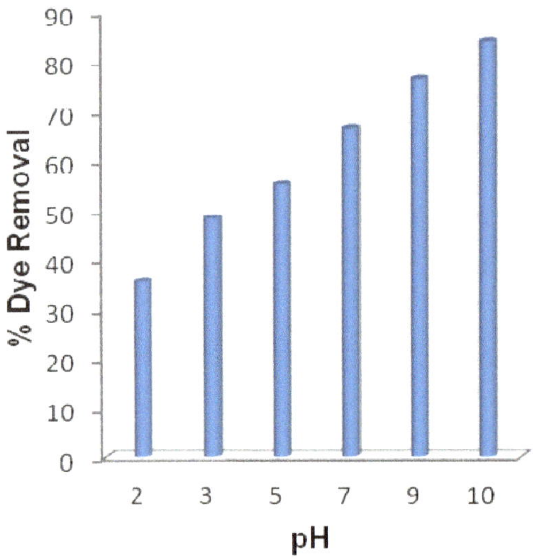

Figure 7a. Effect of pH on the adsorption of malachite green onto unmodified groundnut husk.

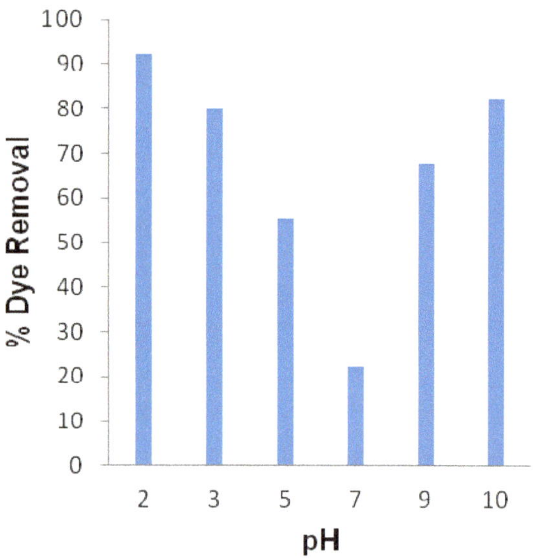

Figure 7b. Effect of pH on the adsorption of malachite green onto EDTA-modified groundnut husk.

affects the adsorptive process through dissociation of functional groups on the surface of the adsorbent. Such behaviour leads to a shift in the equilibrium characteristics of the adsorption process. A similar report on effect of pH was also reported for the adsorption of malachite green onto rattan saw dust (Hameed et al, 2008). But for the chemically modified groundnut husk, the sorption process was minimal at pH 7 and maximum at acidic pH (Figure 7b). At pH 7, the interaction between the carboxylic groups of the modified groundnut husk and water predominates over the interaction between the adsorbent and the dye, hence the reduced adsorption.

However, when the pH of the solution was increased (above pH 7), the uptake of malachite green was also increased. It appears that change in pH solution results in the formation of different ionic species, and different adsorbent surface. When the pH value is lower than 6, the malachite green ions were able to enter into the pore of the adsorbent by forming complex $[MGEDTA]^{3-}$ with the EDTA on the adsorbent surface.

But at pH 7, the zwitter ions form of malachite green in water raise the aggregation of malachite green to form bigger molecule (dimer) and was unable to enter the pore structure of the adsorbent. At pH value higher than 8, the existence of OH^- groups on the adsorbent surface

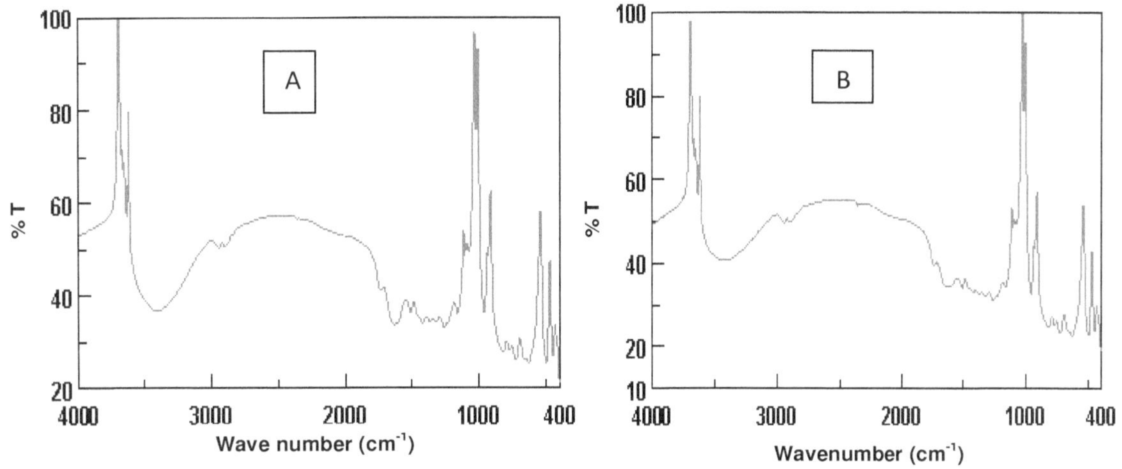

Figure 8. FTIR Spectrum of (A) unmodified groundnut husk and (B) modified groundnut husk.

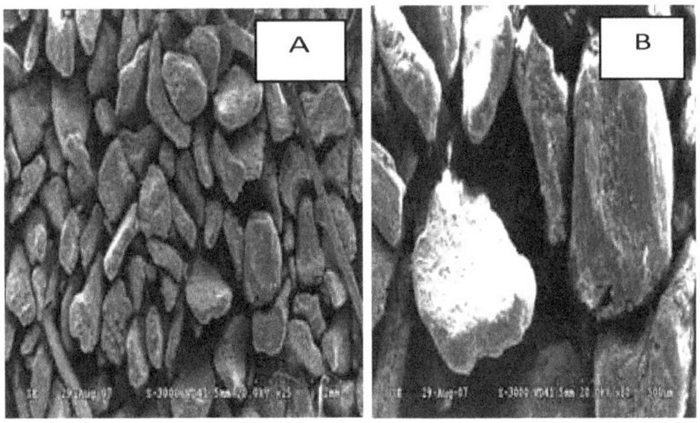

Figure 9. Scanning electron micrograph of (A) unmodified and (B) EDTA-modified groundnut husk.

enhances the adsorption of positively charged dye cations through electrostatic forces of attraction. Hence the increase in adsorption capacity (Arivoli et al., 2007a, b; Kalpana et al., 2007; Yupeng et al., 2005).

The FTIR spectrum and SEM micrographs of groundnut husk

The FTIR spectrum of groundnut husk is shown in Figure 8a. Broad peak at 3400 cm^{-1} is the indicator of -OH and -NH groups. The stretching of the -OH groups bound to methyl groups presented a signal between 2950 and 2887 cm^{-1}. The peaks located at 1737 and 1633 cm^{-1} are characteristics of carbonyl group stretching from aldehydes and ketones. The presence of -OH group, along with carbonyl group, confirms the presence of carboxylic acid groups in the adsorbent. The peaks at 1508 cm^{-1} are associated with the stretching in aromatic rings. The peaks observed at 1071 and 1024 cm^{-1} are due to C-H and C-O bonds. The -OH, - NH, carbonyl and carboxylic groups are important sorption sites (Volesky, 2003). The FTIR spectra of the EDTA modified Groundnut husk, is presented in Figure 8b. As compared to simple groundnut husk, the broadening of -OH peak at 3400 cm^{-1} and carbonyl group peak at 1633 cm^{-1} was observed. This indicates the involvement of hydroxyl and carbonyl groups in the adsorption of malachite green. The SEM micrographs of groundnut husk and EDTA-modified groundnut husk are shown in Figure 9. These micrographs represent a porous structure, with large surface area, and large morphological changes in the modified groundnut husk with increased pore sizes and surface area. The results of surface area analysis further confirmed the porous nature and high surface area of the adsorbent.

Conclusion

The study investigates the dynamics of the adsorption of malachite green on the unmodified groundnut husk and EDTA-modified groundnut husk. The chemically modified groundnut husk shows a higher adsorption capacity than unmodified groundnut husk.

The Langmuir and Freundlich isotherm agree well with the equilibrium isotherm for the two adsorbents. The initial dye concentration and temperature also significantly affect the adsorption capacity of malachite green onto both the modified and unmodified groundnut husks. The pseudo-second-order kinetic model agrees very well with the dynamical behaviour for the adsorption of malachite green onto unmodified and EDTA-modified groundnut husks under several different initial dye concentrations and temperatures.

This study shows that groundnut husk can be used as an adsorbent in the treatment of waste water by sorption system but that its potency can be improved by chemical modification. Groundnut husk is a better and cheaper adsorbent when compared with most adsorbents used for dye removal, and it is found abundantly in nature.

ACKNOWLEDGEMENT

The authors are grateful to the Department of Chemistry and the Central Laboratory of University of Agriculture Abeokuta, Nigeria for providing the facilities and the chemicals used for this work.

REFERENCES

Allen SJ, Koumanova B (2005). Decolourisation of waste water using adsorption. J. Univ. Chem. Technol. Metallur. 40(3):175-192.

Arivoli S, Kalpana K, Sudha R, Rajachandrasekar T (2007a). Comparative Study on Adsorption Kinetics and Thermodynamics of metal ions onto Acid Activated Low Cost carbon. Eur. J. Chem. 4:238-254.

Arivoli S, Venkantraman BR, Rajachandrasekar T, Hema M (2007b). Adsorption of ferrous ion from Aqueous solution by Low Cost Activated carbon obtained from Natural plant Material". Res. J. Chem. Environ. 17:70-78.

Cabuk A, Ilhan S, Filik C, Caliskan F (2005). Lead (II) ion biosorption by pretreated Fungi biomass. Turk. J. Biol. 29:23-28.

Change CF, Yang CH, Shu YO, Chen TI, Shu MS, Liao IC (2001). Effects of temperature, Salinity and Chemical drugs on the in-vitro propagation of the Dinoflagellate parasite, Amylodiniumocellatum. Asian Fish Soc. pp. 31-38.

Freundlich H (1906). Uberdie adsorption in Losungen, Z. Phys. Chem. A 57:385-387.

Furusawa T, Smith JM (1997). Intraparticle mass transport rates in slurries by dynamic adsorption studies, J. Al. Chem. 20(1):88-93.

Green FY (1990). Sigma – Aldrich hand-book of stain dyes and indicators. Aldrich Chemical Company, Milwaukee. pp. 224-228.

HO YS, Mckay G (1998). A comparison of chemisorptions kinetic models applied to pollutant removal on various sorbents. Trans. Inst. Chem. Eng. 76B:332-340.

HO YS, Mckay G (2000). The kinetics of sorption of divalent metal ions onto sphagnum moss peat. Water Res. 30:465-742.

Huang C, Chung YC, Liou MR (1996). Adsorption of Cu (II) and Ni (II) by palletized polymer. J. Harzard. Mater. 45:265-277.

Khaltri SD, Singh M (1997). Colour removal from dye waste water using sugar cane dust as an adsorbent. Adsorp. Sci. Technol. 17(4):269-282.

Mahmoud AS, Ghaly AE, Brooks MS (2007). Removal of dye from textile wastewater using plant oils under different pH and temperature conditions. Am. J. Environ. Sci. 3:205-218.

Malik PK (2003). Use of activated carbon prepared from saw dust and rice husk for adsorption of acid dyes: A Case Study of Acid yellow 36. Dye Pig. 56:239-249.

Mckay GM (1983). The adsorption of dye stuffs from aqueous solution using activated carbon: analytical solution for batch adsorption based on external mass transfer and pore diffusion. Chem. Eng. J. 27: 187-196.

Mohamed M (2004). Acid dye removal: comparison of surfactant–modified mesoporous FSM – 16 with activated carbon derived from rice husk. J. Coll. Interf. Sci. 272:28-34.

Namasivayam C, Yamuna RT (1995). Adsorption of direct red by biogas residual slurry. Environ. Pollut. 89:1-29.

Namasivayam C, Muniasamy N, Gayathri K, Rani M, Reganathan K (1997). Removal of dyes from aqeous solution by cellulosic waste orange peal, Biores. Technol. 57:37-42.

Okieimen FE, Okundia EU, Ogbeifun DE (1991). Sorption of Cadmium and Lead ions on modified groundnut [Arachis hypogea] husk. J. Chem. Technol. Biotechnol. 51:97-103.

Ozacar M, Sengil IA (2003). Adsorption of reactive dye on calcinedalunite from aqueous solutions. J. Hazard. Meter. B98:211-24.

Papic S, Koprivanac N, Bozic AL, Metes A (2004). Removal of some reactive dyes from synthetic wastewater by combined Al (III) coagulation/ carbon adsorption process. Dye Pig. 62:291-298.

Papinulti L, Mouso N, Forchiassin F (2006). Removal and degradation of the fungicide dye malachite green from aqueous solution using the system wheat bran – Fomessclerodermeus. Enzy. Microb. Technol. 39: 848-853.

Raymound EK, Dunald F (1984). Encyclopaedia of chemical technology. New York: wiley; 201-226.

Robinson T, Mcmullan G, Marchant R, Nigam P (2001). Remediation of dyes in textile effluent: A critical review on current treatment technologies with a proposed alternative. Biores. Technol. 77:247-255.

Schnick RA (1998). The impetus to register new therapeutants for aquaculture. Prog. Fish Cultur. 50:190-196.

Sekaran G, Shanmugasundaram KA, Mariappan M, Raghavan KV (1995). Adsorption of dyes by buffing dust of Leather Industry". India J. Chem. Technol. 2:311-355.

Selvarani K (2000). Studies on low cost Adsorbent for the removal of organic and inorganic dyes from water, PhD Thesis. Regional Eng. Coll. Thiruchirapalli. pp. 8-78.

Srivastava S, Sinha R, Roy D (2004). Toxicological effects of malachite green. Aqua. Toxicol. 66(3):319-329.

Vadivelan V, Vasanthakumar K (2005). Equilibrium, Kinetics, Mechanism and process design for the Sorption of methylene blue onto rice husk. J. Coll. Interf. Sci. 286:91-110.

Volesky B (2003). Sorption and biosorption. Montreal-St. Lambert, Quebec, Canada, BV Sorbex Inc. pp. 316-318. ISBN 0-9732983-0-8.

Waranusantigul P,Pokethitiyook P,Kruatrachue M, Upatham ES (2003). Kinetics of basic dye (methylene blue) biosorption by giant duckkweed (Spirodelapolyrrhiza). Environ. Pollut. 125:285-395.

Yupeng G, Jingzhu Z, Hui Z, Shaofeng Y, Zichen W, Hongding X (2005). Use of Rice husk Based porous carbon for the adsorption of Rhodamine B from Aqueous Solution, Dyes and Pig. 66:123-128.

“Photochemical synthesis and spectroscopic studies of 3, 5-alkyl/aryl substituted-tetrahydro-2H-1,3,5-thiadiazine-2-thione and their complexes with $M(CO)_4(\eta^{2:2}$- NBD) [M = Cr, Mo and W] as $M(CO)_4$ transfer reagent”

Ayodhya Singh[1]*, Munesh Kumar[1], Manish Kaushik[2], Anjali Singh[2] and Seema Singh[3]

[1]Department of Chemistry, M. M. H. College, Ghaziabad, U. P., India 201001, India.
[2]Department of Chemistry, D. P. B. S. College, Anoopshahr, Bulandshahr, U. P., India 203001, India.
[3]Department of Chemistry, Kishan P. G. College, Bahraich, U. P. 271801, India, India.

Fifteen new complexes, $[M(CO)_4(L)2]$ [M=Cr; Mo; W]; [L= met-DTTT; et-DTTT; n-pr-DTTT; ph-DTTT and benz-DTTT] have been synthesized by the photochemical exchange reaction of $[(\eta^{2:2}$- NBD)$M(CO)_4]$ (M= Cr, Mo and W) with 3,5-dimethyl-tetrahydro-2H-1,3,5-thiadiazine-2-thione (met-DTTT), 3,5-diethyl-tetrahydro-2H-1,3,5-thiadiazine-2-thione (eth-DTTT), 3,5-di-n-propyl-tetrahydro-2H-1,3,5-thiadiazine-2-thione (n-pr-DTTT), 3,5-diphenyl-tetrahydro-2H-1,3,5-thiadiazine-2-thione (ph-DTTT), and 3,5-dibenzyl-tetrahydro-2H-1,3,5-thiadiazine-2-thione (benz-DTTT). The new complexes have been characterized by elemental analysis, FTIR, 1H NMR spectroscopy. The spectroscopic studies show that DTTT behaves as a monodentate ligand coordinating via the sulfur (C=S) donor atom.

Key words: Metal carbonyl, 3,5-dimethyl-tetrahydro-2H-1,3,5-thiadiazine-2-thione, exchange reactions.

INTRODUCTION

Dazomet (3,5-dimethyl-tetrahydro-2H-1,3,5-thiadiazine-2-thione or met-DTTT) is the fumigant or pesticide currently used to control soil fungi in both the conifer and shrub seedbeds. Control of soil-borne fungi, especially those which cause damping-off, has been good (Goksoyr, 1964). Soil fumigants are pesticides that form gases when applied to soil. Once in the soil, the fumigants work by controlling pests that can disrupt plant growth and crop production. Soil fumigants play a very important role in agriculture, but they also have the potential to pose risk concerns to people (like handlers, workers and bystanders). Soil fumigation can provide benefits to both food consumers and growers. The magnitude of benefits depends on pest pressure, which varies over space and time, and the availability and costs associated with the use of alternatives (Reregistration Eligibility Decision for Dazomet, 2008).

Dazomet is applied before the planting of crops by soil incorporation, thereby causing it to act as a soil fumigant and disinfectant by decomposing to methyl isothiocyanate. It controls soil fungi (such as, Fusarium, Pythium, Rhizoctonia, Sclerotinia, Verticillium and Colletotrichum cocH2:2- NBDes atramentarium), nematodes, germinating weed seeds and soil-dwelling insects (The Pesticide Manual, 2000).

In agriculture, DTTT has found application as fungicide, herbicide and nematocide for cabbage, cucumber, maize, potato and tomato plants. In industrial environments, DTTT has been used as slimicide in paper mills, as a biocide in metal working fluids used in the manufacture of engines, transmissions, aircraft and especially metal products. The chemical DTTT has also been considered for use as a biocide in pump spray-delivered consumer products and custodial supplies. Respiratory exposure may occur from liquid aerosol generated during spray application of materials containing DTTT. Perhaps of a greater significance is the potential for exposure to volatile and semivolatile degradation roducts of DTTT.

* Corresponding author. E-mail: dr.ayodhyasingh@gmail.com.

This 'CS2' incorporated heterocycle is sparingly soluble in water and it has been equivocally proposed to undergo decomposition to formaldehyde, methylisothiocyanate and other compounds in aqueous solution (Drescher and Otto, 1968; Munnecke et al., 1962; Munnecke, 1963; Syamsunder, 1964; Buu-Hoi and Xoung, 1958; Shankarwar et al., 2008; Srivastava and Shrimal, 2002; Subasi et al., 2006; Chakraborty and Patel, 1996; Amar et al., 2002). It has been reported that DTTT gave unstable complexes with Cu(II), Co(II), Ni(II), Fe(III) ions in solution and were not isolated in solid form (Klement et al., 1999). Sulfide carbonyl compounds continue to attract considerable attention not only on account of their fascinating structural chemistry, but also because of their ability to act as electron reservoirs and their potential in catalysis (Martell, 1987). Features of the chemistry of these molecules which are currently of interest include the mechanisms and sites of substitution as well as the modification of reactivity accompanying carbonyl replacement by donor ligands (Zhao et al., 2001). Complexes $[M(CO)_5(DTTT)]$ [M=Cr, Mo, W], $Re(CO)_4Br(DTTT)$ and $[Mn(CO)_2Cp(DTTT)]$ have been reported (Sema et al., 2003). NBD (η^{2}:^{2}-norbornadiene) forms organometallic complexes, where it can act as a two-electron or four-electron donor. Sandwiched compounds such as tetracarbonyl (η^{2}:2-norbornadiene) chromium (0), which is a useful reagent for transferring chromium tetracarbonyl to bidentate phosphine ligands, are known (Markus et al., 2004) 1,5-cyclooctadiene is also the starting material for the synthesis of diamantine and sumanene and it is used as an acetylene transfer agent for instance in reaction with 3,6-di-2-pyridyl-1,2,4,5-tetrazine (Ronald et al., 2001). The thermal substitution of $Mo(CO)_4(\eta^{2}:^{2}$-COD) and $Mo(CO)_4(\eta^{2}:^{2}$-NBD) with, BFEDA gave the same product which was isolated by using the same procedure. Thus, both $Mo(CO)_4(\eta^{2}:^{2}$-NBD) and $Mo(CO)_4(\eta^{2}:^{2}$-COD) can be used as Mo(CO)4 transfer reagent for the synthesis of of $Mo(CO)_4$(BFEDA) (Fatma S.K., 2005; Ayodhya et al., 2010). These literatures prompted us to study the behavior of ($\eta^{2}:^{2}$-NBD)$M(CO)_4$ as transfer reagent in G-6 metal carbonyl complexes.

In this paper, we have used five Schiff bases (Scheme 1) namely 3,5-dimethyl-tetrahydro-2H-1,3,5-thiadiazine-2-thione (met-DTTT), 3,5-diethyl-tetrahydro-2H-1,3,5-thiadiazine-2-thione (eth-DTTT), 3,5-di-n-propyl-tetrahydro-2H-1,3,5-thiadiazine-2-thione (n-pr-DTTT), 3,5-diphenyl-tetrahydro-2H-1, 3,5-thiadiazine-2-thione (ph-DTTT), 3,5-dibenzyl-tetrahydro-2H-1,3,5-thiadiazine-2-thione (benz-DTTT) as ligands with $M(CO)_4(\eta^{2}:^{2}$-NBD) (where M=Cr, Mo and W). M(CO)4(η2:2-NBD) was utilized as $M(CO)_4$ transfer reagent.

In this paper, we report a noble photochemical route to the hitherto unknown fifteen new complexes (1a to 3e). The complexes were characterized by elemental analysis, FT-IR, 1H-NMR spectroscopy. According to this, the ligands (a to e) coordinates to the metal via (C=S) sulfur donor atom in (1a to 3e). The spectroscopic studies suggest monodentate coordination of ligands in $\{M(CO)_4(\text{met-DTTT})_2\}$(1a to 3a); $\{M(CO)_4(\text{eth-DTTT})_2\}$(1b to 3b); $\{M(CO)_4(\text{n-pr-DTTT})_2\}$ (1c to 3c); $\{M(CO)_4(\text{ph-DTTT})_2\}$ (1d to 3d) and $\{M(CO)_4(\text{benz-DTTT})_2\}$ (1e to 3e); where (M=Cr, Mo and W).

MATERIALS AND METHODS

Reactions were carried out under dry argon or in vacuo. All solvents were dried and degassed prior to use. Infrared spectra were recorded on a Perkin-Elmer spectrophotometer (Model-577) in KBr discs and CH2Cl2. All the melting points were determined in an open capillary and are uncorrected. All glassware was oven dried at 120°C, molecular weight of the complexes was determined cryoscopically in benzene. Dichloromethane, 2,5-norbornadiene (η2:2- NBD), n-hexane, n-pentane, benzene, methyl amine hydrochloride, formaldehyde, carbon disulfide, triethyl amine and iso-octane were purchased from E. Merck, while $M(CO)_6$ (M = Cr, Mo, W), 3,5-diethyl-tetrahydro-2H-1,3,5-thiadiazine-2-thione (eth-DTTT), 3,5-di-n-propyl-tetrahydro-2H-1,3,5-thiadiazine-2-thione (n-pr-DTTT), 3,5-diphenyl-tetrahydro-2H-1,3,5-thiadiazine-2-thione (ph-DTTT), 3,5-dibenzyl-tetrahydro-2H-1,3,5-thiadiazine-2-thione (benz-DTTT) were purchased from Aldrich and were used as supplied.

Magnetic susceptibility measurements of the complexes were carried out by Gouy method. UV irradiation was performed with a medium pressure 400W mercury lamp through a quartz bulb.

Synthesis of the 3,5-dimethyl-tetrahydro-2H-1,3,5-thiadiazine-2-thione (met-DTTT). The 3,5-dimethyl-tetrahydro-2H-1,3,5-thiadiazine-2-thione (met-DTTT) was prepared as described in the literature (Wing-Wah et al., 1993). A solution of CH3NH2.HCl (5.0 g, 73 Mmol) in water (6 ml) and triethylamine (8.62 g, 86 mmol) was stirred for 10 min, and then CS2 (2.84 g, 38 mmol) was added drop wise at room temperature with stirring for 20 min. Formaldehyde (2.30 g, 76 mmol) was added drop wise to the above reaction mixture, which was stirred for 30 min until a white precipitate was formed. met-DTTT was obtained as white solid by filtration with 80% yield. The complex was characterized and identified by means of IR and 1H-NMR spectroscopy. IR (in CH2Cl2): v_{sym} (C=S) 655 cm^{-1}, v_{as}(C=S) 600 cm^{-1}, 1H-NMR (in CDCl3) δ= 2.46 (s, 3H, 5-CH3), 3.30 (s, 3H, 3-CH3), 1.27 (s, 2H, 2-CH2), 4.34 (s, 2H, 4-CH2).

Synthesis of the $Mo(CO)_4$(bicyclo[2.1.1] hepta-2,5-diene) or $Mo(CO)_4(\eta^{2:2}$-NBD)

$Mo(CO)_4(\eta^{2:2}$-NBD), $Cr(CO)_4(\eta^{2:2}$-NBD) and $W(CO)_4(\eta^{2:2}$- NBD) were prepared following the method described in the literature (King, 1965; King, 1963, Sarikahya and Senturk, 2001, Warner H. Prinz R., 1965). The preparation of Tetracarbonyl ($\eta^{2:2}$- NBD) molybdenum(0) is typical and is given here.

Tetracarbonyl ($\eta^{2:2}$- NBD)molybdenum(0), was prepared by refluxing $Mo(CO)_6$ and 2,5-norbornadiene ($\eta^{2:2}$- NBD) in iso-octane. The reaction solution was evaporated to dryness. Crystallization from n-hexane solution yields yellow crystals of the complex. The solvent was decanted and crystals were dried in vacuum. The complex was characterized and identified by means of IR spectroscopy. IR (in toluene): A1(2)=2038, A1(1), B1= 1948, B2=1885 cm^{-1}. $Mo(CO)_4(\eta^{2:2}$- NBD) was obtained in 75% yield.

Preparation of (met-DTTT)$_2Mo(CO)_4$ complex from exchange reaction of [($\eta^{2:2}$- NBD)$Mo(CO)_4$]

0.45 g (1.5 mmol) sample of norbornadienetetracarbonylmolybdenum [($\eta^{2:2}$-NBD)$Mo(CO)_4$] was dissolved in 25 ml of n-pentane. Then 0.486 g (3 mmol) of met-

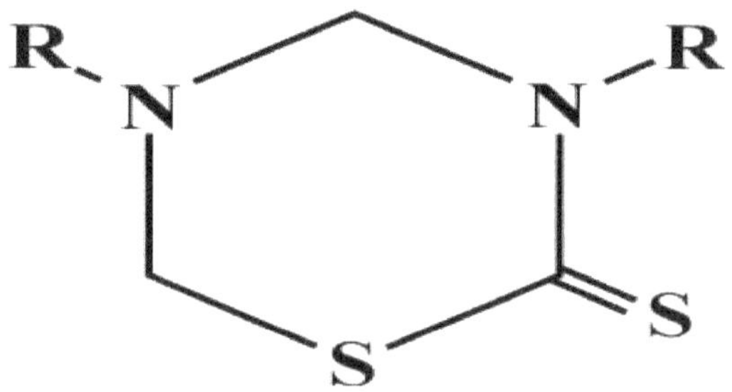

Scheme 1. a. (met-DTTT; R=CH_3); b. (eth-DTTT; R=C_2H_5); c. (n-pr-DTTT; R=C_3H_7); d. (ph-DTTT; R=C_6H_5) and e. (benz-DTTT; R=$C_6H_5CH_2$).

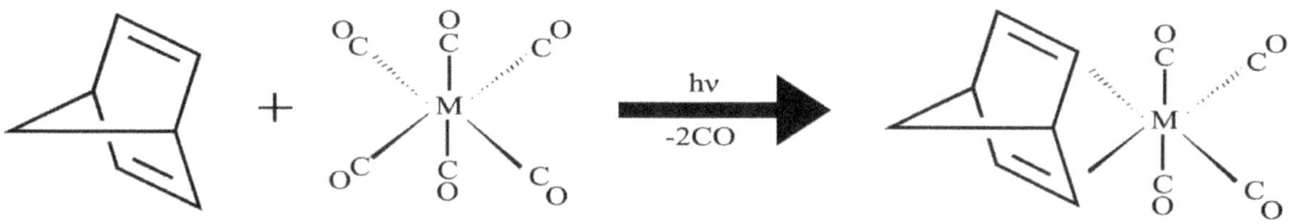

Scheme 2. Thermal reaction of $\eta^{2:2}$-NBD with $M(CO)_6$ (M = Cr, Mo, W).

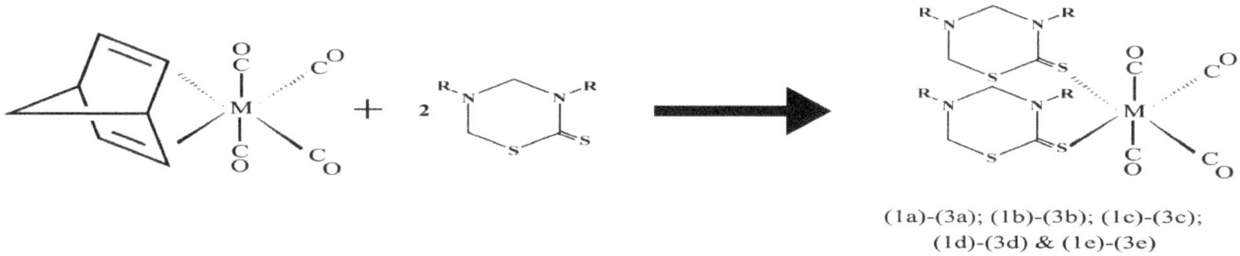

Scheme 3. Ligand Substitution reaction of $\eta^{2:2}$-NBD in $M(CO)_4(\eta^{2:2}$-NBD) with DTTT. (1): M=Cr; (2): M=Mo; (3): M=W; (a): met-DTTT, R=CH_3; (b): eth-DTTT, R=C_2H_5; (c):n-*Pr*-DTTT, *R=C_3H_7*; (d): ph-DTTT, R=C_6H_5 and (e): benz-DTTT, R= $C_6H_5CH_2$.

DTTT was added to the above solution with constant stirring. The mixture was irradiated for 10h. During the irradiation, the color of reaction mixture changed from off-white to dark yellow. After the irradiation the reaction mixture was evaporated under vacuum yielding a deep yellow solid. The solid residue was then washed with n-hexane so as to remove any [($\eta^{2:2}$-NBD)$Mo(CO)_4$], $\eta^{2:2}$- NBD ligand itself, and other n-hexane soluble impurities. The deep yellow residue or remnant was dissolved in 1:1 CH_2Cl_2/n-hexane solution for recrystallization. Yellow microcrystal of [$Mo(CO)_4$(met-DTTT)$_2$] were obtained after waiting 24 h at -15°C with 71% yield.

The other 14 complexes were prepared according to same procedure as described for the preparation of afore mentioned complex by using appropriate [($\eta^{2:2}$-NBD)$M(CO)_4$] (M= Cr, Mo and W) and (met-DTTT), (eth-DTTT), (n-pr-DTTT), (ph-DTTT) and (benz-DTTT).

RESULTS AND DISCUSSION

The transfer reagents $Mo(CO)_4(\eta^{2:2}$-NBD), $Cr(CO)_4(\eta^{2:2}$-NBD) and $W(CO)_4(\eta^{2:2}$-NBD) were prepared by thermal reaction of 2,5-norbornadiene (NBD) with $M(CO)_6$ (M= Cr, Mo, W) according to Scheme 2. Complexes (1a to 3e) were prepared by photochemical reaction as shown in Scheme 3. Analytical data for {$M(CO)_4$(met-DTTT)$_2$}(1a to 3a); {$M(CO)_4$(eth-DTTT)$_2$}(1b to 3b); {$M(CO)_4$(n-pr-DTTT)$_2$} (1c to 3c); {$M(CO)_4$(ph-DTTT)$_2$} (1d to 3d) and {$M(CO)_4$(benz-DTTT)$_2$} (1e to 3e); where M= Cr, Mo and W; complexes are given in Table 1. The photogeneration of $M(CO)_5$ from $M(CO)_6$ (M=Cr, Mo and W) has been extensively studied. These 16-electron $M(CO)_5$ fragments react quickly with any available donor atom to yield a $M(CO)_5L$ species; and where L is a chelating bidentate ligand, rapid continuation to the chelating $M(CO)_4L$ or bridging $M_2(CO)_{10}$ (μ-L) products may occur (Cotton and Wilkinson, 1988; Sema et al., 2003; Ozan et al., 2003; Almond et al, 1997). The photochemical reactions of $Mo(CO)_4(\eta^{2:2}$-NBD) with DTTT proceed in an entirely expected manner by displacement of the NBD ligand to give the monometallic complexes (1a to 3e).

Table 1. Physical and analytical data of Complexes {$M(CO)_4(L)_2$} (L = met-DTTT, eth-DTTT, n-pr-DTTT, ph-DTTT and benz-DTTT) (M = Cr, Mo, W).

Complexes	Empirical formula	Yield %	Found (calculated), %				Mol. wt
			C	H	N	S	
1a	$C_{14}H_{20}O_4N_4S_4Cr$	75	33.12 (34.42)	4.02 (4.13)	11.16 (11.47)	25.28 (26.25)	484 (489)
2a	$C_{14}H_{20}O_4N_4S_4Mo$	71	30.86 (31.58)	3.57 (3.79)	10.37 (10.52)	23.67 (24.08)	529 (533)
3a	$C_{14}H_{20}O_4N_4S_4W$	67	26.25 (27.10)	3.07 (3.25)	8.32 (9.03)	19.28 (20.67)	615 (620)
1b	$C_{18}H_{28}O_4N_4S_4Cr$	73	38.76 (39.69)	5.01 (5.18)	9.98 (10.29)	23.08 (23.55)	541 (545)
2b	$C_{18}H_{28}O_4N_4S_4Mo$	74	35.12 (36.73)	4.12 (4.79)	9.05 (9.52)	21.12 (21.79)	585 (589)
3b	$C_{18}H_{28}O_4N_4S_4W$	68	30.98 (31.96)	4.01 (4.17)	8.00 (8.28)	18.29 (18.96)	673 (677)
1c	$C_{22}H_{36}O_4N_4S_4Cr$	72	43.07 (43.98)	5.76 (6.04)	8.95 (9.33)	20.56 (21.35)	596 (601)
2c	$C_{22}H_{36}O_4N_4S_4Mo$	73	39.94 (40.98)	5.03 (5.63)	8.04 (8.69)	18.44 (19.89)	641 (645)
3c	$C_{22}H_{36}O_4N_4S_4W$	69	35.47 (36.07)	4.54 (4.95)	7.13 (7.65)	16.94 (17.51)	727 (733)
1d	$C_{34}H_{28}O_4N_4S_4Cr$	70	54.57 (55.42)	3.27 (3.83)	7.13 (7.60)	16.89 (17.41)	733 (737)
2d	$C_{34}H_{28}O_4N_4S_4Mo$	72	51.65 (52.30)	3.22 (3.61)	6.97 (7.18)	16.01 (16.43)	777 (781)
3d	$C_{34}H_{28}O_4N_4S_4W$	66	46.12 (47.01)	3.00 (3.25)	6.02 (6.45)	14.02 (14.76)	863 (869)
1e	$C_{38}H_{36}O_4N_4S_4Cr$	76	56.86 (57.56)	4.04 (4.58)	6.54 (7.07)	15.78 (16.17)	787 (793)
2e	$C_{38}H_{36}O_4N_4S_4Cr$	74	53.76 (54.54)	3.94 (4.34)	6.09 (6.69)	14.79 (15.32)	832 (837)
3e	$C_{38}H_{36}O_4N_4S_4Cr$	70	48.76 (49.35)	3.76 (3.92)	5.87 (6.06)	13.22 (13.87)	921 (925)

Table 2a. Selected IR bands (cm^{-1}) and 1H-NMR (ppm) data of ligands.

Ligand	Name of ligands	Abbreviation	(C-S-C)	$v_{as\,(C=S)}$	^{1}H-NMR (in $CDCl_3$) δ ppm
a	3,5-dimethyl-tetrahydro-2H-1,3,5-thiadiazine-2-thione	met-DTTT	837	628	2.46 (s, 3H, 5-CH_3), 3.30 (s, 3H, 3-CH_3), 1.27 (s, 2H, 2-CH_2), 4.34 (s, 2H, 4-CH_2)
b	3,5-diethyl-tetrahydro-2H-1,3,5-thiadiazine-2-thione	eth-DTTT	837	632	0.90 (t, 3H, 5-CH_2CH_3), 2.38 (q, 2H, 5-CH_2CH_3), 1.72 (t, 3H, 3-CH_2CH_3), 3.18 (q, 2H, 3-CH_2CH_3), 1.27 (s, 2H, 2-CH_2), 4.34 (s, 2H, 4-CH_2)
c	3,5-di-n-propyl-tetrahydro-2H-1,3,5-thiadiazine-2-thione	n-pr-DTTT	836	637	1.04 (t, 3H, 5-$CH_2CH_2CH_3$), 2.07 (m, 2H, 5-$CH_2CH_2CH_3$), 4.37 (t, 2H, 5- $CH_2CH_2CH_3$), 1.84 (t, 3H, 3-$CH_2CH_2CH_3$), 2.87 (m, 2H, 3-$CH_2CH_2CH_3$), 5.16 (t, 2H, 3-$CH_2CH_2CH_3$), 1.26 (s, 2H, 2-CH_2), 4.33 (s, 2H, 4-CH_2)
d	3,5-diphenyl-tetrahydro-2H-1,3,5-thiadiazine-2-thione	ph-DTTT	835	635	7.46 (m, 5H, 5-C_6H_5), 7.90 (m, 5H, 3-C_6H_5), 1.26 (s, 2H, 2-CH_2), 4.34 (s, 2H, 4-CH_2)
e	3,5-dibenzyl-tetrahydro-2H-1,3,5-thiadiazine-2-thione	benz-DTTT	836	639	7.46 (m, 5H, 5-$CH_2C_6H_5$), 4.32 (m, 2H, 5-$CH_2C_6H_5$), 7.96 (m, 5H, 3- $CH_2C_6H_5$), 5.12 (m, 2H, 3- $CH_2C_6H_5$), 1.27 (s, 2H, 2-CH_2), 4.34 (s, 2H, 4-CH_2)

The IR Spectral bands and 1H-NMR data of ligands (L = met-DTTT (a); eth-DTTT (b); n-pr-DTTT (c); ph-DTTT (d); benz-DTTT (e))are presented in Table 2a. In this study, photochemical reactions of $M(CO)_4(NBD)$ (M=Cr, Mo and W) with met-DTTT (a); eth-DTTT (b); n-pr-DTTT (c); ph-DTTT (d) and benz-DTTT (e) ligands occurs in this expected manner, and gave hither to a series of complexes (1a) to (1e); (2a) to (2e) and (3a) to (3e) occur via the displacement of NBD from $M(CO)_4(NBD)$ (M=Cr, Mo and W) and co-ordination of metal atom via two Sulfur donor atoms (C=S) of two molecules of the ligand yielding cis-$M(CO)_4L_2$ complexes. Important IR spectral bands $M(CO)_4L_2$ (L = met-DTTT (a); eth-DTTT (b); n-pr-DTTT (c); ph-DTTT (d); benz-DTTT (e) and M=Cr, Mo and W) are presented in Table 2b.

The evidence about the metal-sulfur (M–S) bond formation is the appearance of a new band in all the complexes around 385 to 403 cm^{-1}, which may be assigned to the v(M–S) mode (Hiral et al., 2008, Amin et al., 2001; Jian et al., 1999) can be correlated with the

Table 2b. Selected IR bands (cm^{-1}) and (1H)-NMR (ppm) data of $\{M(CO)_4(L)_2\}$ (L= (a): met-DTTT, R=CH_3; (b): eth-DTTT, R=C_2H_5; (c):n-Pr-DTTT, R=C_3H_7; (d): ph-DTTT, R=C_6H_5 and (e): benz-DTTT, R= $C_6H_5CH_2$) (M = (1): Cr, (2): Mo, (3): W).

Complexes	ν(CO) cm^{-1}	(C-S-C)	$ν_{sym}$ (C=S)	$ν_{as}$ (C=S)	^{1}H-NMR (in $CDCl_3$) δ ppm
1a	2015, 1943, 1910, 1850	836	668	610	2.44 (s, 3H, 5-CH_3), 3.27 (s, 3H, 3-CH_3), 1.26 (s, 2H, 2-CH_2), 4.33 (s, 2H, 4-CH_2)
2a	2010, 1912, 1845, 1795	835	664	607	2.45 (s, 3H, 5-CH_3), 3.28 (s, 3H, 3-CH_3), 1.25 (s, 2H, 2-CH_2), 4.32 (s, 2H, 4-CH_2)
3a	2005, 1851, 1832, 1782	838	666	608	2.44 (s, 3H, 5-CH_3), 3.29 (s, 3H, 3-CH_3), 1.26 (s, 2H, 2-CH_2), 4.33 (s, 2H, 4-CH_2)
1b	2012, 1938, 1904, 1848	838	670	612	0.89 (t, 3H, 5-$CH_2\underline{CH_3}$), 2.32 (q, 2H, 5-$\underline{CH_2}CH_3$), 1.71 (t, 3H, 3-$CH_2\underline{CH_3}$), 3.16 (q, 2H, 3-$\underline{CH_2}CH_3$), 1.26 (s, 2H, 2-CH_2), 4.32 (s, 2H, 4-CH_2)
2b	2007, 1908, 1842, 1790	836	667	610	0.88 (t, 3H, 5-$CH_2\underline{CH_3}$), 2.34 (q, 2H, 5-$\underline{CH_2}CH_3$), 1.70 (t, 3H, 3-$CH_2\underline{CH_3}$), 3.17 (q, 2H, 3-$\underline{CH_2}CH_3$), 1.25 (s, 2H, 2-CH_2), 4.32 (s, 2H, 4-CH_2)
3b	2006, 1854, 1835, 1787	835	664	611	0.89 (t, 3H, 5-$CH_2\underline{CH_3}$), 2.33 (q, 2H, 5-$\underline{CH_2}CH_3$), 1.72 (t, 3H, 3-$CH_2\underline{CH_3}$), 3.17 (q, 2H, 3-$\underline{CH_2}CH_3$), 1.26 (s, 2H, 2-CH_2), 4.33 (s, 2H, 4-CH_2)
1c	2014, 1942, 1907, 1850	837	665	615	1.02 (t, 3H, 5-$CH_2CH_2\underline{CH_3}$), 2.06 (h, 2H, 5- $CH_2\underline{CH_2}CH_3$), 4.32 (t, 2H, 5- $\underline{CH_2}CH_2CH_3$), 1.82 (t, 3H, 3-$CH_2CH_2\underline{CH_3}$), 2.84 (h, 2H, 3-$CH_2\underline{CH_2}CH_3$), 5.10 (t, 2H, 3-$\underline{CH_2}CH_2CH_3$), 1.25 (s, 2H, 2-CH_2), 4.32 (s, 2H, 4-CH_2)
2c	2006, 1910, 1841, 1792	835	663	618	1.01 (t, 3H, 5-$CH_2CH_2\underline{CH_3}$), 2.05 (h, 2H, 5- $CH_2\underline{CH_2}CH_3$), 4.34 (t, 2H, 5- $\underline{CH_2}CH_2CH_3$), 1.81 (t, 3H, 3-$CH_2CH_2\underline{CH_3}$), 2.83 (h, 2H, 3-$CH_2\underline{CH_2}CH_3$), 5.12 (t, 2H, 3-$\underline{CH_2}CH_2CH_3$), 1.24 (s, 2H, 2-CH_2), 4.31 (s, 2H, 4-CH_2)
3c	2002, 1856, 1832, 1788	836	662	617	1.03 (t, 3H, 5-$CH_2CH_2\underline{CH_3}$), 2.06 (h, 2H, 5- $CH_2\underline{CH_2}CH_3$), 4.35 (t, 2H, 5- $\underline{CH_2}CH_2CH_3$), 1.82 (t, 3H, 3-$CH_2CH_2\underline{CH_3}$), 2.84 (h, 2H, 3-$CH_2\underline{CH_2}CH_3$), 5.11 (t, 2H, 3-$\underline{CH_2}CH_2CH_3$), 1.25 (s, 2H, 2-CH_2), 4.33 (s, 2H, 4-CH_2)
1d	2017, 1948, 1908, 1853	836	672	614	7.42 (m, 5H, 5-C_6H_5), 7.87 (m, 5H, 3-C_6H_5), 1.25 (s, 2H, 2-CH_2), 4.33 (s, 2H, 4-CH_2)
2d	2008, 1906, 1844, 1794	837	669	613	7.43 (m, 5H, 5-C_6H_5), 7.88 (m, 5H, 3-C_6H_5), 1.24 (s, 2H, 2-CH_2), 4.31 (s, 2H, 4-CH_2)
3d	2010, 1860, 1831, 1790	838	664	616	7.44 (m, 5H, 5-C_6H_5), 7.89 (m, 5H, 3-C_6H_5), 1.26 (s, 2H, 2-CH_2), 4.33 (s, 2H, 4-CH_2)
1e	2020, 1943, 1902, 1852	837	672	616	7.42 (m, 5H, 5-$CH_2\underline{C_6H_5}$), 4.31 (m, 2H, 5-$\underline{CH_2}C_6H_5$), 7.91 (m, 5H, 3-$CH_2\underline{C_6H_5}$), 5.10 (m, 2H, 3- $\underline{CH_2}C_6H_5$), 1.26 (s, 2H, 2-CH_2), 4.34 (s, 2H, 4-CH_2)
2e	2011, 1910, 1846, 1802	835	669	622	7.43 (m, 5H, 5-$CH_2\underline{C_6H_5}$), 4.32 (m, 2H, 5-$\underline{CH_2}C_6H_5$), 7.92 (m, 5H, 3-$CH_2\underline{C_6H_5}$), 5.11 (m, 2H, 3- $\underline{CH_2}C_6H_5$), 1.27 (s, 2H, 2-CH_2), 4.33 (s, 2H, 4-CH_2)
3e	2014, 1863, 1834, 1791	836	665	619	7.44 (m, 5H, 5-$CH_2\underline{C_6H_5}$), 4.32 (m, 2H, 5-$\underline{CH_2}C_6H_5$), 7.96 (m, 5H, 3-$CH_2\underline{C_6H_5}$), 5.10 (m, 2H, 3- $\underline{CH_2}C_6H_5$), 1.26 (s, 2H, 2-CH_2), 4.33 (s, 2H, 4-CH_2)

decrease in v(CO) modes of (1a to 3e) complexes move to lower wave numbers in comparison with the starting M(CO)6 (M=Cr, Mo and W) molecules (Cotton and Wilkinson, 1988; Mahmud et al., 2007; Maurya et al., 1992). This decrease in the frequency of CO absorption is because of the accumulation of charge density on metal atom (Cr, Mo and W) which stabilizes itself by transferring it back to the nearby CO (a π-acceptor ligand) (Ayodhya, 1999).

Four bands in the range of 2002 to 2020 cm^{-1}, 1851 to 1948 cm^{-1}, 1831 to 1910 cm^{-1} and 1782 to 1853 cm^{-1}, arising from v (CO) vibrations, are seen which presumably have local c2v symmetry of $M(CO)_4$ unit in {$M(CO)_4$(met-DTTT)$_2$}(1a to 3a), {$M(CO)_4$(eth-DTTT)$_2$}(1b to 3b), {$M(CO)_4$(n-pr-DTTT)$_2$} (1c to 3c), {$M(CO)_4$(ph-DTTT)$_2$} (1d to 3d) and {$M(CO)_4$(benz-DTTT)$_2$} (1e to 3e); where M = Cr, Mo and W complexes (Scheme 3). These values are in close resemblance to the values of v (CO) vibration for other sulfur containing di-substituted group-6 metal carbonyls (Cotton and Kraihanzel, 1962; Ozer and Ozkar, 1999; Ayodhya et al., 2010; Timmers and Wacholtz, 1994). The presence of normal ligand bands indicated that these bands were intact in the complexes. The nature and number of CO bands resemble closely to the bands of other known di- and tri-substituted metal carbonyls (Ayodhya et al., 2010; Sema et al., 2003; Darensbourg and Darensbourg, 1970, Abel et al., 1958).

The IR spectra data for (1a to 3e) show a decrease in vas(C=S) on coordination. In the free ligand, the IR active mode vas(C=S) is at 628 cm^{-1} for met-DTTT (Cotton and Kraihanzel, 1962). On coordination, two bands are observed in the range of 666 to 668 cm^{-1} [vsym(C=S)] and 607 to 610 cm^{-1} [vas(C=S)]; 660 to 670 cm^{-1} [vsym(C=S)] and 610 to 612 cm^{-1} [vas(C=S)]; 662 to 665 cm^{-1} [vsym(C=S)] and 615 to 618 cm^{-1} [vas(C=S)]; 664 to 672 cm^{-1} [vsym(C=S)] and 613 to 616 cm^{-1} [vas(C=S)]; and 665 to 672 cm^{-1} [vsym(C=S)] and 616 to 622 cm^{-1} [vas(C=S)] in the case of met-DTTT, eth-DTTT, n-pr-DTTT, ph-DTTT and benz-DTTT respectively. As expected four bands arising from v(CO) vibrations are seen for each complex which presumably have local C2v symmetry of the $M(CO)_4$ unit (Sema et al, 2003) and confirming the cis- complexes (Darensbourg and Darensbourg, 1970).

In addition, magnetic susceptibility measurement shows that (1a to 3e) complexes were diamagnetic. Since these complexes have M(0) [M=Cr, Mo, W] with a low spin d6 configuration. Such diamagnetism might arise from further splitting of the d-orbital in the low symmetry complexes, that is, dxy2, dxz2, dyz2, d(x2-y2)0, and dz20 (Senem et al., 2008, Ayodhya et al., 2010).

Conclusion

The IR spectroscopic results show that the ligands (a to e) coordinate via sulfur (C=S) atom in (1a to 3e) and these ligands behave as monodentate ligand. The IR data are well in accord with cis- coordination of ligands. These results suggest that the ligands

ACKNOWLEDGEMENTS

Sincere thanks are due to Head, Department of Chemistry, D.P.B.S. College, Anoopshahar, Bulandshahar and Head, Department of Chemistry, M.M.H. College, Ghaziabad for providing research facilities. Authors are also thankful to Dr. M.P. Singh, Principal, M.M.H. College, Ghaziabad for his constant encouragement to carry out such research work. Special thanks are also due to CDRI, Lucknow, Intertek, Mumbai and TIFR, Mumbai for allocation of time for various analyses.

REFERENCES

Abel E, Bennett WMA, Burton R Wilkinson G. (1958). Transition-metal complexes of seven-membered ring.

Almond MJ, Sarikahya F, Senturk OS (1997). Photochemical and thermal synthesis of metal carbonyl complexes of tetraalkyl diphosphine disulfides [M(CO)4(R2P(S)P(S)R2)] (M = Mo, W ; R = Me, Et, Pr-n, Bu-n), Polyhedron, 16: 1101-1103.

Amar S, Shrimal AK (April 2002). Schiff base derivatives of tungsten carbonyl , Indian J. Chem., 41A: 785-790.

Amin RR, Elgemeie GEH (2001). The direct electrochemical synthesis of co(ii), ni(ii), and cu(ii) complexes of some pyridinethione derivatives, Synth. React. Inorg. Met-Org. Chem., 31: 431-440.

Ayodhya S (1999). Ph.D. Thesis, C.C.S. University, Meerut.

Ayodhya S, Munesh K, Manish K (2010). Photochemical Synthesis and Spectroscopic Studies of tetraalkyl diphosphine disulfides and their complexes with M(CO)4(η2:2-1,5-cyclooctadiene) [M = Cr, Mo &W] as M(CO)4 transfer reagent, Int. J. of Chem. Appl., 2(1): 151-159 .

Ayodhya S, Munesh K, Manish K (2010). Synthesis and Spectroscopic Studies of 2-Hydroxy-1-napthaldehyde Alkanesulfonylhydrazone(SB1-3) and its complexes with M(CO)6 [M=Cr, Mo & W], Int. J. Appl. Chem., 6(1): 49-58.

Ayodhya S, Munesh K, Manish K, Seema S, Anjali S (2010). Photochemical Reactions of [M(CO)6] (M=Cr, Mo & W) with Quadridentate Schiff- Bases, Oriental J. Chem., 26(2): 601-606.

Buu-Hoi NP, Xoung ND (1958). p-Cyclopentylacetophenone and Its Derivatives, J. Org. Chem., 23: 39-42.

Chakraborty J, Patel RN (1996). Copper-, cobalt- and zinc(II) complexes with monofunctional bidentate Schiff base and monodentate neutral ligands J. Ind. Chem. Soc., 73: 192-195.

Chem D, Martell AE (1987). Dioxygen affinities of synthetic cobalt Schiff base complexes, Inorg. Chem., 26: 1026-1030.

Cotton FA , Wilkinson G (1988). Advance Inorg. Chem., 5th Edn., Wiley Interscience, New York, P. 1047.

Cotton FA, Kraihanzel CS (1962). Vibrational Spectra and Bonding in Metal Carbonyls. I. Infrared Spectra of Phosphine-substituted Group VI Carbonyls in the CO Stretching Region, J. Am. Chem. Soc., 84: 4432-4438.

Darensbourg MY, Darensbourg DJ (1970). Infrared determination of stereochemistry in metal complexes: An application of group theory, J. Chem. Edu., 47(1): 33-35.

Goksoyr J (1964). Chemical and Fungicidal Reactions of 3,5-Dimethyltetrahydro-1,3,5-thiadiazine-2-thione (3,5-D). A Comparison with Sodium N-Methyl Dithiocarbamate and Methyl Isothiocyanate, Acta Chem. Scand., 18: 1341-1352.

Drescher N, Otto S (1968). About the degradation of dazomet in soil, Residue Rev., 23: 49-53.

Fatma SK (May 2005). Thesis Entitled "Synthesis and Characterization of Tetracarbonyl[N,N'- bis(ferrocenylmethylene)ethylenediamine] Molybdenum (0) complexe ", Middle Ease Technical University, pp.

34-35.

Goksoyr J (1964). Chemical and Fungicidal Reactions of 3,5-Dimethyltetrahydro-1,3,5-thiadiazine-2-thione (3,5-D). A Comparison with Sodium N-Methyl

Hiral P, Prajapati Y, Solankee P, Solankee S, Patel G, Kapadia K and Solankee A (2008). Acta ciencia Indica. XXXIV C(2): 327.

Jian F, Wang Z, Bai Z, You X, Fun H , Chinnakali K (1999). Structure of bis(dipropyldithiocarbamate) cadmium(II), [Cd2(n-Pr2dtc)4] (dtc = dithiocarbamate) J. Chem. Crystallogr., 29: 227-231.

King RB (1965). "Organometallic Synthesis", Vol. 1, Academic Press, New York, N. Y., pp. 122-125.

King RB (1963). Complexes of Trivalent Phosphorus Derivatives. II. Metal Carbonyl Complexes of Tris(dimethylamino)-phosphine, Inorg. Chem., 2: 936-944.

King RB (1965). Alkene Metathesis in Organic Synthesis. Organometallic Syntheses, Vol. 1, Academic Press, New York, P. 125.

Klement R, Stock F, Ellias H, Ellias H, Paulur H, Pelikan P, Valke M, Mazar M (1999). Copper(II) complexes with derivatives of salen and tetrahydrosalen: A spectroscopic, electrochemical and structural study, Polyhedron, 18: 3617-3628.

Mahmud T, Iqbal J, Imran M, Mekee V (2007). Synthesis and Spectroscopic Studies of 2-Bromo N, N-Dimethylbenzylamine and Its Complexes with Mo(CO)6 and W(CO)6, J. Appl. Sci., 7(9): 1347-1350.

Markus Strotman, Rudolf Wartchow, Hogler Butenschon (2004). High yield synthesis and structures of some achiral and chiral (diphosphine)tetracarbonylchromium(0) chelate complexes with tetracarbonyl(norbornadiene)chromium(0) as complexation reagent, Arkivoc, 2004(xiii): 57-66.

Maurya RC, Mishra DD, Rao NS (1992). Synthesis and characterization of some ruthenium(II) Schiff base complexes derived from 1-phenyl-3-methyl-4-benzoyl-5-pyrazolone, Polyhedron, 11: 2837-2840.

Munnecke DE (1963). Effect of heating or drying on Armillaria mellea or Trichoderma viride and the relation to survival of A. mellea in soil. Phytopathology, 53: 1140-1143.

Munnecke DE, Domsch KH, Eckert JW (1962). Lateral root inducing compounds from the bacterium Erwinia quercina: Isolation, structure and synthesis, Phytopathology, 52: 1298-1302.

Ozan SS, Sema S, Ummuhan O (2003). Synthesis and Characterization of Metal Carbonyls [M(CO)6(M = Cr, Mo, W), Re(CO)5Br, Mn(CO)3Cp] with 2-Hydroxy-1-napthaldehyde Ethanesulfonylhydrazone (nafesh), Z. Naturforsch., 58b: 1124-1127.

Reregistration Eligibility Decision for Dazomet (July 2008), US-EPA 738-R-08-007.

Ronald NW, Peter AI I (2001). π-Bond Screening in Benzonorbornadienes: The Role of 7-Substituents in Governing the Facial Selectivity for the Diels-Alder Reaction of Benzonorbornadienes with 3,6-Di(2-pyridyl)-s-Tetrazine, Molecules, 6: 353-369.

Sarikahya F, Senturk OS (2001). Photochemical synthesis of metal carbonyl complexes of tetraalkyldiphosphine disulfides, [M2(CO)10(μ-R2P(S)P(S)R2)] (M = Mo, W; R = Me, Et, n-Pr, n-Bu), Syn. React. Inorg. Met.-Org. Nano-Metal Chem., 31: 1843-1851.

Sema S, Ayse E, Ozan SS, Brian TS, Konantin AU, Ummuhan OF, Ugur S (2003). Photochemical reactions of metal carbonyls [M(CO)6 (M=Cr, Mo, W), Re(CO)5Br, Mn(CO)3Cp] with 3,5-dimethyl-tetrahydro-2H-1,3,5-thiadiazine-2-thione (DTTT) and the crystal structure of [W(CO)5(DTTT)], Polyhedron, 22: 1689-1693.

Sema S, Ozan SS, Fadime U (Sarikahya) (2003). Photochemical reactions of M(CO)6 (M = Cr, Mo, W) with Ph2P(S)P(S)Ph2, Tran. Met. Chem., 28: 133-136.

Sema S, Ozan SS, Sarikahya FU (2003). Photochemical reactions of M(CO)6 (M = Cr, Mo, W) with Ph2P(S)P(S)Ph2, Transition Metal Chem., 28: 133-136.

Senem KP, Elif S, Hamdi T (2008). Photochemical Reactions of [M(CO)5THF] (M: Cr, Mo and W) with Tetradentate Schiff-bases, Synthesis and reactivity in Inorg., Metal-org. Nano-Met. Chem., 38: 422-427.

Shankarwar SG, Shankarwar AG, Chandhekar TK (2008). Acta Ciencia Indica, xxxxiv c (02): 219.

Subasi E, Temel H, Senturk OS, Urgur F (2006). Photochemical reactions of metal carbonyls [M(CO)6 (M=Cr, Mo, W)] with N,N'-bis(salicylidene)-1,2-bis- (o-aminophenoxy)ethane, J. Co-ord. Chem., 59: 1807-1811.

Syamsunder K (April, 1964). Chalcones as Inorganic Analytical Reagents. Part I. Spectrophotometric Investigation of Iron 2' : 4' - Dihydroxy Chalcone Complex, Proc. Indian Acad. Sci., 59: 241-247. systems. Part I. The cycloheptatriene–metal complexes and related compounds, J. Chem. Soc., pp. 4559-4563.

The Pesticide Manual (2000). 12th Edition, British Crop Protection Council, UK. thermal synthesis of metal carbonyl complexes of tetraalkyl diphosphine disulfides [M(CO)4(R2P(S)P(S)R2)] (M = Mo, W; R = Me, Et, Pr-n, Bu-n), Polyhedron, 16: 1101-1103.

Timmers FJ, Wacholtz WF (1994). An Advanced Inorganic Laboratory Experiment Using Synthesis and Reactivity of a Cycloheptatriene Molybdenum Complex , J. Chem. Ed., 71: 987.

Werner H, Prinz R (1967). A new way to represent substituted metal carbonyl complexes of chromium and molybdenum, Chem. Ber., 100: 265-270.

Wing-Wah L, Jeong-Han K, Susan ES, Gary BQ, John EC (1993). Metabolism in rats and mice of the soil fumigants metham, methyl isothiocyanate, and dazomet, J. Agric Food Chem., 41(9): 1497-1502.

Zahida O, Saim O (1999). 13C-and 31P-NMR Study of Tetracarbonylbis(diphenylphosphino)alkanemetal(0) Complexes of The Group 6 Elements, Turk. J. Chem., 23: 9-14.

Zhao J, Xhao B, Liu J, Xu W, Wang Z (2001). Spectroscopy study on the photochromism of Schiff Bases N,N′-bis(salicylidene)-1,2-diaminoethane and N,N′-bis(salicylidene)-1,6-hexanediamine, Spectrochim Acta, 57A: 149-154.

Ab-initio study of the ground state structure and properties of Fe^{+2} (thymine)$_2$ $(H_2O)_2$ complex

D. De, S. Dalai and B. R. De*

Department of Chemistry and Chemical Technology, Vidyasagar University, Midnapore, 721102, India.

Hartree-Fock calculations with 6-31G (d) basis set have been done in the gas phase on Fe^{+2} (thymine)$_2$ $(H_2O)_2$ complex with complete geometry optimization. Stable structure for the complex has been found. Single point water phase calculation (PCM) has also been done which shows that the complex is more stable in water implying its physiological action for the removal of excess hazardous Fe^{+2} from the body. Selected optimized geometrical parameters, charge densities on selected atoms have been reported. HOMO-LUMO energies and structures are shown. The LUMO structure shows that the Fe^{+2} play the key role of the complex. The study may help in the new drug discovery.

Key words: Gaussian, gas phase, water phase, Hartree-Fock, charge density.

INTRODUCTION

The specific interactions between the purine and pyrimidine bases are one of the corner stones of the molecular biology (Saengen, 1984; Jeffrey, 1991; Huysken, 1991; Chandra, 1998). It is well known (Snustad et al., 2003) that hemoglobin provides an excellent illustration of quaternary structure. Each unit of hemoglobin is composed of α2 and β2 chains. Hemoglobin is normally formed in the body. If any malfunctioning occurs in its formation (Huheey, 2004) due to genetic or any other causes, several diseases like thalasemia, sickle cell anemia, excess Fe^{+2} accumulations, etc., appear in the body. If this excess Fe^{+2} be removed by complexation with either of the purine or pyrimidine bases, then there may be some possibilities of curing the mentioned diseases. Previously, chelate therapy was used to treat excessive quantities of copper in the body (Huheey, 2004). Keeping this view in mind, the present work has been undertaken to examine theoretically, the reaction between Fe^{+2} and thymine as a case study.

COMPUTATIONAL DETAILS

Complete geometry optimizations for Fe^{+2} (thymine)$_2$ $(H_2O)_2$ complex were done in the gas phase by Hartree-Fock method with 6-31G (d) basis set using Gaussian 03W program (Frisch et al, 2004). Actually, this optimization gives the most stable conformation of the system studied. Moreover for first hand information Hartree-Fock method with 6-31G (d) basis set is sufficient in comparison to more popular DFT method which requires much more computation time. Single point water phase calculations at the gas phase equilibrium geometry were also done using the same basis set by polarizable continuum model (PCM) approach at HF level. Here also optimization in water phase calculation is not done because of computation time problem.

*Corresponding author. E-mail: brdranjan@yahoo.com

RESULTS AND DISCUSSION

Equilibrium geometry of the studied complex with the numbering scheme of the atoms is shown in Figure 1a. The single point water phase calculations using the gas phase equilibrium geometry is shown in Figure 1a, b. The significant variations in these figures are indicated by some selective geometrical parameters given in Table 3. Table 1 reports in the gas phase, the computed total energy (hartree), dipole moment (debye) and ΔE (HOMO-LUMO gap) (hartree), at the equilibrium geometry of the complex along with single point water phase data. Table 2 summarizes the computed net Mulliken charge on the selected hetero atoms of the complex at the equilibrium geometry of the complex in the gas phase including the single point water phase calculations. The hetero atoms are assumed to play the key role for the physiological drug action because these are the most negative centres of the complex. The drug action is assumed to take place through chemical attack.

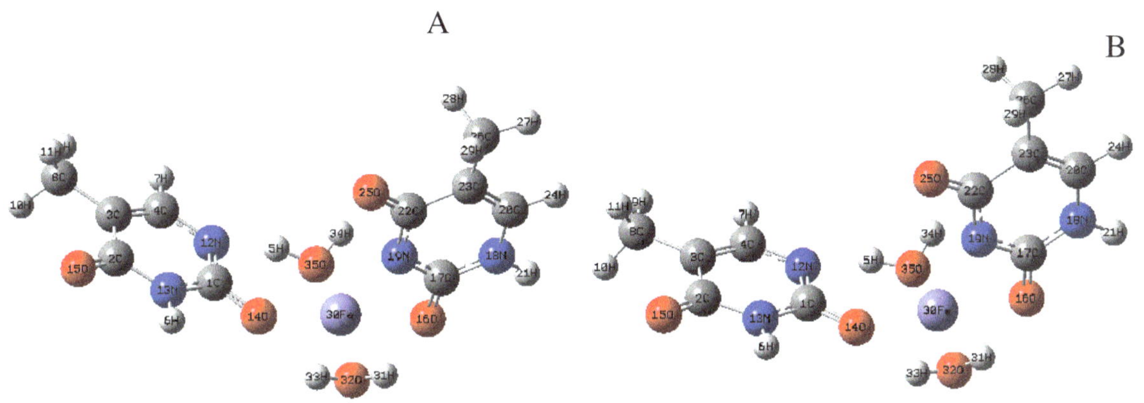

Figure 1. a) Optimized structure of Fe^{+2} $(Thymine)_2$ $(H_2O)_2$ complex in gas phase; b) Single point structure of Fe^{+2} $(Thymine)_2$ $(H_2O)_2$ complex in water phase.

Table 1. Computed total energies (hartree) dipole moments (debye), HOMO, LUMO energies (hartree) and ΔE (HOMO-LUMO gap)(hartree) at the equilibrium geometry of the ground state of Fe^{+2} (thymine $)_2$ $(H_2O)_2$ complex.

Complex in	Total energy	Dipole moment	HOMO	LUMO	ΔE(HOMO-LUMO gap)
Gas phase	-2316.1928	9.9096	-0.2994	0.0971	0.3965
Water phase	-2316.2605	13.1724	-0.3120	0.1133	0.4253

1 hatree = 627.5095 kcal(mole)-1 = 27.2116 eV.

Table 2. Computed net Mullikan charge/electron density on the selected atoms of Fe^{+2} $(thymine)_2$ $(H_2O)_2$ complex at the equilibrium ground state and single point data in water phase.

Atoms	Complex in gas phase	Complex in water phase
H_5	0.5592	0.5585
N_{12}	-0.8485	-0.8473
N_{13}	-0.9034	-0.9090
O_{14}	-0.9069	-0.8916
O_{15}	-0.6212	-0.6919
O_{16}	-0.8066	-0.8402
N_{18}	-0.8667	-0.8704
N_{19}	-0.9982	-0.9973
O_{25}	-0.6047	-0.6922
Fe_{30}	1.3958	1.4030
H_{31}	0.4945	0.5294
O_{32}	-0.9302	-0.9727
H_{33}	0.5128	0.5327
H_{34}	0.4818	0.5213
O_{35}	-0.9952	-1.0231

From Table 1, it is seen that the complex is highly stable in the gas phase at the equilibrium geometry. The stability is increased in the water phase because the total energy of the system in this phase is more negative than in gas phase. Mulliken population analysis is not unique; still it is very important to provide an idea of electron density distribution in a molecular system. The calculated dipole moment reflects an overall charge/electron density distribution in a molecular system. From Table 1 it is seen that the dipole moment of the complex in water phase is 1.5 times greater than that in gas phase as it is expected [polar complex in polar solvent]. The HOMO is more stabilized in water phase but the LUMO is little bit destabilized than the gas phase.

From Table 2, it is seen that the Fe^{+2} (in its closed shell configuration) ion receives 0.6042 and 0.5970 e amount

Table 3. Computed values of some selected geometrical parameters (length in A^0) at the equilibrium ground state of Fe^{+2} $(Thymine)_2$ $(H_2O)_2$ complex.

Atoms	Length	Atoms	length	Atoms	length
C_1-N_{12}	1.3081	O_{16}-Fe_{30}	2.0822	N_{18}-C_{20}	1.3858
C_1-N_{13}	1.3590	C_{17}-Fe_{30}	2.4406	N_{19}-C_{22}	1.3737
C_1-O_{14}	1.2662	N_{19}-Fe_{30}	2.0784	H_{31}-O_{32}	0.9496
C_2-N_{13}	1.3884	Fe_{30}-O_{32}	2.0673	N_{18}-H_{21}	0.9948
C_2-O_{15}	1.2014	Fe_{30}-O_{35}	2.0204	Fe_{30}-N_{12}	3.3400
C_4-N_{12}	1.3662	O_{32}-H_{33}	0.9532	H_5-N_{12}	1.7439
H_5-O_{35}	0.9823	H_{34}-O_{35}	0.9479	H_{33}-O_{14}	2.2501
H_6-N_{13}	0.9969	C_{17}-N_{18}	1.3490	H_{34}-N_{19}	3.1213
O_{14}-Fe_{30}	1.9790	C_{17}-N_{19}	1.3248	H_{31}-N_{19}	2.9124
O_{16}-C_{17}	1.2483	C_{22}-O_{25}	1.2002		

Table 3 (Contd). Computed values of some selected geometrical parameters (angle in degree) at the equilibrium ground state of Fe^{+2} $(Thymine)_2$ $(H_2O)_2$ complex.

Atoms	Angle in degree	Atoms	Angle in degree
O_{14}- Fe $_{30}$-C_{17}	149.32	C_1-O_{14}- Fe $_{30}$-C_{17}	86.15
O_{14}- Fe $_{30}$-N_{19}	116.65	C_1-O_{14}- Fe $_{30}$- N $_{19}$	81.22
O_{14}- Fe $_{30}$-O_{32}	81.14	C_1-O_{14}- Fe $_{30}$-O_{32}	171.22
O_{14}- Fe $_{30}$-O_{35}	92.70	C_1-O_{14}- Fe $_{30}$-O_{35}	-20.88
O_{16}- Fe $_{30}$- N $_{19}$	63.60	O_{14}- Fe $_{30}$-O_{32}-H_{31}	-129.93
O_{16}- Fe $_{30}$-O_{32}	93.71	O_{16}- Fe $_{30}$-O_{32}-H_{33}	167.66
O_{16}- Fe $_{30}$-O_{35}	92.28	C_{17}- Fe $_{30}$-O_{32}-H_{33}	136.84
N $_{19}$- Fe $_{30}$-O_{32}	93.95	N $_{19}$- Fe $_{30}$-O_{32}-H_{31}	-13.55
N $_{19}$- Fe $_{30}$-O_{35}	99.56	O_{35}- Fe $_{30}$-O_{32}-H_{31}	166.38
Fe $_{30}$-O_{32}-H_{33}	103.19	O_{16}-C_{17}-N_{18}-H_{21}	0.19
H_{31}-O_{32}-H_{33}	107.69	O_{25}-C_{22}-C_{23}-C_{26}	-1.09
O_{14}-C_1-N_{12}-C_4	179.55	O_{14}- Fe $_{30}$-O_{35}-H_5	15.59
N_{12}-C_1-O_{14}- Fe $_{30}$	12.27	O_{16}- Fe $_{30}$-O_{35}-H_5	-165.67
N_{13}-C_1-O_{14}- Fe $_{30}$	-168.64	O_{32}- Fe $_{30}$-O_{35}-H_{34}	-158.67
C_3-C_2-N_{13}-C_1	0.11	O_{32}- Fe $_{30}$-O_{35}-H_5	77.83
O_{15}-C_2-N_{13}-H_6	-0.85		

of charge in gas and water phase respectively from two thymine ligands and two water ligands. In almost all the hetero atoms, the charge densities in gas phase is a little less than that in the water phase implying that in water phase, the negative centers become more effective.

HOMO and LUMO structures of the complex both in gas phase and single point water phase are shown in Figures 2a, b, and 3a, b, respectively. The significance of the figures is that complete geometry of the complex is seen which is not reflected from the selected geometrical parameters shown in Table 3. The LUMO structure shows that Fe^{+2} ion plays the key role of the complex both in gas phase and water phase which is also supported from figures. Table 3 shows some selected geometrical parameters of the complex in the gas phase particularly the metal containing zone and other heteroatom containing zone. From Table 3, it is seen that Fe^{+2} containing zones are non planar and this part of the complex is out of plane from both the thymine ligands and two H2O ligands also. Fe –O (35) and Fe-O (32) distances are 2.0204 A0 and 2.0673A0 respectively. Fe-O (14) and Fe-O (16) distances are 1.9796 A0 and 2.0822A0, respectively. Fe-N (12) and Fe-N (19) distances are 3.3400 A0 and 2.0783 A0, respectively. The most stable conformation of the system favors such geometrical parameters.

Conclusions

From the present theoretical study, it is found that the Fe^{+2} $(thymine)_2$ $(H_2O)_2$ complex is a stable complex both

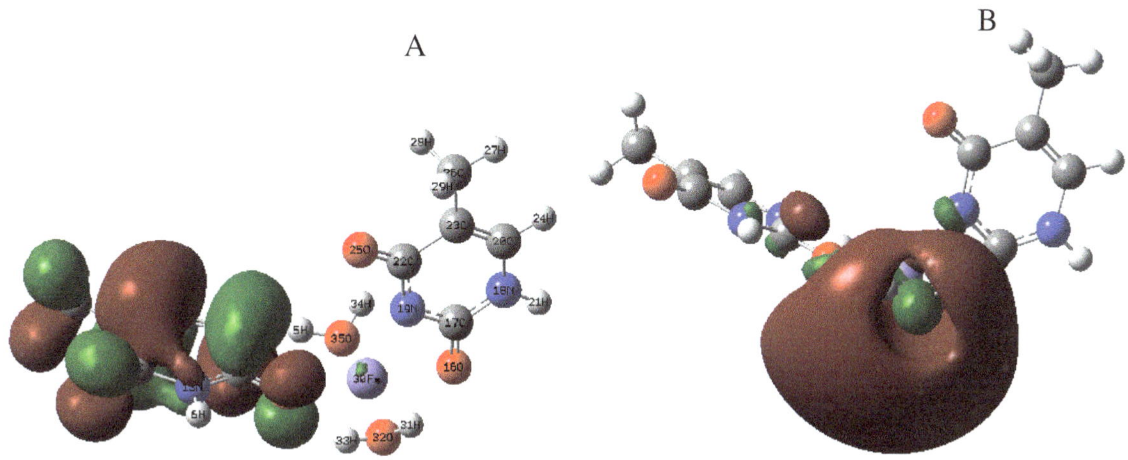

Figure 2. (a) HOMO structure of Fe^{+2} (Thymine)$_2$ $(H_2O)_2$ complex in gas phase; b) LUMO structure of Fe^{+2} (Thymine)$_2$ $(H_2O)_2$ complex in gas phase.

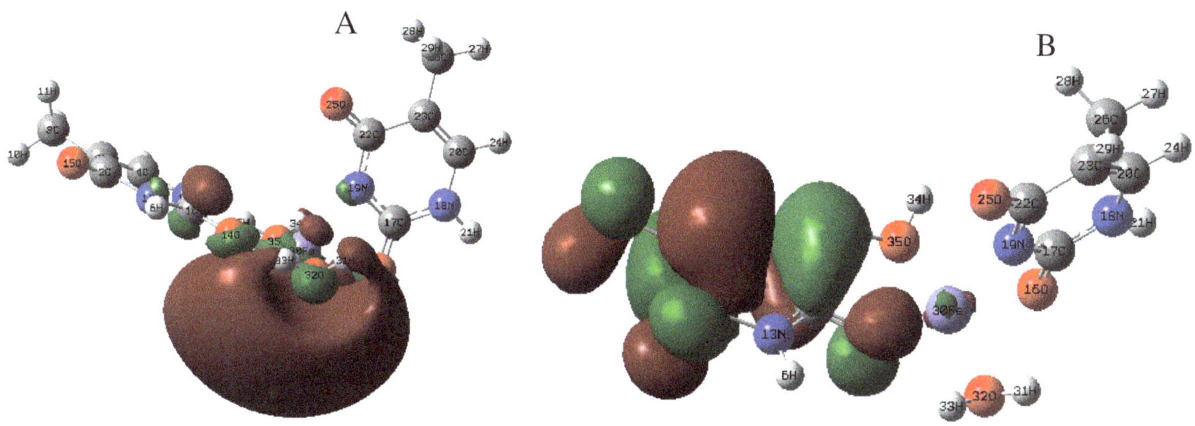

Figure 3. a) LUMO structure of Fe^{+2} (thymine)$_2$ $(H_2O)_2$ complex in water phase. (b) HOMO structure of Fe^{+2} (thymine)$_2$ $(H_2O)_2$ complex in water phase.

in gas phase and in water phase (single point calculation by polarisable continuum model- PCM). This first hand information of the title complex signifies its physiological importance for the removal of excess Fe^{+2} from the body. The LUMO structure shows that Fe^{+2} ion plays the key role of the complex both in gas phase and water phase. The present study may help in the discovery of new drugs.

REFERENCES

Chandra AK (1998). Theoretical study of the interaction between Thyamine and water, Protonation and Deprotonation Enthalpies and Comparison with Uracil, J. Phys. Chem. , 102: 6010-6016.

Frisch MJ, Trucks GW, Schlegel HB, Scuseria GE, Robb MA, Cheeseman JR, Montgomery JA, Vreven Jr. T, Kudin KN, Burant JC, Millam JM, Iyengar SS, Tomasi J, Barone V, Mennucci B, Cossi M, Scalmani G, Rega N, Petersson GA, Nakatsuji H, Hada M, Ehara M, Toyota K, Fukuda R, Hasegawa J, Ishida M, Nakajima T, Honda Y, Kitao O, Nakai H, Klene M, Li X, Knox JE, Hratchian HP, Cross JB, Adamo C, Jaramillo J, Gomperts R, Stratmann RE, Yazyev O, Austin AJ, Cammi R, Pomelli C, Ochterski JW, Ayala PY, Morokuma K, Voth GA, Salvador P, Dannenberg JJ, Zakrzewski VG, Dapprich S, Daniels AD, Strain MC, Farkas O, Malick DK, Rabuck AD, Raghavachari K, Foresman JB, Ortiz JV, Cui Q, Baboul AG, Clifford S, Cioslowski J, Stefanov BB, Liu G, Liashenko A, Piskorz P, Komaromi I, Martin RL, Fox DJ, Keith T, Al-Laham MA, Peng CY, Nanayakkara A, Challacombe M, Gill PMW, Johnson BG, Chen W, Wong MW, Gonzalez C, Pople JA (2004) Gaussian, Inc., Wallingford CT, Huysken Z (1991). In intermolecular forces. An introduction to Modern Methods and Results Springer Verlag, Berlin, Heidelberg, Newyork 24.

Huheey JE (2004). Inorganic Chemistry, Principles of structure and reactivity, Sixth edition, Pearson Education, pp. 902-908, 956-960.

Jeffrey GA (1991). Hydrogen Bonding in Biological structures, Springer Verlag, Berlin, Heidelberg, Newyork.

Saengen W (1984). Principles of Nucleic Acid Structure, Springer Verlag,

Snustad DP (2003). Principles of Genetics, third edition, John Willey and Sons, pp. 303-307.

Co-precipitation synthesis, physical and magnetic properties of manganese ferrite powder

Irfan Elahi*, Rabab Zahira, Kiran Mehmood, Arifa Jamil and Nasir Amin

Department of Physics, University of Agriculture, Faisalabad, Pakistan.

Magnetic manganese ferrite powder was synthesized via chemical co-precipitation techniques using metallic chlorides of manganese and iron. Sodium hydroxide (NaOH) base was used as precipitant agent. The calcinations were performed at 600°C for 5 h. The structural investigation of the prepared sample was performed with X-ray diffractometer (XRD) and scanning electron microscope (SEM). The crystallite size was 26.53 nm, calculated using Sherrer formula. Magnetic properties were studied by vibrating sample magnetometer (VSM) techniques at room temperature.

Key words: Ferrite, co-precipitation, X-ray diffractometer (XRD), scanning electron microscopy (SEM), vibrating sample magnetometer (VSM).

INTRODUCTION

On account of their diverse and fascinating applications in wide technological and scientific fields, the synthesis and characterization of nano-sized magnetic materials have emerged as an important subject during last several years (Awschalom and DiVicenzo, 1995; Raj et al., 1990; Ziolo et al., 1992; Raj et al., 1995; Leslie-Pelecky and Rieke, 1996; Majetich et al., 1997; Kodama, 1999; Sugimoto, 1999; Rosensweig, 1985; Petri-Fink et al., 2005; Zhu et al., 2005; Franc et al., 2003; Ammar et al., 2001; Häfeli et al., 1997; Liu et al., 2000). Among these, ferrites, ferromagnetic having cubic spinel structure (MFe_2O_4) are complex versatile materials have been extensively investigated by the scientists in view of their unique optical, electrical and magnetic properties (Ngo and Pileni, 2001; Singhal et al., 2005; Li et al., 2007; Sun et al., 2004; Deng et al., 2005; Hyeon et al., 2002; Ahmed and Kofinas, 2002; Li et al., 2000; Berbenni et al., 2006; Yang et al., 2004; Meron et al., 2005; Tirosh et al., 2006; Skumiel, 2006; Poddar et al., 2006; Goldman 1999; Abdullah and Yusoff, 1996). Several Physical and Chemical methods like, Hydrothermal (Baruwati et al., 2007; Hu et al., 2008; Verma et al., 2004), Co-precipitation (Suwalka et al., 2007; Mathur et al., 2008; Ataie et al., 1995; Pankov et al., 1993; Aquino et al., 2005), Combustion (Yan et al., 1999), Sol-gel (Mandal et al., 2001; Chatterjee et al., 1993) and some other techniques are being employed for the synthesis of these nano-sized spinel ferrites. Manganese ferrite is exceptionally important member of ferrite family with a variety of applications in modern era of science and engineering. Like other ferrites, Manganese ferrite nano-particles have been synthesized by various methods like ball milling (Mahmoud et al., 2000; Muroi et al., 2001; Ding et al., 1997), reverse micelle synthesis (Liu and Zhang, 2001), thermal decomposition (Balaji et al., 2002), sol-gel method (Nemoto et al., 1995; Gajbhiye and Balaji, 2002), solid-phase reaction (Deraz and El-Shobaky, 2001), pulsed laser deposition (Mahmoud et al., 2003) and thermally activated solid state reaction (Alvani et al., 2005) and have been characterized by X-ray diffractometer (XRD), scanning electron microscopy (SEM), transmission electron microscope (TEM), vibrating sample magnetometer (VSM) and Mössbauer studies.

In the present work, ultra-fine, nano-sized, magnetic manganese ferrite powder has been prepared by co-precipitation technique using metallic chlorides of Mn and Fe as precursors and studied by X-ray diffraction (λ = 1.5404Å), scanning electron microscope and vibrating sample magnetometer.

*Corresponding author. E-mail: si_uaf@yahoo.com.

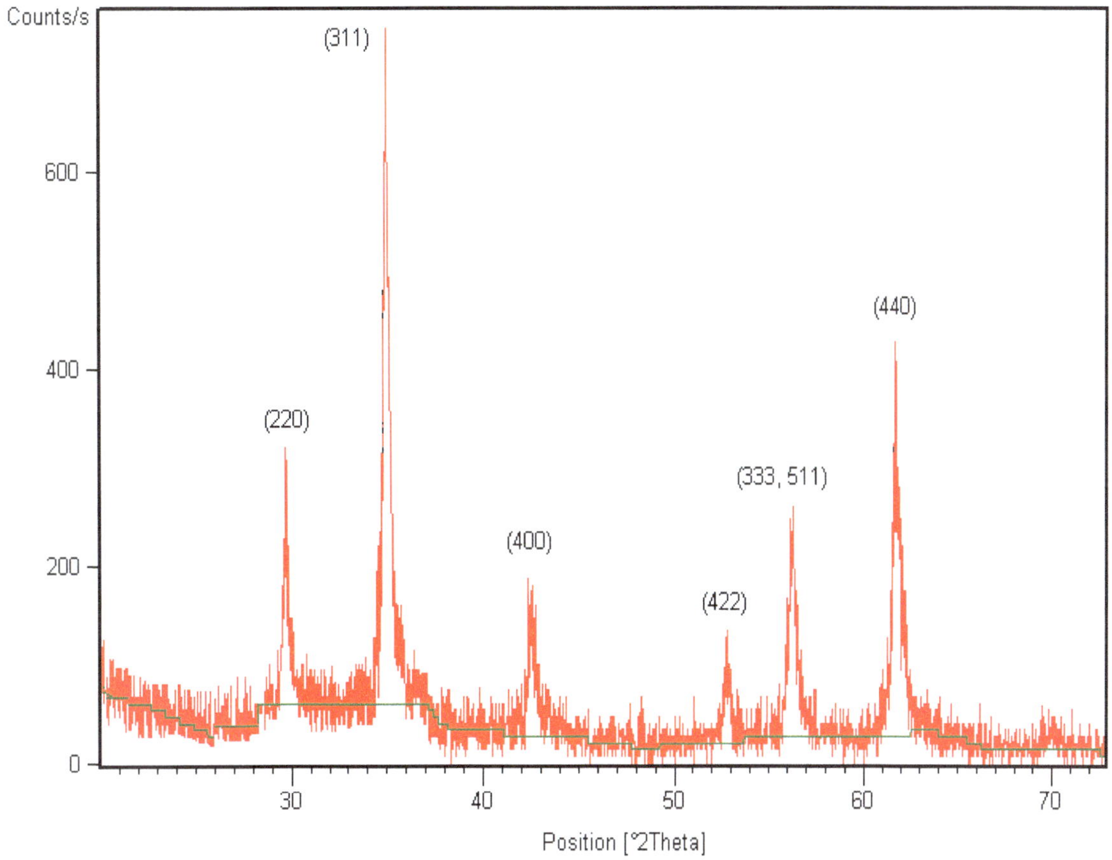

Figure 1. X ray diffraction pattern of manganese ferrite powder.

EXPERIMENTAL PROCEDURE

Nano-sized powder of magnetic manganese ferrite was prepared by co-precipitation method using manganese chloride tetra-hydrated, ferric chloride hexa-hydrated and sodium hydroxide as starting materials. The ratio between Fe and Mn was kept at 2:1, respectively. A 100 ml of salt solution was added drop by drop to the base solution until the metallic to hydroxyl ion (Me/OH) ratio reached to 0.2. The gray precipitates were obtained in this routine. This obtained product was washed and filtered repeatedly with distilled water until the pH of the residual solution reaches 7. This process took about 6 to 7 h. The sample was then dried in an oven keeping temperature at 50°C for 3 h. After attaining the powder form by motor and pestle, calcinations process was carried out at 600°C for 5 h to get the final product of manganese ferrite.

This final manganese ferrite powder was characterized by X-ray diffraction (X' Pert Pro, PANalytical, Netherlands), scanning electron microscope (The Philips XL30) operated at 10 and 20 kV, and vibrating sample magnetometer (BHV-50, Richen Denish Company, Japan). The particle size was calculated by Sherrer's formula given as follows, from the full width at half-maximum of the XRD pattern for the (311) plane of the spinel structure:

$$t = 0.9\,\lambda / B \cos \Theta_B$$

Here t is the mean crystallite size, B is the half width of the relevant diffraction peak, λ the x-ray wavelength and θ is the angle of diffraction (Nan et al., 2006; Cullity, 1978). The broadening (B) of the diffraction peak corresponds to the size of the crystal.

RESULTS AND DISCUSSION

The XRD pattern recorded for the magnetic ferrite powder obtained by co-precipitation and calcined at 600°C is shown in Figure 1. The pattern of powder exhibited all the major peaks related to single phase cubic spinel structure of Manganese Ferrite. The peak analysis shows clear resemblance with the XRD pattern of $MnFe_2O_4$ prepared by different ways (Ahmad et al., 2008). No external peak was detected in the prepared sample. The lattice parameter calculated from XRD information is 8.493Ǻ. The particle size was measured by Scherer's formula using half width of the peak of maximum intensity (311) plane. The value of particle size is calculated as 23.53 nm. The values of lattice constant and the particle size are in line with the previous published results of manganese ferrite prepared by co-precipitation as well as other techniques.

Figure 2 shows morphological pattern of the prepared ferrite powder taken by scanning electron microscope. The SEM figure clearly indicates that microstructure of the grains of the sample is ultra-small. Most of the grains containing a large number of atoms are very small in dimensions, having an average grain size of 10 μm. The size distribution is extraordinarily narrow. Also, there

Figure 2. Scanning electron micrographs at different magnifications of $MnFe_2O_4$ sample calcined at 600°C.

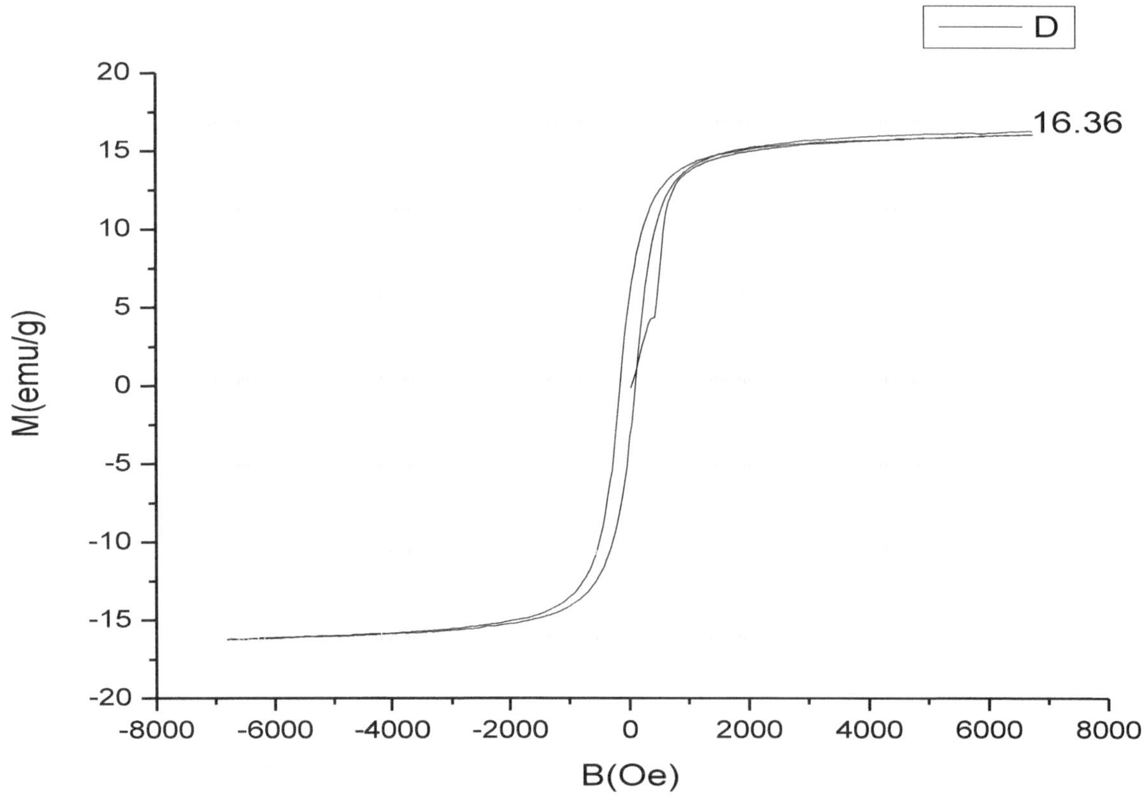

Figure 3. Magnetic hysteresis curve of $MnFe_2O_4$ measured at room temperature by vibrating sample magnetometer.

exist some grains of larger size but having normal growth.

VSM studies

Figure 3 exhibits the hysterics loop of $MnFe_2O_4$ powder taken by vibrating sample magnetometer at room temperature. Instead of having smaller size, the powder exhibited clear hysteresis showing typical magnetic behavior, ferromagnetic, having a non-zero coercivity and remanance. The value of magnetic saturisation 16.36 emu/g, is at B is equal to 6.7 kOe. Low area hysteresis curve shows that the material formed is of soft magnetic material which keeps much importance in magnetic memory purposes.

In this work an effort was made to synthesize manganese ferrite powder by a very simple process of co-precipitation which gained a reasonably incredible success, resulting into small particle size and good magnetic properties as shown by hysteresis curve. Also, further studies on the synthesis of manganese, zinc and manganese ferrite at different temperature conditions are in process.

REFERENCES

Abdullah MH, Yusoff AN (1996). Complex Impedance and Dielectric Properties of an Mg-Zn Ferrite. J. Alloys Compd., 233: 129-135.

Ahmad MA, Okasha N, El-Dek SI (2008). Preparation and characterization of nanometric Mn ferrite via different methods. Nanotechnol., 19 (6): 065003.

Ahmed SR, Kofinas P (2002). Controlled Room Temperature Synthesis of $CoFe_2O_4$ Nanoparticles Through a Block Copolymer Nanoreactor Route. Macromolecules, 35(9): 3338-3341.

Alvani C, Ennas G, La Barbera A, Marongiu G, Padella F, Varsano F (2005). Synthesis and characterization of nanocrystalline $MnFe_2O_4$ advances in thermochemical water splitting. Int. J. Hydrogen Energy, 30(13-14): 1407-1411.

Ammar S, Helfen A , Jouini N, Fiévet F, Rosenman I, Villain F, Molinié P, Danot M (2001). Magnetic properties of ultrafine cobalt ferriteparticles synthesized by hydrolysis in a polyol medium. J. Mater. Chem., 11: 186-192.

Aquino R, Depeyrot J, Sousa M, Tourinho F, Debois E, Perzynski R (2005). Magnetization temperature dependence and freezing of surface spins in magnetic fluids based on ferrite nanoparticles. Phys. Rev. B., 72(18): 184435.

Ataie A, Piramoon MR, Harris IR, Ponton CB (1995). Effect of hydrothermal synthesis environment on the particle morphology, chemistry and magnetic properties of barium hexaferrite. J. Mater. Sci., 30: 5600.

Awschalom DD, DiVicenzo DP (1995). Complex Dynamics of Mesoscopic Magnets. Phys. Today, 48(5): 43.

Balaji G, Gajbhiye NS, Wilde G, Weissmuller J (2002). Magnetic properties of $MnFe_2O_4$ nanoparticles. J. Magn. Magn. Mater., 617(20): 242-245.

Baruwati B, Rana RK, Manorama SV (2007). Further Insights in the Conductivity Behaviour of Nanocrystalline $NiFe_2O_4$. J. Appl. Phys., 101(1): 221.

Berbenni V, Milanese C, Bruni G, Marini A, Pallecchi I (2006). Synthesis and magnetic properties of $ZnFe_2O_4$ obtained by mechanochemically assisted low-temperature annealing of mixtures of Zn and Fe oxalates. Thermochimica Acta. 447(2): 184-189.

Chatterjee A, Das D, Pradhan SK, Chajaravorty D (1993). Synthesis of nanocrystalline nickel-zinc ferrite by the sol-gel method. J. Magn. Magn. Mater., 127(1-2): 214-218.

Cullity BD (1978). Elements of X-ray diffraction. Addison – Wesley Publishing Company, Canada.

Deng H, Li X, Peng Q, Wang X, Chen J, Li Y (2005). Monodisperse Magnetic Single-Crystal Ferrite Microspheres. Angew. Chem. Int. Ed., 44(18): 2782-2785.

Deraz NM, El-Shobaky GA (2001). Solid-solid interaction between ferric oxide and manganese carbonate as influenced by lithium oxide doping. Thermochimica Acta. 375: 137-145.

Ding J, Street R, McCormick PG (1997). Formation of spinel Mn-ferrite during mechanical alloying. J. Magn. Magn. Mater., 171(3): 309-314.

Franc BL, Mandl SJ, Siprashvili Z, Wender P, Contag CH (2003). Breaching biological barriers: protein translocation domains as tools for molecular imaging and therapy. Mol. Imaging, 2(4): 313-323.

Gajbhiye NS, Balaji G (2002). Synthesis, reactivity, and cations inversion studies of nanocrystalline $MnFe_2O_4$ particles. Thermochimica Acta. 385(1-2): 143-151.

Goldman A (1999). Handbook of Ferromagnetic Materials. Kluwer Academic Publishers, Dordrecht (Hingham, MA).

Häfeli U, Schütt W, Teller J, Zborowski M (1997). Scientific and Clinical Applications of Magnetic Carriers. 1st Edition, Plenum, New York. pp. 1-10.

Hu C, Gao Z, Yang X (2008). One-pot low temperature synthesis of MFe_2O_4 (M=Co, Ni, Zn) superparamagnetic nanocrystals. J. Magn. Magn. Mater., 320(8): L70-L73.

Hyeon T, Chung Y, Park J, Lee SS, Kim YW, Park BH (2002). Synthesis of Highly Crystalline and Monodisperse Cobalt Ferrite Nanocrystals. J. Phys. Chem. B., 106(27): 6831-6833.

Kodama RH (1999). Magnetic Nanoparticles. J. Magn. Magn. Mater., 200: 359-372.

Leslie-Pelecky DL, Rieke RD (1996). Magnetic Properties of Nanostructured Materials. Chem. Mater., 8(8): 1770-1783.

Li F, Wang H, Wang L, Wang J (2007). Magnetic properties of $ZnFe_2O_4$ nanoparticles produced by a low-temperature solid-state reaction method. J. Magn. Magn. Mater., 309(2): 295-299.

Li L, Li G, Smith RL, Inomata H (2000). Microstructural Evolution and Magnetic Properties of $NiFe_2O_4$Nanocrystals Dispersed in Amorphous Silica. Chem. Mater., 12(12): 3705-3714.

Liu C, Zou B, Rondinone JA, Zhang ZJ (2000). Chemical Control of Superparamagnetic Properties of Magnesium and Cobalt Spinel Ferrite Nanoparticles through Atomic Level Magnetic Couplings. J. Am. Chem. Soc., 122(26): 6263-6267.

Liu C, Zhang ZJ (2001). Size-dependent superparamagnetic properties of Mn spinel ferrite nanoparticles synthesized from reverse micelles. Chem. Mater., 13(6): 2092-1396.

Mahmoud MH, Hamdeh HH, Ho JC, O'Shea MJ, Walker JC (2000). Mo¨sshauer studics of manganese ferrite fine particles processed by ball-milling. J. Magn. Magn. Mater., 220(2-3): 139-146.

Mahmoud MH,Williams CM, Cai J, Siu I, Walker JC (2003). Investigation of Mn-ferrite films produced by pulsed laser deposition. J. Magn. Magn. Mater., 261(3): 314-318.

Majetich SA, Scott JH, Kirkpatrick EM, Chowdary K, Gallagher K, McHenry ME (1997). Nanostructured Mater., 9: 291.

Mandal K, Mandal SP, Agudo P, Pal M (2001). A study of nanocrystalline (Mn–Zn) ferrite in SiO_2 matrix. Appl. Surf. Sci., 182: 386-389.

Mathur P, Thakur A, Singh M (2008). Effect of nanoparticles on the magnetic properties of Mn–Zn soft ferrite. J. Magn. Magn. Mater., 320(7): 1364-1369.

Meron T, Rosenberg Y, Lereah Y, Markovich G (2005). Synthesis and assembly of high-quality cobalt ferrite nanocrystals prepared by a modified sol-gel technique. J. Magn. Magn. Mater., 292: 11-16.

Muroi M, Street R, McCormick PG, Amighian J (2001). Magnetic properties of ultrafine $MnFe_2O_4$ powders prepared by mechanochemical processing. Physical Review B., 63(18): 184414.

Nan J, Han D, Cui M, Yang M, Pan L (2006). Recycling spent zinc manganese dioxide batteries through synthesizing Zn-Mn ferrite magnetic materials. J. Hazardous Mater., 33: 257.

Nemoto N, Schleich DM, Zhang Y (1995). Preparation of some metal ferrite MFe_2O_4 thin films through a nonaqueous sol method. Mater. Res. Bull., 30(4): 447-452.

Ngo T, Pileni MP (2001). Assemblies of ferrite Nanocrystals: Partial orientation of the easy magnetic axes. J. Phys. Chem. B., 105: 53.

Pankov VV, Pernet M, Germi P, Molard P (1993). Fine hexa ferrite particles for perpendicular recording prepared by the co-precipitation method in presence of an inert component. J. Magn. Magn. Mater. 120(1-3): 69-72.

Petri-Fink A, Chastellain M, Juillerat-Jeanneret L, Ferrari A, Hofmann H (2005). Development of functionalized superparamagnetic Iron oxide nanoparticles for Interaction with human cancer cells. Biomater., 26(15): 2685-2694.

Poddar J, Gass J, Rebar D, Srinath S, Srikanth H, Morrison S, Carpenter E (2006). Magnetocaloric effect in ferrite nanoparticles. J. Magn. Magn. Mater., 307(2): 227-231.

Raj K, Moskowitz R (1990). Commercial applications of ferrofluids. J. Magn. Magn. Mater., 85: 233.

Raj K, Moskowitz B, Casciari R (1995). Advances in ferrofluid technology. J. Magn. Magn. Mater., 149(1-2): 174-180.

Rosensweig RE (1985). Ferrohydrodynamics. Dover Publications, Inc. Mineola, New York. pp. 1-332.

Singhal S, Singh J, Barthwal SK, Chandra K (2005). Structural, magnetic and Mössbauer spectral studies of nano size nickel substituted cobalt ferrites. J. Solid State Chem., 178: 3183-3189.

Skumiel A (2006). J. Magn. Magn. Mater. Suitability of water based magnetic fluid with CoFe. 307: 85-90.

Sugimoto M (1999). The past, present and future of ferrites. J. Am. Ceram. Soc., 82: 269-280.

Sun S, Zeng Robinson DB, Raoux S, Rice PM, Wang SX, Li G (2004). Monodisperse MFe_2O_4 (M = Fe, Co, Mn) Nanoparticles. J. Am. Chem. Soc., 126 (1): 273-279.

Suwalka O, Sharma RK, Sebastian V, Lakshami N, Venugopalan K (2007). A study of nano NiCoZn-ferrite prepared by co-precipitation. J. Magn. Magn. Mater., 313(1): 203.

Tirosh E, Shemer G, Markovich G (2006). Optimizing Cobalt Ferrite Nanocrystal Synthesis Using a Magneto-Optical Probe. Chem. Mater., 18: 465.

Verma S, Joy PA, Khollam YB, Potdar HS, Deshpande SB (2004). Synthesis of nanosized $MgFe_2O_4$ powders by microwave-hydrothermal method. Mater. Lett., 58(6): 1092-1095.

Yan CH, Xu ZG, Cheng FX, Wang ZM, Sun JD, Liao CS, Jia JT (1999). Nanophased $CoFe_2O_4$ prepared by combustion method. Solid State Commun., 111(5): 287-291.

Yang H, Zhang X, Huang C, Yang W, Qiu G (2004). Synthesis of $ZnFe_2O_4$ nanocrystallites by mechanochemical reaction. J. Phys. Chem. Solids, 65(2-3): 1329-1332.

Zhu J, Wu X, Zhang HL (2005). Adult Neural Stem Cell Therapy: Expansion *In vitro*, Tracking *In vivo* and Clinical Transplantation. Current Drug Targets, 6: 97-110.

Ziolo RF, Giannelis EP, Weinstein BA, O'Horo MP, Ganguly BN, Mehrotra V, Russel MW, Huffman DR (1992). Science, 257(5067): 219-223.

Permissions

All chapters in this book were first published in AJPAC, by Academic Journals; hereby published with permission under the Creative Commons Attribution License or equivalent. Every chapter published in this book has been scrutinized by our experts. Their significance has been extensively debated. The topics covered herein carry significant findings which will fuel the growth of the discipline. They may even be implemented as practical applications or may be referred to as a beginning point for another development.

The contributors of this book come from diverse backgrounds, making this book a truly international effort. This book will bring forth new frontiers with its revolutionizing research information and detailed analysis of the nascent developments around the world.

We would like to thank all the contributing authors for lending their expertise to make the book truly unique. They have played a crucial role in the development of this book. Without their invaluable contributions this book wouldn't have been possible. They have made vital efforts to compile up to date information on the varied aspects of this subject to make this book a valuable addition to the collection of many professionals and students.

This book was conceptualized with the vision of imparting up-to-date information and advanced data in this field. To ensure the same, a matchless editorial board was set up. Every individual on the board went through rigorous rounds of assessment to prove their worth. After which they invested a large part of their time researching and compiling the most relevant data for our readers.

The editorial board has been involved in producing this book since its inception. They have spent rigorous hours researching and exploring the diverse topics which have resulted in the successful publishing of this book. They have passed on their knowledge of decades through this book. To expedite this challenging task, the publisher supported the team at every step. A small team of assistant editors was also appointed to further simplify the editing procedure and attain best results for the readers.

Apart from the editorial board, the designing team has also invested a significant amount of their time in understanding the subject and creating the most relevant covers. They scrutinized every image to scout for the most suitable representation of the subject and create an appropriate cover for the book.

The publishing team has been an ardent support to the editorial, designing and production team. Their endless efforts to recruit the best for this project, has resulted in the accomplishment of this book. They are a veteran in the field of academics and their pool of knowledge is as vast as their experience in printing. Their expertise and guidance has proved useful at every step. Their uncompromising quality standards have made this book an exceptional effort. Their encouragement from time to time has been an inspiration for everyone.

The publisher and the editorial board hope that this book will prove to be a valuable piece of knowledge for researchers, students, practitioners and scholars across the globe.

List of Contributors

Abida Perveen
Department of Chemistry, University of Karachi, Kaarchi-75270 Pakistan

Teimoor Nezamoleslam
Department of Chemistry, University of Karachi, Kaarchi-75270 Pakistan

Iftikhar Imam Naqvi
Department of Chemistry, University of Karachi, Kaarchi-75270 Pakistan

Soheila Sedaghat
Department of Chemistry, Islamic Azad University, Tehran North branch, 19136-74711, Iran

I. R. Jack
Department of Chemistry, Rivers State University of Science and Technology, P. M. B. 5080, Port Harcourt, Nigeria

K. O. Orubite
Department of Chemistry, Rivers State University of Science and Technology, P. M. B. 5080, Port Harcourt, Nigeria

U. A. Kuevi
Laboratoire de Chimie Théorique et de Spectroscopie Moléculaire (LACTHESMO) Université d'Abomey-Calavi, 03 BP: 3409 Cotonou, Bénin Republic

Y. G. S. Atohoun
Laboratoire de Chimie Théorique et de Spectroscopie Moléculaire (LACTHESMO) Université d'Abomey-Calavi, 03 BP: 3409 Cotonou, Bénin Republic

J. B. Mensah
Laboratoire de Chimie Théorique et de Spectroscopie Moléculaire (LACTHESMO) Université d'Abomey-Calavi, 03 BP: 3409 Cotonou, Bénin Republic

Ali Moghimi
Department of Chemistry, Varamin (Pishva) Branch Islamic Azad University, Varamin, Iran

Paul Ocheje Ameh
Department of Chemistry, Ahmadu Bello University, Zaria, Kaduna State, Nigeria

Mary Adelaide Oladipo
Department of Pure and Applied Chemistry, Ladoke Akintola University of Technology, P. M. B 4000, Ogbomoso, Oyo State, Nigeria

Olugbenga Solomon Bello
Department of Pure and Applied Chemistry, Ladoke Akintola University of Technology, P. M. B 4000, Ogbomoso, Oyo State, Nigeria

Adewumi Idowu Adeagbo
Department of Chemistry, Emmanuel Alayande College of Education Oyo, Oyo State, Nigeria

Hassan Rezaei
Environmental Sciences Department, University of Pune - Pune, India

Satish D. Kulkarni
Environmental Sciences Department, University of Pune - Pune, India

Praveen G. Saptarshi
Environmental Sciences Department, University of Pune - Pune, India

Malathi Challa
Department of Chemistry, MSRIT, Bangalore-54, Karnataka, India

T. Srinivasulu Reddy
Department of Chemistry, S. K. University, Anantapur - 515003, Andra Pradesh, India

Xuehui Zhao
Key Laboratory of Green Packaging and Application Biological Nanotechnology of Hunan Province, Hunan University of Technology, Zhuzhou 412008, P. R. China

Hongmei peng
Key Laboratory of Green Packaging and Application Biological Nanotechnology of Hunan Province, Hunan University of Technology, Zhuzhou 412008, P. R. China

Liangcai Yu
Key Laboratory of Theoretical Chemistry and Molecular Simulation of Ministry of Education, Hunan Provincial University Key Laboratory of QSAR/QSPR, School of Chemistry and Chemical Engineering, Hunan University of Science and Technology, Xiangtan Hunan, 411201, P. R. China

Robert Byamukama
Chemistry Department, Makerere University, P. O. Box 7062, Kampala, Uganda

George Ogweng
Chemistry Department, Kyambogo University, P. O. Box 1, Kyambogo- Kampala, Uganda

Monica Jordheim
Department of Chemistry, University of Bergen, Allegt.41, 5007 Bergen, Norway

Øyvind M. Andersen
Department of Chemistry, University of Bergen, Allegt.41, 5007 Bergen, Norway

Bernard T. Kiremire
Chemistry Department, Makerere University, P. O. Box 7062, Kampala, Uganda

O. Ombaka
Department of Chemistry, Chuka University College, Box 109, Chuka, Kenya

J. M. Gichumbi
Department of Chemistry, Chuka University College, Box 109, Chuka, Kenya

Mohammad Hakimi
Chemistry Department, Payame Noor University, 19395-4697 Tehran, Iran

Keyvan Moeini
Chemistry Department, Payame Noor University, 19395-4697 Tehran, Iran

Zahra Mardani
Chemistry Department, Payame Noor University, 19395-4697 Tehran, Iran

Sanaz Khorasani
Molecular Sciences Institute, School of Chemistry, University of the Witwatersrand, Johannesburg, South Africa

I. Mobasherpour
Ceramics Department, Materials and Energy Research Center, P. O. Box 31787/316, Karaj, Iran

E. Salahi
Ceramics Department, Materials and Energy Research Center, P. O. Box 31787/316, Karaj, Iran

M. Pazouki
Energy Department, Materials and Energy Research Center, P. O. Box 31787/316, Karaj, Iran

A. Bahari
Department of Physics, University of Mazandaran, Iran

M. Asgharzadeh
Department of Physics, University of Mazandaran, Iran

M. Najafikhah
Department of Physics, University of Mazandaran, Iran

Shankar Ramchandra Thopate
Department of Chemistry, Dr. John Barnabas Post Graduate School for Biological Studies, Ahmednagar College, Ahmednagar, Maharashtra. 414001. India

Santosh Rajaram Kote
Department of Chemistry, Dr. John Barnabas Post Graduate School for Biological Studies, Ahmednagar College, Ahmednagar, Maharashtra. 414001. India

Ruth Hojvat
Facultad de Ciencias Exactas y Naturales, Universidad de Belgrano, Zabala 1837 CABA (CP 1426DQG) Argentina
Escuela de Tecnología y Escuela de Ciencias Agrarias, Universidad Nacional del Noroeste de la Provincia de Buenos Aires, Sede Junin, R. Sáenz Peña 456, Junín B6000, Pcia de Buenos Aires, Argentina

Dora A. Barbiric
Facultad de Ingeniería, Universidad de Buenos Aires, Av. Paseo Colón 850, C1063ACV, Buenos Aires, Argentina

Eduardo A. Castro
Instituto de Investigaciones Fisicoquímicas Teóricas y Aplicadas, Universidad Nacional de La Plata, Suc. 4, C.C. 16, 1900La Plata, Buenos Aires, Argentina

C. Boukratem
Institute of Chemistry. University of Oran, 31000, Algeria

Y. Bouhadda

Department of Chemistry, University of Mascara, 29000, Algeria

T. Fergoug
Department of Chemistry, University of Mascara, 29000, Algeria

M. A. Zenasni
Department of Sciences, Faculty of Sciences and Technology, University of Bechar, 08000 Bechar, Algeria

S. Benfarhi
Department of Chemistry, Faculty of Sciences, University of Batna, 05000 Batna, Algeria

A. Mansri
Department of Chemistry, Faculty of Sciences, University of Tlemcen, 13000 Tlemcen, Algeria

H. Benmehdi
Department of Sciences, Faculty of Sciences and Technology, University of Bechar, 08000 Bechar, Algeria

B. Meroufel
Department of Sciences, Faculty of Sciences and Technology, University of Bechar, 08000 Bechar, Algeria

J. Desbrieres
University of Pau, Helioparc Pau Pyrenees, 64053 Pau Cedex 09, France

R. Dedriveres
University of Pau, Helioparc Pau Pyrenees, 64053 Pau Cedex 09, France

Abdou A. Abdou
Nuclear Materials Authority (NMA) Egypt

Prithwiraj Byabartta
Departmento de Quimica Inorganica-Instituto de Ciencia deMateriales de Aragon, Universidad de Zaragoza-CSIC, Zaragoza-50009, Spain
Department of Chemistry, Jogesh Chandra Chaudhuri College, 30-Prince Anwar Shah Road, Kolkata-700033, India

Ali Moghimi
Department of Chemistry, Varamin (Pishva) Branch Islamic Azad University, Varamin, Iran

Majid Farahmandjou
Department of Physics, Varamin Pishva Branch, Islamic Azad University, Varamin, Iran

Asmaa A. Ibrahim
Department of Chemistry, Faculty of Science, South Valley University, Aswan 81528, Egypt

Ali Moghimi
Department of Chemistry, Varamin (Pishva) Branch, Islamic Azad University, Varamin, Iran

Okome Mintsa Madeleine
Department of Chemistry-Biochemistry, Faculty of Medicine, Health Sciences University, Owendo, Gabon

Nguele Jean-Calvin
Department of Chemistry-Biochemistry, Faculty of Medicine, Health Sciences University, Owendo, Gabon

Eba François
Laboratory of Chemistry, Department of Sciences, École Normale Supérieure, Libreville, Gabon

Amblès André
Laboratory of Synthesis and Reactivity of Natural Substances, UMR-CNRS 6514, Faculty of Sciences, University of Poitiers, 40, avenue du Recteur Pineau, 86022 Poitiers-Cedex, France

Joffre Janie
Laboratory of Synthesis and Reactivity of Natural Substances, UMR-CNRS 6514, Faculty of Sciences, University of Poitiers, 40, avenue du Recteur Pineau, 86022 Poitiers-Cedex, France

Tadjou DO REGO
Laboratoire National de Pharmacognosie, Centre Beninois de la Recherche Scientifique et Technique (CBRST), BP 06 Oganla, Porto-Novo, Benin
Laboratory of Physic and Synthesis Organic Chemistry (LaCOPS), University of Abomey-Calavi, Faculty of Sciences and Technics (FAST), BP: 4521 Cotonou, Benin

Serge Fredys Rodrigue O. SENOU
Laboratoire National de Pharmacognosie, Centre Beninois de la Recherche Scientifique et Technique (CBRST), BP 06 Oganla, Porto-Novo, Benin
Laboratory of Physical Chemistry, University of Abomey-Calavi (UAC), Faculty of Sciences and Technics (FAST), Cotonou, Benin

Fidèl DIMON
Laboratory of Physical Chemistry, University of Abomey-Calavi (UAC), Faculty of Sciences and Technics (FAST), Cotonou, Benin

Salome KPOVIESSI
Laboratory of Physic and Synthesis Organic Chemistry (LaCOPS), University of Abomey-Calavi, Faculty of Sciences and Technics (FAST), BP: 4521 Cotonou, Benin

Alban HOUNGBEME
Laboratoire National de Pharmacognosie, Centre Beninois de la Recherche Scientifique et Technique (CBRST), BP 06 Oganla, Porto-Novo, Benin

Coffi AZANDEGBE ENI
Laboratory of Physical Chemistry, University of Abomey-Calavi (UAC), Faculty of Sciences and Technics (FAST), Cotonou, Benin

Georges C. ACCROMBESSI
Laboratory of Physic and Synthesis Organic Chemistry (LaCOPS), University of Abomey-Calavi, Faculty of Sciences and Technics (FAST), BP: 4521 Cotonou, Benin

Jacques POUPAERT
Louvain Drug Research Institute (LDRI), School of Pharmacy, Université catholique de Louvain, B1 7203 Avenue Emmanuel Mounier 72, B-1200 Brussels, Belgium

Fernand A. GBAGUIDI
Laboratoire National de Pharmacognosie, Centre Beninois de la Recherche Scientifique et Technique (CBRST), BP 06 Oganla, Porto-Novo, Benin
Laboratory of Physic and Synthesis Organic Chemistry (LaCOPS), University of Abomey-Calavi, Faculty of Sciences and Technics (FAST), BP: 4521 Cotonou, Benin

N. Sharmila
Department of Chemistry, Acharya Nagarjuna University, Nagarjunanagar - 522 510, A.P., India

R. S. C. Rao
Laila Impex R & D Centre, Jawahar Auto Nagar, Vijayawada, AP-India

B. Hari Babu
Department of Chemistry, Acharya Nagarjuna University, Nagarjunanagar - 522 510, A.P., India

G. Trimurtulu
Laila Impex R & D Centre, Jawahar Auto Nagar, Vijayawada, AP-India

P. V. V. Satyanarayana
Department of Chemistry, Acharya Nagarjuna University, Nagarjunanagar - 522 510, A.P., India

S. A. Ahmed
Department of Chemistry, University of Agriculture, Abeokuta, Nigeria

N. O. Obi-Egbedi
Department of Chemistry, University of Ibadan, Ibadan, Nigeria

N. W. Odozi
Department of Chemistry, University of Ibadan, Ibadan, Nigeria

I. Iweibo
Department of Chemistry, University of Ibadan, Ibadan, Nigeria

M. D. Adeoye
Department of Chemical Sciences, Fountain University, Osogbo, Nigeria

Mohammed Oday Ezzat
Centre for Drug Research, Universiti Sains Malaysia, 11800 Minden, Penang, Malaysia

Mohd Nizam Mord
Centre for Drug Research, Universiti Sains Malaysia, 11800 Minden, Penang, Malaysia

Anjali Soni
Department of Chemistry, Government V.Y.T.PG., Autonomous College Durg (C.G.), 491001, India

Alka Tiwari
Department of Chemistry, Government V.Y.T.PG., Autonomous College Durg (C.G.), 491001, India

A. K. Bajpai
Bose Memorial Research Laboratory, Department of Chemistry, Government Autonomous Science College, Jabalpur (M.P.), 482001, India

I. A. ADEJORO
Department of Chemistry, University of Ibadan, Ibadan, Nigeria

Adebayo A. ADENIYI
Department of Chemistry, University of Ibadan, Ibadan, Nigeria

Diepreye Ere
Department of Chemical Sciences, Niger Delta University, Wilberforce Island, Bayelsa State, Nigeria

Allen T. Ekubo
Department of Chemical Sciences, Niger Delta University, Wilberforce Island, Bayelsa State, Nigeria

Bamgbose Janet Titilayo
Department of Chemistry, University of Agriculture, Abeokuta, Ogun State, Nigeria

Bamigbade Akeem Adesina
Department of Chemistry, University of Agriculture, Abeokuta, Ogun State, Nigeria

Nkiko Mojisola Olubunmi
Department of Chemistry, University of Agriculture, Abeokuta, Ogun State, Nigeria

Ahmed Sikiru Akinyeye
Department of Chemistry, University of Agriculture, Abeokuta, Ogun State, Nigeria

Ikotun Adebomi Ayodeji
Department of Chemistry, Bowen University, Iwo, Osun State, Nigeria

Ayodhya Singh
Department of Chemistry, M. M. H. College, Ghaziabad, U. P., India 201001, India

Munesh Kumar
Department of Chemistry, M. M. H. College, Ghaziabad, U. P., India 201001, India

Manish Kaushik
Department of Chemistry, D. P. B. S. College, Anoopshahr, Bulandshahr, U. P., India 203001, India

Anjali Singh
Department of Chemistry, D. P. B. S. College, Anoopshahr, Bulandshahr, U. P., India 203001, India

Seema Singh
Department of Chemistry, Kishan P. G. College, Bahraich, U. P. 271801, India, India

D. De
Department of Chemistry and Chemical Technology, Vidyasagar University, Midnapore, 721102, India

S. Dalai
Department of Chemistry and Chemical Technology, Vidyasagar University, Midnapore, 721102, India

B. R. De
Department of Chemistry and Chemical Technology, Vidyasagar University, Midnapore, 721102, India

Irfan Elahi
Department of Physics, University of Agriculture, Faisalabad, Pakistan

Rabab Zahira
Department of Physics, University of Agriculture, Faisalabad, Pakistan

Kiran Mehmood
Department of Physics, University of Agriculture, Faisalabad, Pakistan

Arifa Jamil
Department of Physics, University of Agriculture, Faisalabad, Pakistan

Nasir Amin
Department of Physics, University of Agriculture, Faisalabad, Pakistan